Microfluidics
Synthesis, Device Fabrication and Applications

Edited by

Inamuddin[1], Tariq Altalhi[2], Abdalla Salah Mustafa[3], Mohammad Luqman[4]

[1]Advanced Functional Materials Laboratory, Department of Applied Chemistry, Zakir Husain College of Engineering and Technology, Faculty of Engineering and Technology, Aligarh Muslim University, India

[2]Department of Chemistry, College of Science, Taif University, 21944 Taif, Kingdom of Saudi Arabia

[3]SRB Environmental Services, Rabigh, Saudi Arabia

[4]Department of Chemical Engineering, College of Engineering, Taibah University, Yanbu, Kingdom of Saudi Arabia

Published by **Materials Research Forum LLC**
Millersville, PA 17551, USA

Published as part of the book series
Materials Research Foundations
Volume 191 (2026)
ISSN 2471-8890 (Print)
ISSN 2471-8904 (Online)

Print ISBN 978-1-64490-420-6
eBook ISBN 978-1-64490-421-3

Distributed worldwide by

Materials Research Forum LLC
105 Springdale Lane
Millersville, PA 17551
USA
https://mrforum.com

Manufactured in the United States of America
10 9 8 7 6 5 4 3 2 1

Table of Contents

Preface

Microfluidic technology has emerged as a powerful tool due to its ability to precisely manipulate and control very small quantities of fluids. Advances in materials science, fabrication techniques, and biomedical engineering have significantly broadened its applications in wearable technologies, point-of-care diagnostics, drug delivery systems, biosensors, and lab-on-a-chip platforms. These rapid developments have created the need for an updated resource that highlights recent innovations and practical applications in the field.

Microfluidics: Synthesis, Device Fabrication and Applications book provides a comprehensive overview of recent progress and emerging trends in microfluidic technology, including material synthesis strategies, device fabrication methods, and multidisciplinary applications. The book discusses the integration of functional materials into microfluidic systems and covers fabrication techniques such as photolithography, 3D printing, laser ablation, hot embossing, etching, and other advanced methods.

It further explores applications of microfluidics in biomedical diagnostics, biosensors, environmental monitoring, healthcare, and energy systems. In addition, the book addresses current challenges, technological advancements, and future perspectives in the development of high-performance microfluidic platforms with improved sensitivity, miniaturization, and rapid analytical capabilities.

The most relevant target audience of this book are researchers, academicians, scientists, postgraduate students, doctoral scholars, and professionals in chemistry, materials science, biomedical engineering, nanotechnology, biotechnology, pharmaceutical sciences, and microfabrication technologies. The book will also benefit industrial researchers, healthcare professionals, device manufacturers, and engineers working in diagnostics, sensing technologies, and advanced microfluidic systems and devices. In addition, it may serve as a reference text for undergraduate, postgraduate, and PhD-level courses related to microfluidics, biomedical engineering, nanotechnology, and interdisciplinary research applications. This book contains 13 chapters, which are summarized below:

Chapter 1 elaborates on the evolution and importance of microfluidics, setting the stage for how micro-scale fluid channels are fabricated using various techniques and materials. It compares fabrication methods and material choices, discusses key challenges, highlights recent innovations and commercialization efforts, and explains how microfluidics is shaping the future of lab-on-a-chip technologies.

Chapter 2 reviews conventional and advanced fabrication methods for microfluidic devices, including soft lithography, molding techniques, and emerging additive manufacturing approaches for complex three-dimensional structures. It highlights material considerations, scalability, and resolution limits, and introduces next-generation techniques such as AI-assisted optimization, 4D printing, and ultrafast laser ablation, emphasizing their role in future analytical, biosensing, and point-of-care diagnostic applications.

Chapter 3 presents an overview of microfluidic platforms for food analysis. It describes the materials, fabrication methods, and detection approaches used in microfluidic analytical devices. Applications include the detection of pathogens, toxins, contaminants, and allergens, as well as the quantification of nutrients and additives, highlighting the potential of these devices for rapid, low-cost, on-site food quality and safety monitoring.

Chapter 4 provides a comprehensive overview of electron beam lithography (EBL), beginning with its fundamental principles and progressing to practical applications. It explains the physics of electron–material interactions and the role of resist systems in defining nanostructures, followed by key processing steps such as exposure, development, and etching. Various electron beam sources and their effects on resolution and throughput are analyzed. The advantages of EBL, including high resolution and maskless patterning, are compared with its limitations, particularly speed and cost. Emerging trends and future advancements, such as novel resist materials and next-generation e-beam systems, are also discussed.

Chapter 5 presents a combined bibliometric analysis of scientific publications and patents to examine materials, fabrication methods, applications, and detection techniques for microfluidic devices used in water quality monitoring. Detection methods for various analytes are discussed in detail. Paper-based microfluidic devices (μPADs) are highlighted for their portability, low cost, and in situ analysis capabilities. Emphasis is placed on advantages, limitations, emerging trends, and research gaps in the field.

Chapter 6 explores advanced nanomaterials, including metallic, carbon-based, and composite nanomaterials, that enhance microfluidic biosensor performance by improving sensitivity, specificity, and diagnostic speed. It covers fundamental biosensing principles, sensor types, and key performance parameters. Functionalization strategies, bioreceptor immobilization, and nanomaterial integration into microfluidic platforms are discussed, along with applications in early disease diagnosis, environmental monitoring, and food safety, highlighting prospects for point-of-care diagnostics.

Chapter 7 focuses on the crucial role of advanced materials in enhancing the efficiency of microfluidic electrochemical sensors (MES). MES combine precise fluid handling with electrochemical analysis for rapid point-of-care diagnostics. Applications span biomedical diagnostics, environmental monitoring, and food safety. Future directions include addressing challenges related to scalability and reproducibility to enable integration with wearable sensors and Internet of Things (IoT) frameworks.

Chapter 8 provides a comprehensive overview of microfluidic immunosensors. Key microfluidic platforms, including glass, paper, and polydimethylsiloxane (PDMS), are described alongside fluid-flow mechanisms such as capillary action, electrokinetic transport, and pressure-driven flow. The chapter concludes by discussing challenges related to sensitivity and scalability and outlines future trends for advancing real-world applications.

Chapter 9 outlines microfluidic platforms for modern bioanalysis, describing continuous-flow, droplet-based, digital, and paper-based systems. It focuses on applications such as DNA/RNA testing, protein assays, and point-of-care diagnostics. Advances in drug delivery, organ-on-a-chip models, and omics integration are highlighted, along with advantages, limitations, and future potential.

Chapter 10 describes microfluidic strategies for biomolecule separation, covering fundamental principles, materials, fabrication techniques, and diverse separation mechanisms. Applications in diagnostics, proteomics, genomics, and environmental testing are discussed.

The chapter also summarizes advantages, limitations, and emerging trends such as droplet systems, paper microfluidics, and AI-enabled automation.

Chapter 11 discusses the fundamental principles and advantages of microfluidic flow chemistry and explores its application across various organic reactions. It highlights the role of microfluidics in enabling sustainable and scalable chemical synthesis and discusses industrial applications and future perspectives.

Chapter 12 reviews research progress on microfluidic materials for energy applications, including their properties and applications in capacitors, fuel cells, and batteries. Current challenges and future development directions for microfluidic energy systems are also discussed.

Chapter 13 discusses microfluidic devices that enable fast, automated, and remote analysis using very small fluid volumes. Applications in healthcare, environmental monitoring, and food safety for real-time, on-site detection are highlighted. Key microfluidic formats, including digital, centrifugal, paper-based, and wearable systems, are presented, along with challenges related to device robustness, fluid handling, and waste management.

Materials Research Foundations 191 (2026) 1-35 https://doi.org/10.21741/9781644904213-1

Chapter 1

Microfluidic Device Fabrication: Techniques, Materials, Integration Strategies and Commercialization Prospects

Vipul D. Prajapati[1*], Varad Gore[2]

[1]Department of Pharmaceutics, S.S.R. College of Pharmacy (Permanently Affiliated to Savitribai Phule Pune University), Sayli-Silvassa Road, Silvassa, U. T. of D.N.H. and D.D., India

[2]Department of Pharmaceutical Quality Assurance, S.S.R. College of Pharmacy (Permanently Affiliated to Savitribai Phule Pune Universitypph), Sayli-Silvassa Road, Silvassa, U. T. of D.N.H. and D.D., India

*vippra2000@yahoo.com

Abstract

Microfluidic devices allow precise manipulation of fluids at the microscale, transforming biomedical, chemical, and environmental applications. This paper reviews key fabrication techniques, including photolithography, soft lithography, three-dimensional printing, laser micromachining, and paper-based methods, comparing their resolution, cost, speed, and scalability. It discusses materials such as polymers, glass, silicon, and hydrogels, emphasizing factors like biocompatibility, transparency, and strength. The integration of fabrication processes with material properties, along with challenges and optimization strategies, is analysed. Recent advances in sustainable fabrication, smart materials, and artificial intelligence-assisted design are explored. The study concludes with insights into patents, emerging technologies, and commercial potential in microfluidics.

Keywords

Microfluidic, Fabrication Techniques, Photolithography, Soft Lithography, Three-Dimensional Printing, Laser Micromachining, Paper-based Methods, Biocompatibility, Transparency

List of Abbreviations

LOC : Lab-on-a-chip
MEMS : Microelectromechanical systems
PDMS : Polydimethylsiloxane
UV : Ultraviolet
DNA : Deoxyribonucleic Acid
PMMA : Poly methyl methacrylate
COC : Cyclic olefin copolymer
PC : Polycarbonate
μPADs : Microfluidic paper-based analytical devices

Materials Research Foundations 191 (2026) 1-35 | https://doi.org/10.21741/9781644904213-1

FDM : Fused deposition modelling
SLA : Stereolithography
DLP : Digital light processing
PEGDA : Poly ethylene glycol diacrylate
PET : Polyethylene Terephthalate
CAD : Computer Aided Design
CNC : Computer Numerical Control
RIE : Reactive ion etching
DRIE : Deep reactive ion etching
POC : Point-of-care
PEG : Poly ethylene glycol
PEGDA : Polyethylene glycol diacrylate
PCR : Polymerase Chain Reaction
PI : Polyimide
AI : Artificial intelligence
IC : Integrated circuit
3D : Three Dimensional
2D : Two Dimensional

Contents

Materials Research Foundations 191 (2026) 1-35 | https://doi.org/10.21741/9781644904213-1

1. Introduction

Microfluidic devices allow for the manipulation and control of small fluid volumes, typically in the microlitre to nanolitre range, within microscale channels. By assembling several components, such as chambers, valves, and channels, onto a tiny platform, these instruments allow for accurate fluid manipulation. The main attraction of microfluidics is its ability to miniaturize laboratory processes, also referred to as lab-on-a-chip (LOC) systems, and provide rapid, low-volume, and cost-effective solutions for chemical and biological analysis [1].

1.1 Historical Development of Microfluidics

The history of microfluidics is fundamentally shaped by the development of semiconductors and microelectromechanical systems (MEMS). The field started to take shape in the late 1970s when Terry et al. used fabrication techniques borrowed from microelectronics to create the first miniature analytical equipment, a silicon-based gas chromatograph [2]. This modernization marked a shift toward scaling down laboratory processes to the microscale.

Microfluidics emerged as a distinct area in the 1980s and early 1990s, using MEMS technologies to create devices that could handle fluids at sub-millimeter scales. Notably, Andreas Manz and his associates were the first to come up with the concept of micro total analysis systems (µTAS), which aimed to combine sample preparation, reaction, separation, and detection on a single chip [3]. This was a major turning point, as it introduced the idea of the "lab-on-a-chip", now a central perception in microfluidics.

George Whiteside and his team also pioneered soft lithography in the 1990s, particularly with polydimethylsiloxane (PDMS). Soft lithography democratized access to microfabrication by making it possible to create microfluidic devices rapidly and cheaply without the requirement for expensive cleanroom facilities [4].

Government research funding, progressions in material science, and the evolution of MEMS fabrication methods helped fuel the rapid progress of microfluidics in the 2000s [5]. From fundamental studies, microfluidics has developed into a broad field with applications in industry, environmental science, diagnostics, and medicine. Because of their transparency, flexibility, and biocompatibility, polymers like PDMS eventually replaced silicon or glass as the primary materials used in early devices [6].

Microfluidics is currently expanding due to advancements in three-dimensional (3D) printing, nanofabrication, and the incorporation of electrical, optical, and biological components. It has become a fundamental technology in producing POC diagnostics, organ-on-chip platforms, and high-throughput screening tools for drug development [7]. A summarized history of various microfluidics is depicted in Figure 1.

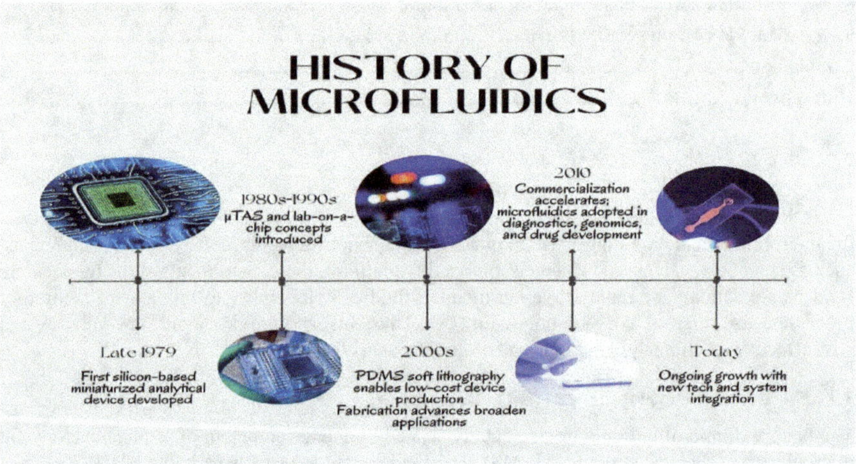

Figure 1. *Schematic flow chart exhibiting the historical timelines of microfluidics fabrication techniques.*

1.2 Importance of Microfluidics in Modern Science

Microfluidics has become a transformative tool in various scientific domains. In biotechnology field it allows for single-cell analysis, high-throughput screening, and integration with proteomic and genomic platforms [8]. The technology enables lower reagent usage, decreased contamination, faster analysis, and the possibility of automation. Furthermore, its applications range from environmental sensing and point-of-care diagnostics, to drug discovery and artificial organ systems [9].

1.3 Scope and Structure of the Report

The manufacturing of microfluidic devices is covered in broad length in this paper, starting with a thorough examination of the many fabrication methods now in use and concluding with a thorough examination of the materials utilized to make the devices. The mail goal is to help all the engineers and researchers for choosing the best materials and techniques for a given application. Future directions in microfluidics, current innovations, and the integration of techniques with materials is also covered in the following sections.

2. Fabrication Techniques for Microfluidic Devices

Various types of fabrication techniques for the development of microfluidic devices with history, main mechanism in short, along with their status till date for suitable purposes is briefly discussed in the following section with the help of keen reference work.

2.1 Photolithography

In 1958, Bell Labs' Jay W. Lathrop and James R. Nall developed the method for fabricating microfluidic devices using photolithography. By altering semiconductor methods to create microscopic fluidic channels, they created the LOC technology. It involves employing light to transfer geometric designs from a mask to a photoresist-coated substrate. After cleaning a silicon or glass wafer, a photoresist is typically spin-coated onto it. The photoresist is then exposed by the Ultra-Violet (UV) light source through a patterned photomask, which, depending on whether the resist is positive or negative, either softens or hardens the exposed areas. After removing the unexposed (or exposed) areas during the development process, a precise pattern is left behind it that can be used as a mold or etched onto the substrate.

Photolithography enables extremely high-resolution features—down to the submicron level—which is necessary for applications like DNA analysis chips, single-cell assays, and microreactors that call for accuracy and miniaturization [10]. Nevertheless, the method is heavily reliant on Ultra-Violet (UV) exposure devices, costly optical masks, and cleanroom settings. Furthermore, although multilayer stacking can enable some degree of vertical integration, photolithography is mainly restricted to planar, two-dimensional (2D) devices.

The compatibility with non-traditional materials has improved with recent developments. For example, he modified the surface of PDMS, a typically soft, flexible elastomer that is not naturally suited for lithography, using oxygen plasma to improve photoresist adhesion, thereby introducing a direct photolithography technique [11]. This opens the new avenues for hybrid devices that combine rigid substrates and the soft materials.

Materials Research Foundations 191 (2026) 1-35	https://doi.org/10.21741/9781644904213-1

Photolithography is the gold standard for producing precise master molds for soft lithography in both research and industry applications. Additionally, it is a fundamental method for developing high-performance microfluidic devices. Photolithography-based microfluidic fabrication is particularly accurate, reproducible, and mass-producible for LOC devices. However, it requires cleanroom facilities, is costly, time-consuming, and only works with stiff materials like silicon and glass.

2.2 Soft Lithography

In the mid-1990s, Younan Xia and George M. Whitesides invented the soft lithography. In 1998, they employed PDMS for the first time to create microfluidic devices, offering a versatile, reasonably priced, and readily available alternative to traditional photolithography. Making a master mold with the required microchannel patterns on it is the first stage in the procedure. Usually, photolithography is used to create this mold. PDMS is poured onto the mold, let to harden, and then peeled off to form a negative copy of the original structure. The PDMS device can then be bonded to glass or another PDMS layer to form enclosed microfluidic channels. Soft lithography is favoured in biological and biomedical applications due to PDMS's optical transparency, oxygen permeability, low cost, and ease of handling. It does not require cleanroom facilities for replication steps, making it accessible for academic laboratories and rapid prototyping [12].

The soft lithography also has some drawbacks. Because of its mechanical softness and ability to absorb tiny hydrophobic molecules, PDMS can make some chemical or pharmaceutical experiments more challenging. Furthermore, it is difficult to fabricate complex 3D architectures or multilayer structures using standard soft lithography alone.

To address this, advanced methods like two-photon polymerization and multi-layer soft lithography have been developed. Lin et al. introduced a hybrid approach that combines conventional photolithography with two-photon polymerization to generate master moulds featuring complex 3D microstructures. This enables precise patterning while maintaining cost-effectiveness and scalability [13].

Soft lithography's low cost, ease of production, and compatibility with flexible and biocompatible materials like PDMS make it an excellent choice for biological applications and rapid prototyping. It's more accessible because it doesn't require expensive cleanroom facilities. It does, however, have certain drawbacks, including a lesser resolution than photolithography, the possibility of soft material deformation, and PDMS's limited long-term durability and chemical resistance.

In summary, soft lithography remains one of the most accessible and adaptable microfabrication techniques, especially for biological assays, rapid prototyping, and educational use. Its compatibility with hybrid strategies continues to push the boundaries of soft materials in microfluidic engineering.

2.3 Injection Moulding

In the early 2000s, John Wesley Hyatt and Isaiah Hyatt modified injection molding for the purposes of fabricating microfluidic devices. It was perfect for commercial applications since it developed. Now onwards it is possible to produce microfluidic devices in large quantities at low-cost utilizing thermoplastics like PMMA and COC. In the process, thermoplastic polymers, such as PMMA, COC, or PC, are melted and then injected under high pressure into a mold cavity that

Materials Research Foundations 191 (2026) 1-35 https://doi.org/10.21741/9781644904213-1

has pre-made microstructures. Once the molded part has cooled and solidified, it is released from the mold. Thousands of identical microfluidic chips with excellent dimensional precision can be created using this method. The cost-effectiveness of injection molding at large scale is one of its main benefits. While the initial fabrication of metal moulds (usually made of steel or aluminium) is time-consuming and expensive, the unit cost drops significantly for large-volume production, making it an ideal for disposable devices such as point-of-care test cartridges [14].

Injection moulding also offers fast cycle times, strong mechanical properties, and compatibility with automation. Thermoplastics used in this process can be selected for specific properties—like chemical resistance or optical clarity—allowing for tailored solutions in areas like genomics, fluid mixing, or biosensing [15].

Injection molding has few drawbacks despite its many advantages. Because mold creation is expensive and time-consuming, the method is less appropriate for rapid prototyping or trial iterations. Furthermore, without extremely specialized tools, obtaining micro-scale features (below ~50 μm) might be challenging. To prevent flaws during demoulding, design restrictions like draft angles and flow lines must also be taken into account.

To overcome these barriers, hybrid approaches are emerging. For instance, combining injection moulding with hot embossing or surface micromachining enables improved resolution and design flexibility.

Injection moulding enables fast, low-cost mass production of durable microfluidic devices but requires expensive molds upfront and is less flexible for quick design changes or very fine features. In conclusion, injection moulding is the method of choice for scalable, robust, and repeatable microfluidic device fabrication in commercial applications.

2.4 Hot Embossing

Researchers like Stephen A. Soper pioneered hot embossing for microfluidic manufacturing in the mid-1990s. It offers the more affordable option to photolithography by precisely creating microchannels in the thermoplastics using the heated molds. After heating a polymer substrate over its glass transition temperature (Tg), it is uniformly compressed and pressed against a hard master mold, usually composed of silicon, nickel, or steel. When the substrate cools below Tg, it is demolded, leaving behind a precise negative of the mold's microstructures. This method is especially valued for its high fidelity, optical transparency, and material versatility, making it a strong alternative to soft lithography for creating rigid, disposable microfluidic chips [16]. It is particularly useful for fabricating bioanalytical devices, optical sensors, and fluidic cartridges that demand stability and mechanical durability. Hot embossing has lower initial setup costs than the injection molding, which makes it ideal for medium-volume production. But because each cycle involves pressing, heating, and cooling steps, which reduces its throughput, it is typically slower. One major challenge lies in demoulding, which can lead to feature distortion or damage if not managed carefully—especially in structures with high aspect ratios.

According to Wu and Gu, hot embossing achieves structural features as small as 100 nano-meters, depending on mould quality and process conditions, and is compatible with multi-layer bonding techniques for full microfluidic system assembly [17].

Innovations in embossing include ultrasonic embossing and roll-to-roll hot embossing, which improve speed and enable continuous fabrication. These hold particular promise for high-volume,

Materials Research Foundations 191 (2026) 1-35 https://doi.org/10.21741/9781644904213-1

large-area applications such as lab-on-foil systems and microfluidic paper-based analytical devices (μPADs).

For medium-scale polymer device production, hot embossing provides accurate microchannel fabrication with acceptable surface quality at a reasonable cost. However, it is less appropriate for quick prototyping or very high-volume manufacturing, has longer cycle times than injection molding, and necessitates a specialist equipment.

In summary, hot embossing is a precise, material-flexible method ideal for rigid microfluidic devices with optical or mechanical requirements, and is a key enabler for transitioning from lab-scale to real-world deployment.

2.5 Three-dimensional Printing and Additive Manufacturing

Researchers like Michael McAlpine showed that 3D printing could rapidly produce intricate, adaptable microfluidic structures without the need for molds or masks, which led to the development of 3D printing for the manufacturing of microfluidic devices in 2013. 3D printing has made it possible to create complex 3D geometries, such as spirals, embedded channels, and interconnects, that are difficult or impossible to achieve with lithographic techniques, directly and without the use of a mask, in contrast to traditional fabrication methods that rely on molds or masks.

Fused Deposition Modeling (FDM), Stereolithography (SLA), and Two-Photon Polymerization are some of the 3D printing technologies utilized in microfluidics. SLA and digital light processing (DLP) are particularly useful for achieving higher resolution, down to tens of micro-meters, while FDM is favoured for its low cost and accessibility [18].

Even with its versatility, 3D printing in microfluidics still has difficulties, particularly when it comes to finding translucent, biocompatible materials, smooth channel surfaces, and fine detail resolution. Biological experiments may be impacted by surface roughness in printed channels, which can impede fluid flow. Furthermore, conventional 3D printers might not consistently create microchannels smaller than about 100 μm, which restricts their use in single-cell or nanoscale applications.

To address these issues and improve feature resolution and reproducibility, researchers used direct laser path programming and multi-resin printing. Using PEGDA-based resins, Ahmed et al. demonstrated stereolithographic techniques that enable integrated components like pumps and valves as well as embedded channel characteristics to be manufactured in a single step [19].

Additionally, hybrid approaches that combine 3D printing with soft lithography—such as 3D printed moulds for PDMS casting—enable complex, multi-layered devices while retaining the material compensations like transparency and biocompatibility [20].

3D printing makes it possible to quickly, easily, and affordably fabricate complex and customized microfluidic devices without the need for molds. However, it often has worse resolution and surface quality than traditional methods, and the choice of materials may be more limited, which might affect the device's performance and its life. In summary, 3D printing is altering microfluidic prototyping by offering customization, speed, and design liberty, especially in research and early-stage development.

Materials Research Foundations 191 (2026) 1-35 https://doi.org/10.21741/9781644904213-1

2.6 Laser Micromachining

In the 1990s, researchers like Stephen A. Soper showed how laser micromachining could be used for quick, accurate, maskless patterning on polymers and glass, providing a versatile substitute for more conventional techniques. This led to the development of laser micromachining for microfluidic manufacturing. Laser micromachining is a direct-write fabrication technique that uses high-energy laser beams to engrave or ablate microscale designs onto substrates for microfluidic devices. Unlike photolithography or molding, laser micromachining provides speed, flexibility, and mask-less customization without the use of masks or molds. This makes it perfect for rapid prototyping, one-off designs, and small-batch production.

By directing a laser beam—typically CO_2, UV, or femtosecond lasers—on a target material, such as glass, PMMA, PET, or PDMS, the process creates channels, holes, or chambers by locally heating and removing material. The laser's cutting accuracy and material compatibility are determined by its wavelength, pulse length, and energy intensity. For instance, femtosecond lasers are perfect for bio-integrated applications because of their high resolution and little heat damage [21].

Laser micromachining is particularly advantageous for rapid iteration, as design modifications can be made in CAD software and implemented instantly without fabricating new moulds. It is also effective for creating multilayer structures, through-bore holes, or reservoirs, which are difficult with planar lithographic techniques.

The surface roughness, debris generation, and heat degradation in certain polymers are some of the method's drawbacks. These artifacts may need post-processing, like polishing or plasma cleaning, or they may obstruct fluid flow. Furthermore, it is still difficult to consistently achieve very fine resolution (<20 μm) across wide areas, particularly on softer materials like PDMS [22].

Recent innovations include integrating laser micromachining with other methods, such as using it to prepare moulds for soft lithography or to post-process 3D printed microfluidic devices for improved its detail overall functionalities.

Laser micromachining allows fast, precise, and mask-free fabrication of microfluidic devices on various materials without cleanroom needs. However, it can cause surface roughness, has limited resolution compared to photolithography, and may damage sensitive materials due to heat. In summary, laser micromachining is a versatile, contactless, and fast tool for microfluidic device fabrication, especially valuable for the custom or multilayer systems.

2.7 Micro-Milling

Researchers like Andreas Manz demonstrated how micro-milling, which was first used for microfluidic production in the early 2000s, could rapidly produce intricate microchannels by manually machining metals and plastics without the need for cleanroom facilities. Micro-milling is a mechanical subtractive manufacturing process that uses tiny, fast rotating cutting tools to remove material from a solid block in order to create microfluidic devices. It is carried out on CNC (Computer Numerical Control) milling machines, which may be configured to create intricate microchannel geometries in metals and molds for PDMS casting as well as thermoplastics like PMMA, COC, and PC. Unlike lithography-based techniques, micro-milling makes it possible to create 3D geometries with non-planar features, such rounded or semi-circular cross-sections, and varied depths, which are challenging to accomplish with photolithography or soft lithography [23].

Materials Research Foundations 191 (2026) 1-35 https://doi.org/10.21741/9781644904213-1

It is particularly advantageous for creating master moulds for hot embossing or PDMS soft lithography, allowing reusability and modification.

Micro-milling provides high design freedom and doesn't require a cleanroom, making it cost-effective for low-to-medium volume fabrication and rapid prototyping. Precision tools with diameters as small as 10 μm can achieve microscale resolution, and newer machines offer sub-10 μm positional accuracy [24].

But there are difficulties with the process. Over time, tool wear can reduce resolution, and surface roughness may necessitate post-processing to enhance bonding qualities or optical clarity. Furthermore, vibrations or tool deflections may have an impact on attaining homogeneity in narrow channels, particularly when working with the soft or heat-sensitive materials.

Recent developments in hybrid techniques use micro-milled moulds combined with soft lithography to fabricate circular or 3D curved microchannels, are useful in vascular or biomimetic systems [25].

Micro-milling offers fast, flexible fabrication of complex 3D microfluidic structures without the need for cleanrooms, making it great for prototyping. However, it has limited resolution compared to lithography, can cause tool wear, and may produce rougher surfaces requiring additional finishing.

In conclusion, micro-milling is a flexible, precise, and mould-friendly technique ideal for creating complex or 3D microfluidic architectures, especially when reusable tooling or non-planar features are required.

2.8 Etching Methods (Wet and Dry Etching)

Etching methods of wet and dry types were established in the 1960s for semiconductors and used for microfluidics in the late 1980s by researchers like Andreas Manz to build precise microchannels on silicon and glass substrates. Microfluidic structures are commonly made using wet and dry etching techniques, especially on substrates made of silicon and glass. These methods selectively remove material to form channels, wells, and through-holes by using patterned masks to direct chemical reactions or plasma processes.

Wet etching involves immersing the substrate in a chemical solution (e.g., hydrofluoric acid for glass or potassium hydroxide for silicon). It is simple, cost-effective, and ideal for creating isotropic or anisotropic characteristics depending on the crystal alignment of the material. However, wet etching lacks precision for vertical sidewalls and often suffers from undercutting, which limits its use in firmly packed microchannel networks [26].

Dry etching, more especially reactive ion etching (RIE) and deep reactive ion etching (DRIE), uses plasma composed of chemically reactive ions to etch material in a vacuum chamber. These techniques are perfect for creating straight-walled microchannels in silicon or glass because they provide highly anisotropic profiles, superior aspect ratios, and exact control over etch depth. With lateral resolution in the micron range, DRIE, which is frequently utilized in MEMS production, allows for the construction of devices with depths more than hundreds of micrometers.

Etching is often used in tandem with photolithography, where a patterned photoresist acts as an etch mask. It is especially valuable for fabricating robust, chemically impervious, and high-

Materials Research Foundations 191 (2026) 1-35 https://doi.org/10.21741/9781644904213-1

temperature compatible microfluidic systems, such as those used in gas chromatography or high-voltage electrophoresis.

He also demonstrated the integration of wet and dry etching techniques with nano porous membranes to create multiscale microfluidic platforms for tuning biochemical interactions [27].

Etching methods provide high-resolution, precise microchannel fabrication essential for microfluidics and are well-established techniques. However, they often require hazardous chemicals or plasma equipment, involve complex processes, and are mostly limited to rigid materials like silicon and glass, making them less suitable for flexible or rapid prototyping.

In summary, wet and dry etching provide precision, material compatibility, and deep etch capabilities, making them indispensable for high-performance, silicon-based microfluidic applications.

2.9 Paper-Based Microfabrication

The Harvard group of George M. Whitesides announced paper-based microfluidic fabrication in 2007. They used patterned paper to make inexpensive, pump-free devices that were perfect for portable diagnostics and resource-constrained applications. Paper-based microfluidics is a new low-cost production method that creates reaction zones and microchannels using porous paper substrates such as cellulose or nitrocellulose. Instead of using conventional channel walls, paper uses capillary action to passively transfer fluids without the need for external pumps or power sources. In resource-constrained contexts, this approach is particularly advantageous for point-of-care diagnostics due to its affordability, portability, and simplicity of use. Paper-based devices are made using a variety of fabrication techniques, including photolithography, laser cutting, screen printing, inkjet printing, and wax printing.

The most popular of these is wax printing, which involves depositing wax in predetermined patterns and melting it to pierce the paper and create hydrophobic barriers that keep aqueous solutions in the intended channels [28].

Paper-based microfluidics supports multiplexed assays, colorimetric detection, and sample filtration, making it ideal for diagnostic tools for glucose, pregnancy, infectious diseases, and even environmental toxins. Devices such as lateral flow assays and origami-style 3D paper chips have been successfully installed in the field.

While the method is highly ascendable and eco-friendly, it does have limitations. Resolution is relatively low (often >200 μm), and the variability of paper porosity can affect flow rates and reproducibility. Moreover, device durability and mechanical robustness are lower than polymer-based systems, which may affect storage and handling in humid environments.

The application of paper microfluidics is growing fast as the result of developments in the multilayer paper-based devices and the incorporation of optical and electrochemical sensors. These days, hybrid approaches include microheaters and flexible electronics, opening up new possibilities for wearable and mobile health diagnostics [29].

Paper-based microfluidic fabrication is simple, affordable, and ideal for portable, disposable diagnostic devices, especially in resource-constrained contexts. It uses easily accessible materials, relies on capillary action for fluid transport, and doesn't require additional pumps. In contrast to more sophisticated production techniques, it has drawbacks such decreased durability and

Materials Research Foundations 191 (2026) 1-35
https://doi.org/10.21741/9781644904213-1

compatibility with specific reagents, less control over flow rates, and worse precision and resolution.

In summary, paper-based microfabrication is a cost-effective, accessible, and scalable technology ideal for diagnostic applications where simplicity, portability, and affordability are critical.

2.10 Hybrid and Emerging Techniques

In order to get around the drawbacks of individual techniques, researchers combined different techniques, such as 3D printing, soft lithography, and laser machining, to create hybrid and emergent microfluidic manufacturing techniques, which started to garner attention in the 2010s. For cutting-edge applications, these developments sought to improve gadget functionality, complexity, and material compatibility. The state-of-the-art in microfluidic device engineering is represented by hybrid and developing fabrication techniques, which combine the benefits of numerous traditional techniques, including as photolithography, soft lithography, 3D printing, and etching. As a result, limitations can be overcome and multiscale, multifunctional, high-performance devices can be produced.

Hybrid approaches are particularly useful for creating complex geometries and multilayer architectures. For instance, Shankles integrated photolithography and electron beam lithography to fabricate microfluidic platforms with nano porous membranes, allowing selective transport of molecules and fine-tuned control of biochemical reactions. This method allows for control over a variety of length scales within a single device, which is crucial for applications such as selective filtering or cell signaling studies [30].

Emerging techniques often leverage 3D printing in combination with PDMS moulding, known as 3D printed transfer moulding. Glick et al. (2016) demonstrated a method where complex moulds were fabricated using 3D printing and then used to cast multi-layer PDMS microfluidic devices. This allowed for the creation of non-planar geometries, flexible valves, and vias for vertical interconnections, which are difficult to fabricate using planar lithographic techniques alone [31].

Other innovations include multi-resin 3D printing, which allows for spatial tuning of material properties within a single chip. Ahmed used this technique to fabricate integrated components like valves and pumps with flexible and rigid regions—fabricated simultaneously in one print job—streamlining device development [32].

For complicated microfluidic devices, hybrid and developing techniques provide more design flexibility, enhanced functionality, and the ability to combine materials and processes. They can, however, be expensive, technically challenging, and necessitate certain tools and knowledge.

In summary, hybrid and emerging methods enable greater design complexity, functional integration, and multi-material compatibility, pushing microfluidics toward advanced applications like organ-on-chip systems, high-throughput screening, and smart diagnostics. Table 1 depicts about various microfabrication techniques, including their invention periods (where available), inventors or originators (if known), and their pros and cons.

Materials Research Foundations 191 (2026) 1-35 https://doi.org/10.21741/9781644904213-1

Table 1. *A summary of comparison of the various microfabrication techniques with their invention dates, inventors (if known), and their pros and cons.*

Technique	Year Invented	Inventor(s)	Pros	Cons	References
Photolithography	1958	Jay W. Lathrop & James R. Nall (Bell Labs)	High-resolution patterning; scalable for mass production	Expensive equipment; limited to flat substrates; complex process.	[10]
Soft Lithography	~1995	Younan Xia & George M. Whitesides	Low-cost; suitable for flexible substrates; high resolution (~30 nm)	Less durable molds; not ideal for high-volume production.	[11]
Injection Molding	2000	John Wesley Hyatt & Isaiah Hyatt	High-volume production; consistent quality	High initial tooling cost; limited to thermoplastics.	[12]
Hot Embossing	1990s	Stephen A. Soper	High fidelity replication; suitable for thermoplastics	Long cycle times; not ideal for high-throughput production.	[13]
3D Printing / Additive Manufacturing	2013	Michael McAlpine	Rapid prototyping; complex geometries; minimal waste	Slower than traditional methods; limited material choices.	[14–17]
Laser Micromachining	1990s	Stephen A. Soper	High precision; contactless process	High equipment cost; thermal effects can damage materials.	[18,19]
Micro-Milling	early 2000s	Andreas Manz	High precision; suitable for hard materials	Tool wear; limited to certain geometries.	[20]
Etching (Wet & Dry)	1980s	Andreas Manz	Anisotropic etching; suitable for microfabrication	Wet etching lacks precision; dry etching equipment is expensive.	[20]
Paper-Based Microfabrication	2007	George M. Whitesides	Low-cost; biodegradable; suitable for point-of-care diagnostics	Limited resolution; not suitable for complex devices.	[22–24]
Hybrid & Emerging Techniques	2000s	Various researchers	Combines advantages of multiple methods; versatile	Complex integration; may require specialized equipment.	[25-26]

Materials Research Foundations 191 (2026) 1-35 https://doi.org/10.21741/9781644904213-1

3. Comparison of Fabrication Techniques

3.1 Resolution and Accuracy

Resolution and accuracy are critical parameters in microfluidic fabrication, determining the smallest features that can be reliably reproduced. Photolithography and dry etching (e.g., reactive ion etching) achieve the highest resolution, with features under 1 μm possible, making them ideal for nanofluidic applications or intricate device architectures [33]. However, these methods require cleanroom conditions and are costly.

3D printing, especially using digital light processing (DLP) or stereolithography (SLA), offers resolutions around 50–100 μm, with some LCD-based printers reaching 35 μm lateral features and 22 μm vertical layers [34]. However, surface roughness and feature fidelity remain challenges in additive manufacturing.

Soft lithography using PDMS offers moderate resolution (~10–100 μm), sufficient for most biological assays. Its elastomeric nature may lead to slight deformation or swelling over time, reducing dimensional accuracy [35].

Micro-milling and laser micromachining can achieve resolutions as low as 10–25 μm depending on the tool or laser type, but often struggle with surface finish and repeatability for complex geometries [36].

Paper-based methods offer the lowest resolution, typically >200 μm, though advances in inkjet printing and wax plotting have improved this to ~60–100 μm [37].

3.2 Production Time and Throughput

When scaling microfluidic devices from development to commercial or clinical application, production time and throughput are crucial factors. These variables are influenced by the complexity of the process, the equipment required, and the degree of automation in addition to the speed of manufacturing.

Injection moulding and hot embossing are the leading techniques in terms of throughput for large-scale manufacturing. Once a mould is prepared, thousands of microfluidic chips can be fabricated in minutes per cycle. This makes them the preferred methods for mass production of diagnostic cartridges and disposable test strips [38]. However, mould fabrication itself is time-consuming and expensive, making them less suitable for prototyping.

Soft lithography is frequently used in small-batch manufacturing and prototyping. It has a moderate throughput and can produce several dozen PDMS chips in a day. Although setup time is accelerated by the absence of cleanroom facilities, automation is limited by the manual demoulding and bonding processes.

3D printing, especially inkjet and DLP-based systems, offers rapid turnaround for prototyping, with print times ranging from minutes to hours depending on complexity. While it excels in customization, simultaneous multi-device printing is limited, which constrains large-scale throughput [39].

Laser micromachining and micro-milling offer fast, on-demand fabrication for individual devices. They are excellent for design iteration but less efficient for bulk manufacturing unless paired with replication techniques.

Paper-based microfluidics offers extremely high throughput at low cost, especially when fabricated using roll-to-roll processing or continuous ink plotting, enabling thousands of units per hour [40].

3.3 Cost and Scalability

When choosing a fabrication method for microfluidic devices, cost and scalability are important factors, especially when moving from lab-scale research to industrial or clinical applications.

Injection moulding offers excellent scalability due to its ability to produce thousands of identical parts quickly with low per-unit cost. However, the initial investment in mould design and tooling is high, which is only justified for large-scale production runs [41]. It is ideal for mass production of disposable devices, such as point-of-care diagnostic cartridges.

Soft lithography is highly cost-effective for low-volume production and rapid prototyping. The master mould can be created via photolithography, and subsequent PDMS replicas can be produced without specialized equipment. While not as ascendable as injection moulding, it remains a cornerstone in academic labs for its low upfront cost and flexibility [42].

3D printing technologies, especially low-cost LCD or DLP printers, provide affordable and customizable platforms for iterative design. While print materials can be more expensive and slower than moulding, 3D printing offers exceptional flexibility and eliminates the need for moulds, making it ideal for small-scale or personalized applications [43].

Microfluidics based on paper is excellent in terms of scalability and affordability. These devices are particularly helpful in environments with limited resources and for field diagnostics since techniques like wax printing and inkjet printing enable roll-to-roll or batch production at incredibly low costs [44]. Table 2 depicts the comparison of various fabrication techniques used for analyte detection, including the target analytes, quantification limits, runtime, key conclusions, and potential applications.

Materials Research Foundations 191 (2026) 1-35 https://doi.org/10.21741/9781644904213-1

***Table 2.** Summary of the comparative features of various microfluidic fabrication techniques with references.*

Technique	Analytes / Application Areas	Quantification Limit / Resolution	Runtime / Throughput	Key Conclusion	Applications	Refs.
Photolithography	Biomolecules, chemicals	High resolution (sub-micron)	Moderate to slow	Precise and reproducible; limited by planar 2D patterning	MEMS, biosensors, microfluidics	[10]
Soft Lithography	Cells, proteins, DNA	~50–100 µm resolution	Fast and cost-effective	Flexible, ideal for bio-patterning and microfluidics	Biosensors, tissue engineering	[11]
Injection Molding	Bulk fluidic systems	Medium resolution	High throughput	Suited for mass production	Industrial diagnostics, lab-on-chip	[12]
Hot Embossing	Biological/chemical sensing	~100 nm – 1 µm	Moderate	Good for thermoplastics; requires heat/pressure	Disposable microfluidic devices	[13]
3D Printing / Additive Manufacturing	Large molecule assays	~10–100 µm	Rapid prototyping	Custom geometries possible, low resolution limits use	Prototyping, organ-on-chip systems	[14–17]
Laser Micromachining	Bioanalytes (e.g., in PDMS)	~10–20 µm (depends on laser)	Fast	Good for 3D structuring; minimal chemical processing	Implantable sensors, 3D devices	[18,19]
Micro-Milling	General chemical analytes	~10 µm	Moderate	Good for circular channels and molds	Vascular models, microreactors	[20]
Etching (Wet/Dry)	MEMS and microfluidics	Sub-100 nm possible	Slow	High resolution but involves hazardous chemicals	Semiconductor devices, sensors	[20]
Paper-Based Microfabrication	Glucose, proteins, pathogens	~nM–µM depending on design	Very fast	Inexpensive and disposable	Point-of-care diagnostics	[22–24]
Hybrid/ Emerging	Complex biochemical systems	Tunable; sub-100 nm	Moderate	Merges benefit of multiple methods (e.g., soft lithography + e-beam)	Advanced biochips, energy devices	[25-26]

4. Materials for Microfluidic Devices

Various types of microfluidic devises are available in the market for analytical estimation of various types of molecules in various areas. Figure 2 exhibits the salient outlook of various materials used for the same purpose.

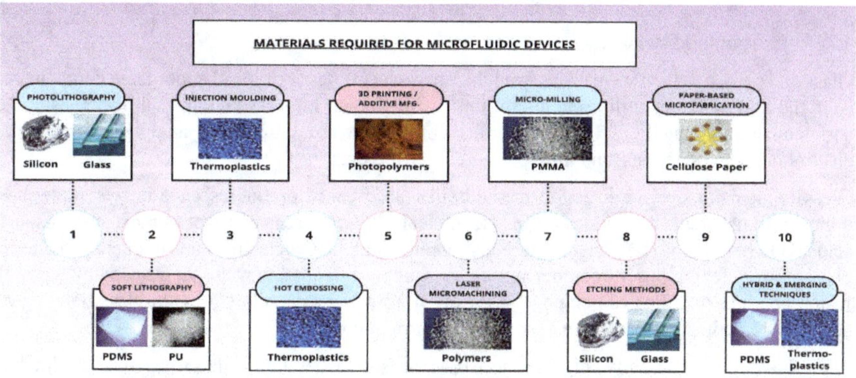

Figure 2. *An illustration depicting about various materials required for microfluidic devices.*

4.1 Polymeric Materials

The most popular materials in microfluidics are the polymeric components due to their low cost, optical transparency, flexibility, and compatibility with various production methods. The most well-known of these are cyclic olefin copolymer (COC), polymethyl methacrylate (PMMA), and polydimethylsiloxane (PDMS).

Because it is transparent, gas-permeable, biocompatible, and simple to mold using soft lithography, PDMS is mostly preferred in the academic settings. It is perfect for biological tests and prototyping. However, PDMS suffers from issues like absorption of hydrophobic molecules and mechanical deformation under pressure [45].

In addition to having superior optical qualities, PMMA is more robust and resistant to chemicals than PDMS. It is frequently utilized in hot embossing, injection molding, and laser micromachining. Although PMMA is easy to produce due to its low glass transition temperature, it is less resistant to organic solvents and is more likely to shatter under stress [46].

Because of its superior optical clarity, minimal water absorption, and chemical inertness, COC is being utilized more and more in commercial diagnostics. It has superior barrier qualities over PMMA or PDMS and works well with hot embossing, injection molding, and laser ablation. COC was demonstrated for its effective use in multi-layer microfluidic systems using freeform fabrication and solvent bonding [47].

Materials Research Foundations 191 (2026) 1-35 https://doi.org/10.21741/9781644904213-1

Fabrication compatibility;

- PDMS: Ideal for soft lithography and hybrid bonding with glass/PMMA.
- PMMA: Suited for laser micromachining, micro-milling, embossing.
- COC: Best for high-volume injection moulding and solvent bonding.

4.2 Inorganic Materials

Glass and silicon, which were taken from the microelectronics sector as the inorganic compounds, were the first materials utilized in microfluidics due to their high resolution, chemical resistance, and temperature stability. These materials continue to be crucial for analytical chemistry, electrokinetic applications, and high-pressure microfluidics.

Glass is ideal for optical detection methods like absorbance or fluorescence assays since it is optically transparent. It provides a consistent, clean surface that is resistant to most solvents and acids and chemically inert for bioanalytical applications. Microchannels are frequently made in glass by wet etching, laser micromachining, or sandblasting, and they are then sealed by thermal fusion bonding or adhesive bonding [48]. Recent advancements allow glass-to-glass bonding at bench scale, reducing reliance on expensive cleanroom processes.

Silicon, while opaque, is widely used in MEMS-based microfluidics due to its semiconducting properties and compatibility with photolithography, dry etching, and thin film deposition. It is ideal for pressure-resistant and thermal applications, such as PCR microchips or microheaters. Silicon's major limitation is lack of optical transparency, which restricts its use in detection-intensive applications unless paired with a transparent layer.
In fluidics, silicon and glass can perform better than polymers in terms of thermal performance and vibrational synchronization, particularly when manipulating particles or cells at high energies [49].

Fabrication compatibility:

- **Glass**: Suitable for wet etching, laser micromachining, thermal bonding, and cleanroom fabrication.
- **Silicon**: Best for photolithography, dry (Reactive Ion Etching (RIE) / Deep Reactive Ion Etching (DRIE)) etching, and integration with electronics (MEMS/NEMS).

4.3 Paper-Based Substrates

Paper has become a revolutionary material thanks to microfluidic paper-based analytical devices (μPADs), commonly referred to as low-cost, point-of-care (POC) microfluidic devices. It is widely used because of its abundance, biodegradability, portability, and capacity to facilitate passive fluid movement via capillary action, which eliminates the need for pumps or other power sources.

Cellulose filter paper and chromatography paper are frequently used paper types because of their consistent porosity and hydrophilicity. Particularly in situations with limited resources, these substrates are perfect for colorimetric detection, immunoassays, and environmental diagnostics.

Paper-based microfluidics frequently uses screen, laser, inkjet, or wax printing to produce hydrophobic barriers that direct fluid flow inside hydrophilic zones. For example, wax printing

Materials Research Foundations 191 (2026) 1-35
https://doi.org/10.21741/9781644904213-1

creates impenetrable barriers by depositing molten wax designs that pierce the paper fibers [50]. Newer techniques like origami-based 3D μPADs expand capabilities by enabling multi-step reactions and multiplexed detection.

Paper-based substrates are scalable and reasonably priced, yet they have drawbacks. Among these include low resolution (usually greater than 100 μm), variable flow rates brought on by fiber variability, and restricted durability in adverse environments. Paper's porous porosity also makes it more difficult to integrate with optical or electrical sensors and limits accurate volume control.

Fabrication compatibility:

- It works best with roll-to-roll processing, inkjet printing, screen printing, laser cutting, and wax printing.
- Incompatible with conventional cleanroom-based photolithography or high-temperature processing.

4.4 Hydrogels and Biological Materials

Hydrogels are soft, water-swollen polymers that closely resemble the extracellular matrix of biological tissues. They are perfect for biomimetic microfluidic systems such as organ-on-chip models and cell culture platforms. Hydrogels, which can be made of natural or synthetic polymers such as collagen, alginate, gelatin, agarose, or polyethylene glycol (PEG), are perfect for promoting cell adhesion, migration, and proliferation because of their high-water content, biocompatibility, and adjustable mechanical characteristics [51].

Natural hydrogels with built-in bioactivity, such collagen or gelatin, work well for structures that are packed with cells. PEG and other synthetic hydrogels provide more control over mechanical strength and rate of degradation, which makes them appropriate for perfusion systems and long-term cultures.

Hydrogels can be structured using micro-moulding, photopolymerization, or microfluidic flow-focusing. For example, photopolymerizable PEG-diacrylate (PEGDA) can be patterned using UV lithography, enabling the fabrication of accurate microchannel geometries and hydrogel scaffolds for 3D cell culture [52].

Hydrogels are also being used to create bio-inks for cell-rich microfluidic devices using 3D printing techniques like stereolithography. This combination enhances the structural and biological properties of microfluidic devices used in drug screening, tissue engineering, and disease modelling.

However, hydrogels generally have low mechanical strength and can deform or dry out over time. They are also incompatible with high-temperature processing, limiting their integration with rigid materials like glass or silicon.

Fabrication compatibility:

- Compatible with micro-moulding, photolithography (with PEGDA), soft lithography, and 3D bioprinting.
- Not compatible with high-temperature or high-pressure techniques like injection moulding or hot embossing.

Materials Research Foundations 191 (2026) 1-35 https://doi.org/10.21741/9781644904213-1

4.5 Composite and Hybrid Materials

Composite and hybrid materials offer customized solutions for high-performance, multipurpose microfluidic devices by combining two or more different materials to capitalize on their unique characteristics. Advanced applications such as wearable sensors, flexible electronics, organ-on-a-chip systems, and stimuli-responsive diagnostics are using these materials more and more.

A mutual hybrid pairing is PDMS-glass or PDMS-PMMA, where PDMS offers flexibility and biocompatibility, while the rigid counterpart (glass or PMMA) contributes mechanical support and optical transparency. This pairing allows for easy bonding via plasma activation, and is frequently used in applications requiring microscopy-based detection, like cell migration assays or drug screening platforms [53].

Nanocomposites—such as PDMS mixed with nanoparticles (e.g., graphene, gold, silver, or carbon nanotubes)—improve electrical conductivity, sensitivity, or chemical resistance. These are essential in developing electrochemical biosensors, wearable devices, and microelectrode-integrated systems.

Hybrid fabrication also enables multilayer architectures, where different materials are processed layer-by-layer using a combination of 3D printing, soft lithography, and lamination. For example, Glick demonstrated a double-sided microfluidic chip using a 3D printed mould for PDMS casting, integrating complex vertical interconnects for cell culture studies [54].

These merged systems provide the benefit of customized mechanical, optical, or chemical properties, but their fabrication often requires complex bonding techniques, material compatibility assessments, and optimized process parameters.

Fabrication compatibility:

- Compatible with soft lithography, 3D printing, lamination, plasma bonding, and moulding techniques.
- Requires careful thermal, chemical, or surface activation matching between layers.

5. Material Properties and Selection Criteria

5.1 Biocompatibility and Bio-functionality

Biocompatibility is a crucial factor to take into account when selecting materials for microfluidic systems utilized in biomedical applications such as organ-on-chip, drug delivery, and diagnostics. Materials must not release hazardous leachable or trigger immunological reactions. Because of their exceptional biocompatibility and capacity to promote cell survival and adhesion, PDMS, COC, PMMA, and specific hydrogels like PEG and collagen are frequently utilized [55].

Hydrogels, due to their high-water content and bioactivity, are particularly effective in mimicking extracellular matrices, which makes them ideal for tissue engineering and cell culture applications. Surface treatments like oxygen plasma or UV-ozone also enhance biocompatibility and reduce nonspecific adsorption in thermoplastics like PC and COC [56].

Materials Research Foundations 191 (2026) 1-35 https://doi.org/10.21741/9781644904213-1

5.2 Optical and Thermal Properties

Materials with low autofluorescence and high optical clarity are required for optical detection techniques in microfluidics, such as fluorescence and absorbance. Because of their transparency in the UV-visible spectrum, materials for fabrication such as glass, COC, PMMA, and PDMS are frequently used [57].

Materials with excellent heat resilience are required for thermal cycling applications like PCR chips. Because silicon and glass are resistant to thermal distortion, they are both great options. Thermally active devices can be rapidly prototyped using thermoplastics such as carbon-based polymers (COC) and polyimide (PI) due to their modest heat resistance [58].

5.3 Mechanical and Chemical Stability

A microfluidic device's motorized integrity guarantees that it will retain its structural integrity when subjected to pressure-driven flow or manipulation. Excellent motorized strength is provided by silicon, glass, and rigid thermoplastics like PC and COC, which are appropriate for systems linking high pressure or prolonged flow [59].

Chemical confrontation is essential for working with solvents, surfactants, or biological reagents. While PDMS is versatile, it tends to absorb small hydrophobic molecules, which can affect reproducibility. In contrast, COC, PC, and PMMA show better chemical inertness, making them ideal for biosensing and sample preparation involving organic solvents [59].

5.4 Surface Functionalization and Modification

Surface modification tailor's materials for specific functions, such as cell adhesion, electrochemical sensing, or reducing nonspecific adsorption. Techniques include plasma treatment, salinization, polymer grafting, or protein coating. For instance, plasma oxidation of PDMS introduces hydroxyl groups, enhancing hydrophilicity and bonding to glass [60].

Functionalization is chiefly important in biosensors and cell-based devices, where surface chemistry governs selectivity and sensitivity. Some materials like SU-8 and silk fibroin naturally offer functional groups for binding biomolecules, expanding their utility in optofluidic and analytical devices [61]. Figure 3 depicts about the various properties of the materials employed in the Fabrication of Microfluidic materials, while table 3 summarizing about various materials used in fabrication of microfluidic techniques.

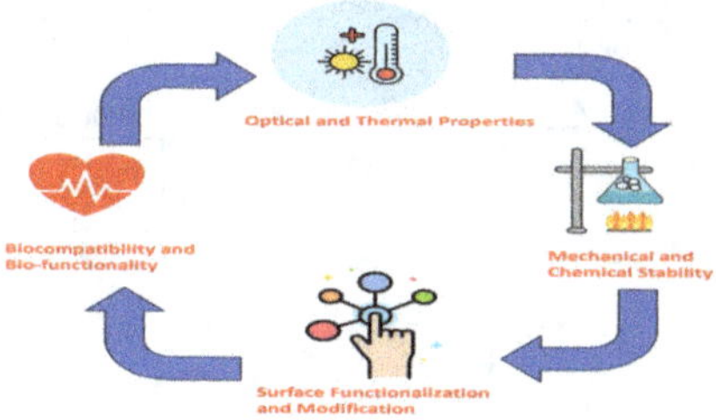

Figure 3. *An Illustration about the various properties of the materials employed in the Fabrication of Microfluidic materials.*

Table 3. *Comparison of various materials with properties and their selection criteria for fabrication of microfluidic devices.*

Name of Material	Biocompatibility	Optical / Thermal Properties	Mechanical / Chemical Stability	Surface Functionalization	References
PDMS	Excellent for cell culture; gas-permeable	High optical clarity; limited thermal stability	Flexible but absorbs hydrophobic molecules	Easy with plasma oxidation; supports salinization and protein coating	[45–47]
PMMA	Good for biosensors and drug assays	Excellent transparency; low Tg (~105°C)	Brittle; chemically inert; moderate mechanical strength	Supports UV and chemical activation; less versatile than PDMS	[55]
COC	Highly biocompatible thermoplastic	Very low autofluorescence; high thermal resistance	High rigidity; excellent solvent resistance	Good with UV ozone; supports microfluidic bonding techniques	[56]
Glass	Inert and non-toxic; ideal for long-term assays	Transparent and thermally stable	Mechanically robust; chemically inert	Plasma or silane coating enables biofunctionalization	[57,58]
Silicon	Biocompatible but opaque; suitable for MEMS	Opaque; excellent heat conduction	High mechanical strength; chemically stable	Compatible with lithographic coatings and surface salinization	[56,57]
Hydrogels (PEG, Gelatin)	Excellent; mimic extracellular matrix	Translucent; limited to moderate thermal stability	Soft; prone to dehydration; limited chemical stability	Highly tailorable via crosslinking, functional peptides, or bioactive molecules	[58]
Paper	Opaque or translucent; no thermal resistance	Fragile; low chemical resistance	Limited, but wax printing and ink functionalization possible	Opaque or translucent; no thermal resistance	[22-24]
Composites (e.g. PDMS + CNTs)	Tunable bio-functionality depending on fillers	Varies by base material; enhancements possible	Strength, conductivity, and chemical properties can be engineered	Functionalization depends on filler and base material; often advanced chemistries needed	[45–47]
SU-8	Good biocompatibility; used in viral detection systems	Transparent; stable up to 200°C	Chemically and thermally stable	Supports biomolecule immobilization via amine or epoxy groups	[68]

Materials Research Foundations 191 (2026) 1-35 https://doi.org/10.21741/9781644904213-1

6. Integration of Fabrication Techniques and Materials

6.1 Technique-Material Compatibility Matrix

The integration of suitable materials with fabrication techniques is necessary for microfluidic devices to be structurally sound, economical, and useful. For instance, PDMS is the ideal complement to soft lithography due to its flexibility and simplicity of demoulding. High-throughput replication is made possible by the greater compatibility of COC and PMMA with hot embossing or injection molding [62]. In contrast, glass and silicon, used for precise applications, are best processed using wet/dry etching or MEMS-based photolithography.

Hybrid systems are acquiring momentum. For instance, 3D printed moulds for PDMS allow incorporation of custom geometries with soft materials [63], while sacrificial materials like photosensitive polycarbonates can be patterned via lithography and decomposed to create hollow channels in a range of substrates [64].

6.2 Optimization Strategies

Optimization involves tuning fabrication parameters for stronger bonding, better resolution, and reduced defects. For example, in PDMS multilayer systems, adjusting curing temperature and incorporating vent channels helps reduce trapped air and improve bonding strength [65].

Design optimization can also be software-driven. Computer-aided design (CAD) and simulation tools allow prediction of fluid dynamics, thermal effects, and electromechanical behaviour, ensuring the selected materials and techniques deliver the desired output [66].

6.3 Common Challenges and Practical Solutions

Challenges in integration include material incompatibility, delamination, and inconsistent bonding. For instance, PDMS often fails to bond reliably with thermoplastics or glass unless treated by oxygen plasma or adhesive layers [67].

Another issue is thermal mismatch, which can lead to distorting or cracking when combining materials like PDMS and silicon. Solutions include temperature-controlled curing and the application of specialty resins or intermediate adhesion layers like SU-8 [68].

6.4 Case Studies and Examples

A successful example is the incorporation of PDMS microchannels with a Surface Plasmon Resonance (SPR) sensor using low-cost materials like PMMA and Tygon tubing. This setup allowed real-time biosensing while keeping fabrication accessible [69].

Another case involves the use of polyimide-based channels etched with dry plasma methods, creating microfluidics well-matched with biosensor integration due to polyimide's thermal and chemical stability [70]. 3D printed elastomeric membranes with through-holes have also enabled compact, high-throughput platforms for DNA and protein immobilization, demonstrating how incorporation can solve miniaturization encounters in diagnostics [71].

7. Recent Advances and Future Directions

7.1 Sustainable and Eco-Friendly Fabrication

Eco-friendly materials and fabrication techniques in microfluidics have become more popular as a result of the drive for sustainability. Interestingly, paper-based microfluidics is at the forefront, providing portable, affordable, and biodegradable options for environmental monitoring and diagnostics. These tools are perfect for point-of-care applications in environments with limited resources because they are simple to fabricate using wax or inkjet printing and produce little chemical waste [72]. Additionally, microfluidic fuel cells present a sustainable path to power portable systems by converting chemical energy directly within the chip [73].

7.2 Smart and Responsive Materials

For dynamic fluid flow and sensing, smart microfluidics incorporates sensitive materials, such as hydrogels that react to pH, temperature, or chemicals. Without electronics, these systems can function as self-regulating sensors, pumps, or valves. For example, in organ-on-a-chip devices, sensitive hydrogels can swell or contract to regulate flow, allowing for passive mixing and actuation.[74]. These materials are also key to targeted drug delivery, offering smart microcapsules for on-demand release [75].

7.3 Artificial Intelligence-Assisted Design and Fabrication

The integration of artificial intelligence (AI) and machine learning into microfluidics enables predictive modelling, design automation, and optimization. AI reduces trial-and-error in device development by helping with flow simulation, material selection, and error correction. AI is also essential for creating intelligent diagnostic systems that combine real-time sensing, processing, and interpretation [76]. This trend is especially evident in wearable biosensors and portable diagnostics [77].

7.4 Toward Fully Integrated Lab-on-a-Chip Systems

The goal of recent initiatives is to develop fully integrated LOC systems that incorporate data processing, actuators, sensors, and fluidics on a single platform. These systems independently carry out intricate analysis, from sample input to result output. Advances in hybrid microfluidic-integrated circuit (IC) chips are enabling previously unattainable speeds and sizes for high-throughput cytometry and biosensing [78]. Such technologies are foundational to organ-on-chip, personalized medicine, and mobile health diagnostics [79]. The recent advances and future directions in microfluidics along with various pros and cons, future purpose views, and applications are exhibited in **table 4.**

Table 4. *A summary of* recent advances and future directions in microfluidics *including various* advantages, disadvantages, future directions, and applications.

Section	Description	Advantages	Disadvantages	Future Directions	Applications	Refs.
Sustainable and Eco-Friendly Fabrication	Focuses on eco-friendly materials and techniques such as paper-based microfluidics and microfluidic fuel cells that promote sustainability.	Low-cost and biodegradable Minimal chemical waste Simple fabrication (wax or inkjet printing) Suitable for low-resource settings	Limited durability Restricted to aqueous fluids Lower precision than polymer-based chips	Development of recyclable polymers- Integration with renewable energy systems Improvement in multi-layer paper devices	Environmental monitoring Point-of-care diagnostics- Portable biosensors	[72,73]
Smart and Responsive Materials	Uses stimuli-responsive materials (e.g., hydrogels) that react to pH, temperature, or chemicals for self-regulating systems.	No need for external electronics Enables self-actuating systems High sensitivity and adaptability Useful for dynamic control	Complex material synthesis Slower response time Limited mechanical stability	Integration with electronic sensing Use in autonomous organ-on-chip models Enhanced drug-release precision	Organ-on-a-chip- Smart sensors targeted drug delivery	[74,75]
AI-Assisted Design and Fabrication	Employs AI and machine learning for predictive modeling, automated design, and process optimization in microfluidics.	Reduces trial-and-error Enhances precision and efficiency Accelerates prototyping enables smart diagnostics	Requires large datasets High computational cost Algorithm dependency may reduce flexibility	AI-driven real-time adaptive control Cloud-connected fabrication systems Autonomous chip design	Wearable biosensors Portable diagnostic devices Automated fluidic system optimization	[76,77]
Toward Fully Integrated Lab-on-a-Chip Systems	Combines sensing, actuation, data processing, and fluidics into a single, autonomous platform for complete analysis.	Compact and multifunctional High throughput and speed Minimal sample and reagent use enables on-site analysis	Complex fabrication Expensive integration process Power and data handling limitations	Hybrid microfluidic IC chip development Miniaturized multi-analyte platforms Integration with AI and IoT systems	Personalized medicine Organ-on-a-chip- Mobile and telemedicine diagnostics	[78,79]

8. Patents

Microfluidic device patents are now essential to the innovation and commercialization of LOC technology. These devices, which deal with fluids at the microscale, are widely used in drug research, diagnostics, and environmental monitoring. Both technological maturity and growing market importance are reflected in the spike in patent activity over the last 20 years. Key trends in microfluidic patents, such as invention domains, design approaches, and commercialization channels, are examined in such synthesis.

Microfluidics enables rapid, low-cost, and high-throughput analysis, particularly for biomedical and pharmaceutical applications. Because of their transformative potential, protecting these technologies through patents is vital for fostering industry partnerships and securing investment. Yetisen and Volpatti emphasize that patent protection complements academic publishing and helps researchers transition discoveries into market-ready solutions. They outline the legal frameworks for filing patents in the US, EU, and Japan and recommend balancing consultancy, licensing, and academic output to maximize impact [80].

The primary areas of innovation in microfluidics are also shown by patent analysis. Additionally, Yesil-Celiktas found patent patterns in microfluidic applications connected to enzymes. These include new materials like ionic liquids, integrated microreactors, and innovative biomolecule immobilization techniques that indicate the convergence of biology, chemistry, and engineering [81].

Design strategies are another critical focus in patent literature. Lee et al. (2018) analysed over 200 U.S. patents and extracted 36 reusable design strategies, helping engineers generate new perceptions more effectively. These strategies not only enhance creativity but also reduce development time for commercial microfluidic platforms [82].

Further emphasizing the global impact of microfluidic IP, Qiao et al. mapped the diffusion of microfluidic technologies across disciplines using patent citation networks. Their findings underscore the interdisciplinary reach of microfluidics and how patents facilitate technological transfer across sectors like biotechnology and materials science [83]. In summary, from inception to commercialization, patents are critical to the development of microfluidic devices. In addition to documenting innovation and maintaining competitive advantages, they provide a blueprint for future research and product development. In order to convert microfluidic breakthroughs into scalable solutions, strategic patenting will remain crucial as the sector develops.

9. Market commercialization

Due to improvements in design, scalability, and application scope, the commercialization of microfluidic manufacturing methods and devices has grown significantly during the past 20 years. Low-cost, compact, and incredibly effective platforms for environmental monitoring, drug research, and diagnostics are provided by microfluidic systems, also known as LOC devices. These characteristics, together with developing fabrication techniques, are driving microfluidics' entry into general markets.

However, technical viability alone doesn't ensure market success. Volpatti and Yetisen stress that commercialization requires more than device performance—it must meet real-world needs and integrate into existing workflows. Profitable businesses have concentrated on specialized markets like environmental sensing and point-of-care diagnostics, utilizing microfluidics' advantages in automation and miniaturization [84].

Microfluidic diagnostics commercial success stories demonstrate these tactics. The development of point-of-care diagnostic devices was examined by Chin et al., who observed that although numerous ideas have shown promise, very few have been successfully brought to market. Regulatory obstacles, manufacturing costs, and the requirement for strong fluidic and electrical component integration are some of the obstacles. Their research emphasizes that improving device performance, manufacturing scalability, and user experience are typically necessary for commercial viability [85].

Microfluidics has a well-established economic potential. According to Kim, as early as the 2010s, microfluidic diagnostics was poised to take a sizable chunk of the in vitro diagnostics business, which was valued at over $40 billion. The technology's advantages—faster turnaround, reduced reagent consumption, and portability—fit very nicely with the healthcare industry's drive for decentralization and customization [86].

Looking forward, Bohr et al. predict that the integration of microfluidics with wearable devices and personalized drug delivery systems will further expand market opportunities. Emerging manufacturing techniques like 3D printing and miniaturized chip-based drug production may facilitate on-demand pharmaceutical fabrication at clinical sites [87]. With respect to commercialization of various techniques for desire polymeric microfluidic device fabrication, various features are required to understand based on the reported study [1,41,53,88-90], that is comparatively depicted in table 5.

Table 5*. Summary of comparative features of various fabrication technology for polymeric microfluidic devices.*

<table>
<tr><th rowspan="2">Machine/Technology name</th><th colspan="6">Comparative features on various heads [1, 41, 53, 88-90]</th></tr>
<tr><th>Cost of the machine</th><th>Compatible for desired purpose using various materials</th><th>Technologies capability to detect smallest change in fabrication</th><th>Complex in the design of microfluidic device</th><th>Operating speed for the process</th><th>Material passing for the operation</th></tr>
<tr><td>3D Printable system</td><td>Comparatively less in cost</td><td rowspan="3">Challenge is there in selection based on materials properties for desire output</td><td>Slightly moderate</td><td>Highest</td><td>Slow</td><td>Slow</td></tr>
<tr><td>Micro-level embossing technique</td><td>Moderately in cost</td><td>Highest</td><td>Lowest</td><td>Fast</td><td>Rapid</td></tr>
<tr><td>Micro-level injectable molding</td><td>Highly expensive</td><td>Highest</td><td>Lowest</td><td>Very fast</td><td>Highly rapid</td></tr>
<tr><td>Laser-based micro-level machine</td><td>Comparatively less in cost</td><td rowspan="2">Highly utilized for various materials</td><td>Moderate</td><td>Moderate</td><td>Moderate</td><td>Moderate in operation</td></tr>
<tr><td>CNC-based micro-level machine</td><td>Moderately in cost</td><td>Moderate</td><td>Moderate</td><td>Moderate</td><td>Moderate in operation</td></tr>
<tr><td>PDMS-based machine</td><td>Very cheap in comparison to other techniques</td><td>Very less utilized based on materials properties</td><td>Highest</td><td>Lowest</td><td>Very slow</td><td>Very slow</td></tr>
</table>

Conclusion

In conclusion, microfluidic devices to be commercialized, fabrication technology must be matched with actual market needs. Microfluidics has the potential to revolutionize several industries, including medicines and diagnostics, as fabrication techniques become more affordable and scalable and integration issues are addressed.

Microfluidic devices have transformed the way fluids are handled at the microscale, making a significant impact in diagnostics, medication delivery, environmental monitoring, and LOC systems. As discussed in this research, their development is dependent on the evolution of several fabrication techniques, ranging from traditional photolithography to developing 3D printing and paper-based approaches, each having unique advantages in terms of cost-effectiveness, scalability, and resolution. The requirements of a specific application must guide the choice of materials, such as glass, polymers, or hybrid composites, achieving a balance between durability, biocompatibility, and functional integration.

The combination of fabrication processes with appropriate materials is critical for developing efficient and dependable microfluidic platforms. AI-assisted design, smart materials, and sustainable manufacturing processes point to a bright future of fully integrated, environmentally friendly lab-on-chip systems. Patent trends emphasize continued scientific advancement, while commercialization efforts reflect a maturing market that is more focused on scalable, real-world solutions in healthcare and industry.

In summary, microfluidics represents a dynamic and rapidly evolving field. Its interdisciplinary nature, coupled with ongoing innovation in fabrication and material science, ensures it will remain at the forefront of modern scientific and industrial development.

References

[1] Y.J. Juang, Y.J. Chiu, Fabrication of polymer microfluidics: An overview, Polymers. 14 (2022) 1-18. https://doi.org/10.3390/polym14102028

[2] R. Yuan, J. Lee, H.W. Su, E. Levy, T. Khudiyev, J. Voldman, Y. Fink, Microfluidics in structured multimaterial fibers, Proc. Natl. Acad. Sci. U.S.A. 115 (2018) E10830-E10838. https://doi.org/10.1073/pnas.1809459115

[3] H. Schroeder, M. Adler, K. Gerigk, B.M. Chorus, F. Gotz, C.M. Niemeyer, User configurable microfluidic device for multiplexed immunoassays based on DNA-directed assembly, Anal. Chem. 81 (2009) 1275-1279. https://doi.org/10.1021/ac802228k

[4] G.M. Whitesides, The origins and the future of microfluidics, Nature. 442 (2006) 368-373. https://doi.org/10.1038/nature05058

[5] N. Convery, N. Gadegaard, 30 years of microfluidics, Micro Nano Eng. 2 (2019) 76-91. https://doi.org/10.1016/j.mne.2019.01.003

[6] P.N. Nge, C.I. Rogers, A.T. Woolley, Advances in microfluidic materials, functions, integration, and applications, Chem. Rev. 113 (2013) 2550-2583. https://doi.org/10.1021/cr300337x

[7] A. Aryasomayajula, P. Bayat, P. Rezai, P.R. Selvaganapathy, Microfluidic devices and their applications, in: B. Bhusan (Eds.), Springer handbook of nanotechnology, Springer Berlin Heidelberg, Springer, Switzerland, 2017, pp. 487-536. https://doi.org/10.1007/978-3-662-54357-3_16

[8] R. Barry, D. Ivanov, Microfluidics in biotechnology, J. Nanobiotechnol. 2 (2004) 1-5. https://doi.org/10.1186/1477-3155-2-2

[9] K. Karimi, A. Fardoost, N. Mhatre, J. Rajan, D. Boisvert, M. Javanmard, A thorough review of emerging technologies in micro- and nanochannel fabrication: Limitations, applications, and comparison, Micromachines. 15 (2024) 1-29. https://doi.org/10.3390/mi15101274

[10] W. Chen, R.H.W. Lam, J. Fu, Photolithographic surface micromachining of polydimethylsiloxane (PDMS), Lab Chip. 12 (2012) 391-395. https://doi.org/10.1039/C1LC20721K

[11] J. Christoffersson, C.F. Mandenius, Fabrication of a microfluidic cell culture device using photolithographic and soft lithographic techniques, Methods Mol. Biol. 1994 (2019) 227-233. https://doi.org/10.1007/978-1-4939-9477-9_21

[12] Y. Lin, C. Gao, D. Gritsenko, R. Zhou, J. Xu, Soft lithography based on photolithography and two-photon polymerization, Microfluid. Nanofluid. 22 (2018) 1-27. https://doi.org/10.1007/s10404-018-2118-5

[13] J. Wu, M. Gu, Microfluidic sensing: State of the art fabrication and detection techniques, J. Biomed. Opt. 16 (2011) 1-11. https://doi.org/10.1117/1.3607430

[14] P. Juskova, A. Ollitrault, M. Serra, J.L. Viovy, L. Malaquin, Resolution improvement of 3D stereolithography through direct laser trajectory programming: Application to microfluidic deterministic lateral displacement device, Anal. Chim. Acta. 1000 (2018) 239-247. https://doi.org/10.1016/j.aca.2017.11.062

[15] I. Ahmed, K. Sullivan, A. Priye, Multi-resin masked stereolithography (MSLA) 3D printing for rapid and inexpensive prototyping of microfluidic chips with integrated functional components, Biosensors. 12 (2022) 1-17. https://doi.org/10.3390/bios12080652

[16] S. Smith, M. Sypabekova, S. Kim, Double-sided tape in microfluidics: A cost-effective method in device fabrication, Biosensors. 14 (2024) 1-28. https://doi.org/10.3390/bios14050249

[17] P.G. Shankles, A.C. Timm, M.J. Doktycz, S.T. Retterer, Fabrication of nanoporous membranes for tuning microbial interactions and biochemical reactions, J. Vac. Sci. Technol. B Nanotechnol. Microelectron. 33 (2015) 1-8. https://doi.org/10.1116/1.4932671

[18] A.K. Sen, A. Raj, U. Banerjee, S.R. Iqbal, Soft lithography, molding, and micromachining techniques for polymer micro devices, Methods Mol. Biol. 1906 (2019) 13-54. https://doi.org/10.1007/978-1-4939-8964-5_2

[19] B.K. Gale, A.R. Jafek, C.J. Lambert, B.L. Goenner, H. Moghimifam, U.C. Nze, S.K. Kamarapu, A review of current methods in microfluidic device fabrication and future commercialization prospects, Inventions. 3 (2018) 1-25. https://doi.org/10.3390/inventions3030060

[20] J. Xu, M. Harasek, M. Gfohler, From soft lithography to 3D printing: Current status and future of microfluidic device fabrication, Polymers. 17 (2025) 1-22. https://doi.org/10.3390/polym17040455

[21] S.A. Jaywant, M.A.A. Rehmani, T. Nayak, K. Mehmood, Fabrication and characterization of 3D printed microfluidics, in: J. Billingsley, P. Brett (Eds.), Mechatronics and Machine Vision in Practice 4, Springer Nature Switzerland AG, Springer Cham., Switzerland, 2021, pp. 119-124. https://doi.org/10.1007/978-3-030-43703-9_10

[22] G. Jenkins, Y. Wang, Y.L. Xie, Q. Wu, W. Huang, L. Wang, X. Yang, Printed electronics integrated with paper-based microfluidics: New methodologies for next-generation health care, Microfluid. Nanofluid. 19 (2015) 251-261. https://doi.org/10.1007/s10404-014-1496-6

[23] T. Thomas, A. Agrawal, Design and fabrication of microfluidic devices: A cost-effective approach for high throughput production, J. Micromech. Microeng. 34 (2024) 1-21. https://doi.org/10.1088/1361-6439/ad104b

[24] R.M. Roller, M. Lieberman, Beyond wax printing: The future of paper analytical device fabrication, Sens. Actuators B: Chem. 392 (2023) 1-14. https://doi.org/10.1016/j.snb.2023.134059

[25] Y.J. Chuah, Z.T. Heng, J.S. Tan, L.M. Tay, C.S. Lim, Y. Kang, D.A. Wang, Surface modifications to polydimethylsiloxane substrate for stabilizing prolonged bone marrow stromal cell culture, Colloids Surf. B: Biointerfaces. 191 (2020) 1-10. https://doi.org/10.1016/j.colsurfb.2020.110995

[26] P.M.V. Midwoud, A. Janse, M.T. Merema, G.M.M. Groothuis, E. Verpoorte, Comparison of biocompatibility and adsorption properties of different plastics for advanced microfluidic cell and tissue culture models, Anal. Chem. 84 (2012) 3938-3944. https://doi.org/10.1021/ac300771z

[27] Q. Chen, G. Li, Y. Nie, S. Yao, J. Zhao, Investigation and improvement of reversible microfluidic devices based on glass-PDMS-glass sandwich configuration, Microfluid. Nanofluid. 16 (2014) 83-90. https://doi.org/10.1007/s10404-013-1222-9

[28] S.A.M. Shaegh, A. Pourmand, M. Nabavinia, H. Avci, A. Tamayol, P. Mostafalu, H.B. Ghavifekr, E.N. Aghdam, M.R. Dokmeci, A. Khademhosseini, Y.S. Zhang, Rapid prototyping of whole-thermoplastic microfluidics with built-in microvalves using laser ablation and thermal fusion bonding, Sens. Actuators B: Chem. 255 (2018) 100-109. https://doi.org/10.1016/j.snb.2017.07.138

[29] H. Cong, N. Zhang, Perspectives in translating microfluidic devices from laboratory prototyping into scale-up production, Biomicrofluidics. 16 (2022) 1-15. https://doi.org/10.1063/5.0079045

[30] A.A.D.B. Yaseen, Integrated microfluidic MEMS and their biomedical applications, in: A.N. Laskovski (Eds.), Biomedical engineering, trends in electronics, communications and software, InTech, Croatia, 2011, pp. 1-20.

[31] A. Amini, R.M. Guijt, T. Themelis, J.D. Vos, S. Eeltink, Recent developments in digital light processing 3D-printing techniques for microfluidic analytical devices, J. Chromatogr. A. 1692 (2023) 1-14. https://doi.org/10.1016/j.chroma.2023.463842

[32] F. Sommonte, N. Denora, D.A. Lamprou, Combining 3D printing and microfluidic techniques: A powerful synergy for nanomedicine, Pharmaceuticals. 16 (2023) 1-24. https://doi.org/10.3390/ph16010069

[33] S.M. Scott, Z. Ali, Fabrication methods for microfluidic devices: An overview, Micromachines. 12 (2021) 1-38. https://doi.org/10.3390/mi12030319

[34] G. Gonzalez, I. Roppolo, C.F. Pirri, A. Chiappone, Current and emerging trends in polymeric 3D printed microfluidic devices, Addit. Manuf. 55 (2022) 1-22. https://doi.org/10.1016/j.addma.2022.102867

[35] D. Kefallinou, M. Grigoriou, D.T. Boumpas, E. Gogolides, A. Tserepi, Fabrication of a 3D microfluidic cell culture device for bone marrow-on-a-chip, Micro Nano Eng. 9 (2020) 1-11. https://doi.org/10.1016/j.mne.2020.100075

[36] J.L. Erkal, A. Selimovic, B.C. Gross, S.Y. Lockwood, E.L. Walton, S. McNamara, R.S. Martin, D.M. Spence, 3D printed microfluidic devices with integrated versatile and reusable electrodes, Lab Chip. 14 (2014) 2023-2032. https://doi.org/10.1039/C4LC00171K

[37] W. Ying, C. Zhou, H. Sun, W. Huang, Paper-based flexible electronic devices: Processing, integration, and applications, npj Flex. Electron. 9 (2025) 1-30. https://doi.org/10.1038/s41528-025-00446-z

[38] T.D. Do, U.T. Pham, L.P. Nguyen, T.M. Nguyen, C.N. Bui, S. Oliver, P. Pham, T.Q. Tran, B.T. Hoang, M.T.H. Pham, D.T.N. Pham, D.T. Nguyen, Fabrication of a low-cost microfluidic device for high-throughput drug testing on static and dynamic cancer spheroid culture models, Diagnostics. 13 (2023) 1-16. https://doi.org/10.3390/diagnostics13081394

[39] V.V. Shapovalov, S.V. Chapek, A.A. Tereshchenko, A.N. Bulgakov, A.P. Bagliy, V.V. Volkov, P.V. Konarev, M.A. Soldatov, S.A. Soldatov, A.A. Guda, A.V. Soldatov, 3D-printed microfluidic system for the in situ diagnostics and screening of nanoparticles synthesis parameters, Micro Nano Eng. 20 (2023) 1-12. https://doi.org/10.1016/j.mne.2023.100224

[40] S. Altundemir, A.K. Uguz, K. Ulgen, A review on wax printed microfluidic paper-based devices for international health, Biomicrofluidics. 11 (2017) 1-26. https://doi.org/10.1063/1.4991504

[41] T. Ching, X. Nie, S.Y. Chang, Y.C. Toh, M. Hashimoto, Techniques and materials for the fabrication of microfluidic devices, in: M. Mozafari (Eds.), Principles of human organs-on-chips, Elsevier Ltd., United Kingdom, 2023, pp. 1-36. https://doi.org/10.1016/B978-0-12-823536-2.00014-6

[42] T. Nguyen, T. Sarkar, T. Tran, S.M. Moinuddin, D. Saha, F. Ahsan, Multilayer soft photolithography fabrication of microfluidic devices using a custom-built wafer-scale PDMS slab aligner and cost-efficient equipment, Micromachines. 13 (2022) 1-10. https://doi.org/10.3390/mi13081357

[43] H. Shafique, V. Karamzadeh, G. Kim, Y. Morocz, A. Sohrabi-Kashani, M.L. Shen, D. Juncker, High-resolution low-cost LCD 3D printing of microfluidics, Lab Chip. 24 (2024) 2774-2790. https://doi.org/10.1039/D3LC01125A

[44] D.E. Damon, Y.S. Maher, M. Yin, F.P.M. Jjunju, I.S. Young, S. Taylor, S. Maher, A.K.B. Tawiah, 2D wax-printed paper substrates with extended solvent supply capabilities allow enhanced ion signal in paper spray ionization, Analyst. 141 (2016) 3866-3873. https://doi.org/10.1039/C6AN00168H

[45] I. Ramos, M. Goncalves, I.M. Goncalves, V. Carvalho, E. Fernandes, R. Lima, D. Pinho, PDMS surface wettability modification and its applications: A systematic review, J. Mol. Liq. 434 (2025) 1-20. https://doi.org/10.1016/j.molliq.2025.127978

[46] T. Limongi, F. Guzzi, E. Parrotta, P. Candeloro, S. Scalise, V. Lucchino, F. Gentile, L. Tirinato, M.L. Coluccio, B. Torre, M. Allione, M. Marini, F. Susa, E.D. Fabrizio, G. Cuda, G. Perozziello, Microfluidics for 3D cell and tissue cultures: Microfabricative and ethical aspects updates, Cells. 11 (2022) 1-16. https://doi.org/10.3390/cells11101699

[47] B.L. Stark, M. Gamboa, A. Esparza, T.J.C. Word, D. Bermudez, L. Carlon, D.A. Roberson, B. Joddar, S.N. Diaz, Materials characterization of stereolithography 3D printed polymer to develop a self-driven microfluidic device for bioanalytical applications, ACS Appl. Bio Mater. 7 (2024) 7883-7894. https://doi.org/10.1021/acsabm.4c00059

[48] Y. Liu, A. Hansen, R.K. Shaha, C. Frick, J. Oakey, Bench scale glass-to-glass bonding for microfluidic prototyping, Microsyst. Technol. 26 (2020) 3581-3589. https://doi.org/10.1007/s00542-020-04819-4

[49] C. Zhang, D. Xing, Y. Li, Micropumps, microvalves, and micromixers within PCR microfluidic chips: Advances and trends, Biotechnol. Adv. 25 (2007) 483-514. https://doi.org/10.1016/j.biotechadv.2007.05.003

[50] R.A. Ruiz, J. L. Gonzalez, M.V. Alvarado, N.W. Martinez, A.W. Martinez, Beyond wax printing: Fabrication of paper-based microfluidic devices using a thermal transfer printer, Anal. Chem. 94 (2022) 8833-8837. https://doi.org/10.1021/acs.analchem.2c01534

[51] K. Mao, X. Min, H. Zhang, K. Zhang, H. Cao, Y. Guo, Z. Yang, Paper-based microfluidics for rapid diagnostics and drug delivery, ACS Omega. 9 (2024) 45537-455474. https://doi.org/10.1021/acsomega.4c07776

[52] K.M. Leong, A.Y. Sun, M.L. Quach, C.H. Lin, C.A. Craig, F. Guo, T.R. Robinson, M.M. Chang, A.O. Olanrewaju, Democratizing access to microfluidics: Rapid prototyping of open microchannels with low-cost LCD 3D printers, ACS Omega. 9 (2024) 45537-45544. https://doi.org/10.1021/acsomega.4c07776

[53] A. Shakeri, S. Khan, T.F. Didar, Conventional and emerging strategies for the fabrication and functionalization of PDMS-based microfluidic devices, Lab Chip. 21 (2021) 3053-3075. https://doi.org/10.1039/D1LC00288K

[54] J. Zheng, S. Ju, X. Zhou, S. Tian, Fabrication of microfluidic devices by 3D printing: Technology, materials, applications and prospects, Results Eng. 28 (2025) 153-156. https://doi.org/10.1016/j.rineng.2025.107270

[55] M.M. Faghih, M.K. Sharp, Solvent-based bonding of PMMA-PMMA for microfluidic applications, Microsyst. Technol. 25 (2019) 3547-3558. https://doi.org/10.1007/s00542-018-4266-7

[56] P.M. van Midwoud, A. Janse, M.T. Merema, G.M.M. Groothuis, E. Verpoorte, Bibliometric study of plastics microfluidic chip from 1994 to 2022: A review, Heliyon. 11 (2025) 1-12.

[57] Q. Chen, G. Li, Y. Nie, S. Yao, J. Zhao, Fabrication of microfluidic devices based on glass-PDMS-glass technology, Microelectron. Eng. 84 (2007) 1265-1269. https://doi.org/10.1016/j.mee.2007.01.276

[58] A. Zulfiqar, A. Pfreundt, W.E. Svendsen, M. Dimaki, Fabrication of polyimide based microfluidic channels for biosensor devices, J. Micromech. Microeng. 25 (2015) 1-8. https://doi.org/10.1088/0960-1317/25/3/035022

[59] Y. Wei, T. Wang, Y. Wang, S. Zeng, Y.P. Ho, H.P. Ho, Rapid prototyping of multi-functional and biocompatible Parafilm®-based microfluidic devices by laser ablation and thermal bonding, Micromachines. 14 (2023) 1-15. https://doi.org/10.3390/mi14030656

[60] Y. Ma, X. Sun, Z. Cai, M. Tu, Y. Wang, Q. Ouyang, X. Yan, G. Jing, G. Yang, Transformation gap from research findings to large-scale commercialized products in microfluidic field, Mater. Today Bio. 29 (2024) 1-19. https://doi.org/10.1016/j.mtbio.2024.101373

[61] J.J. Amsden, P. Domachuk, A. Gopinath, R.D. White, L.D. Negro, D.L. Kaplan, F.G. Omenetto, Rapid nanoimprinting of silk fibroin films for biophotonic applications, Adv. Mater. 22 (2010) 1746-1749. https://doi.org/10.1002/adma.200903166

[62] M.W. Ashraf, S. Tayyaba, N. Afzulpurkar, Micro electromechanical systems (MEMS) based microfluidic devices for biomedical applications, Int. J. Mol. Sci. 12 (2011) 3648-3704. https://doi.org/10.3390/ijms12063648

[63] E. Park, S. Lim, Dynamic phase control with printing and fluidic materials' interaction by inkjet printing an RF sensor directly on a stereolithographic 3D printed microfluidic structure, Lab Chip. 21 (2021) 4364-4378. https://doi.org/10.1039/D1LC00419K

[64] B.A. Peeni, M.L. Lee, A.R. Hawkins, A.T. Woolley, Sacrificial layer microfluidic device fabrication methods, Electrophoresis. 27 (2006) 4888-4895. https://doi.org/10.1002/elps.200600399

[65] M.F.A. Franco, R. Kouider, R.K. Al-Karany, E.M.C. Correa, A. Al-Kassir, Recent advances in polymer science and fabrication processes for enhanced microfluidic applications: An overview, Micromachines. 15 (2024) 1-53. https://doi.org/10.3390/mi15091137

[66] E.E. Tsur, Computer-aided design of microfluidic circuits, Annu. Rev. Biomed. Eng. 22 (2020) 285-307. https://doi.org/10.1146/annurev-bioeng-082219-033358

[67] M. Li, S. Li, J. Wu, W. Wen, W. Li, G. Alici, A simple and cost-effective method for fabrication of integrated electronic-microfluidic devices using a laser-patterned PDMS layer, Microfluid. Nanofluid. 12 (2012) 751-760. https://doi.org/10.1007/s10404-011-0917-z

[68] M.I. Hajam, M.M. Khan, Microfluidics: A concise review of the history, principles, design, applications, and future outlook, Biomater. Sci. 12 (2024) 218-251. https://doi.org/10.1039/D3BM01463K

[69] D.S. Wang, S.K. Fan, Microfluidic surface plasmon resonance sensors: From principles to point-of-care applications, Sensors. 16 (2016) 1-18. https://doi.org/10.3390/s16081175

[70] S. Metz, R. Holzer, P. Renaud, Polyimide-based microfluidic devices, Lab Chip. 1 (2001) 29-34. https://doi.org/10.1039/b103896f

[71] D. Brassard, L. Clime, K. Li, M. Geissler, C. Miville-Godin, E. Roy, T. Veres, 3D thermoplastic elastomer microfluidic devices for biological probe immobilization, Lab Chip. 11 (2011) 4099-4107. https://doi.org/10.1039/c1lc20714h

[72] A.J. Thompson, L.J. Ma, T. Major, M. Jeakle, O.L. Csorba, M.J. Goudie, H. Handa, A.R. Pena, J.A. Potkay, Assessing and improving the biocompatibility of microfluidic artificial lungs, Acta Biomater. 112 (2020) 190-201. https://doi.org/10.1016/j.actbio.2020.05.008

[73] R. Boden, S. Ogden, K. Hjort, Microdispenser with continuous flow and selectable target volume for microfluidic high-pressure applications, J. Microelectromech. Syst. 23 (2014) 452-458. https://doi.org/10.1109/JMEMS.2013.2279976

[74] M.O.O. Oncel, S. Odabas, L. Uzun, D. Hur, B. Garipcan, A facile surface modification of poly(dimethylsiloxane) with amino acid conjugated self-assembled monolayers for enhanced osteoblast cell behavior, Colloids Surf. B: Biointerfaces. 196 (2020) 1-11. https://doi.org/10.1016/j.colsurfb.2020.111343

[75] A. Bhattacharyya, C.M. Klapperich, Thermoplastic microfluidic device for on-chip purification of nucleic acids for disposable diagnostics, Anal. Chem. 78 (2006) 788-792. https://doi.org/10.1021/ac051449j

[76] H.F. Tsai, S. Podder, P.Y. Chen, Microsystem advances through integration with artificial intelligence, Micromachines. 14 (2023) 1-30. https://doi.org/10.3390/mi14040826

[77] D.U.P. Madukwe, M.I.M. Ogburia, N. Nduka, J. Nzeobi, Smart microfluidics: Synergy of machine learning and microfluidics in the development of medical diagnostics for chronic and emerging infectious diseases, Crit. Rev. Biomed. Eng. 51 (2023) 41-58. https://doi.org/10.1615/CritRevBiomedEng.2023047211

[78] V. Iyer, D.A. Issadore, F. Aflatouni, The next generation of hybrid microfluidic/integrated circuit chips: Recent and upcoming advances in high-speed, high-throughput, and multifunctional lab-on-IC systems, Lab Chip. 23 (2023) 2553-2576. https://doi.org/10.1039/D2LC01163H

[79] S. Park, Y. Zhang, S. Lin, T.H. Wang, S. Yang, Advances in microfluidic PCR for point-of-care infectious disease diagnostics, Biotechnol. Adv. 29 (2011) 830-839. https://doi.org/10.1016/j.biotechadv.2011.06.017

[80] A.K. Yetisen, L.R. Volpatti, Patent protection and licensing in microfluidics, Lab Chip. 14 (2014) 2217-2227. https://doi.org/10.1039/c4lc00399c

[81] O.Y. Celiktas, Patenting trends in enzyme related microfluidic applications, Biochem. Eng. J. 92 (2014) 53-62. https://doi.org/10.1016/j.bej.2014.06.017

[82] J.W. Lee, S.R. Daly, A.Y.H. Saad, C.M. Seifert, J. Lutz, Using design strategies from microfluidic device patents to support idea generation, Microfluid. Nanofluid. 22 (2018) 1-22. https://doi.org/10.1007/s10404-018-2089-6

[83] Q. Zheng, L.C. Huang, F.F. Wu, W. Dan, Z. Hui, Analyzing technological knowledge diffusion among technological fields using patent data: The example of microfluidics, Int. J. Innov. Technol. Manag. 14 (2017) 3103-3111. https://doi.org/10.1142/S0219877017400041

[84] L.R. Volpatti, A.K. Yetisen, Commercialization of microfluidic devices, Trends Biotechnol. 32 (2014) 347-350. https://doi.org/10.1016/j.tibtech.2014.04.010

[85] C.D. Chin, V. Linder, S.K. Sia, Commercialization of microfluidic point-of-care diagnostic devices, Lab Chip. 12 (2012) 2118-2134. https://doi.org/10.1039/c2lc21204h

[86] L. Kim, Overview of the microfluidic diagnostics commercial landscape, Methods Mol. Biol. 949 (2013) 65-83. https://doi.org/10.1007/978-1-62703-134-9_5

[87] A. Bohr, S. Colombo, H. Jensen, Future of microfluidics in research and in the market, in: H.A. Santos, D. Liu, H. Zhang (Eds.), Microfluidics for pharmaceutical applications, Elsevier Inc., Netherlands, 2019, pp. 425-465. https://doi.org/10.1016/B978-0-12-812659-2.00016-8

[88] X. Liu, A. Sun, J. Brodsky, I. Gablech, T. Lednicky, P. Voparilova, O. Zitka, W. Zeng, P. Neuzil, Microfluidics chips fabrication techniques comparison, Sci. Rep. 14 (2024) 1-13. https://doi.org/10.1038/s41598-024-80332-2

[89] C.W. Tsao, Polymer microfluidics: Simple, low-cost fabrication process bridging academic lab research to commercialized production, Micromachines. 7 (2016) 1-11. https://doi.org/10.3390/mi7120225

[90] G.S. Fiorini, D.T. Chiu, Disposable microfluidic devices: Fabrication, function, and application, Biotechniques. 38 (2005) 429-446. https://doi.org/10.2144/05383RV02

Materials Research Foundations 191 (2026) 36-71 https://doi.org/10.21741/9781644904213-2

Chapter 2

State-of-the-Art Microfluidic Manufacturing: Processes, Functional Materials, and Analytical Applications

Farnaz Saadi, Afsaneh Mollahosseini*

Research Laboratory of Spectroscopy & Micro and Nano Extraction, Department of Chemistry, Iran University of Science and Technology (IUST), Narmak, Tehran, Iran

amollahosseini@iust.ac.ir*

Abstract

In the last few decades, the microfabrication of devices has transitioned away from standard photolithography and silicon-based micromachining to more advanced techniques like 3D printing and computational model-based design optimization. This evolution was fueled predominantly by the increasing need for portable, miniaturized platforms for high-performance analysis and point-of-care diagnostics to solve complex scientific problems. This chapter provides a snapshot of contemporary fabrication approaches, emphasizing the synergy between technological innovation, functional material designs for maximum optimization, and application specific design limitations. The chapter also touches on existing trends and future directions in the micro-fabrication process, particularly in the development of next-generation lab-on-a-chip technologies that have tremendous prospects in diagnostics and analysis-based science.

Keywords

Microfluidic Analytical Platforms, Point-of-Care Diagnostics, Miniaturized Analytical Systems, Bioanalytical Micro Devices, Lab-on-A-Chip Systems, Sustainable Microfluidic Materials

Contents

Materials Research Foundations 191 (2026) 36-71 https://doi.org/10.21741/9781644904213-2

1. Introduction

Microfluidics has attracted much attention in the analytical chemistry arena due to its ability to facilitate exact measurements in the microscale range [1,2], A specific use of these systems can be found in chemical analysis [1], fluidic sensors [3], and online analytical systems [3,4]. Microfluidics has emerged as a disruptive force in a range of scientific fields that include analytical chemistry, biomedical engineering, and point-of-care diagnostics. Fundamentally, the discipline centers on the controlled fluidic regimes in the channels of the order of micrometers in diameter [1] that provide complete control of fluid flow, reagent mixing [2], and on-chip chemistry [3,5]. These characteristics have enabled the manufacture of integrations that are miniaturized and efficient in configuration which were appropriate for chemical analysis and high throughput. In the early days, microfluidic device fabrication was highly influenced by semiconductor production, and photolithography was a cornerstone technique [6]. These techniques were necessary for the production of chemical microfluidic sensors and on-chip chemical analysis systems [7] in chip form that played a major part in the optimization of chemical detection systems [8]. With the incorporation of the wet and dry etching techniques, the method allowed for the production of chemical or high-resolution micro-structures on silicon or glass substrates. In spite of the precision and reproducibility of the process involved, the method was faced with several limitations that proved to include high cost of operation, dependence on a clean room environment, and a lack of material versatility [9,10]. These materials, among which PDMS stands out, have found extensive application in developing microfluidic chemical sensors because of their bio compatibility [11], optical transparency [12], and chemical stability [13,14] to form portable chemical sensing systems for analysis in real time [15]. These materials paved the way for the invention of techniques like hot embossing [16] and soft-lithography [17,18]. PDMS found

Materials Research Foundations 191 (2026) 36-71 https://doi.org/10.21741/9781644904213-2

extensive application in the academic world due to its mechanical compliance, optical transparency, and bio-compatibility [19]. The availability of soft lithography made the process of prototyping even more democratized through the ability to repattern microchannel designs in a timely manner without the need for elaborate infrastructure [17,20]. The beginning of the 21st century saw a major turning point the use of 3D printing [21] techniques, including stereo-lithography (SLA) [22], digital light processing (DLP) [23], and inkjet-based printing [24], became increasingly integrated into microfluidic production. Such technologies facilitate the direct building of microfluidic elements on an increasingly finer scale and geometrical complexity without the need for in between molds. Importantly, these technologies facilitate the production of embedded and non-planar structures and greatly enhance system level integration [25]. In parallel to these advancements in hardware, computational tools have played an increasingly important role in facilitating the design and fabrication process. Platforms driven by algorithms now aid in the optimization of structural designs and functional material selection through simulated performance-based selection, reducing the duration of the development cycle and enhancing device functionality [26]. Ultrafast laser ablation is an efficient fabrication method. It has found the most use in the fabrication of high-density microfluidic devices for chemical sensing due to the precision it offers in patterning the material to enable efficient chemical recognition and detection. It structures the material directly using ultrafast athermal ablation [27,28]. This process causes minimal thermal. This precision renders it most desirable for applications involving sensitivity in fields of application in areas such as opto-fluidics, as well as biomedical implants [29]. Functional material selection for functional properties remains a highly important consideration in the case of microfabrication. Whereas PDMS has the advantage of ease of use and elasticity, it has the disadvantage of solvent absorption and permeability to gases [30,31]. Thermo-plastics such as PMMA and COC have the advantage of chemical resistance and mechanical stability and are well-suited to scalable production using precision injection molding [32,33]. For applications requiring high chemical inertness and optical transparency, classical materials such as glass and silicon cannot be replaced. As microfluidic technologies continue to mature to enable new applications ranging from pathogen detection to gene editing workflows the discipline is tending towards hybrid fabrication approaches that blend the capabilities of several approaches [34]. In this chapter we aim to present an organized review of existing fabrication strategies, determine their relevance in the domain of analytical chemistry, and outline areas that are expected to lead to future innovation.

2. Classification of Microfluidic Fabrication Techniques

Microfluidic device fabrication involves a broad range of techniques that are supported by distinct physico-chemical properties, material constraints, and resolution capabilities. For ease of systematization of the techniques involved, these process approaches have traditionally been classified into five major categories: subtractive, formative, additive, hybrid, and emerging artificial intelligence-based techniques. Each category has distinct merits and drawbacks that determine their suitability for particular architectures and functions of the microfluidic devices.

2.1 Precision Subtractive Manufacturing

Subtractive fabrications involve Material removal to form definition in microstructure. Photolithography, usually used with wet etching (for example using hydrofluoric acid for glass) [35] or dry etching (RF plasma etching (RIE)/ deep reactive Ion etching (DRIE)), will be the standard for achieving submicron resolution [36]. Photolithography is a pattern transfer process that begins as a photoresist layer, usually a UV-sensitive polymer, is deposited on a planar substrate such as silicon or glass [35] (Fig. 1a). A photomask, containing micro-patterns that are predefined by the user, is carefully aligned whereupon ultraviolet light is propagated through the mask [35–37] (Fig. 1b,c). The exposed regions of the photoresist will be chemically altered to a more soluble (positive resist) or a less soluble (negative resist) state after development in a developer solution [38,39] (Fig. 1d). The developed material is a patterned photoresist that will act as a mask for the subsequent etching steps which will form very specific microscale geometries with lateral resolution (less than microne), which are used in many high-precision microfluidic channels [40]. Wet etching is then performed to dissolve any exposed substrate material (the substrate interacts chemically with the etching solution) [41]. Wet etchants typically consist of hydrofluoric acid (HF) for glass or potassium hydroxide (KOH) for silicon [42]. Wet etching is an isotropic process, which means that it etches backwards, sideways, and forwards simultaneously [40], which causes lateral undercutting beneath the photoresist mask [43]. This lateral undercutting can limit Aspect ratio control, but provides a net uniform removal in depth as required when the ultimate geometry of interest is at shallow depth or if the geometry does not require oversight and depth [43]. In contrast, dry etching employs plasma-assisted techniques that enable anisotropic etching [44], or etching in which the vertical and lateral etch rates differ [45] (Fig. 2). This category includes Reactive Ion Etching (RIE) and Deep Reactive Ion Etching (DRIE). In RIE, reactive gas species (e.g., CF_4, SF_6, O_2, etc.) [46] are introduced into a low-pressure chamber for Plasma generation [46]. The generated ions are accelerated toward the substrate, drawn through an electric field [47]. Subjected to both physical sputtering and chemical etching, RIE is particularly precise as it allows for good control of etch depth and sidewall profile [48,49]. It is valuable for forming vertical microchannels with high fidelity in thermally stable materials [50]. DRIE is a specific high-aspect-ratio variant of RIE that often employs the Bosch process [51,52]. This method consists of etching (with SF_6) and passivating (with C_4F_8) in alternating cycles [53–55], which allows for the growth of long, vertical, Steep-sided features, which can achieve depth/width aspect ratios greater than 20:1 [56]. DRIE is especially useful in the fabrication of MEMS [57] and Microfluidic chip fabrication structures such as high-resolution vertical channels [58] or through-silicon vias (TSVs) [58,59]. Though DRIE can be high-precision, it is expensive and infrastructure intensive [60,61]. DRIE requires a controlled environment and high-cost photomasks [60]. For these reasons, MEMS integration is a good fit [62], while flexible [63] or disposable [64] platforms are usually not [60].

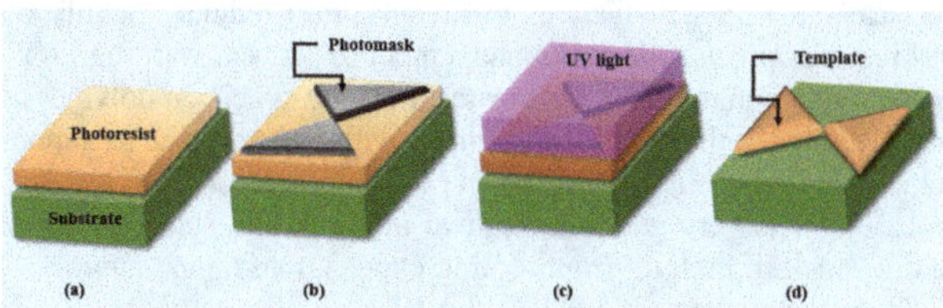

Fig. 1. Deposition of a photoresist layer on a flat substrate such as silicon or glass (a), A photomask containing micropatterns predefined by the user is precisely aligned (b), Ultraviolet light is emitted (c), The areas containing photoresist change shape according to the pattern (d). (Drawn by the author)

Materials Research Foundations 191 (2026) 36-71 https://doi.org/10.21741/9781644904213-2

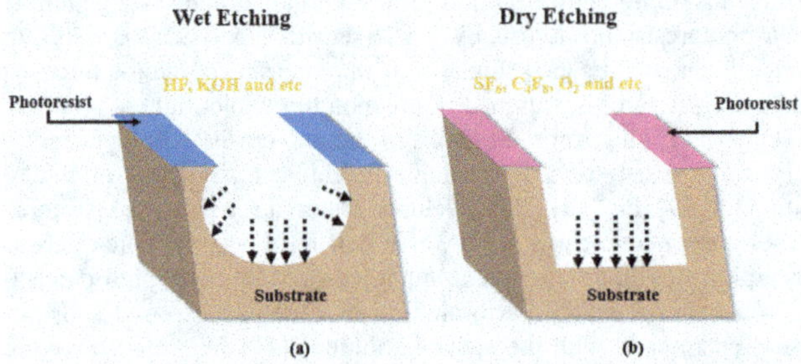

Fig. 2. Wet etching is performed to dissolve any exposed substrate material and etch back, side, and front simultaneously, creating a lateral undercut beneath the photomask (a), Dry etching uses plasma techniques that enable etching where the vertical etching rate varies (b). (Drawn by the author)

2.2 Shape-Formatting Methods

Here, we discuss the multiple shape-forming methods utilized in the fabrication of microfluidics, which have played a pivotal role in the production and design of chemical analysis systems. Unlike subtractive processes, techniques of the shape-forming method comprise deformation or molding of materials into the desired patterns without mass subtraction [65,66]. Among these is soft-lithography, which utilizes elastomeric polymers, most commonly polydimethylsiloxane (PDMS), in replicating microchannel features directly from a master mold, usually made through photolithography [67–69] (Fig. 3). Among its merits are optical transparency, bio-compatibility, gas permeability, and the ease of bonding, which allows PDMS to be perfectly adapted to biological and chemical microenvironments [70]. Nevertheless, PDMS also has intrinsic disadvantages, including its ability to absorb small hydrophobic molecules and poor resistance to organic solvents [71]. Such shortcomings have driven the use of other thermoplastic materials like poly methyl methacrylate (PMMA) [72], also known as PMMA, and cyclic olefin copolymer (COC) [73] in hot embossing. In hot embossing, a thermoplastic substrate is heated above the glass transition temperature and forced under pressure against a master mold, resulting in permanent replication of the features when cooled [74,75] (Fig. 4). This process allows for the creation of durable and chemically inert micro features but results in a reduced fidelity of the mold after multiple uses and requires precise thermal management to prevent warping [76,77]. In industrial production, precision injection molding is a key technique in precision polymer manufacturing. It entails the filling of a highly machined metal mold with hot polymer to support high-capacity production of disposable diagnostic cartridges and point-of-care test (POCT) componentry [78,79] (Fig. 5). The process enjoys very good mechanical integrity and stability in terms of shape but suffers from the drawback of high up-front tooling expenses and poor support for rapid iterative design [80,81]. These techniques are vital to the process of manufacturing chemical analysis systems and apparatus. These shape-forming techniques play a central role in the production of rapid diagnostic devices and chemical sensors that capitalize on novel polymeric materials to enhance sensitivity and chemical environment resistance [18].

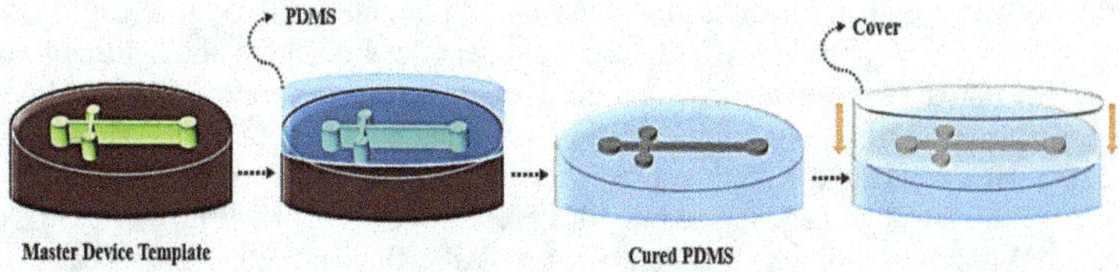

Fig. 3. The steps for manufacturing a microfluidic chip using soft lithography include molding PDMS from a master pattern, curing the polymer, and finally bonding the cover layer to form fluid channels. (Drawn by the author)

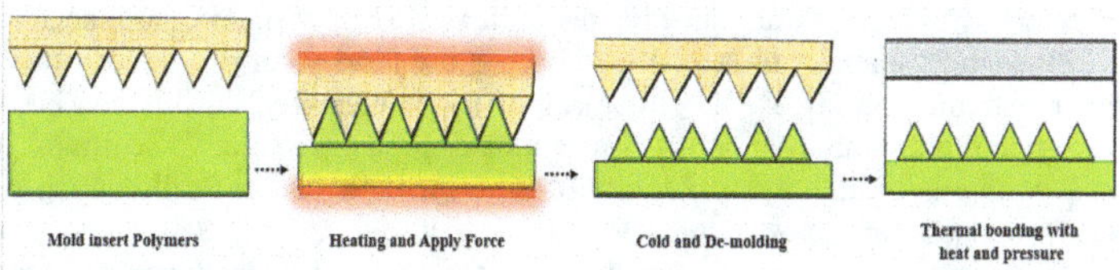

Fig. 4. The steps of the Hot Embossing process include placing the polymer in a mold, applying heat and pressure to form the pattern, and finally cooling and removing the mold to create microstructures on the polymer substrate. (Drawn by the author)

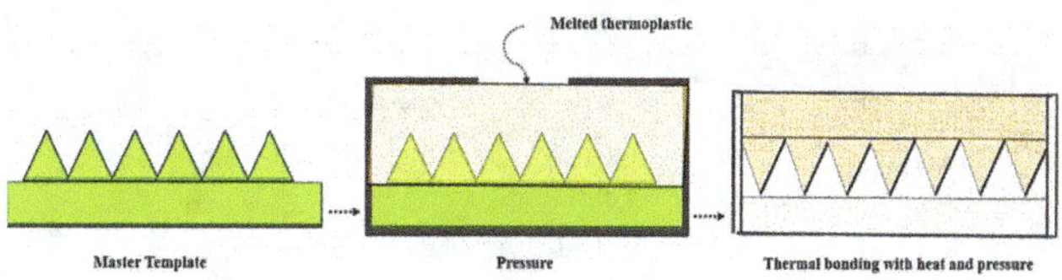

Fig. 5. The steps of the injection molding process include injecting molten polymer into a master mold, applying pressure to precisely shape the microstructures, and finally thermal bonding to form the final layer of the microfluidic chip. (Drawn by the author)

2.3 Fabrication Additive

Fabrication Additive otherwise referred to as 3D printing, has introduced a new approach to manufacturing by allowing for the creation of complex geometry in a layer-by-layer construction from Digital design data [82]. Stereo lithography (SLA) and digital light processing (DLP) techniques deploy a photo polymerizable resin and light projection technology to fabricate microscale structures with feature resolution limits approaching 25 µm [83–85]. In SLA, a UV laser is used to selectively cure resin at predetermined coordinates, layer-by-layer [86], which allow for fine vertical resolution and limited distortion [87] of features [88,89] (Fig. 6). DLP, on the other hand, utilizes a digital micro mirror device [90] and flashes the entire cross sectional images to the resin at once allowing for higher speed of fabrication [91] while still maintaining microscale accuracy [92,93] (Fig. 7). Both of these optical technologies provide high fidelity

Materials Research Foundations 191 (2026) 36-71
https://doi.org/10.21741/9781644904213-2

feature fidelity and are ideal for the complex design of internal channels necessary in embedded microfluidic networks and these platforms are most beneficial for the manufacture of non-planar structures, integrated reservoirs, and embedded microfluidic networks [94–96]. Inkjet based printing prints droplets of functional inks typically mixing polymers or biomolecules [97] onto a substrate to form a controlled pattern which is useful for integration with biological or chemical sensors [98,99], on the other hand it provides lower resolution [100,101] (Fig. 8). Fused Deposition Modeling (FDM) prints through extruding thermoplastic filaments through a heated nozzle to create successively deposited layers [102]; again, while advantageous in terms of cost and range of materials [103], actual resolution is still often limited to >100 μm [104], and often statutory surface roughness [105] can affect microfluidic flow characteristics [106,107] (Fig. 9). Two-photon polymerization (2PP), where femtosecond laser pulses at focus are employed to initiate the voxel level polymerization to achieve submicron feature fabrication in real three dimensional space, has emerged as a breakthrough method [108,109] (Fig. 10). The nonlinear absorption process of 2PP enables spatially to be confined only at the focal point and thereby facilitates the fabrication of extremely complex 3D structures of nanometer precision. It has been used in fabrication of nanofluidic devices, optical waveguides, and cell interactive scaffolds [110,111]. Fabrication Additive is limited by availability of biocompatible and optically transparent resins and surface roughness and post processing needs. However, its ease of redesigning and functionally tailoring ensures fabrication additive a potent directed research prototype tool and for fabrication customized diagnostic platforms. [25,113].

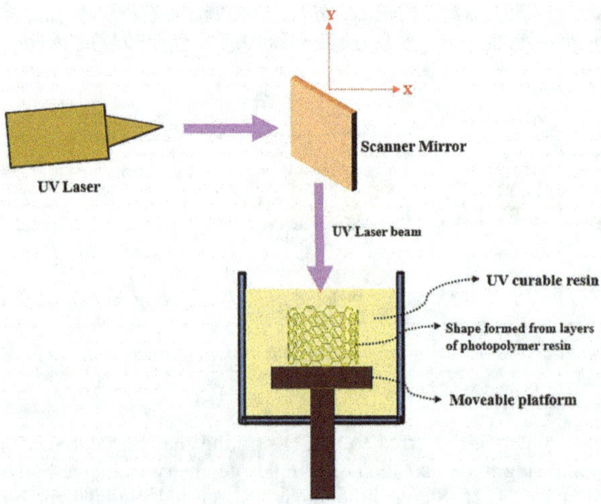

Fig. 6. Stereolithography (SLA) process steps involve layer-by-layer curing of photopolymer resin using an ultraviolet laser to create highly accurate microscopic structures in the 3D printing process. (Drawn by the author)

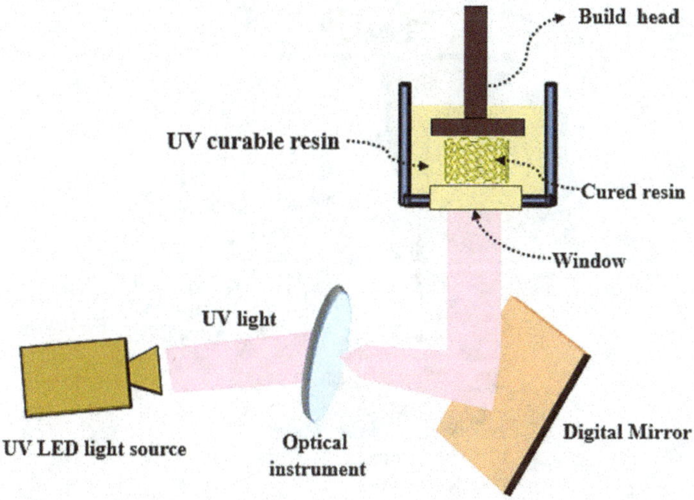

Fig. 7. Schematic view of the 3D printing process using Digital Light Processing (DLP), in which a microfluidic structure is created with high precision and speed by simultaneously irradiating the image of each layer onto photopolymer resin. (Drawn by the author)

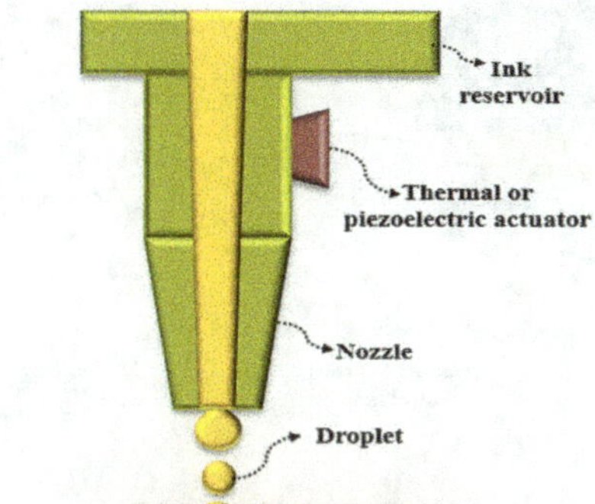

Fig. 8. Inkjet-based printing is a process used to fabricate microfluidic devices, involving the spraying of controlled droplets of ink or polymer onto a substrate to create precise patterns and integrate biological or chemical components into the chip. (Drawn by the author)

Materials Research Foundations 191 (2026) 36-71 https://doi.org/10.21741/9781644904213-2

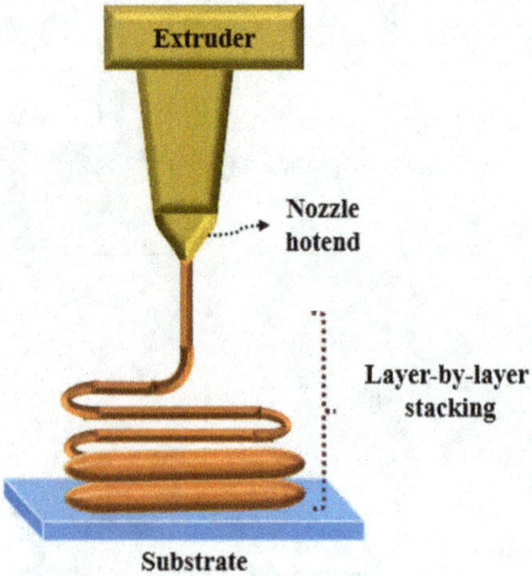

Fig. 9. Fused Deposition Modeling (FDM) process involves melting and extruding thermoplastic filament from a hot nozzle and depositing material layer by layer to build 3D microfluidic structures. (Drawn by the author)

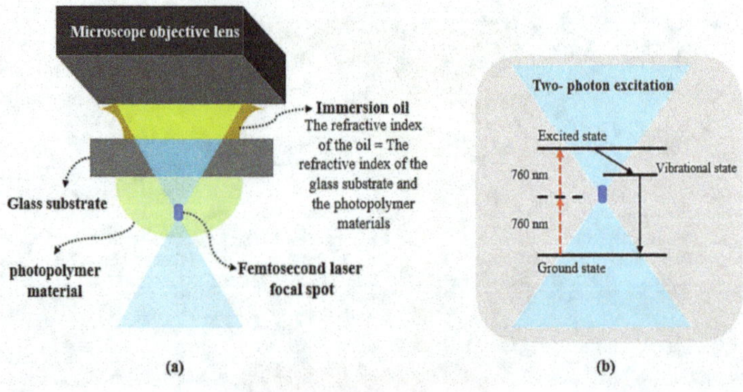

Fig. 10. Two-Photon Polymerization (2PP) process:Focusing of a femtosecond laser beam by a microscope objective lens through an immersion oil with a refractive index matching that of the glass and photopolymer, which results in highly localized polymerization at the focus point(a); Two-Photon Excitation Mechanism, in which the simultaneous absorption of two photons provides the energy required to transition the molecule from the ground state to the excited state, initiating the polymerization reaction in a very small volume of material(b). (Drawn by the author)

Materials Research Foundations 191 (2026) 36-71 https://doi.org/10.21741/9781644904213-2

2.4 Multimodal Manufacturing Approaches

As the complexity of applications involving the use of microfluidics increases, the single technique approaches tend to fall short. Hybrid methodologies have thus emerged to harness the complementary capabilities of dissimilar techniques [114,115]. For example, the coupling of SLA-based printing and PDMS-based soft-lithography enables the production of devices that incorporate intricate structural elements along with biocompatible fluidic interfaces [116–118]. Likewise, laser micromachining may be used after precision injection molding to add high-resolution features or to functionalize device surfaces selectively [119] (Fig. 11). They provide for multimodal workflows such as embedded electronics [120], optical sensors [121], and surface patterning [122] into a microfluidic chip. The merging of formative, subtractive, and additive processes also allows for the production of reconfigurable, multi-layered platforms customized for application specific requirements [123].

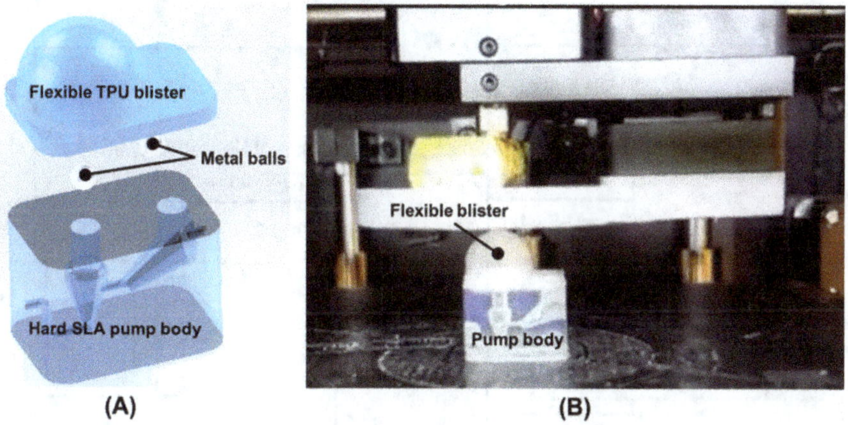

Figure 11. (A) Schematic illustration of the 3D-printed, hard-soft micro pump that was powered by a finger. It consists of one hard SLA-printed pump body, two metal balls and one flexible TPU blister. (B) 3D printing of the flexible TPU blister on the hard SLA-printed pump body which was pre-filled with two metal balls before FDM 3D printing [123]. Reproduced from Ruiz et al., "Fabrication of Hard–Soft Microfluidic Devices Using Hybrid 3D Printing," Micromachines, 2020, 11(6), 567, licensed under Creative Commons Attribution 4.0 International License.

2.5 Integrating Artificial Intelligence (AI), 4D printing and femtosecond lasers in advanced manufacturing technologies

A new framework is emerging in fabrication science in the form of the incorporation of artificial intelligence and machine learning algorithms into the production and design pipeline [124]. These Computational frameworks support predictive modeling, parameter optimization through automation, and real-time process correction through performance feedback [125]. AI-guided systems have the ability to adjust printing parameters, the durations of etching, or curing profiles in real-time to make up for Material heterogeneities or Environmental fluctuations [125] (Fig. 12).

Another emerging area of work involves 4D printing wherein the printed micro-structures change temporally as a function of external stimuli including pH, heat, or magnetic fields [126,127]. Dynamic systems promise enormous potential when utilized in the fabrication of responsive valves, self-healing channels [128], and Programmable assay modules [129–131]. Ultrafast Laser Ablation also continues to redefine the boundaries of precision manufacturing. Using ultrashort laser pulses, the process reduces Thermal damage and allows high-fidelity patterning of delicate substrates, such as glass, hydrogels, and biodegradable polymers [132,133] (Fig. 13). It is ideally suited for the ability to provide Micron-level control and refractive index control for both bio-implantable devices and for the synthesis of opto-fluidic systems [134].

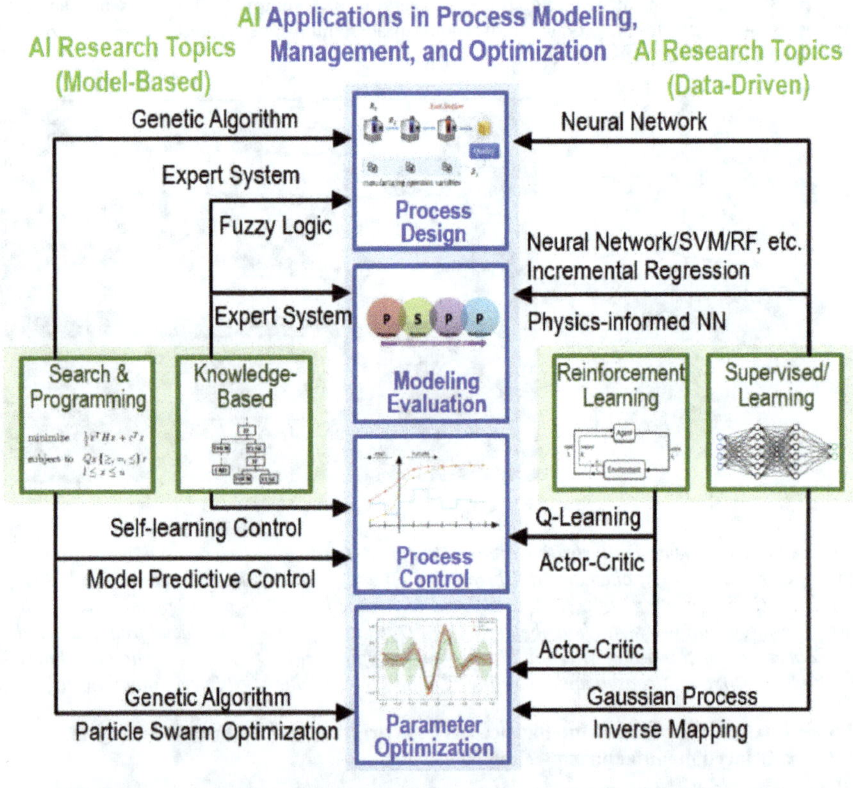

Figure 12. AI research for process design, modeling, and optimization [125]. Reproduced from Gao et al., "Artificial Intelligence in Manufacturing: State of the Art, Perspectives, and Future Directions," CIRP Annals, 2024, 73(2), 723–749, licensed under Creative Commons Attribution 4.0 International License.

Materials Research Foundations 191 (2026) 36-71 https://doi.org/10.21741/9781644904213-2

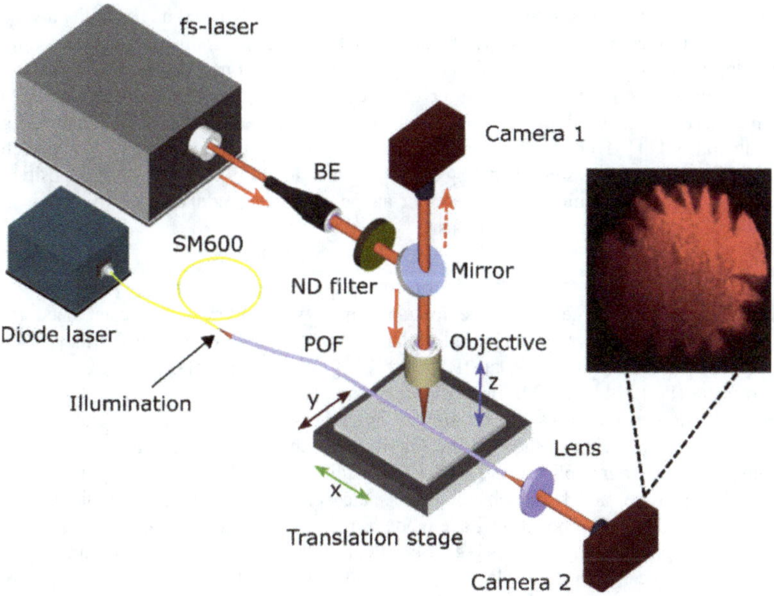

Figure 13. The ultrafast laser micromachining system, in which the laser beam is expanded using a beam expander (BE) and guided through a variable neutral-density (ND) filter to adjust the laser power to the fabrication system; Camera 1 is used for alignment purposes (a). The inset image in (a) shows the distal end profile of the POF with the shadows of the machined structures after the femtosecond machining process. Top view (b) and cross-sectional view (c) micrographs of a microwindow are also shown. Reproduced from Chandrasekharan et al., "Ultrafast laser ablation of a multicore polymer optical fiber for multipoint light emission," Optics Express, 2021, 29(13), 20765–20775, licensed under Creative Commons Attribution 4.0 International License

3. Materials in Microfluidic Fabrication

Microfluidic device performance is heavily impacted by material selection, as materials influence optical transparency, chemical resistance, thermal and mechanical durability, and biocompatibility. Functional material selection determines not only which manufacturing techniques will suffice but also influences the device's compatibility with downstream processes such as cell culture, chemical detection, and nucleic acid amplification. This section provides a structured summary of the foremost classes of materials utilized in microfluidic devices, everything from elastomers and thermoplastics to rigid substrates and new biodegradable alternatives.

3.1 Polydimethylsiloxane (PDMS)

Polydimethylsiloxane (PDMS) continues to be a favorite material for university microfluidics due to its elasticity, mold ability, and compatibility with soft-lithography [135,136]. It is amenable to

Materials Research Foundations 191 (2026) 36-71 https://doi.org/10.21741/9781644904213-2

detection with light due to its clarity within the UV-visible spectrum [30], and its flexibility allows for the creation of pneumatic valves [137] and tunable structures. Industrial or chemically demanding applications are, however, limited by PDMS limitations. Permeation by small hydrophobic analytes results in analyte loss or cross contamination [30], and susceptibility to organic solvents results in swelling and distortion of micro-channels, potentially jeopardizing assay reproducibility [11]. To overcome such deficiencies, various surface treatments have been studied and engineered, such as sol-gel coatings [138], silanization [139], and Parylene deposition [11,140]. These improve chemical robustness but at the same time add complexity and may compromise device stability on a long-term basis.

3.2 Thermoplastics: PMMA, COC, and PC

Thermoplastics prove a suitable alternative for elastomers where there is a greater need for mechanical strength, chemical resistance, and scalability [141]. Some of the most widely used thermoplastics are polymethyl methacrylate (PMMA) [142], cyclic olefin copolymer (COC) [143], and polycarbonate (PC) [144]. PMMA is highly appreciated for its clarity and acceptable bio-compatibility with tissues. It is accessible for forming by the hot embossing, laser micromachining, and precision injection molding methods [145]. Fragility and susceptibility to solvents may limit its effectiveness under harsh applications [146]. COC exhibits better thermal stability, resistance against auto fluorescence, and better chemical resistance [147]. These properties, along with compatibility with roll-to-roll production, have made it one of the top material choices for usage within optics and diagnostics applications [148,149]. PC is highly resistant to impacts and is heat-resistant, and therefore can be used for devices requiring autoclaving [150]. Its propensity to turn yellow with UV light and reduced optical clarity will restrict its applications in optical-sensitive systems [150].

3.3 Glass

Glass is still a key material for high-performance uses in which optical clarity, chemical inertness, and mechanical stability are paramount [151]. Fused silica and borosilicate are uniquely suitable for high-temperature and high-voltage system applications [152,153]. Some common methods of fabrication include wet/dry etching [154], femtosecond laser ablation [152], and powder blasting [155]. Glass is also suitable for electro-kinetic and electrochemical integration due to its insulating properties and pressure stability [154]. Although there are all of these benefits, fragility, expense, and tedious processing have restricted its application to disposable uses or the masses. In any case, it is still uniquely valuable to precision analytical devices [152].

3.4 Silicon

Silicon has been at the forefront of early developments in microfluidic systems, particularly the ones utilizing MEMS technologies [2]. Its photolithography and deep reactive ion etching compatibility allows for the fabrication of high-resolution micro structures with very good dimensional control [57,156]. Silicon's thermal conductivity is high and is useful in temperature-sensitive applications such as PCR [157], and its semiconducting nature enables electronic components such as field-effect transistors and capacitive sensors to be integrated [158]. Its transparency in the visible spectrum, however, limits its application in optical detection systems [159]. In addition, its production involves cleanroom facilities and expensive tools, which confine its application to multifunctional or highly integrated devices [160,161].

Materials Research Foundations 191 (2026) 36-71 https://doi.org/10.21741/9781644904213-2

3.5 Paper and Cellulose-Based Materials

Micro PADs, paper-based microfluidics, represent a low-cost, disposable option that is suitable for point-of-care diagnostics where resource limited settings are prevalent [162,163]. Micro PADs exploit the phenomenon of capillary action within cellulose or nitrocellulose membranes without the need for external pumps [164]. Resolution and design complexity are limited from a physical perspective, but the devices are low-cost, easily manufactured (e.g., wax printing or inkjet printing), and highly scalable. They have proven suitable for use in immunoassays, colorimetric assays, and environmental monitoring [165,166]. Examples of recent breakthroughs have seen hybrid polymer [167] integration and multilayer 3D paper devices increase their applications and enable quantitative analyses [168]. As an outgrowth of cellulose-based platforms, naturally porous materials like wood have also seen application in microfluidics [169,170]. Wood's natural anisotropic structure facilitates directional capillary driven flow comparable to substrates in paper, but with complementary mechanical rigidity and tunable hydrodynamics. Pores of varying sizes and flow rates due to the diversity of wood types (softwoods versus hardwoods) allow for selective suitability in diagnostic and sensing tasks. Surface modification techniques such as laser engraving [171], chemical functionalization [172], or lamination [171] enable manipulation of pathways for fluids and hydrophobicity. These attributes place wood derived microfluidics as an upcoming biodegradable option for low-cost environmental monitoring [172] and point-of-care diagnostics in sustainable platform development [171].

3.6 Bio-degradable and Bio-resorbable Materials

Biopolymers such as polylactic acid (PLA), polycaprolactone (PCL), and silk fibroin are being targeted ever more for use in sustainable microfluidic devices [173]. They are especially suitable for single-use point-of-care devices [173], transient implant devices [174,175], and environmentally responsible sensing platforms. They have the potential to degrade under physiological or environmental conditions to reduce long-term waste generation. However, they lack thermal and mechanical stability [176] along with compatibility with traditional manufacturing methods [177]. To overcome all of them, various methods of processing like low-temperature embossing [176], inkjet printing [178], and laser sintering [179] are being designed to encompass biodegradable substrate support along with imparting structural integrity.

3.7 Comparative Evaluation of Fabrication Techniques

To select the appropriate fabrication process, one has to balance resolution, cost, scalability, and design complexity properly. A comparative summary of the most common methods is provided in Table 1: Photolithography has unmatched resolution, submicron-scale feature capabilities, and support for integration at the level of MEMS. Its drawback is the high-cost setups and rigidity. Soft-lithography is suitable for rapid prototyping due to its low cost and ease of implementation, even though it is low in solvent resistance and long-term dimensional stability. Hot embossing and precision injection molding are both suitable for large-volume production with good reproducibility but have high setup costs to manufacture the mold. Fabrication Additive enables manufacturing of intricate, non-flat shapes and quick iteration in the absence of molds, but at the expense of optical quality and fewer materials today. Hybrid and algorithmic strategies merge various methods to deliver superior performance, lower turnaround time, and evolving system architectures. With computational design tools evolving, they increasingly play pivotal roles in next-generation, multifaceted platform development for microfluidics.

Table 1. Comparative Assessment of Microfluidic Fabrication Techniques (Prepare by author)

Fabrication Method	Resolution	Material Compatibility	Scalability	Cost Efficiency	Suitable Applications	Advantages	Limitations
Photolithography + Etching	Submicron (<1 µm)	Silicon, Glass	Low	High Capital Costs	MEMS integration, electrophoresis, flow control	Ultra-high resolution; excellent reproducibility	Expensive infrastructure; rigid substrates only
Soft Lithography	5–50 µm	PDMS, hydrogels	Medium	Low to Moderate	Cell culture, droplet generation, organ-on-chip	Biocompatible; flexible; rapid prototyping	Poor chemical resistance; gas permeability
Hot Embossing	20–100 µm	PMMA, COC, PC	Medium to High	Moderate	Disposable diagnostics, optical biosensors	Good optical properties; chemical stability	Mold degradation over time; thermal control needed
Injection Molding	50–150 µm	Thermoplastics (PMMA, COC, PS, etc.)	Very High	High Initial Cost	Commercial cartridges, lateral flow systems	High throughput; reproducible	Not ideal for prototyping; tooling cost
SLA / DLP 3D Printing	25–100 µm	Limited resins	Low to Medium	Low to Moderate	Custom geometries, embedded sensors	Moldless design; complex 3D structures	Surface roughness; limited transparency
FDM / Inkjet 3D Printing	>100 µm	Thermoplastics, functional inks	Medium	Low	Rapid testing, fluidic adapters	Affordable; open-access systems	Low resolution; poor bonding
Two-Photon Polymerization	<1 µm	Photopolymers	Low	Very High	Nano-fluidics, micro-optics	Exceptional resolution; complex freeform structures	Time-consuming; limited scalability
Laser Micromachining	1–10 µm	Glass, thermoplastics, hydrogels	Medium	High Equipment Cost	Opto-fluidics, implantable biosensors	Non-contact; minimal thermal damage	Expensive equipment; low throughput
Hybrid	Depends on technique	Multi-material systems	Medium	Moderate	Multifunctional, adaptive systems	Combines precision with flexibility	Workflow complexity; integration challenges
AI-Assisted Fabrication	Design-stage	All compatible platforms	Scalable (indirect)	Reduced design cycles	Optimization, real-time quality control	Predictive modeling; automated process refinement	Requires training datasets; less mature in some workflows

Materials Research Foundations 191 (2026) 36-71 | https://doi.org/10.21741/9781644904213-2

4. Applications in Analytical Chemistry Microfluidic

Technologies has reshaped analytical chemistry by means of compact, integrated systems, lab-on-a-chip, and point-of-care systems that can conduct complicated, multistep assays in spatially confined areas. Microfluidic systems minimize sample and reagent volumes, enhancing the sensitivity of analysis; laminar flow regimes ensure reproducibility, and spatial confinement allows selective analyte addressing, making them perfectly suitable for trace-level, quantitative analysis in complicated matrices. Microfluidic devices are being applied widely in environmental monitoring, pharmaceutical drug screening, and diagnostics, where they are able to perform rapid and low limit of detection (LOD) analysis of analytes, toxins, pathogens, and biomarkers, from small volumes of samples. Microfluidic devices are highly beneficial where real-time analysis with minimal consumption of a sample is needed, during food safety analysis, detection of pollutants, and clinical diagnostics for personalized medicine. These devices allow the performance of analysis with a higher throughput using less reagent and minimum volumes of a sample. Microfluidic devices bring about considerable improvement in efficiency of analysis with minimized Reagent Efficiency, faster reaction kinematics, and accurate spatial and temporal control. Microfluidic devices are perfectly compatible with electrochemical, optical, and mass spectrometric detection techniques leading the way to the development of Miniaturized, automated, and high-throughput analyzers. This section discusses major applications of microfluidic devices in the context of analytical chemistry with a focus on core functionalities and recent advancements involve the development of droplet-based microfluidics and digital microfluidics, which have greatly facilitated control over tiny volumes of fluids with a very high precision. Digital microfluidics employs arrays of electrodes to manipulate fluid droplets, facilitating exact dispensation, mixing, and kinetic control of on-chip reactions, whereas droplet-based microfluidics is able to separate a sample into individual droplets, optimizing the conditions for individual reactions.

4.1 Electrochemical detection

Electrochemical detection is particularly suitable for microfluidic formats because of low power needs, the ability to monitor in real time, and scalability [180]. Micro-fabricated electrodes, usually made from gold, platinum, or carbon, integrated into microfluidic channels allow for accurate quantitation of redox-active analytes, heavy metals [181], neurotransmitters [182], and metabolites. Screen printing [181], inkjet deposition [182], and laser ablation [183] are methods used to produce highly patterned electrodes on glass or thermoplastic substrates. Geometry and microchannel flow dynamics control electrochemical performance. Laminar flow properties of micro channels for enhanced mass transportation stability ensure better signal reproducibility and temporal resolution [181]. Nanostructured electrodes (e.g., graphene, nanotubes) and redox-active coatings are newly emerged components that double the sensitivity and boost the strategy for specific recognition [184]. These components improve the kinetic of electron transfers and add a large surface area to volume for multi-analyte detection in redox-active models like neuro DS [185] and environmentally important pollutants [186]. Currently, highly efficient immunosensors [187] and biosensors have resulted from advancements that involve the use of electrochemical methods like electrochemical impedance spectroscopy (EIS) and amplified voltammetry for detection of analytes and low-abundance biomarkers with a higher degree of precision [187]. Electrochemical impedance spectroscopy provides tremendous insights into sensor surface-biomolecule interaction for a real-time monitoring of binding events with the surface [188].

Materials Research Foundations 191 (2026) 36-71 https://doi.org/10.21741/9781644904213-2

Improved voltammetry methods are able to improve the resolution of currents so that they are highly recommended for early disease detection and for ecological monitoring [189]. Electrochemical impedance spectroscopy is being highly researched for early detection of neurodegenerative disease by performing amyloid-beta interaction detection on a real time basis [190], while differential pulse voltammetry has delivered a high degree of efficiency for quantification of tumor markers like Prostate-Specific Antigen (PSA) and Carcinoembryonic Antigen (CEA) at ultra-low concentration levels [191,192].

4.2 Optical detection and Spectroscopy

Microfluidics is inherently compatible with detection-optics, with benefits of minimal optics pathway length, minimal sample volume, and optics transparencies of the material. Microfluidic applications have had easy adoption for a broad spectrum of light-based techniques fluorescence [193], absorbance [194], chemi-luminescence [195], surface enhanced Raman scattering (SERS) [196], and surface Plasmon resonance (SPR) [197]. Functional material selection is critical to the attainment of optical performance. Optical clarity and minimal native fluorescence are offered by polymers such as PMMA and COC, which allow for the integration of optical devices such as fibers, micro lenses, and waveguides [198]. New fabrication methods such as Ultrafast Laser Ablation have also made the development of on-chip photonic components such as diffraction gratings and interferometers possible [199]. All of these technologies enable a wide range of applications from those that involve quantitative fluorescence assays and real-time PCR to miniaturized clinical diagnostics and environmental monitoring biosensors. Among them, surface-enhanced Raman scattering is capable of delivering single-molecule sensitivity for trace pollutants and bio-marker detection, but signal amplification on a reproducible scale is difficult to achieve due to nanostructure fabrication defects and excitation pathway alignment within microfluidic channels [200,201].

4.3 Mass Spectrometry Integration

Microfluidic coupling with mass spectrometry has greatly enhanced the efficiency of analysis and reduced analyte degradation and sample carryover [202]. Microfluidic electrospray ionization (μESI) interfaces that enable direct, Inline connection to MS detector interfacing allow on-stream reaction monitoring and highly sensitive detection of analytes [203]. μESI is less suppressive of ions and has a higher tolerance to matrix complexity compared to conventional ESI, enabling consistent detection of analytes with clinical and forensic matrices that are highly complex [204,205]. Apart from continuous-flow devices, droplet-based microfluidics is a versatile option for the precise handling of samples. Droplet-based microfluidics provides precise control of the mixing of the reagents at the time of MS introduction, which effectively prevents cross-contamination [206]. Micro-fabricated glass or silicon chips with on-chip nano emitters have also come forth for unbroken, monolithic MS interfacing [207]. This is particularly valuable for proteomics [208], metabolomics, and Pharmaceutical screening, where accurate Sample metering and low dead volume are crucial for quantitation accuracy [209]. Furthermore, microfluidic integration with quantitative mass spectrometry improves performance and suppresses ion suppressions by enabling consistent, steady nano spray generation and reliable temporal control for the delivery of analyte into the ionization source [202,208].

Materials Research Foundations 191 (2026) 36-71
https://doi.org/10.21741/9781644904213-2

4.4 Sample Preparation and Pre-concentration

One of the strongest benefits of microfluidic devices is their capability for on-board integration of sample preparation with the analysis device [210]. Critical functions such as filtration, mixing of the reagents, and separation can be integrated within a single chip [211]. Integration eliminates off-chip handling of the sample and variability due to manual pre-processing, a major point of concern for trace detection of biological and environmental analytes [212]. Reproducible analysis demands efficient mixing of the reagents, especially for the confined microfluidic environment [213]. Passive micro mixers based on staggered herringbone or serpentine channel geometries maximize diffusion-based mixing [214], while active mixing methods based on electro kinetic or acoustic driving fields optimize performance for viscous or multi-phase fluids [215]. For determination of low-abundant analytes, on-chip pre-concentration techniques such as isotachophoresis [216], solid-phase extraction (SPE) [217], and immunoaffinity [218] capture are being applied with growing frequency. These techniques are especially relevant to applications that generate a requirement for sensitivity, such as for monitoring the environment, forensic analysis, and other applications.

4.5 Immunoassays and Multiplex Detection

Microfluidic immunoassays take advantage of the exquisite specificity of antibody-antigen interactions in micro engineered environments to enable scalable, automated detection [219,220]. Channel surface functionalization provides for target analyte capture, with integrated labels and reagents delivering either optical or electrochemical outputs [11]. Multiplex analysis of several biomarkers is made possible using parallel microchannel arrays and internal networks of valves vital for applications ranging from sepsis diagnostics to hormonal profiling and assessment of the immune status [221]. Some recent developments involve hybrid platforms bringing thermoplastics together with other substances, for example, paper and nitrocellulose, to create low-cost lateral flow-based devices with faster analysis performance [222]. Multiplex immunoassay platforms based on these are being used ever more widely for the simultaneous profiling of cytokines [223], tumor markers [224], and hormones to allow clinicians to track disease progression, response to treatment, and immunity responses in near-real-time at high analysis resolution [225].

5. Future Trends and Technological Outlook

Anticipated advances in microfluidic technology are likely to be driven by highly interdisciplinary approaches drawing upon advanced materials science, intelligent computation and scalable manufacturing processes. These advances are expected to be propelled by artificial intelligence (AI) and machine learning (ML) approaches which will help develop and modify the design and Fabrication processes extensively so that fewer prototype iterations will be needed, optimization will be automatic and real-time changes to processes will be facilitated. In analytical chemistry, AI derived algorithms will provide rapid data interpretation, improved predictive modelling and more correct overall analyses of chemical and biological substances by advanced processing of sensor data. Meanwhile, a world-wide drive towards sustainability is giving rise to the use of biodegradable and bio-derived polymers for microfluidic devices. By using production processes which are solvent-less and low temperature, it will be convenient to evolve productive devices with reduced eco systems damaging qualities. Such bio-derived materials will permit fully disposable microfluidic systems which will be able to retain their analytical precision while miniaturization the waste discharge of plastics, particularly in Point-of-care diagnostics and environmental monitoring. Another potential avenue which is developing, is that of 4D microfluidics where the microfluidic devices are created using the operation of smart materials which are responsive to change as a result of an external stimulus such as temperature, pH, electricity or magnetic field. This means that self-regulating, adaptive microfluidic components such as programmable valves and or tunable flow channels will also be possible. Finally, in terms of scalability of manufacture, roll-to-roll fabrication is on track to being a viable process for the high throughput manufacture of flexibly wearable and disposable diagnostic devices. Furthermore, nanointegrated fluidic systems combined with plasmonic nanostructures, quantum dots and carbon tubes will offer ultra-sensitive detection platforms, capable of single molecule detection. Taken together, these technology directions indicate that the future of microfluidics lies with the simultaneous also incorporation of smart materials, intelligent design and sustainable large scale production.

Conclusion

The combination of advanced microfluidic fabrication techniques and analytical chemistry has revolutionized the field of chemical sensing and diagnostics considerably. The Shape-forming methods, Additive manufacturing, and algorithm-controlled multimodal assembly allow for accurate geometrical control, chemical compatibility, and high-order functional integration, facilitating the design of miniaturized, rugged devices for quantitative chemical and biochemical analyses. Analytically, microfluidic devices improve dramatically the analytical figures of merit, namely, the limit of detection (LOD), limit of quantification (LOQ), signal-to-noise ratio, and reproducibility. Their suitability for electrochemical, optical, and mass spectrometry detection means that it is possible to analyze clinically and environmentally relevant analytes such as biomarkers, toxins, and metabolites simultaneously, in low volumes, and in multiplexed form. Digital and droplet-microfluidic formats lead to ultrasensitive and high-throughput analysis and statistically sound quantification. On-chip integrated sample pretreatment, on the other hand, improves assay automation and stringency in point-of-care and remote diagnostics.

References

[1] E. Livak-Dahl, I. Sinn, M. Burns, Microfluidic chemical analysis systems, Annu. Rev. Chem. Biomol. Eng. 2 (2011) 325-353. https://doi.org/10.1146/annurev-chembioeng-061010-114215

[2] K.F. Lei, Introduction: the origin, current status, and future of microfluidics, Microfluidics: Fundam., Devices Appl. 1 (2018) 1-18. https://doi.org/10.1002/9783527800643.ch1

[3] R.K. Mousah, D.S. Zageer, Microfluidic devices to monitor water pollution, (2024). https://doi.org/10.14719/pst.4092

[4] E.K. Sackmann, A.L. Fulton, D.J. Beebe, The present and future role of microfluidics in biomedical research, Nat. 507 (2014) 181-189. https://doi.org/10.1038/nature13118

[5] G.M. Whitesides, The origins and the future of microfluidics, Nat. 442 (2006) 368-373. https://doi.org/10.1038/nature05058

[6] Y.-J. Juang, Y.-J. Chiu, Fabrication of polymer microfluidics: An overview, Polymers 14 (2022) 2028. https://doi.org/10.3390/polym14102028

[7] S.M. Scott, Z. Ali, Fabrication methods for microfluidic devices: An overview, Micromachines 12 (2021) 319. https://doi.org/10.3390/mi12030319

[8] D.S. Dkhar, R. Kumari, S.J. Malode, N.P. Shetti, P. Chandra, Integrated lab-on-a-chip devices: Fabrication methodologies, transduction system for sensing purposes, J. Pharm. Biomed. Anal. 223 (2023) 115120. https://doi.org/10.1016/j.jpba.2022.115120

[9] M. Castano-Alvarez, D.F.P. Ayuso, M.G. Granda, M.T. Fernández-Abedul, J.R. García, A. Costa-García, Critical points in the fabrication of microfluidic devices on glass substrates, Sensors Actuat. B Chem. 130 (2008) 436-448. https://doi.org/10.1016/j.snb.2007.09.043

[10] C. Iliescu, B. Chen, J. Miao, On the wet etching of Pyrex glass, Sensors Actuat. A Phys. 143 (2008) 154-161. https://doi.org/10.1016/j.sna.2007.11.022

[11] A. Shakeri, S. Khan, T.F. Didar, Conventional and emerging strategies for the fabrication and functionalization of PDMS-based microfluidic devices, Lab Chip 21 (2021) 3053-3075. https://doi.org/10.1039/D1LC00288K

[12] K. Raj M, S. Chakraborty, PDMS microfluidics: A mini review, J. Appl. Polym. Sci. 137 (2020) 48958. https://doi.org/10.1002/app.48958

[13] J. Friend, L. Yeo, Fabrication of microfluidic devices using polydimethylsiloxane, Biomicrofluidics 4 (2010) 022811. https://doi.org/10.1063/1.3259624

[14] A.W. Auner, K.M. Tasneem, D.A. Markov, L.J. McCawley, M.S. Hutson, Chemical-PDMS binding kinetics and implications for bioavailability in microfluidic devices, Lab Chip 19 (2019) 864-874. https://doi.org/10.1039/C8LC00796A

[15] C.M. Penso, J.L. Rocha, M.S. Martins, P.J. Sousa, V.C. Pinto, G. Minas, M.M. Silva, L.M. Goncalves, PtOEP-PDMS-based optical oxygen sensor, Sensors 21 (2021) 5645. https://doi.org/10.3390/s21165645

[16] H. Becker, U. Heim, Hot embossing as a method for the fabrication of polymer high aspect ratio structures, Sensors Actuat. A Phys. 83 (2000) 130-135. https://doi.org/10.1016/S0924-4247(00)00296-X

[17] J.C. McDonald, D.C. Duffy, J.R. Anderson, D.T. Chiu, H. Wu, O.J. Schueller, G.M. Whitesides, Fabrication of microfluidic systems in poly(dimethylsiloxane), Electrophoresis 21 (2000) 27-40. https://doi.org/10.1002/(SICI)1522-2683(20000101)21:1<27::AID-ELPS27>3.0.CO;2-C

[18] J. Wu, M. Gu, Microfluidic sensing: state of the art fabrication and detection techniques, J. Biomed. Opt. 16 (2011) 080901. https://doi.org/10.1117/1.3607430

[19] D.A. Chang-Yen, R.K. Eich, B.K. Gale, A monolithic PDMS waveguide system fabricated using soft-lithography techniques, J. Lightwave Technol. 23 (2005) 2088. https://doi.org/10.1109/JLT.2005.849932

[20] N. Bhattacharjee, A. Urrios, S. Kang, A. Folch, The upcoming 3D-printing revolution in microfluidics, Lab Chip 16 (2016) 1720-1742. https://doi.org/10.1039/C6LC00163G

[21] J. Collingwood, K. De Silva, K. Arif, High-speed 3D printing for microfluidics: Opportunities and challenges, Mater. Today: Proc. (2023). https://doi.org/10.1016/j.matpr.2023.05.683

[22] P. Juskova, A. Ollitrault, M. Serra, J.-L. Viovy, L. Malaquin, Resolution improvement of 3D stereo-lithography through direct laser trajectory programming: Application to microfluidic deterministic lateral displacement device, Anal. Chim. Acta 1000 (2018) 239-247. https://doi.org/10.1016/j.aca.2017.11.062

[23] W. Li, M. Wang, H. Ma, F.A. Chapa-Villarreal, A.O. Lobo, Y.S. Zhang, Stereolithography apparatus and digital light processing-based 3D bioprinting for tissue fabrication, iScience 26 (2023) 102345. https://doi.org/10.1016/j.isci.2023.106039

[24] S. Chen, Z. He, S. Choi, I.V. Novosselov, Characterization of inkjet-printed digital microfluidics devices, Sensors 21 (2021) 3064. https://doi.org/10.3390/s21093064

[25] R. Su, F. Wang, M.C. McAlpine, 3D printed microfluidics: advances in strategies, integration, and applications, Lab Chip 23 (2023) 1279-1299. https://doi.org/10.1039/D2LC01177H

[26] I. Kundacina, O. Kundacina, D. Miskovic, V. Radonic, Advancing microfluidic design with machine learning: a Bayesian optimization approach, Lab Chip (2025) in press. https://doi.org/10.1039/D4LC00872C/v2/response1

[27] E. Mottay, X. Liu, H. Zhang, E. Mazur, R. Sanatinia, W. Pfleging, Industrial applications of ultrafast laser processing, MRS Bull. 41 (2016) 984-992. https://doi.org/10.1557/mrs.2016.275

[28] F. Poitrasson, F.-X. d'Abzac, Femtosecond laser ablation inductively coupled plasma source mass spectrometry for elemental and isotopic analysis: are ultrafast lasers worthwhile?, J. Anal. At. Spectrom. 32 (2017) 1075-1091. https://doi.org/10.1039/C7JA00084G

[29] F. He, Y. Liao, J. Lin, J. Song, L. Qiao, Y. Cheng, K. Sugioka, Femtosecond laser fabrication of monolithically integrated microfluidic sensors in glass, Sensors 14 (2014) 19402-19440. https://doi.org/10.3390/s141019402

[30] M.W. Toepke, D.J. Beebe, PDMS absorption of small molecules and consequences in microfluidic applications, Lab Chip 6 (2006) 1484-1486. https://doi.org/10.1039/b612140c

[31] C. Rumens, M. Ziai, K. Belsey, J. Batchelor, S. Holder, Swelling of PDMS networks in solvent vapours; applications for passive RFID wireless sensors, J. Mater. Chem. C 3 (2015) 10091-10098. https://doi.org/10.1039/C5TC01927C

[32] B. Guan, J.-H. Pai, M. Cherrill, B. Michalatos, C. Priest, Injection moulding of micropillar arrays: a comparison of poly(methyl methacrylate) and cyclic olefin copolymer, Microsyst. Technol. 28 (2022) 2083-2091. https://doi.org/10.1007/s00542-022-05350-4

[33] M.F. Alexandre-Franco, R. Kouider, R. Kassir Al-Karany, E.M. Cuerda-Correa, A. Al-Kassir, Recent Advances in Polymer Science and Fabrication Processes for Enhanced Microfluidic Applications: An Overview, Micromachines 15 (2024) 1137. https://doi.org/10.3390/mi15091137

[34] J.B. Nielsen, R.L. Hanson, H.M. Almughamsi, C. Pang, T.R. Fish, A.T. Woolley, Microfluidics: innovations in materials and their fabrication and functionalization, Anal. Chem. 92 (2019) 150-168. https://doi.org/10.1021/acs.analchem.9b04986

[35] D. Kauzlarick, Fundamentals of microfabrication, the science of miniaturization, [Book Review], IEEE Eng. Med. Biol. Mag. 22 (2003) 109-111. https://doi.org/10.1109/MEMB.2003.1195708

[36] S. Das, V.C. Srivastava, Microfluidic-based photocatalytic microreactor for environmental application: a review of fabrication substrates and techniques, and operating parameters, Photochem. Photobiol. Sci. 15 (2016) 714-730. https://doi.org/10.1039/c5pp00469a

[37] T.W. Lee, S. Jeon, J. Maria, J. Zaumseil, J.W. Hsu, J.A. Rogers, Soft-contact optical lithography using transparent elastomeric stamps and application to nanopatterned organic light-emitting devices, Adv. Funct. Mater. 15 (2005) 1435-1439. https://doi.org/10.1002/adfm.200500109

[38] A. Sengupta, S. Herminghaus, C. Bahr, Liquid crystal microfluidics: surface, elastic and viscous interactions at microscales, Liquid Crystals Rev. 2 (2014) 73-110. https://doi.org/10.1080/21680396.2014.963716

[39] V.M. Martinez, T.F. Edgar, Control of lithography in semiconductor manufacturing, IEEE Control Syst. Mag. 26 (2006) 46-55. https://doi.org/10.1109/MCS.2006.252811

[40] N. Salih, N. Nafarizal, C. Soon, M. Sahdan, A. Tijjani, U. Hashim, Glass etching for cost-effective microchannels fabrication, 2014 IEEE Int. Conf. Semiconductor Electron. (ICSE2014), IEEE, 2014, pp. 432-435. https://doi.org/10.1109/SMELEC.2014.6920890

[41] M.M. Hasan, F. Rahman, Aqueous Wet Etching of Silicon With Alkali Metal Hydroxides, Int. Manuf. Sci. Eng. Conf., ASME, 2024, p. V002T05A015. https://doi.org/10.1115/MSEC2024-131691

[42] M. Hassaan, U. Saleem, A. Singh, A.J. Haque, K. Wang, Recent Advances in Positive Photoresists: Mechanisms and Fabrication, Materials 17 (2024) 2552. https://doi.org/10.3390/ma17112552

[43] U. Abidin, B.Y. Majlis, J. Yunas, Fabrication of pyramidal cavity structure with micron-sized tip using anisotropic KOH etching of silicon (100), J. Teknol. (Sci. & Eng.) 74 (2015) 1-10. https://doi.org/10.11113/jt.v74.4846

[44] K. Hu, Q. Wang, Method for fabricating micro/nanostructures via scanning probe microscope and anisotropic wet etching, J. Nanosci. Nanotechnol. 13 (2013) 3966-3971. https://doi.org/10.1166/jnn.2013.7129

[45] J. Zhang, W.C. Hon, L.L. Leung, K. Chen, CMOS-compatible micromachining techniques for fabricating high-performance edge-suspended RF/microwave passive components on silicon substrates, J. Micromech. Microeng. 15 (2004) 328-335. https://doi.org/10.1088/0960-1317/15/2/012

[46] M. Huff, Recent advances in reactive ion etching and applications of high-aspect-ratio microfabrication, Micromachines 12 (2021) 991. https://doi.org/10.3390/mi12080991

[47] Y. Wu, H. He, Reactive ion etching and deep reactive ion etching processes, 2nd Int. Conf. Mechanical, Electronics, and Electrical Autom. Control (METMS 2022), SPIE, 2022, pp. 34-43. https://doi.org/10.1117/12.2634681

[48] M.D. Henry, C. Welch, A. Scherer, Techniques of cryogenic reactive ion etching in silicon for fabrication of sensors, J. Vac. Sci. Technol. A 27 (2009) 1211-1216. https://doi.org/10.1116/1.3196790

[49] Z. Ren, P. Heard, J.M. Marshall, P.A. Thomas, S. Yu, Etching characteristics of LiNbO3 in reactive ion etching and inductively coupled plasma, J. Appl. Phys. 103 (2008) 033301. https://doi.org/10.1063/1.2838180

[50] M. Ashraf, S.V. Sundararajan, G. Grenci, Low-power, low-pressure reactive-ion etching process for silicon etching with vertical and smooth walls for mechanobiology application, J. Micro/Nanolithogr. MEMS MOEMS 16 (2017) 034501. https://doi.org/10.1117/1.JMM.16.3.034501

[51] F. Laermer, A. Urban, Milestones in deep reactive ion etching, 13th Int. Conf. Solid-State Sensors, Actuators Microsystems, TRANSDUCERS'05, IEEE, 2005, pp. 1118-1121. https://doi.org/10.1109/SENSOR.2005.1497272

[52] H.V. Jansen, M.J. De Boer, S. Unnikrishnan, M. Louwerse, M.C. Elwenspoek, Black silicon method X: a review on high speed and selective plasma etching of silicon with profile control: an in-depth comparison between Bosch and cryostat DRIE processes as a roadmap to next generation equipment, J. Micromech. Microeng. 19 (2009) 033001. https://doi.org/10.1088/0960-1317/19/3/033001

[53] G. Sun, T. Gao, X. Zhao, H. Zhang, Fabrication of micro/nano dual-scale structures by improved deep reactive ion etching, J. Micromech. Microeng. 20 (2010) 075028. https://doi.org/10.1088/0960-1317/20/7/075028

[54] M. Blauw, T. Zijlstra, E. van der Drift, Balancing the etching and passivation in time-multiplexed deep dry etching of silicon, J. Vac. Sci. Technol. B 19 (2001) 2930-2934. https://doi.org/10.1116/1.1415511

[55] H. Rhee, H. Kwon, C.-K. Kim, H. Kim, J. Yoo, Y.W. Kim, Comparison of deep silicon etching using SF6/C4F8 and SF6/C4F6 plasmas in the Bosch process, J. Vac. Sci. Technol. B 26 (2008) 576-581. https://doi.org/10.1116/1.2884763

[56] K.-S. Kim, Y.-C. Lee, J.-H. Ahn, J.Y. Song, C.D. Yoo, S.-B. Jung, Effect of Process Parameters on TSV Formation Using Deep Reactive Ion Etching, Korean J. Metals Mater. 48 (2010) 1028-1034. https://doi.org/10.3365/KJMM.2010.48.11.1028

[57] M.S. Gerlt, N.F. Läubli, M. Manser, B.J. Nelson, J. Dual, Reduced etch lag and high aspect ratios by deep reactive ion etching (DRIE), Micromachines 12 (2021) 542. https://doi.org/10.3390/mi12050542

[58] Z. Wang, F. Jiang, W. Zhang, Si dry etching for TSV formation and backside reveal, Proc. 5th Electron. Syst.-integration Technol. Conf. (ESTC), IEEE, 2014, pp. 1-3. https://doi.org/10.1109/ESTC.2014.6962840

[59] C.W. Tang, H.T. Young, K.M. Li, Innovative through-silicon-via formation approach for wafer-level packaging applications, J. Micromech. Microeng. 22 (2012) 045019. https://doi.org/10.1088/0960-1317/22/4/045019

[60] K. Karimi, A. Fardoost, N. Mhatre, J. Rajan, D. Boisvert, M. Javanmard, A Thorough Review of Emerging Technologies in Micro-and Nanochannel Fabrication: Limitations,

Applications, and Comparison, Micromachines 15 (2024) 1274. https://doi.org/10.3390/mi15101274

[61] R. Westerman, L. Martinez, D. Pays-Volard, K. Mackenzie, T. Lazerand, Deep silicon etching: current capabilities and future directions, Micromach. Microfab. Process Technol. XIX, SPIE, 2014, pp. 38-51. https://doi.org/10.1117/12.2046694

[62] B. Wenk, J. Collet, V. Gaff, Technology platform for high performance MEMS inertial & vibration sensors, 2024 IEEE Int. Symp. Inertial Sensors Syst. (INERTIAL), IEEE, 2024, pp. 1-4. https://doi.org/10.1109/INERTIAL60399.2024.10502059

[63] M. Prince, X. Ma, P. Docker, M. Ward, P. Prewett, The development of a novel Bio-MEMS filtration chip for the separation of specific cells in fluid suspension, Proc. Inst. Mech. Eng. Part H 221 (2007) 113-128. https://doi.org/10.1243/09544119JEIM190

[64] J. Thevenoud, B. Mercier, T. Bourouina, F. Marty, M. Puech, N. Launay, DRIE technology: from micro to nanoapplications, Alcatel Micro Machining Syst., ESIEE, 2011, pp. 1-8.

[65] C. Tomassoni, O.A. Peverini, G. Venanzoni, G. Addamo, F. Paonessa, G. Virone, 3D printing of microwave and millimeter-wave filters: Additive manufacturing technologies applied in the development of high-performance filters with novel topologies, IEEE Microw. Mag. 21 (2020) 24-45. https://doi.org/10.1109/MMM.2020.2979153

[66] S. Dritsas, Y. Vijay, M. Dimopoulou, N. Sanadiya, J.G. Fernandez, An additive and subtractive process for manufacturing with natural composites, Robotic Fabrication Archit., Art Des. 2018, Springer, 2019, pp. 181-191. https://doi.org/10.1007/978-3-319-92294-2_14

[67] D.C. Duffy, J.C. McDonald, O.J. Schueller, G.M. Whitesides, Rapid prototyping of microfluidic systems in poly(dimethylsiloxane), Anal. Chem. 70 (1998) 4974-4984. https://doi.org/10.1021/ac980656z

[68] S. Natarajan, D. Chang-Yen, B. Gale, Large-area, high-aspect-ratio SU-8 molds for the fabrication of PDMS microfluidic devices, J. Micromech. Microeng. 18 (2008) 045021. https://doi.org/10.1088/0960-1317/18/4/045021

[69] Y. Ma, J. Thiele, L. Abdelmohsen, J. Xu, W.T. Huck, Biocompatible macro-initiators controlling radical retention in microfluidic on-chip photo-polymerization of water-in-oil emulsions, Chem. Commun. 50 (2014) 112-114. https://doi.org/10.1039/C3CC46733C

[70] X. Liu, S. Li, Fabrication of a three-layer PDMS pneumatic microfluidic chip for micro liquid sample operation, SLAS Technol. 25 (2020) 151-161. https://doi.org/10.1177/2472630319870126

[71] B. Cortese, M.C. Mowlem, H. Morgan, Characterisation of an irreversible bonding process for COC-COC and COC-PDMS-COC sandwich structures and application to microvalves, Sensors Actuat. B Chem. 160 (2011) 1473-1480. https://doi.org/10.1016/j.snb.2011.07.040

[72] H. Persson, S. Park, M. Mohan, K.K. Cheung, C.A. Simmons, E.W. Young, Rapid assembly of PMMA microfluidic devices with PETE membranes for studying the endothelium, Sensors Actuat. B Chem. 356 (2022) 131342. https://doi.org/10.1016/j.snb.2021.131342

[73] P.S. Nunes, P.D. Ohlsson, O. Ordeig, J.P. Kutter, Cyclic olefin polymers: emerging materials for lab-on-a-chip applications, Microfluid. Nanofluid. 9 (2010) 145-161. https://doi.org/10.1007/s10404-010-0605-4

[74] J. Sun, J. Zhuang, Y. Liu, H. Xu, J. Horne, E.K. Wujcik, H. Liu, J.E. Ryu, D. Wu, Z. Guo, Development and application of hot embossing in polymer processing: a review, ES Mater. Manuf. 6 (2019) 3-17.

[75] Y. He, J.-Z. Fu, Z.-C. Chen, Research on optimization of the hot embossing process, J. Micromech. Microeng. 17 (2007) 2420-2425. https://doi.org/10.1088/0960-1317/17/12/005

[76] T.-Y. Lin, T. Do, P. Kwon, P.B. Lillehoj, 3D printed metal molds for hot embossing plastic microfluidic devices, Lab Chip 17 (2017) 241-247. https://doi.org/10.1039/C6LC01430E

[77] M. Asif, R.N. Tait, P. Berini, Hot embossing of microfluidics in cyclic-olefin co-polymer using a wafer aligner-bonder, Microsyst. Technol. 27 (2021) 3899-3906. https://doi.org/10.1007/s00542-020-05188-8

[78] K. Roskos, A.I. Hickerson, H.-W. Lu, T.M. Ferguson, D.N. Shinde, Y. Klaue, A. Niemz, Simple system for isothermal DNA amplification coupled to lateral flow detection, PLoS One 8 (2013) e69355. https://doi.org/10.1371/journal.pone.0069355

[79] C.H. Ahn, J.-W. Choi, G. Beaucage, J.H. Nevin, J.-B. Lee, A. Puntambekar, J.Y. Lee, Disposable smart lab on a chip for point-of-care clinical diagnostics, Proc. IEEE 92 (2004) 154-173. https://doi.org/10.1109/JPROC.2003.820548

[80] S.H. Choi, D.S. Kim, T.H. Kwon, Microinjection molded disposable microfluidic lab-on-a-chip for efficient detection of agglutination, Microsyst. Technol. 15 (2009) 309-316. https://doi.org/10.1007/s00542-008-0689-x

[81] I. Maia, C. Rocha, P. Pontes, V. Cardoso, J. Miranda, A. Moita, G. Minas, A.L. Moreira, R. Lima, Heat transfer and fluid flow investigations in PDMS microchannel heat sinks fabricated by means of a low-cost 3D printer, IntechOpen, London, UK, 2020. https://doi.org/10.5772/intechopen.89735

[82] A. Kantaros, T. Ganetsos, D. Piromalis, 3D and 4D printing as integrated manufacturing methods of industry 4.0, 3D and 4D Printing as Integr. Manuf. Methods Ind. 4 (2023) 12-22. https://doi.org/10.3844/ajeassp.2023.12.22

[83] K. Griffin, D. Pappas, 3D printed microfluidics for bioanalysis: A review of recent advancements and applications, TrAC Trends Anal. Chem. 158 (2023) 116892. https://doi.org/10.1016/j.trac.2022.116892

[84] X. Li, M. Wang, T.P. Davis, L. Zhang, R. Qiao, Advancing tissue culture with light-driven 3D-printed microfluidic devices, Biosensors 14 (2024) 301. https://doi.org/10.3390/bios14060301

[85] C. Lee, Y. Chen, P. Wang, D.C. Wallace, P.J. Burke, A three-dimensional printed inertial microfluidic platform for isolation of minute quantities of vital mitochondria, Anal. Chem. 94 (2022) 6930-6938. https://doi.org/10.1021/acs.analchem.1c03244

[86] H.K. Park, M. Shin, B. Kim, J.W. Park, H. Lee, A visible light-curable yet visible wavelength-transparent resin for stereolithography 3D printing, NPG Asia Mater. 10 (2018) 82. https://doi.org/10.1038/s41427-018-0021-x

[87] Z. Miller, B. Stidham, T. Fairbanks, C. Maldonado, The use of stereolithography (SLA) additive manufacturing in space-based instrumentation, 2023 IEEE Aerospace Conf., IEEE, 2023, pp. 1-10. https://doi.org/10.1109/AERO55745.2023.10115988

[88] B.A. Basilia, A.J.C. Boniel, G.S.B. Borilla III, P.A.N. de Yro, 3D Printing of Urethane Dimethacrylate with Nanoclay for Enhanced Curing Properties via Stereolithography, Adv. Sci. Technol. 155 (2024) 59-64. https://doi.org/10.4028/p-EeP7MJ

[89] T. Wohlers, T. Gornet, N. Mostow, I. Campbell, O. Diegel, J. Kowen, R. Huff, B. Stucker, I. Fidan, A. Doukas, History of additive manufacturing, 2016.

[90] J. Zhang, Q. Hu, S. Wang, J. Tao, M. Gou, Digital light processing based three-dimensional printing for medical applications, Int. J. Bioprint. 6 (2019) 242. https://doi.org/10.18063/ijb.v6i1.242

[91] X. Peng, L. Yue, S. Liang, S. Montgomery, C. Lu, C.M. Cheng, R. Beyah, R.R. Zhao, H.J. Qi, Multi-color 3D printing via single-vat grayscale digital light processing, Adv. Funct. Mater. 32 (2022) 2112329. https://doi.org/10.1002/adfm.202112329

[92] L. Wu, Y. Song, Recent innovations in interfacial strategies for DLP 3D printing process optimization, Mater. Horiz. 12 (2025) 401-417. https://doi.org/10.1039/D4MH01160K

[93] F. Gholami, L. Yue, M. Li, A. Jain, A. Mahmood, M. Fratarcangeli, R. Ramprasad, H.J. Qi, Fast and efficient fabrication of functional electronic devices through grayscale digital light processing 3D printing, Adv. Mater. 36 (2024) 2408774. https://doi.org/10.1002/adma.202408774

[94] A. Amini, R.M. Guijt, T. Themelis, J. De Vos, S. Eeltink, Recent developments in digital light processing 3D-printing techniques for microfluidic analytical devices, J. Chromatogr. A 1692 (2023) 463842. https://doi.org/10.1016/j.chroma.2023.463842

[95] J.G.L. Santiago, O.M.R. Martínez, P.A. Sundaram, 3D-printed microfluidic and millifluidic devices for cell culture and mechanotransduction studies: A review, J. Micromanuf. 7 (2024) 216-230. https://doi.org/10.1177/25165984241286651

[96] C. Tzivelekis, P. Sgardelis, K. Waldron, R. Whalley, D. Huo, K. Dalgarno, Fabrication routes via projection stereolithography for 3D-printing of microfluidic geometries for nucleic acid amplification, PLoS One 15 (2020) e0240237. https://doi.org/10.1371/journal.pone.0240237

[97] A. Maddaus, P. Curley, M.A. Griswold, B.D. Costa, S. Hou, K.J. Jeong, E. Song, L.F. Deravi, Design and fabrication of bio-hybrid materials using inkjet printing, Biointerphases 11 (2016) 4. https://doi.org/10.1116/1.4966164

[98] Y. Sui, C.A. Zorman, Inkjet printing of metal structures for electrochemical sensor applications, J. Electrochem. Soc. 167 (2020) 037571. https://doi.org/10.1149/1945-7111/ab721f

[99] E.K. Arora, V. Sharma, A. Ravi, A. Shahi, S. Jagtap, A. Adhikari, J.K. Dash, P. Kumar, R. Patel, Polyaniline-based ink for inkjet printing for supercapacitors, sensors, and electrochromic devices, Energies 16 (2023) 6716. https://doi.org/10.3390/en16186716

[100] D.A. Lukyanov, O.V. Levin, Inkjet Printing with (Semi) conductive Conjugated Polymers: A Review, ChemEngineering 8 (2024) 53. https://doi.org/10.3390/chemengineering8030053

[101] E. Tekin, P.J. Smith, U.S. Schubert, Inkjet printing as a deposition and patterning tool for polymers and inorganic particles, Soft Matter 4 (2008) 703-713. https://doi.org/10.1039/b711984d

[102] G.I. Salentijn, P.E. Oomen, M. Grajewski, E. Verpoorte, Fused deposition modeling 3D printing for (bio) analytical device fabrication: procedures, materials, and applications, Anal. Chem. 89 (2017) 7053-7061. https://doi.org/10.1021/acs.analchem.7b00828

[103] A. Kumar, Z. Alam, D.A. Khan, S. Jha, Nanofinishing of FDM-fabricated components using ball end magnetorheological finishing process, Mater. Manuf. Process. 34 (2019) 232-242. https://doi.org/10.1080/10426914.2018.1512136

[104] R.F. Quero, G.D. da Silveira, J.A.F. da Silva, D.P. de Jesus, Understanding and improving FDM 3D printing to fabricate high-resolution and optically transparent microfluidic devices, Lab Chip 21 (2021) 3715-3729. https://doi.org/10.1039/D1LC00518A

[105] N. Jayanth, P. Senthil, C. Prakash, Effect of chemical treatment on tensile strength and surface roughness of 3D-printed ABS using the FDM process, Virtual Phys. Prototyp. 13 (2018) 155-163. https://doi.org/10.1080/17452759.2018.1449565

[106] N.P. Macdonald, J.M. Cabot, P. Smejkal, R.M. Guijt, B. Paull, M.C. Breadmore, Comparing microfluidic performance of three-dimensional (3D) printing platforms, Anal. Chem. 89 (2017) 3858-3866. https://doi.org/10.1021/acs.analchem.7b00136

[107] R.K. Lade Jr, E.J. Hippchen, C.W. Macosko, L.F. Francis, Dynamics of capillary-driven flow in 3D printed open microchannels, Langmuir 33 (2017) 2949-2964. https://doi.org/10.1021/acs.langmuir.6b04506

[108] L. Florea, E. Blasco, V. Mattoli, New Frontiers in Materials and Technologies for 3D Two Photon Polymerization, Adv. Funct. Mater. 33 (2023) 39. https://doi.org/10.1002/adfm.202305697

[109] V.F. Paz, M. Emons, K. Obata, A. Ovsianikov, S. Peterhänsel, K. Frenner, C. Reinhardt, B. Chichkov, U. Morgner, W. Osten, Development of functional sub-100 nm structures with 3D two-photon polymerization technique and optical methods for characterization, J. Laser Appl. 24 (2012) 4. https://doi.org/10.2351/1.4712151

[110] J.-F. Xing, M.-L. Zheng, X.-M. Duan, Two-photon polymerization microfabrication of hydrogels: an advanced 3D printing technology for tissue engineering and drug delivery, Chem. Soc. Rev. 44 (2015) 5031-5039. https://doi.org/10.1039/C5CS00278H

[111] Z. Faraji Rad, P.D. Prewett, G.J. Davies, High-resolution two-photon polymerization: the most versatile technique for the fabrication of microneedle arrays, Microsystems Nanoeng. 7 (2021) 71. https://doi.org/10.1038/s41378-021-00298-3

[112] M. Vaezi, H. Seitz, S. Yang, A review on 3D micro-additive manufacturing technologies, Int. J. Adv. Manuf. Technol. 67 (2013) 1721-1754. https://doi.org/10.1007/s00170-012-4605-2

[113] S. Waheed, J.M. Cabot, N.P. Macdonald, T. Lewis, R.M. Guijt, B. Paull, M.C. Breadmore, 3D printed microfluidic devices: enablers and barriers, Lab Chip 16 (2016) 1993-2013. https://doi.org/10.1039/C6LC00284F

[114] A. Waldbaur, H. Rapp, K. Länge, B.E. Rapp, Let there be chip-towards rapid prototyping of microfluidic devices: one-step manufacturing processes, Anal. Methods 3 (2011) 2681-2716. https://doi.org/10.1039/c1ay05253e

[115] M.I. Mohammed, M.N.H.Z. Alam, A. Kouzani, I. Gibson, Fabrication of microfluidic devices: Improvement of surface quality of CO2 laser machined poly(methylmethacrylate) polymer, J. Micromech. Microeng. 27 (2016) 015021. https://doi.org/10.1088/0960-1317/27/1/015021

[116-118] P.F. Costa, H.J. Albers, J.E. Linssen, H.H. Middelkamp, L. Van Der Hout, R. Passier, A. Van Den Berg, J. Malda, A.D. Van Der Meer, Mimicking arterial thrombosis in a 3D-printed microfluidic in vitro vascular model based on computed tomography angiography data, Lab Chip 17 (2017) 2785-2792. https://doi.org/10.1039/C7LC00202E

M.G. Mohamed, H. Kumar, Z. Wang, N. Martin, B. Mills, K. Kim, Rapid and inexpensive fabrication of multi-depth microfluidic device using high-resolution LCD stereolithographic 3D printing, J. Manuf. Mater. Process. 3 (2019) 26. https://doi.org/10.3390/jmmp3010026

J. Kajtez, S. Buchmann, S. Vasudevan, M. Birtele, S. Rocchetti, K. Pless, L. Adden, R. Tena, R. Kattler, B. Röhr, D. Hyll, S. Rizzi, E. Guimarães, H. Jaksch, N. Picht, K. Egger, M. Schwab, C. Heckl, F. Barzaghi, N. Galli, D. Meier, G. Laurenti, J. Honegger, L. Tomasek, D. Verkhratsky, B. Wirth, D. Custers, A. Antczak, F. Potthoff, M. Fussenegger, T. Schwab, M. Galli, Human 3D neuronal models for studying neural development and disease, Nat. Commun. 13 (2022) 3070.

[119] M. Li, P. Xiang, J. He, Z. Dong, Z. Zhang, J. Qiu, Y. Fan, Rapid prototyping of microfluidic devices using dry-film photoresist and direct-write lithography, Microfluid. Nanofluid. 23 (2019) 41.

[120] H. Bhattacharjee, R.M. Urrios, S. Kang, A. Folch, The upcoming 3D-printing revolution in microfluidics, Lab Chip 16 (2016) 1720-1742. https://doi.org/10.1039/C6LC00163G

[121] Y. Gong, X. Wang, T. Meng, S. Zeng, J. Li, Rapid prototyping of microfluidic devices using low-cost 3D printing technology, Microsyst. Technol. 26 (2020) 295-303.

[122] A. Malmberg, C. Edman, R. Andersson, M. Larsson, Rapid prototyping of microfluidic devices in glass using picosecond laser ablation and thermal bonding, Microfluid. Nanofluid. 21 (2017) 122.

[123] S. Zhang, F. Liu, D. Zhao, Microfluidics fabrication with high resolution using femtosecond laser direct writing, Opt. Laser Technol. 115 (2019) 35-45. https://doi.org/10.1016/j.optlastec.2019.02.046

[124] S. Maruo, O. Nakamura, S. Kawata, Three-dimensional microfabrication with two-photon-absorbed photopolymerization, Opt. Lett. 22 (1997) 132-134. https://doi.org/10.1364/OL.22.000132

[125] J. Sun, L. Zhang, S. Xu, Y. Fu, Advances in two-photon polymerization 3D printing for micro/nano biomedical devices, Micromachines 11 (2020) 798.

[126] K. Malinauskas, S. Žukauskas, A. Hasegawa, S. Hayasaki, R. Mizeikis, S. Juodkazis, S. Juodkazis, Ultrafast laser processing of materials: from science to industry, Light Sci. Appl. 5 (2016) e16133. https://doi.org/10.1038/lsa.2016.133

[127] C. Liu, C. Chen, X. Hu, L. Lin, Two-photon polymerization: A promising 3D printing technology for advanced functional microstructures, J. Mater. Sci. Technol. 105 (2022) 165-187.

[128] X. Huang, Y. Li, J. Lu, W. Fang, X. Li, High-resolution 3D printing of microfluidic devices with complex geometries via two-photon polymerization, Lab Chip 21 (2021) 1123-1132.

[129-131] A. Urrios, H. Bhattacharjee, S. Kang, A. Folch, 3D-printing for microfluidics: a review, Lab Chip 16 (2016) 1720-1742. https://doi.org/10.1039/C6LC00163G

L. Li, X. Peng, D. Zhu, J. Zhang, P. Xiao, Recent progress in polymers with dynamic covalent bonds, Macromolecular Chemistry and Physics, 224(20) (2023) 2300224. https://doi.org/10.1002/macp.202300224

M. Wang, Z. Zhang, G. Li, A. Jing, Room-temperature self-healing conductive elastomers for modular assembly as a microfluidic electrochemical biosensing platform for the detection of colorectal cancer exosomes, Micromachines, 14(3) (2023) 617. https://doi.org/10.3390/mi14030617

[132] K. Sugioka, Y. Cheng, Ultrafast lasers-reliable tools for advanced materials processing, Light: Science & Applications, 3(4) (2014) e149-e149. https://doi.org/10.1038/lsa.2014.30

[133] H.K. Chandrasekharan, E.P. McShane, K. Dhaliwal, R.R. Thomson, M.G. Tanner, Ultrafast laser ablation of a multicore polymer optical fiber for multipoint light emission, Optics Express, 29(13) (2021) 20765-20775. https://doi.org/10.1364/OE.424494

[134] K. Sugioka, Y. Cheng, Femtosecond laser processing for optofluidic fabrication, Lab on a Chip, 12(19) (2012) 3576-3589. https://doi.org/10.1039/c2lc40366h

[135] A. Gokaltun, M.L. Yarmush, A. Asatekin, O.B. Usta, Recent advances in nonbiofouling PDMS surface modification strategies applicable to microfluidic technology, Technology, 5(01) (2017) 1-12. https://doi.org/10.1142/S2339547817300013

[136] D.B. Weibel, W.R. DiLuzio, G.M. Whitesides, Microfabrication meets microbiology, Nature Reviews Microbiology, 5(3) (2007) 209-218. https://doi.org/10.1038/nrmicro1616

[137] X. Sun, R.T. Kelly, K. Tang, R.D. Smith, Ultrasensitive nanoelectrospray ionization-mass spectrometry using poly(dimethylsiloxane) microchips with monolithically integrated emitters, Analyst, 135(9) (2010) 2296-2302. https://doi.org/10.1039/c0an00253d

[138] F. Bollino, E. Tranquillo, Surface modification of implants by sol-gel coating technology: Advantages and applications, Materials Proceedings, 2(1) (2020) 7. https://doi.org/10.3390/CIWC2020-06827

[139] C. Carrera-Figueiras, Y. Pérez-Padilla, M.A. Estrella-Gutiérrez, E.G. Uc-Cayetano, J.A. Juárez-Moreno, A. Avila-Ortega, Surface science engineering through sol-gel process, Applied Surface Science, IntechOpen, 2019. https://doi.org/10.5772/intechopen.83676

[140] R. Kemkemer, Z. Zenghao, Y. Linxiao, K. Athanasopulu, K. Frey, Z. Cui, H. Su, L. Luo, Surface modification of Polydimethylsiloxane by hydrogels for microfluidic applications, Current Directions in Biomedical Engineering, 5(1) (2019) 93-96. https://doi.org/10.1515/cdbme-2019-0024

[141] S. Schneider, E.J. Brás, O. Schneider, K. Schlünder, P. Loskill, Facile patterning of thermoplastic elastomers and robust bonding to glass and thermoplastics for microfluidic cell culture and organ-on-chip, Micromachines, 12(5) (2021) 575. https://doi.org/10.3390/mi12050575

[142] V. Sunkara, D.-K. Park, H. Hwang, R. Chantiwas, S.A. Soper, Y.-K. Cho, Simple room temperature bonding of thermoplastics and poly(dimethylsiloxane), Lab on a Chip, 11(5) (2011) 962-965. https://doi.org/10.1039/C0LC00272K

[143] P.M. van Midwoud, A. Janse, M.T. Merema, G.M. Groothuis, E. Verpoorte, Comparison of biocompatibility and adsorption properties of different plastics for advanced microfluidic cell and tissue culture models, Analytical Chemistry, 84(9) (2012) 3938-3944. https://doi.org/10.1021/ac300771z

[144] A. Alrifaiy, O.A. Lindahl, K. Ramser, Polymer-based microfluidic devices for pharmacy, biology and tissue engineering, Polymers, 4(3) (2012) 1349-1398. https://doi.org/10.3390/polym4031349

[145] M.S. Zafar, Prosthodontic applications of polymethyl methacrylate (PMMA): An update, Polymers, 12(10) (2020) 2299. https://doi.org/10.3390/polym12102299

[146] F. Ghasemi, A. Jahani, A. Moradi, M.H. Ebrahimzadeh, N. Jirofti, Different modification methods of poly methyl methacrylate (PMMA) bone cement for orthopedic surgery applications, Archives of Bone and Joint Surgery, 11(8) (2023) 485.

[147] G. Khanarian, H. Celanese, Optical properties of cyclic olefin copolymers, Optical Engineering, 40(6) (2001) 1024-1029. https://doi.org/10.1117/1.1369411

[148] B. Bruijns, A. Veciana, R. Tiggelaar, H. Gardeniers, Cyclic olefin copolymer microfluidic devices for forensic applications, Biosensors, 9(3) (2019) 85. https://doi.org/10.3390/bios9030085

[149] J.Y. Shin, J.Y. Park, C. Liu, J. He, S.C. Kim, Chemical structure and physical properties of cyclic olefin copolymers (IUPAC Technical Report), Pure and Applied Chemistry, 77(5) (2005) 801-814. https://doi.org/10.1351/pac200577050801

[150] G. Bozzano, M. Dente, R. Del Rosso, Poly(bisphenol A carbonate) recycling: high pressure hydrolysis can be a convenient way, in: Material Recycling-Trends and Perspectives, InTech Rijeka, Croatia, 2012, pp. 115-132. https://doi.org/10.5772/32974

[151] A. Gruzd, A. Tokarev, I. Tokarev, D. Kuksenkov, S. Minko, All-nanoparticle monolayer broadband antireflective and self-cleaning transparent glass coatings, ACS Applied Materials & Interfaces, 13(5) (2021) 6767-6777. https://doi.org/10.1021/acsami.0c18776

[152] J.W. Chan, T. Huser, S. Risbud, D. Krol, Structural changes in fused silica after exposure to focused femtosecond laser pulses, Optics Letters, 26(21) (2001) 1726-1728. https://doi.org/10.1364/OL.26.001726

[153] M. Gromada, A. Tłuczek, R. Cygan, Silica-based ceramic cores for high-pressure turbine airfoil blades in aircraft engines, Processing and Application of Ceramics, 18(3) (2024) 299-306. https://doi.org/10.2298/PAC2403299G

[154] M. Hejazian, E. Balaur, L. Flueckiger, L. Hor, C. Darmanin, B. Abbey, Microfluidic mixing and jetting devices based on SU8 and glass for time-resolved molecular imaging experiments, in: Microfluidics, Biomems, and Medical Microsystems XVII, SPIE, 2019, pp. 59-65. https://doi.org/10.1117/12.2508437

[155] C.J. Easley, J.A. Humphrey, J.P. Landers, Thermal isolation of microchip reaction chambers for rapid non-contact DNA amplification, Journal of Micromechanics and Microengineering, 17(9) (2007) 1758. https://doi.org/10.1088/0960-1317/17/9/002

[156] D.K. Maurya, W.Y. Ng, K.A. Mahabadi, Y.N. Liang, I. Rodríguez, Fabrication of lab-on-chip platforms by hot embossing and photo patterning, Biotechnology Journal: Healthcare Nutrition Technology, 2(11) (2007) 1381-1388. https://doi.org/10.1002/biot.200700107

[157] T. Claudio, N. Stein, D.G. Stroppa, B. Klobes, M.M. Koza, P. Kudejova, N. Petermann, H. Wiggers, G. Schierning, R.P. Hermann, Nanocrystalline silicon: lattice dynamics and enhanced thermoelectric properties, Physical Chemistry Chemical Physics, 16(47) (2014) 25701-25709. https://doi.org/10.1039/C3CP53749H

[158] A. Wu, L. Wang, E. Jensen, R. Mathies, B. Boser, Modular integration of electronics and microfluidic systems using flexible printed circuit boards, Lab on a Chip, 10(4) (2010) 519-521. https://doi.org/10.1039/B922830F

[159] Y.-Y. Zhao, H. Kim, V.-N. Nguyen, S. Jang, W.J. Jang, J. Yoon, Recent advances and prospects in organic molecule-based phototheranostic agents for enhanced cancer phototherapy, Coordination Chemistry Reviews, 501 (2024) 215560. https://doi.org/10.1016/j.ccr.2023.215560

[160] J. Wang, D. Xue, X. Li, Silicon monolithic microflow sensors: a review, Journal of Micromechanics and Microengineering, 31(10) (2021) 104002. https://doi.org/10.1088/1361-6439/ac1f3a

[161] W. Xi, F. Kong, J.C. Yeo, L. Yu, S. Sonam, M. Dao, X. Gong, C.T. Lim, Soft tubular microfluidics for 2D and 3D applications, Proceedings of the National Academy of Sciences, 114(40) (2017) 10590-10595. https://doi.org/10.1073/pnas.1712195114

[162] A.K. Yetisen, M.S. Akram, C.R. Lowe, Based microfluidic point-of-care diagnostic devices, Lab on a Chip, 13(12) (2013) 2210-2251. https://doi.org/10.1039/c3lc50169h

[163] X. Li, D.R. Ballerini, W. Shen, A perspective on paper-based microfluidics: Current status and future trends, Biomicrofluidics, 6(1) (2012). https://doi.org/10.1063/1.3687398

[164] E. Noviana, T. Ozer, C.S. Carrell, J.S. Link, C. McMahon, I. Jang, C.S. Henry, Microfluidic paper-based analytical devices: from design to applications, Chemical Reviews, 121(19) (2021) 11835-11885. https://doi.org/10.1021/acs.chemrev.0c01335

[165] Y.-J. Juang, S.-K. Hsu, Fabrication of paper-based microfluidics by spray on printed paper, Polymers, 14(3) (2022) 639. https://doi.org/10.3390/polym14030639

[166] A. Vafadar, S. Takallu, S.K. Alashti, S. Rashidi, S. Bahrani, A. Tajbakhsh, E. Mirzaei, A. Savardashtaki, Advancements in microfluidic platforms for rapid biomarker diagnostics of infectious diseases, Microchemical Journal, (2024) 112296. https://doi.org/10.1016/j.microc.2024.112296

[167] W. Zhou, M. Dou, S.S. Timilsina, F. Xu, X. Li, Recent innovations in cost-effective polymer and paper hybrid microfluidic devices, Lab on a Chip, 21(14) (2021) 2658-2683. https://doi.org/10.1039/D1LC00414J

[168] P. Liu, B. Li, L. Fu, Y. Huang, M. Man, J. Qi, X. Sun, Q. Kang, D. Shen, L. Chen, Hybrid three-dimensionally printed paper-based microfluidic platform for investigating a cell's apoptosis and intracellular cross-talk, ACS Sensors, 5(2) (2020) 464-473. https://doi.org/10.1021/acssensors.9b02205

[169] H.A. Silva-Neto, I.V. Arantes, A.L. Ferreira, G.H. do Nascimento, G.N. Meloni, W.R. de Araujo, T.R. Paixão, W.K. Coltro, Recent advances on paper-based microfluidic devices for bioanalysis, TrAC Trends in Analytical Chemistry, 158 (2023) 116893. https://doi.org/10.1016/j.trac.2022.116893

[170] A. Andar, M.-S. Hasan, V. Srinivasan, M. Al-Adhami, E. Gutierrez, D. Burgenson, X. Ge, L. Tolosa, Y. Kostov, G. Rao, Wood microfluidics, Analytical Chemistry, 91(17) (2019) 11004-11012. https://doi.org/10.1021/acs.analchem.9b01232

[171] P. Mesquita, L. Gong, Y. Lin, Low-cost microfluidics: Towards affordable environmental monitoring and assessment, Frontiers in Lab on a Chip Technologies, 1 (2022) 1074009. https://doi.org/10.3389/frlct.2022.1074009

[172] P. Aryal, C. Hefner, B. Martinez, C.S. Henry, Microfluidics in environmental analysis: advancements, challenges, and future prospects for rapid and efficient monitoring, Lab on a Chip, 24(5) (2024) 1175-1206. https://doi.org/10.1039/D3LC00871A

[173] M. Dadras Chomachayi, A. Jalali-Arani, F.R. Beltrán, M.U. De la Orden, J. Martínez Urreaga, Biodegradable nanocomposites developed from PLA/PCL blends and silk fibroin nanoparticles: study on the microstructure, thermal behavior, crystallinity and performance, Journal of Polymers and the Environment, 28 (2020) 1252-1264. https://doi.org/10.1007/s10924-020-01684-0

[174] C. Cui, Q. Fu, L. Meng, S. Hao, R. Dai, J. Yang, Recent progress in natural biopolymers conductive hydrogels for flexible wearable sensors and energy devices: materials, structures, and performance, ACS Applied Bio Materials, 4(1) (2020) 85-121. https://doi.org/10.1021/acsabm.0c00807

[175] J. Mettakoonpitak, K. Khongsoun, N. Wongwan, S. Kaewbutdee, A. Siripinyanond, A. Kuharuk, C.S. Henry, Simple biodegradable plastic screen-printing for microfluidic paper-

based analytical devices, Sensors and Actuators B: Chemical, 331 (2021) 129463. https://doi.org/10.1016/j.snb.2021.129463

[176] A. Mtibe, M.P. Motloung, J. Bandyopadhyay, S.S. Ray, Synthetic biopolymers and their composites: Advantages and limitations-An overview, Macromolecular Rapid Communications, 42(15) (2021) 2100130. https://doi.org/10.1002/marc.202100130

[177] M.S.B. Reddy, D. Ponnamma, R. Choudhary, K.K. Sadasivuni, A comparative review of natural and synthetic biopolymer composite scaffolds, Polymers, 13(7) (2021) 1105. https://doi.org/10.3390/polym13071105

[178] Y.K. Jo, D. Lee, Biopolymer microparticles prepared by microfluidics for biomedical applications, Small, 16(9) (2020) 1903736. https://doi.org/10.1002/smll.201903736

[179] N. Gowthaman, H. Lim, T. Sreeraj, A. Amalraj, S. Gopi, Advantages of biopolymers over synthetic polymers: social, economic, and environmental aspects, in: Biopolymers and Their Industrial Applications, Elsevier, 2021, pp. 351-372. https://doi.org/10.1016/B978-0-12-819240-5.00015-8

[180] D. Park, J.J. Faraway, Sequential design for response curve estimation, Journal of Nonparametric Statistics, 9(2) (1998) 155-164. https://doi.org/10.1080/10485259808832740

[181] Y. Hong, M. Wu, G. Chen, Z. Dai, Y. Zhang, G. Chen, X. Dong, 3D printed microfluidic device with microporous Mn2O3-modified screen printed electrode for real-time determination of heavy metal ions, ACS Applied Materials & Interfaces, 8(48) (2016) 32940-32947. https://doi.org/10.1021/acsami.6b10464

[182] Y. Sun, J. Ma, Y. Wang, S. Qiao, Y. Feng, Z. Li, Z. Wang, Y. Han, Z. Zhu, A wearable patch sensor for simultaneous detection of dopamine and glucose in sweat, Analytica, 4(2) (2023) 170-181. https://doi.org/10.3390/analytica4020014

[183] A. Sawires, R. Trouillon, Flexible electrodes for the electroanalysis of neurotransmitters, in: 2024 22nd IEEE Interregional NEWCAS Conference (NEWCAS), IEEE, 2024, pp. 263-267. https://doi.org/10.1109/NewCAS58973.2024.10666300

[184] L. Gao, S. Huang, X. Li, Z. Liu, X. Huang, Y. Li, H.J. Chen, X. Lin, J. Liu, X. Xie, Modification of metal microelectrodes with vertical graphene structure for enhanced electrochemical detection of dopamine, Advanced Materials Interfaces, (2024) 2400767. https://doi.org/10.1002/admi.202400767

[185] C. Yang, C.B. Jacobs, M.D. Nguyen, M. Ganesana, A.G. Zestos, I.N. Ivanov, A.A. Puretzky, C.M. Rouleau, D.B. Geohegan, B.J. Venton, Carbon nanotubes grown on metal microelectrodes for the detection of dopamine, Analytical Chemistry, 88(1) (2016) 645-652. https://doi.org/10.1021/acs.analchem.5b01257

[186] R. Kim, N. Hong, Y. Nam, Gold nanograin microelectrodes for neuroelectronic interfaces, Biotechnology Journal, 8(2) (2013) 206-214. https://doi.org/10.1002/biot.201200219

[187] J. Kim, M. Park, Recent progress in electrochemical immunosensors, Biosensors, 11(10) (2021) 360. https://doi.org/10.3390/bios11100360

[188] C. Brett, Electrochemical impedance spectroscopy for characterization of electrochemical sensors and biosensors, ECS Transactions, 13(13) (2008) 67. https://doi.org/10.1149/1.3004030

[189] A. Bonanni, M. Pumera, Y. Miyahara, Influence of gold nanoparticle size (2-50 nm) upon its electrochemical behavior: An electrochemical impedance spectroscopic and voltammetric study, Physical Chemistry Chemical Physics, 13(11) (2011) 4980-4986. https://doi.org/10.1039/c0cp01209b

[190] Y.-R. Wang, S.C. Chien, Y.C. Lin, J.-C. Chen, Portable electrochemical impedance spectroscopy device implementation for Alzheimer's disease-related protein, Electrochemical Society Meeting Abstracts, 243, The Electrochemical Society, Inc., 2023, pp. 1892-1892. https://doi.org/10.1149/MA2023-01341892mtgabs

[191] Y. Sun, J. Fan, L. Cui, W. Ke, F. Zheng, Y. Zhao, Fluorometric nanoprobes for simultaneous aptamer-based detection of carcinoembryonic antigen and prostate specific antigen, Microchimica Acta, 186 (2019) 1-10. https://doi.org/10.1007/s00604-019-3281-4

[192] A.O. Idris, N. Mabuba, O.A. Arotiba, An exfoliated graphite-based electrochemical immunosensor on a dendrimer/carbon nanodot platform for the detection of carcinoembryonic antigen cancer biomarker, Biosensors, 9(1) (2019) 39. https://doi.org/10.3390/bios9010039

[193] H. Gai, Y. Li, E.S. Yeung, Optical detection systems on microfluidic chips, in: Microfluidics: Technologies and Applications, 2011, pp. 171-201. https://doi.org/10.1007/128_2011_144

[194] N.M.M. Pires, T. Dong, U. Hanke, N. Hoivik, Recent developments in optical detection technologies in lab-on-a-chip devices for biosensing applications, Sensors, 14(8) (2014) 15458-15479. https://doi.org/10.3390/s140815458

[195] S. Ahmadsaidulu, O. Banik, P. Kumar, S. Kumar, E. Banoth, Microfluidic point-of-care diagnostics for multi-disease detection using optical techniques: a review, IEEE Transactions on NanoBioscience, 23(1) (2023) 140–147

[196] B.B. Xu, Z.C. Ma, H. Wang, X.Q. Liu, Y.L. Zhang, X.L. Zhang, R. Zhang, H.B. Jiang, H.B. Sun, A SERS-active microfluidic device with tunable surface plasmon resonances, Electrophoresis, 32(23) (2011) 3378–3384.

[197] D.-S. Wang, S.-K. Fan, Microfluidic surface plasmon resonance sensors: from principles to point-of-care applications, Sensors, 16(8) (2016) 1175.

[198] R. Chantiwas, M.L. Hupert, S.R. Pullagurla, S. Balamurugan, J. Tamarit-López, S. Park, P. Datta, J. Goettert, Y.-K. Cho, S.A. Soper, Simple replication methods for producing nanoslits in thermoplastics and the transport dynamics of double-stranded DNA through these slits, Lab on a Chip, 10(23) (2010) 3255–3264.

[199] C. Byram, S.S.B. Moram, D. Banerjee, R. Beeram, J. Rathod, V.R. Soma, Review of ultrafast laser ablation for sensing and photonic applications, Journal of Optics, 25(4) (2023) 043001.

[200] M. Bhavya, N. Rhakho, S.R. Jena, S. Yadav, A. Altaee, M. Saxena, A.K. Samal, Detection of PFAS via surface-enhanced Raman scattering: challenges and future perspectives, Sustainable Chemistry for the Environment, 3 (2023) 100031.

[201] R. Shi, X. Liu, Y. Ying, Facing challenges in real-life application of surface-enhanced Raman scattering: design and nanofabrication of SERS substrates for rapid field test of food contaminants, Journal of Agricultural and Food Chemistry, 66(26) (2017) 6525–6543.

[202] X. Feng, B.F. Liu, J. Li, X. Liu, Advances in coupling microfluidic chips to mass spectrometry, Mass Spectrometry Reviews, 34(5) (2015) 535–557.

[203] D. Gao, H. Liu, Y. Jiang, J.-M. Lin, Recent advances in microfluidics combined with mass spectrometry: technologies and applications, Lab on a Chip, 13(17) (2013) 3309–3322.

[204] D.S. Kulyk, C.F. Miller, A.K. Badu-Tawiah, Reactive charged droplets for reduction of matrix effects in electrospray ionization mass spectrometry, Analytical Chemistry, 87(21) (2015) 10988–10994.

[205] Y.-Q. Huang, J.-Q. You, B.-F. Yuan, Y.-Q. Feng, Sample preparation and direct electrospray ionization on a tip column for rapid mass spectrometry analysis of complex samples, Analyst, 137(19) (2012) 4593-4597. https://doi.org/10.1039/c2an35856e

[206] D.J. Steyer, R.T. Kennedy, High-throughput nanoelectrospray ionization-mass spectrometry analysis of microfluidic droplet samples, Analytical Chemistry, 91(10) (2019) 6645-6651. https://doi.org/10.1021/acs.analchem.9b00571

[207] W. Kim, M. Guo, P. Yang, D. Wang, Microfabricated monolithic multinozzle emitters for nanoelectrospray mass spectrometry, Analytical Chemistry, 79(10) (2007) 3703-3707. https://doi.org/10.1021/ac070010j

[208] J. Lee, S.A. Soper, K.K. Murray, Microfluidic chips for mass spectrometry-based proteomics, Journal of Mass Spectrometry, 44(5) (2009) 579-593. https://doi.org/10.1002/jms.1585

[209] Y. Yang, Y. Su, X. Wang, W. Gao, X. Lu, H. Lam, R. Tian, Fritted tip capillary column with negligible dead volume facilitated ultrasensitive and deep proteomics, Analytica Chimica Acta, 1201 (2022) 339615. https://doi.org/10.1016/j.aca.2022.339615

[210] L.M. Rey Gomez, R. Hirani, A. Care, D.W. Inglis, Y. Wang, Emerging microfluidic devices for sample preparation of undiluted whole blood to enable the detection of biomarkers, ACS Sensors, 8(4) (2023) 1404-1421. https://doi.org/10.1021/acssensors.2c02696

[211] S. Wang, Q. Xu, Y. Cai, Q. Wang, Y. Liu, D. Wang, Biological particle separation techniques based on microfluidics, Interdisciplinary Medicine, 2(2) (2024) e20240003. https://doi.org/10.1002/INMD.20240003

[212] G. Luka, A. Ahmadi, H. Najjaran, E. Alocilja, M. DeRosa, K. Wolthers, A. Malki, H. Aziz, A. Althani, M. Hoorfar, Microfluidics integrated biosensors: A leading technology towards lab-on-a-chip and sensing applications, Sensors, 15(12) (2015) 30011-30031. https://doi.org/10.3390/s151229783

[213] A.D. Stroock, S.K. Dertinger, A. Ajdari, I. Mezic, H.A. Stone, G.M. Whitesides, Chaotic mixer for microchannels, Science, 295(5555) (2002) 647-651. https://doi.org/10.1126/science.1066238

[214] T.J. Kwak, Y.G. Nam, M.A. Najera, S.W. Lee, J.R. Strickler, W.-J. Chang, Convex grooves in staggered herringbone mixer improve mixing efficiency of laminar flow in microchannel, PloS One, 11(11) (2016) e0166068. https://doi.org/10.1371/journal.pone.0166068

[215] S. Orbay, A. Ozcelik, J. Lata, M. Kaynak, M. Wu, T.J. Huang, Mixing high-viscosity fluids via acoustically driven bubbles, Journal of Micromechanics and Microengineering, 27(1) (2016) 015008. https://doi.org/10.1088/0960-1317/27/1/015008

[216] W. Zhou, B. Zhang, Y. Liu, C. Wang, W. Sun, W. Li, Z. Chen, Advances in capillary electrophoresis-mass spectrometry for cell analysis, TrAC Trends in Analytical Chemistry, 117 (2019) 316-330. https://doi.org/10.1016/j.trac.2019.05.011

[217] N. Delaunay, V. Pichon, M.-C. Hennion, Immunoaffinity solid-phase extraction for the trace-analysis of low-molecular-mass analytes in complex sample matrices, Journal of Chromatography B: Biomedical Sciences and Applications, 745(1) (2000) 15-37. https://doi.org/10.1016/S0378-4347(00)00116-X

[218] N. Delaunay-Bertoncini, M.-C. Hennion, Immunoaffinity solid-phase extraction for pharmaceutical and biomedical trace-analysis-coupling with HPLC and CE-perspectives, Journal of Pharmaceutical and Biomedical Analysis, 34(4) (2004) 717-736. https://doi.org/10.1016/S0731-7085(03)00559-4

[219] L. Mou, X. Jiang, Materials for microfluidic immunoassays: A review, Advanced Healthcare Materials, 6(15) (2017) 1601403. https://doi.org/10.1002/adhm.201601403

[220] K.N. Han, C.A. Li, G.H. Seong, Microfluidic chips for immunoassays, Annual Review of Analytical Chemistry, 6(1) (2013) 119-141. https://doi.org/10.1146/annurev-anchem-062012-092616

[221] S.H. Ko, S.J. Kim, L.F. Cheow, L.D. Li, K.H. Kang, J. Han, Massively parallel concentration device for multiplexed immunoassays, Lab on a Chip, 11(7) (2011) 1351-1358. https://doi.org/10.1039/c0lc00349b

[222] F. Chen, Q. Hu, H. Li, Y. Xie, L. Xiu, Y. Zhang, X. Guo, K. Yin, Multiplex detection of infectious diseases on microfluidic platforms, Biosensors, 13(3) (2023) 410. https://doi.org/10.3390/bios13030410

[223] K. McKinski, H. Tang, K. Wang, M. Birchler, M. Wright, Comparison of highly sensitive, multiplex immunoassay platforms for streamlined clinical cytokine quantification, Bioanalysis, 17(1) (2025) 17-29. https://doi.org/10.1080/17576180.2024.2442190

[224] D.E. Bender, M.O. Schaettler, K.C. Sheehan, T.M. Johanns, G.P. Dunn, Cytokine profiling in plasma from patients with brain tumors versus healthy individuals using 2 different multiplex immunoassay platforms, Biomarker Insights, 16 (2021) 1-12. https://doi.org/10.1177/11772719211006666

[225] W. Zhao, Y. Zhou, Y.-Z. Feng, X. Niu, Y. Zhao, J. Zhao, Y. Dong, M. Tan, Y. Xianyu, Y. Chen, Computer vision-based artificial intelligence-mediated encoding-decoding for multiplexed microfluidic digital immunoassay, ACS Nano, 17(14) (2023) 13700-13714. https://doi.org/10.1021/acsnano.3c02941

Materials Research Foundations 191 (2026) 72-95 — https://doi.org/10.21741/9781644904213-3

Chapter 3

Microfluidic Devices for Food Analysis

Ramya Mohandass[1], Aravind Thiruvenkadam[1], Ashish Kapoor[2,*]

[1]Molecular Genetics Laboratory, Department of Genetic Engineering, SRM Institute of Science and Technology, Kattankulathur, Chengalpattu, Tamil Nadu 603203, India

[2]Department of Chemical Engineering, Harcourt Butler Technical University, Kanpur, Uttar Pradesh 208002, India

*ashishkapoorchem@gmail.com

Abstract

The increasing outbreaks of various foodborne diseases have made food hygiene and quality monitoring a priority concern worldwide. Although traditional analysis methods are reliable, they have some disadvantages like slow rate of processing, high cost, and the need for sophisticated infrastructure. Microfluidic platforms have proved to be a worthwhile substitute for traditional methods by offering a rapid, cost-effective, and portable means of food analysis. Microfluidic platforms function based on miniaturized platforms to enable the control of minute quantities of fluids with minimal error. This chapter provides a general overview of microfluidics for food analysis, from the basics of fluid dynamics at the microscale to different propulsion and mixing platforms, and fabrication techniques like soft lithography, paper-based platforms, and 3D printing. This study examines key microfluidic platforms in the context of their applications for detecting pollutants, pathogens, and quality indicators. We further investigate the incorporation of microfluidic devices with various techniques, such as optical methods (fluorescence, colorimetry, SPR) and electrochemical techniques (voltammetry, amperometry, impedance spectroscopy). This chapter emphasizes the possibilities of microfluidic technologies to transform food diagnostics by allowing high-throughput, real-time, on-site testing, particularly in low-resource environments.

Keywords

Lab-on-a-Chip, Biosensors, Isothermal Amplification, Food-Borne Pathogens, Allergens, Spoilage Detection

Contents

Materials Research Foundations 191 (2026) 72-95 https://doi.org/10.21741/9781644904213-3

1. Introduction

Food quality and safety are significant international problems affecting public health, trade, and consumer confidence. The World Health Organization (WHO) has estimated that food-borne disease results in 600 million cases and 420,000 fatalities annually [1]. Urbanization and globalization have marked the significance of rapid, sensitive, and cost-effective analytical technology among complex supply chains and emerging food-borne diseases [2]. Conventional food analysis methods such as chromatography, mass spectrometry, and microbiological culturing are precise but tend to need large-scale instrumentation, specialized expertise, and quite a long time. Food companies and clinical diagnostic laboratories require in-place analysis, real-time analysis, and high-throughput analysis, which none of these can provide [3]. Microfluidic technology has been an emergent alternative to conventional techniques for the improvement of food diagnostic capabilities. Microfluidics refers to the technology and science of handling small amounts of liquids (typically in the range of microliters to nanoliters) within a miniaturized device for chemical and biological tests [4]. Microfluidic devices were initially conceptualized for applications in biomedicine, but these have rapidly transferred into other areas such as environmental monitoring, drugs, and, more recently, food analysis. These tools bring various disciplines together and merge a variety of laboratory tasks such as sample preparation, separation, mixing, and detection into one chip. Their advantages include low reagent use, faster processing, improved analytical sensitivity, and the possibility of portability and automation [5]. Microfluidics in food analysis applications enables in situ, real-time detection of contaminating agents like pathogens, toxins, allergens, heavy metals, and pesticides [6]. The microfluidic platforms' miniaturized characteristics provide compatibility with a wide range of detection technologies, such as optical, electrochemical, and colorimetric techniques [7]. Microfluidic devices are combined with biosensing components such as antibodies, aptamers, or nucleic acids to provide high sensitivity and specificity. Significantly, the devices could be tailored to various tests of food quality parameters like pH, viscosity, nutritional content, and shelf-life parameters [8]. Recent developments in paper-based microfluidics, 3D printing, and flexible substrates have also more opened up the use of microfluidics for application in underdeveloped regions, hence enabling application in small-scale food production industries and on-site monitoring [9]. Microfluidic devices have also been integrated with isothermal amplification methods, like loop-mediated isothermal amplification (LAMP), recombinase polymerase amplification (RPA), for pathogen detection in food without the need for thermal cycling [10]. This chapter intends to give an overview of microfluidics in food analysis, such as the concept of microfluidic design, fabrication

Materials Research Foundations 191 (2026) 72-95 https://doi.org/10.21741/9781644904213-3

techniques, and interfacing them with other analytical methods. Case studies and recent advances in the field will be highlighted to show the diversity and potential of microfluidic technology.

2. Fundamentals of microfluidic systems for food analysis

Microfluidics technology is the process of manipulating small volumes of fluids in channels with dimensions less than microscale. At this scale, fluids tend to follow laminar flow rather than turbulent flow, leading to more predictable kinetic behavior. The need for external pumps can be eliminated as fluids adapt surface tension for the movement within the channel at this scale, unlike the gravitational force at larger scales. Fluids exhibit capillary force at microscale, which facilitate their transport through the pores [11]. This capillary action provides the foundation for paper-based microfluidic systems. Paper-based microfluidics (μPADs) exploit cellulose paper as the main material, with a low-cost disposable option over conventional microfluidic devices. The μPADs are prepared by applying hydrophobic barriers by wax printing or inkjet etching to create microchannels for the capillary movement of the samples. These microchannels direct fluid flow without additional forces [12,13]. This simple mechanism allows for the development of portable, on-site analysis of food samples like a lateral flow μPAD for the detection of egg allergen (ovalbumin) in spiked food samples [14]. Likewise, a μPAD-ELISA hybrid assay was fabricated to detect clenbuterol in milk, which employed the porousness of the paper for antibody trapping and made it possible to detect at 0.2 ppb. This decreased the use of reagent as well as analysis time in comparison to conventional ELISA [15]. The economical and simple construction make μPADs preferrable for POC needs, specifically in infrastructure limited places. Moreover, their application in biological samples enables screening of foodborne pathogens at a rapid pace [16,17].

Polymers like polydimethylsiloxane (PDMS), poly(methyl methacrylate) (PMMA), and polycarbonate (PC) are also widely used to create microfluidic devices for different uses in food hygiene [18]. Polymer-based microfluidic devices are often fabricated by a method called soft lithography, which is the process of developing microfluidic device by pouring polymer material in a flexible patterned mold. This method is quicker and economical than the photolithography technique [19]. There are also other fabrication techniques like hot embossing, injection molding, lased ablation, 3D printing, micromachining, and micro-cutting. The selection of fabrication technique is determined by the substrate material and the required production volume [20]. PDMS is a widely used polymer because of its optical transparency, biocompatibility, and moldability. Its transparency is ideal for optical detection methods in food analysis, as seen in a colorimetric microfluidic device to detect nitrite in meat [21]. PMMA is an another commonly used polymer material because of its mechanical strength and the fact that it is easy to pattern using laser ablation. Laser ablation technique employs laser to carve the cavity in a polymer to create a microfluidic chip [19]. These properties make them suitable for creating complex devices with integrated components, such as a PMMA/paper hybrid device to analyze cyclamate, an artificial sweetner, using a diazotization reaction [22]. 3D printing techniques, such as fused deposition modeling (FDM) and stereolithography (SLA), has enabled the creation of intricate microfluidic structures with high resolution and biocompatibility. The techniques allow several materials and functions to be integrated into a single device, enhancing versatility and application of microfluidic systems in food analysis [23]. This capability was demonstrated in a 3D-printed microfluidic device with multiplex LAMP for the recognition of three foodborne pathogens, such as *Salmonella* typhimurium, *Listeria monocytogenes*, and *Escherichia coli* in various food samples. The device was 100% accurate, which demonstrated its ability for rapid and trustworthy food safety tests [24]. Table 1 provides the summary of commonly used materials, their fabrication approach, and their respective food applications.

Table 1 *Various fabrication materials and methods used in microfluidic devices for food analysis*

Material	Common fabrication method	Properties	Example in food application	Reference
PDMS	Soft lithography	Optically transparent, biocompatible, easy prototyping	Nitrite detection in meat	[21]
PMMA	Laser ablation, CNC machining	Good mechanical strength, low cost, easy patterning	Cyclamate detection in food samples	[22]
PC	Hot embossing, Injection molding	High thermal stability, durable, good solvent resistance	Detection of *S. aureus*, *B. cereus* and *L. monocytogenes* in food samples	[25]
Paper	Wax printing, Inkjet etching	Very low cost, disposable, capillary-driven flow	Egg allergen (ovalbumin) detection in spiked food	[14]
3D-printed resins	Stereolithography, Fusion Deposition Modelling	Easy prototyping, customizable designs	Detection of Salmonella, Listeria, E. coli in milk, chicken, lettuce	[24]

The design of a microfluidic device determines the efficient manipulation of samples and reagents. Actuators, pumps, valves, mixers, and detectors are some of the essential components in a microfluidic device. Actuators control the dynamics of fluids and samples by converting energy into mechanical motion in microfluidic systems [26]. Pumps provide the driving force for the fluids, which can be integrated with or without valves [27]. Valves control the entry and exit of fluids and regulate their flow within the channels. Since fluid flow is laminar in microfluidics, the mixing of different fluids by diffusion alone is slow and limited when compared to turbulent mixing. In these types of reactions, micromixers are employed for speeding up the diffusion-driven mixing [28,29]. Micromixers are divided into two types such as active and passive micromixers. Active micromixers necessitates external forces, while passive micromixers employs intricate geometries in the channel to enhance the diffusion [30]. For instance, a micromixer was used in an electrochemical 3D microfluidic chip for real-time glucose sensing in liquid analytes [31]. Detectors are analytical components used to quantify the fluids based on different methods. The types of detections employed in microfluidics are discussed in detail in the upcoming sections.

Different mechanisms for propulsion, such as pressure-driven, electrokinetic, and centrifugal flows, can be employed to regulate fluid dynamics in microfluidics. Pressure-driven systems use external pumps to create flow within microchannels, while electrokinetic mechanisms allow electric fields to drive charged species within the fluid. Centrifugal microfluidics works through rotation in order to propel fluids along channels [32]. Microchip capillary electrophoresis enabled electrokinetic mechanism to separate and detect ciprofloxacin in milk samples [33]. Centrifugal microfluidic platforms utilize rotational motion for fluid management, eliminating the need for external pumps and allowing extensive integration. This was demonstrated in a centrifugal

Materials Research Foundations 191 (2026) 72-95 https://doi.org/10.21741/9781644904213-3

microfluidic chip designed to simultaneously identify milk from camels, cows, horses, goats, and yaks [34].

Droplet microfluidics (DMF) is an approach that has found widespread usage in food analysis. It exploits the single droplet as a reaction chamber, which reduces the consumption of reagents and enables high-throughput analysis of food constituents [35]. Encapsulation of different food samples in droplets facilitates parallel analysis of different parameters, which facilitates rapid assay developments. The versatility of droplet microfluidics also facilitates compatibility with other analytical devices, like PCR detection of pathogens [36]. DMF employs electrowetting or dielectrophoresis to move droplets on a planar surface, which permits programmable and accurate liquid handling. Their properties allowed them to develop microfluidic platforms that can efficiently detect nitrite in food samples [37], copper ions in wine [38], and ciprofloxacin in milk [39]. Additionally, recent developments point to the incorporation of nanomaterials that provide increased sensitivity and performance in food analysis [40,41]. Advances such as smartphone compatibility enable real-time monitoring alongside analysis capabilities, such that such technologies are well suited for field testing [42,43]. Such devices are particularly useful in point-of-care (POC) diagnostics due to their portability alongside the capacity of integrating to other bioanalytical methods [44].

3. Detection mechanisms in microfluidic food analysis

Microfluidic devices employ different types of detection methods to achieve rapid and sensitive identification of contaminants, pathogens, or adulterants for food analysis applications. The outline of detection methods used in microfluidic food analysis is presented in Table 2

***Table 2** Overview of Detection Methods in Microfluidic Food Analysis*

Detection method	Principle	Example Food Analyte	Mechanism of Signal Generation	Advantages	Reference
Optical	Measures changes in light (fluorescence, color, absorbance)	2,4-D pesticide	Pesticide binds with quantum dots providing fluorescence	High sensitivity, portable readout	[45]
Electrochemical	Measures changes in electrical properties (current, voltage, impedance)	*Listeria monocytogenes*	Pathogen binding alters electrode impedance across AC frequencies	Label-free, real-time, portable	[46]
Mass-based	Measures minute changes in mass on a sensor's surface	Bacteria	Surface acoustic wave sensors detect frequency shifts caused by bacterial mass binding to sensor surface	Real-time, label-free detection with high sensitivity	[47]
Magnetic	Uses magnetic beads to capture targets	*Salmonella* typhimurium	Bacterial DNA is captured by magnetic beads and amplified by PCR on-chip	High specificity	[48]

Materials Research Foundations 191 (2026) 72-95 | https://doi.org/10.21741/9781644904213-3

3.1 Optical detection

Optical sensing dominates microfluidic food safety platforms due to its non-destructive nature, multiplexing capability, and compatibility with miniaturized detection systems. Recent advances focus on enhancing sensitivity in complex matrices, simplifying readout, and enabling field deployment. Optical detection is efficient in visualizing results by converting a molecular binding reaction or a chemical reaction into a measurable change in light. Fluorescence-based techniques are particularly advantageous within microfluidic environments due to their high sensitivity and property to locate low amounts of analytes. The primary component is fluorophore molecule, which has the property of absorbing light at one wavelength and emitting it at a higher wavelength. A microfluidic paper chip was integrated with cadmium telluride quantum dots (CdTe QDs) to detect the pesticide 2,4-dichlorophenoxyacetic acid (2,4-D) in spiked bean sprouts by fluorescence quenching. In the absence of pesticide, the CdTe Qds are highly fluorescent. When the sample containing the pesticide is introduced to the device, the 2,4-D binds to the quantom dots facilitating an electron transfer from the quantom dots to the pesticide molecule. This quenches the fluorescence and the intensity of the light decreases. This method displayed results within 18 minutes, while achieving an LOD of 93 nM [45]. Another approach uses the natural autofluorescence of certain molecules, similar to the L-tryptophan/levofloxacin-based fluorescence device for E. coli detection in different food samples. This system measured the ratiometric fluorescence signal from L-typtophan and levofloxacin with an LOD of 102–158 CFU/mL [49]. Colorimetric sensors are simple, low-cost and can be interpreted visually without any complex equipment. A microfluidic thread-based device was developed to detect carbaryl in food. Researchers extracted 1-naphthol from carbaryl, reacted it with 4-methoxybenzenediazonium tetrafluoroborate (MBDF) on the device to create a visible color change. The results are analyzed using smartphone imaging with detection limits of 0.020–0.027 mg/kg and recoveries of 92.3–105.9% [50]. Surface plasmon resonance (SPR) is a highly sensitive technique that involves the sensing of shifts in optical resonance wavelength with respect to variations in the local refractive index. Researchers have integrated a 3D-printed microfluidic platform with SPR biosensor, to detect four pesticides in strawberry extract using surface-enhanced Raman scattering (SERS). The sensor consists of a thin gold film functionalized with target-specific sensing is based on the changes in refractive index. This device provided increased selectivity, recovery, and an LOD range of 44–88 μg/kg [51]. Chemiluminescence is light emitted as a result of a chemical reaction, which does not require external light or heat for excitation. A microfluidic chip was designed to quantify the presence of monosaccharide sugars such as glucose and fructose using chemiluminescence emission from the luminol-tetrachloroaurate system. The sample is introduced to the device containing a solution of luminol and tetrachloroaurate, where the sugars react with the solution to release energy in the form of light. The intensity is measured and the LOD in consumables was 0.65 μM for glucose and 0.69 μM for fructose [52].

3.2 Electrochemical detection

Electrochemical sensing is very optimal for food analysis because its operation is unaffected by sample colour or turbidity, which is a challenge for optical detection when analysing samples like milk or juice. They integrate seamlessly with microfluidics for portable, quantitative food contaminant detection, using miniaturized electrodes, low power requirements, and real-time signal output. These methods work by measuring changes in electrical properties, like current, voltage, or impedance which results from a chemical reaction or binding at electrode surface.

Materials Research Foundations 191 (2026) 72-95 https://doi.org/10.21741/9781644904213-3

Voltammetric techniques refer to the measurement of current that results from applying a varying voltage to an electrode. Differential pulse voltammetry (DPV) is a sensitive form of this technique which is effective in detecting toxins. A microfluidic system was developed to detect fungal mycotoxin, Ochratoxin-A in food samples using DPV. The binding of toxin to the electrode reduces the ability of nearby redox probe to transfer electrons to the electrode's surface. DPV applies a series of voltage pulses and measures there resulting peak current. The toxin binding reduces this current peak and the reduction is proportional to ochratoxin-A concentration, allowing sensitive down to 350 pg/mL in food samples [53]. Amperometry refers to a technique where a constant voltage is applied to a working electrode, and the resultant current is determined over time. This is effective in monitoring a specific analyte for a continuous period. A μPAD with thermoplastic electrode was used to detect tryptamine, a spoilage indicator, in water and cheese samples using amperometric detection. The voltage of oxidation potential of tryptamine is constantly applied to the electrode and when the sample containing tryptamine passes through the electrode, it is oxidised and release electrons generating current. The sensor measures the current and the device achieved an LOD range of 10 to 75 μmol/L, with the thermoplastic electrode additionally being foul-resistant [54]. Electrochemical Impedance Spectroscopy (EIS) observes the impedance (resistance + reactance) of a system across a range of AC frequencies. Unlike voltammetry or amperometry, which measure current at a fixed or varying voltage, EIS applies a small alternating voltage and analyzes how the system resists or stores electrical energy [55]. A miniaturized, portable EIS platform was designed for the detection of the foodborne pathogen *Listeria monocytogenes* in milk samples. The platform utilizes microelectrodes with pathogen-specific antibody and the binding of *Listeria* to the antibody creates an insulating layer for the electrode. This increases the impedance to the flow of charge between electrode and the solution. EIS applies a small AC voltage through a range of frequencies, and measures the resulting impedance at each frequency. The charge transfer resistance is proportional to the number of bacterial cells captures, allowing for high sensitivity and an LOD of 5.5 cfu/mL [46]. Capacitive sensing measures how well a system can store electrical charge between two conductive surfaces (electrodes), which are separated by a dielectric material. When a fluid flows in a microfluidic device, droplets move, or cells pass by, they alter the dielectric environment between the electrodes and cause a measurable change in capacitance [56]. For example, an aptasensor coupled interfacial capacitive sensing to enable one-step detection of aflatoxin B1, apotent mycotoxin. An array of electrodes is coated with toxin-specific aptamers, which binds with aflatoxin B1 when the corresponding sample is introduced. This binding displaces ions and alters the structure of the electrical double layer at the electrode's surface, causing a measurable change in the interfacial capacitance. The device achieved sub-femtomolar sensitivity (LOD: 620 aM) and high selectivity, with detection completed in 20 seconds [57].

3.3 Mass-based detection

Mass-based detection techniques exploit the quantifiable mass of macroscale and microscale particles for different analyte detection. They directly measure the mass of analytes as they bind to the sensor without the need for any colorimetric tags. Quartz Crystal Microbalance (QCM) is a highly sensitive piezoelectric sensor that measures tiny mass changes by detecting shifts in the resonant frequency of a quartz crystal. When molecules (like proteins, DNA, or pathogens) attach with the crystal surface, the added mass makes the crystal to vibrate more slowly. This frequency shift (Δf) is proportional to the mass change (Δm) on the surface, often in the nanogram or picogram range [58]. QCM is an effective strategy for real-time identification of microorganisms

Materials Research Foundations 191 (2026) 72-95 | https://doi.org/10.21741/9781644904213-3

in food samples [59]. Surface Acoustic Waves (SAWs)-based sensors utilizes mechanical waves created on the surface of a piezoelectric substrate using interdigitated transducers (IDTs). The properties of these waves alters after the changes to surface caused by binding of analytes, which can be detected [60]. A study presented a SAW-based microfluidic system to rapidly concentrate and detect bacteria in raw samples like milk. The milk sample is introduced into a microfluidic chamber containing a packed bed of tiny polystyrene beads. The SAW device is activated, generating acoustic waves that create Bjerknes forces. These forces trap and concentrate the bacterial cells from the milk onto the beads, effectively separating them from the complex matrix. A fluorescent stain is then introduced, which labels the captured bacteria on the beads, making them optically detectable. The acoustic field is then altered to release the stained bacteria from the beads to a detection zone where their fluorescence is measured. This method combines acoustic concentration with optical detection to achieve a very low LOD of 34 CFU/mL, completing the entire analysis in just one hour [47].

3.4 Magnetic detection

Magnetic detection techniques in microfluidic devices are gaining traction in the food analysis domain. One of the essential advantages of this system is to isolate the analyte from complex food matrices. The employment of magnetic particles facilitates various functions, including separation, enrichment, and detection of analytes. Magnetic-bead-based platforms enable the concentration of target particles, thereby significantly improving detection limits and providing a higher degree of specificity [61]. A study developed a 3D-printed magnetic extraction method using magnetic silica beads (MSBs) for isolating DNA from bulk bacterial samples, combined with PCR in microfluidic system for *Salmonella* detection in food samples. After lysis with proteinase K, the sample flowed through a serpentine channel where MSBs, held by a magnetic field, captured the DNA. The beads were then washed, dried, and eluted to yield purified DNA, which was amplified using a microfluidic PCR chip and commercial qPCR. The method achieved over 90% extraction efficiency and recognized *Salmonella* with an LOD of 10^2 CFU/mL [48]. A handheld device was developed for detecting *E. coli* in food and clinical specimen using immunomagnetic separation and a LOC system. It combines a 3D microfluidic network with a giant magnetoresistive (GMR) biosensor that detects magnetic field changes from superparamagnetic beads bound to target antigens. The device uses a [Cu/Co] multilayer with 3.2% magnetoresistance and high sensitivity, coated with silicon nitride for effective antibody immobilization. An SU-8-based microfluidic circuit guides samples, and 3D-printed packaging with standard tubing enables safe, simple use and straightforward replacement [62]. Another study created magnetic nanoparticles (AgMNP) by attaching silver nanoparticles to Fe_3O_4, serving as both SERS enhancers and magnetic fluid particles. These AgMNPs enable rapid isolation and detection of malachite green, a harmful ingredient from fish skin [63].

4. Applications in food analysis

Microfluidic technologies have developed as significant assets in food analysis, offering fast, sensitive, and affordable alternatives for a variety of applications. Table 3 provides an overview of various food analysis applications and their working mechanism.

Materials Research Foundations 191 (2026) 72-95 https://doi.org/10.21741/9781644904213-3

Table 3 *Microfluidic applications for Food Safety and Quality*

Target analyte	Food matrix	Microfluidic platform	Mechanism of detection	Outcomes	References
Pathogen - *Staphylococcus aureus* and *Vibrio parahaemolyticus*	Shrimp	PDMS/paper hybrid microfluidic chip with multiplex LAMP	Food sample is added to the device and distributed to LAMP reaction site by capillary action. The presence of pathogen results in LAMP products, which is visualized as test line in lateral flow strip.	Rapid detection (~30 min), multiplex detection, and prevents contaminatio n	[64]
Toxin - aflatoxin B1 and DON	Spiked corn extract	Comb-shaped microfluidic electrochemical aptasensor	Food sample is added to the device to contact with aptamer. The binding of toxin with the aptamer generates a measurable electrochemical change. The intensity of change indicates the concentration of the toxin.	Ultralow LOD (pg/mL to ng/mL) and multiplex detection	[65]
Chemical contaminants – Carbaryl	Rice, wheat, bean, and cabbage	Microfluidic thread-based analytical device	The sample is hydrolysed to convert carbaryl into 1-naphthol, which is then extracted by effervescence-assisted liquid phase microextraction. The extract is reacted with 4-methoxybenzenediazonium tetrafluoroborate to produce orange azo dye in the device and intensity is visualized in smartphone.	Cost effective, high stability, and portable	[50]
Allergens – Peanut allergen (Ara h1)	Cookies, milk, and bread	Microfluidic paper-based electrochemical device	The allergen specific aptamer was bound to an electrode surface through Au-S bonds. When the food sample is added, the allergen binds with the aptamer and alters the electron transfer at the electrode surface. The decrease in current is measured to detect the allergen.	Real-time, rapid analysis, good specificity and stability	[66]
Nutrients and additives - Ascorbic acid (Vitamin C), Nitrate and Sunset yellow	Not specified	Microfluidic paper-based sensor	Three fluorescent carbon dots, specific to each compound is present in distinct zones in the device. The food sample is added to the device and interacted with the respective carbon dots to provide either fluorescence quenching or enhancement. The change in fluorescence is visualized under UV light.	Reliable, easy to mass produce, and highly sensitive	[67]

Materials Research Foundations 191 (2026) 72-95 https://doi.org/10.21741/9781644904213-3

4.1 Pathogen detection

In food analysis, microfluidics technology has shown great promise, especially for pathogen detection. The drawbacks of culture-based techniques are time-consumption, labour-intensive, and lack of portability. They are addressed by combining sample preparation, amplification, and sensing into portable platforms. The incorporation of LAMP-based methods are the key advancements in the use of microfluidics in food safety. By employing LAMP technique within microfluidic devices, researchers have achieved rapid identification of multiple infectious organisms, without the need for complex equipment and laboratory settings. The risk of reagent handling with contamination is higher in a field testing environment, which is a challenge for conventional molecular techniques. Researchers developed a sample-to-answer μPAD platform with LAMP to detect *Vibrio vulnificus* and *Salmonella* Typhimurium. The sample is added in the sample pad, which serves as an amplification platform, and the DNA is amplified in the presence of the pathogen. This design cleverly adapted the heating used for amplification to open the wax channel allowing the amplified products into lateral flow strip. The strip visualizes the result as a test line, producing a interpretable colour band for the presence of pathogen. [68]. Another challenge in multiplex assays is ensuring the equal distribution of sample across different reaction chamber and manual addition poses the contamination risk mentioned before. A study developed a method of self-priming μPAD integrated with multiplex LAMP to detect foodborne pathogens, *Staphylococcus aureus* and *Vibrio parahaemolyticus*. Self-priming is a process in which the microfluidic device is kept in a vacuum, where the air within the chip is removed and the sample is loaded. When the chip is brought back to the regular environment, the sample is directed to the respective chambers due to the kinetic energy provided by the air re-entering the chambers of the device. This method provided a contamination-free sample loading approach and ensured the equal distribution of samples to multiple reaction chambers [64]. Another study incorporated the similar self-priming LAMP system into a PMMA-based microfluidic chip. This device provided a rapid, visual detection of six food-borne pathogens directly from food samples [69]. Another inventive solution includes introduction of finger pump to initiate the sample flow in the device. A study integrated a simple microfluidic device with multiplex LAMP to find the existence of two enterotoxin genes of *Staphylococcus aureus* in ruminant milk. The sample is added to the dilution chamber, which is then pumped with a finger to the amplification chamber, where the LAMP process continues. Once the reaction is completed, the amplified product is pumped again to a lateral flow strip, where the presence of toxins is detected by the visualization of the test line. This method is also effective in avoiding contamination and ensuring equal distribution, which are critical for pathogen detection in food samples [70]. The integration of SERS with a LAMP-based microfluidic device provides an efficient optical detection platform for pathogen detection. Such device was developed to detect *Listeria monocytogenes* and *Listeria innocua* by using gold nanoparticles, 1-naphthalenethiol, and glutathione. The detection is based on the alteration in the SERS signal of 1-naphthalenethiol regulated by the clustering of AuNPs initiated by the formation of pyrophosphate-glutathione complexes with free magnesium ions during the LAMP reaction [71]. Moreover, the fabrication and customization of microfluidic chips facilitate the simultaneous identification of multiple pathogens in food matrices, a task that traditional methods may struggle to perform effectively. This multiparameter strategy is reflected in a study demonstrating a fast and sensitive immunoassay-based RPA technique for the concurrent recognition of multiple microorganisms. Since food items tend to harbor multiple contaminating species, single-target assays lack efficiency in managing the current issues of food safety. Therefore, novel microfluidic

Materials Research Foundations 191 (2026) 72-95 https://doi.org/10.21741/9781644904213-3

strategies that have multiplexing capacity provide large gains towards pathogen detection [72–74]. For example, a paper/PDMS hybrid microfluidic device was used for the identification of norovirus in food using RPA. The analysis took only 35 min, with an LOD of 200 PFU/mL [75]. Aptamers have an advantage of overcoming the need for time consuming DNA extraction step, which many molecular tests require. Aptamers could be coupled with microfluidic platforms to enable whole cell detection of various food-borne pathogens in real-time. For instance, *Zhang et al.* developed an aptamer for the whole cell of *E. coli* and employed the aptamer to capture the bacteria in a microfluidic device. The device effectively distinguished *E. coli* from other species, offering a quick, economical, and portable approach for microbial identification in food hygiene, ecological, and healthcare applications [76].

4.2 Toxin Detection

Mycotoxins are toxic fungal secondary metabolites, that are synthesized by *Fusarium, Aspergillus* and other species. They are usually seen as co-contaminants in many agricultural commodities and have the potential to inflict serious harm to food hygiene and public wellness. The detection of these tiny, highly stable components at low concentrations is very crucial in food analysis. Various methodologies based on microfluidics and biosensing have been introduced for the quick analysis, sensitivity, and multiplex detection of mycotoxins in foodstuffs. Food products are often contaminated with several mycotoxins at once, making it necessary to have multiplex detection in a single run. *Nolan et al.* provided a method involving mass-sensitive microarray (MSMA) biosensors that utilize solid-mounted resonator technology. The method allowed the simultaneous recognition of major mycotoxins, such as T2-toxin (T2), zearalenone (ZEN), and fumonisin B1 (FumB1). The microarray chip contains multiple tiny resonators, where each of them are coated with different toxin-specific antibodies. When a food extract is introduced and the mycotoxins bind to their corresponding antibodies, a tiny amount of mass is added to the resonator, which causes a shift in its vibration frequency. The sensor measures the frequency change for each resonator, which is directly proportional to the amount of each mycotoxin present. The MSMA achieved detection limits of 1.3 ng/mL for T2, 6.8 ng/mL for ZEN, and 2.0 ng/mL for FumB1 in singleplex reaction, while the multiplex reaction showed comparable sensitivities, demonstrating its effectiveness as a field-deployable and cost-efficient platform [77]. As mentioned earlier, detecting mycotoxins at lower concentrations requires a high sensitive method. A study presented a comb-shaped microfluidic electrochemical aptasensor for concurrent identification of aflatoxin B1 (AFB1) and deoxynivalenol (DON). The device contains two sets of microelectrodes corresponding to two aptamers specific for each AFB1 and DON. As the mycotoxins are selectively bound to their respective aptamers, the electrical properties at the electrode surface is altered. This prevents the electron transfer of a redox probe in the solution and the change in the signal is measured, which is proportional to the concentration of the bound mycotoxin. This system had ultralow LODs of 7.74 pg/mL and 0.172 ng/mL for AFB1 and DON, respectively, offering exceptional sensitivity and specificity for on-site, simultaneous detection [65]. Another study approached the challenge by designing a CRISPR/Cas12a microfluidic biosensor for detecting low-concentration mycotoxins in complex food matrices. This platform combined with quantum dots and photonic crystals to amplify fluorescent signals. The biosensor exhibited a wider analytical range and ultra-low LOD in the femtogram per milliliter range [78]. Many of the biosensors are developed for single use, which could increase the operating cost during frequent testing. Researchers developed a reusable microfluidic device with molecularly imprinted polymers (MIPs) to detect zearalenone (ZEN) using quantum dot fluorescence. This reusable

Materials Research Foundations 191 (2026) 72-95 https://doi.org/10.21741/9781644904213-3

system enabled selective and reusable recognition of ZEN in real food samples with high recoveries (91–105%) and minimal manual intervention, demonstrating high portability and rapid turnaround time [79].

4.3 Chemical Contaminant detection

Pesticide residues, heavy metals, and antibiotic residues are the most common chemical contaminant present in the fruits, vegetables, and food products. They pose a severe harm to the public health, if they are present above the threshold. Pesticide residues, especially from organophosphorus (OP) compounds and carbamates, carry considerable risks to human health and environmental well-being [80]. The traditional pesticide tests require laboratory equipment, man-power, and takes a considerable amount of time. Microfluidic devices offer a rapid diagnostic tools for detecting these contaminant with high sensitivity and portability. Deepak *et al.* designed a portable microfluidic device for colorimetric detection of three OPs, including malathion, chlorpyrifos, and dimethoate. The device is based on the acetylcholinesterase (AChE) inhibition assay. In the micromixers, the enzyme AChE is mixed with its substrate, where it breaks down the substrate and produces a bright yellow product. If an OP pesticide is present in the food sample, it inhibits the activity of the AChE enzyme and less yellow product is formed. A low-cost optoelectronic circuit is integrated into the device to measure the colour intensity and the device had an LOD range of 0.07-2.35 ppm [81]. Similarly, Pang *et al.* developed portable microfluidic system for OP pesticide detection using molybdenum-based reagent for colour change. The device achieved an LOD of 0.035 mg/L and a 30-second average detection time [82]. Another microfluidic platform was developed to enable selective detection of dimethoate, an OP compound. Oil-soluble CdSe quantum dots are used as the fluorescent material and a surface molecularly imprinted polymer (SMIP) is created on the paper chip which has dimethoate-specific recognition sites. The binding of dimethoate molecules to the imprinted sites quenches the fluorescence of the quantum dots. The degree of fluorescence quenching is proportional to the concentration of dimethoate in the sample. Glass fiber paper was used in this device to minimize background fluorescence and the device achieved an LOD of 0.13 μmol/L with high recovery (97.6-104.1%) [83]. Fungicides, insecticides, and herbicides have increasingly been detected in food products, enabling researchers to develop corresponding microfluidic devices for their rapid and sensitive detection [84,85]. Heavy metals, such as lead (Pb^{2+}), mercury (Hg^{2+}), cadmium (Cd^{2+}), and arsenic (As^{3+}) pose serious environmental and public safety risks due to their persistence, bioaccumulation, and toxicity. Prolonged exposure to even trace amounts can lead to chronic illnesses, including cancer, kidney damage, and neurological disorders [86,87]. This created the need for extremely sensitive detection methods for comprehensive screening of multiple heavy metals. An aptamer-based microfluidic sensor was designed for the concurrent identification of Hg^{2+} and Pb^{2+} ions using fluorescence. The system utilizes fluorescence resonance energy transfer (FRET) with graphene oxide (GO) and contains two unique DNA aptamers labeled with FAM dye and HEX dye for Hg^{2+} and Pb^{2+}, respectively. Initially, these aptamers are adsorbed onto a sheet of GO, which quenches their fluorescence. When the metal ions bind specifically to their corresponding aptamers, the aptamer changes its 3D shape and detach from the GO surface. Once the FAM and HEX dyes move away from the GO quencher, they fluoresce brightly under excitation light. The device demonstrated high selectivity and sensitivity, with an LOD of 0.70 ppb and 0.53 ppb for Hg^{2+} and Pb^{2+}, respectively [87]. Another μPAD was developed for the simultaneous recognition of four heavy metals (Pb^{2+}, Hg^{2+}, Cd^{2+}, and As^{3+}) using aptamer-based fluorescence in real food samples like apple and lettuce. The device was low-cost, portable, and

Materials Research Foundations 191 (2026) 72-95 https://doi.org/10.21741/9781644904213-3

LODs were 4.20 nM for Pb^{2+}, 1.70 nM for Hg^{2+}, 2.04 nM for Cd^{2+}, and 1.65 nM for As^{3+} [88]. Antibiotic residues in food products such as milk, eggs, and meat pose serious health risks, including the emergence of antibiotic resistance and allergic responses, necessitating rapid, low-cost, and sensitive detection methods. The researchers developed a disposable μPAD for identifying and discriminating eight antibiotics in milk and eggs. The detection is based on colorimetric changes resulting from the differential interaction of antibiotics with metallochromic complexes. The μPAD contains an array of five different metal-indicator complexes, where a metal ion is bound to a colored indicator molecule. When the antibiotic molecules interact with the metal-indicator complexes, the original indicator is displaced and resulted in a color change. Each antibiotic interacts with the array of complexes in a different way, generating a unique color pattern. The color changes are captured and analyzed using chemometric data analysis, which allows the discrimination and identification of the different antibiotics present in the sample. This sensor, analyzed via linear discriminant analysis, achieved 100% accuracy at 5.0 mg/L and could detect trace levels down to 1.0 mg/L [89]. A novel microfluidic system was developed that incorporates hybrid magnetoliposomes (h-MLs) to determine aminoglycoside antibiotics (AAGs) in milk and meat samples. The h-MLs are specially designed vesicles that contains hydrophobic magnetic nanoparticles (Fe3O4@AuNPs-C12SH) and encapsulated N-acetylcysteine. The magnetic nanoparticles allow the h-MLs to be retained in reaction zone using an external electromagnet and N-acetylcysteine is a reagent that is released to participate in the detection reaction. The food sample containing AAGs and another reagent, o-phthalaldehyde, are passed through the chip. The AAGs react with the o-phthalaldehyde and the N-acetylcysteine to form a fluorescent isoindole derivate. The device is placed in a conventional fluorimeter and the intensity of the fluorescence is measured, which is proportional to the concentration of AAGs in the sample [90].

4.4 Allergen detection

Food allergies are serious and increasing global health issue with a significant need for sensitive, fast, and reliable detection methodologies for food allergens in foods and in environmental places. The conventional laboratory assay like ELISA has been in use for decades but suffer from various drawbacks like long assay duration, reagent consumption, and the requirement of specialized laboratory space and skilled human resources [91]. Traditionally, it has been the gold standard in laboratory-based methods. Microfluidic-based technologies have opened new avenues for meeting these challenges through miniaturization, automation, and integration of analytical assays. The researchers developed a method that combines microfluidic separation with mass spectrometry to quantify peanut allergens in chocolate-derived foods. Peanut proteins are extracted from the food sample using a buffer and the peptide fragments are separated using a microfluidic chromatography system. The protein is analyzed using mass spectrometry with a technique called Multiple Reaction Monitoring (MRM), which targets specific peptide fragments derived from the peanut allergens. The signal from the sample is compared to a known standard to quantify the amount of peanut protein in the original food product. Results agreed well with ELISA analysis, and it was shown that microfluidic LC-MS/MS platforms can be used as alternative tools for the identification of allergic components in processed food, in addition to offering simplified sample preparation and matrix-tolerant quantification [92]. The need for low-cost, portable diagnostic tool is addressed by the development of a μPAD for the identification of Ara h1, a major peanut allergen, in cookies and milk. The device uses nanocomposites made of black phosphorus and gold to create a highly conductive sensor surface, which amplifies the electrochemical signal. The sensor is functionalized with allergen-specific aptamers that alters the flow of electrons at the electrode interface when Ara h1 binds to the aptamers. The change in the electrical current is measured by DPV technique, allowing for the quantification of Ara h1 with an LOD of 11.8 ng/mL and high recovery rates [66]. The same allergen, Ara h1 was detected in another study using a fluorescence-assisted microfluidic assay with QD-aptamer-GO complexes. The biosensor achieved detection within 10 minutes, relying on fluorescence quenching and recovery by graphene oxide upon target binding [93]. In food processing plants and cooking units, air-borne allergens poses a serious risk of contaminating the food products. This threat is difficult to monitor with traditional methods. To address this problem, *Chen et al.* developed a herringbone-patterned microfluidic ELISA chip to recognize allergens in bioaerosols. The aerosol particles are collected and introduced into the microfluidic chip, where the herringbone structure on the channel floor creates micro-vortices. This brings the allergen particles into contact with the detection antibodies immobilized on the chip's surface. A standard sandwich ELISA is performed, but the enhanced mixing by micro-vortices allows the entire assay to be completed much faster. A fluorescent or colorimetric signal is measured, allowing the detection of four different allergens (ovalbumin, ovomucoid, lysozyme, and tropomyosin) with high sensitivity (LODs of 0.31–7.8 ng/mL) and no cross-reactivity. This system is applicable for on-site air quality monitoring and allergen exposure control [94]. Allergen detection in complex, coloured liquid samples is a challenge for optical detection due to interference. A discardable electrochemical microfluidic device was created to detect ovalbumin, utilizing an immunoassay combined with magnetic beads. Magnetic beads coated with anti-ovalbumin antibodies selectively capture ovalbumin in the sample. A second antibody, labeled with the enzyme horseradish peroxidase (HRP), is introduced and binds to the captured ovalbumin, forming a sandwich complex on the magnetic bead. A magnet is used to hold the beads while the rest of the interfering wine matrix is washed away. A chemical substrate is poured over the beads and the HRP enzyme on the beads catalyzes a reaction that produces an electrical current, which is measured by an electrode. This sandwich-type immunoassay successfully quantified ovalbumin and achieved ultrasensitive detection at 0.2 fg/mL [95]. Microfluidics has been applied not only to food analysis but also to personal allergy diagnostics. In a study, *Polonschii et al.* created a novel POC Surface Plasmon Resonance (SPR) platform combining microfluidics and panel peptide-ligand biosensors to differentiate between sensitization and actual allergy. The platform was tested using peanut and hazelnut allergens and had the ability to distinguish different sera responses according to discrete peptide epitopes [96].

4.5 Nutrient and Additive Quantification

In food analysis, microfluidic systems are used to deliver precise and reproducible quantifications of additives, nutrients, and bioactive compounds, apart from contaminant detections. These platforms offer a simple, cost-effective alternative to complex laboratory techniques like chromatography. One study presented a technique based on microchip gel electrophoresis for protein content analysis, which discriminated the protein profiles of different plant-based beverage types [97]. μPADs have shown significant potential in food quality and nutritional monitoring because of their simplicity, ease of portability, and affordability. *Deng et al.* developed an inkjet-printed μPAD to detect ascorbic acid (AA) or Vitamin C, nitrite (NO_2^-), and sunset yellow (SY) based on fluorescent carbon dots. This device used hydrophobic/hydrophilic patterning for effective fluid flow and achieved on-site visual detection levels of 40 mmol/L for AA, 6 mmol/L for NO_2^-, and 60 μmol/L for SY [67]. Similarly, a colorimetric μPAD was created for Vitamin C (AA) detection using the silver nitrate reduction reaction. The paper in the μPAD is pre-loaded

Materials Research Foundations 191 (2026) 72-95 https://doi.org/10.21741/9781644904213-3

with silver nitrate ($AgNO_3$), which reacts with AA and reduced to solid silver nanoparticles (Ag^0). This reaction produces a visible colored spot on the paper, which is captured in a smartphone and analyzed in Image analysis software to provide a quantitative result. The results were validated against HPLC using fresh orange juice as a real sample and displayed strong correlation [98]. For the detection of vitamin D_3, a fat-soluble vitamin, a new microchip-based solid-phase extraction (SPE) system was constructed on a novel magnetic adsorbent, which offered high surface area and high affinity. The adsorbent captures and concentrates Vitamin D3 molecules while other fats and proteins are washed away. It was coupled with LC-MS/MS detection to attain low detection limit (1.4 ng/mL), and high recovery (95.2–101%) from food and biological samples [99]. For lipid analysis, a μPAD was constructed based on cholesterol oxidase, horseradish peroxidase (HRP), and 3,3′,5,5′-tetramethylbenzidine (TMB) for visual detection of free cholesterol. The colorimeter changes were recorded by smartphone and interpreted by ImageJ, showing the low detection limit of the sensor and its potential for broad clinical and nutritional use [100]. Similarly, the quantification of iodide in seaweed and dietary supplements was performed in a microfluidic device. The device used hydrogen peroxide to oxidize TMB in the presence of iodide to generate a blue-colored product. Also, incorporating powdered silica into the device enhanced the signal clarity, with an LOD of 3 μM [101]. The employment of microfluidic methods has proven to be highly promising for improving bioavailability as well as the therapeutic action of bioactive compounds in functional foods. For example, fucoxanthin, a xanthophyll carotenoid with advantageous health characteristics, has poor water solubility and instability. Hence, *Tian et al.* utilized a microfluidic coaxial electrospinning method to encapsulate fucoxanthin into zein-gelatin core-shell nanofibers. In this system, two separate polymer solutions are pumped through a coaxial needle, where the inner solution contains fucoxanthin mixed with a zein polymer and the outer solution contains a gelatin polymer. A high voltage is applied to the needle tip, which stretches the liquid jets into thin, continuous fibers. The fucoxanthin-zein mixture forms the core and the gelatin forms a protective outer shell. The resultant nanofibers exhibited an encapsulation efficiency of 98.58%, enhanced photostability, UV stability, and heat stability, together with high biocompatibility and cellular protective actions in HepG2 cells [102]. Likewise, the synergistic combination of astaxanthin and anthocyanin in nanoparticle form was found to reverse lipid accumulation and oxidative stress in high-fat, high-sugar diet-induced *Caenorhabditis elegans*. The nanoparticles promoted lifespan extension, improved motility, and enhanced reproductive function [103].

Conclusion and future challenges

In food analysis, microfluidic technology has emerged as a revolutionary tool, bringing rapid, precise, and cheap solutions for detecting a variety of contaminants and quality indicators [104]. Microfluidics enables miniaturization and automation of challenging analytical processes that normally require large equipment and skilled operators by integrating multiple steps on one compact platform. Microfluidic systems are ideal for high-throughput screening and POC applications since this integration reduces reagent and sample volumes and operates at significantly shorter analysis times [105]. The versatility of microfluidics has been demonstrated in the identification of chemical contaminants and biological threats, such as toxins, allergens, and pathogens [81,86,106]. Improvements in paper-based platforms, 3D printing, and new biosensing approaches, coupled with applications of aptamers, quantum dots, and nanomaterials, have further increased the versatility of such devices for applications in food hygiene and quality control.

Materials Research Foundations 191 (2026) 72-95 https://doi.org/10.21741/9781644904213-3

Effective incorporation of techniques such as LAMP, SERS, EIS, and fluorescence-based assays with microfluidic platforms has improved the sensitivity of detection and enabled multiplex analysis [46,71,84]. Significantly, the portability, cost-effectiveness, and scalability of microfluidic systems render them uniquely deployable in resource-limited settings, small-scale food production, and field monitoring. Connectivity with smartphones and IoT connection are also likely to evolve easy-to-use diagnostic devices that facilitate real-time detection and decision-making [43,107]. Conclusively, while these benefits exist, regulatory validation, large-scale manufacturing, and robustness across a wide range of food matrices remain challenges. Future work must be aimed at standardizing microfluidic platforms, enhancing their reproducibility, and further broadening their ability to detect novel food adulterants and emerging contaminants. Interdisciplinary collaboration will also be crucial in pushing innovation and commercialization. In summary, microfluidic devices hold the key to the future of food analysis, with potential for a transition towards faster, safer, and more convenient diagnostics that can meet the increasing needs of international food safety and quality assurance.

References

[1] M. Manoswini, A.G. Majumdar, B. Pany, B.R. Sahu, P.S. Mohanty, Rapid detections of food pathogens using metal, semiconducting nanoparticles, and their hybrids: A review, Emergent Mater. 6 (2023) 15-30. https://doi.org/10.1007/s42247-022-00441-4

[2] M. Thorsen, J. Hill, J. Farber, F. Yiannas, I.M.C.M. Rietjens, P. Venter, R. Lues, P. Bremer, Megatrends and emerging issues: Impacts on food safety, Compr. Rev. Food Sci. Food Saf. 24 (2025) e70170. https://doi.org/10.1111/1541-4337.70170

[3] V. Ahuja, A. Singh, D. Paul, D. Dasgupta, P. Urajová, S. Ghosh, R. Singh, G. Sahoo, D. Ewe, K. Saurav, Recent advances in the detection of food toxins using mass spectrometry, Chem. Res. Toxicol. 36 (2023) 1834-1863. https://doi.org/10.1021/acs.chemrestox.3c00241

[4] Y. Zhang, J. Li, S. Jiao, Y. Li, Y. Zhou, X. Zhang, B. Maryam, X. Liu, Microfluidic sensors for the detection of emerging contaminants in water: A review, Sci. Total Environ. 929 (2024) 172734. https://doi.org/10.1016/j.scitotenv.2024.172734

[5] S. Siavashy, M. Soltani, S. Rahimi, M. Hosseinali, Z. Guilandokht, K. Raahemifar, Recent advancements in microfluidic-based biosensors for detection of genes and proteins: Applications and techniques, Biosens. Bioelectron.:X 19 (2024) 100489. https://doi.org/10.1016/j.biosx.2024.100489

[6] A. Nilghaz, S.M. Mousavi, M. Li, J. Tian, R. Cao, X. Wang, Paper-based microfluidics for food safety and quality analysis, Trends Food Sci. Technol. 118 (2021) 273-284. https://doi.org/10.1016/j.tifs.2021.08.029

[7] H. Tavakoli, S. Mohammadi, X. Li, G. Fu, X.J. Li, Microfluidic platforms integrated with nano-sensors for point-of-care bioanalysis, TrAC, Trends Anal. Chem. 157 (2022) 116806. https://doi.org/10.1016/j.trac.2022.116806

[8] B. Dorst, J. Mehta, K. Bekaert, E. Rouah-Martin, W. Coen, P. Dubruel, R. Blust, J. Robbens, Recent advances in recognition elements of food and environmental biosensors: A review, Biosens. Bioelectron. 26 (2010) 1178-1194. https://doi.org/10.1016/j.bios.2010.07.033

[9] S. Boobphahom, M.N. Ly, V. Soum, N. Pyun, O.S. Kwon, N. Rodthongkum, K. Shin, Recent advances in microfluidic paper-based analytical devices toward high-throughput screening, Molecules 25 (2020) 2970. https://doi.org/10.3390/molecules25132970

[10] D. Gao, X. Guo, Y. Yang, H. Shi, R. Hao, S. Wang, Z.J. Li, R. Zhao, H. Song, Microfluidic chip and isothermal amplification technologies for the detection of pathogenic nucleic acid, J. Biol. Eng. 16 (2022) 33. https://doi.org/10.1186/s13036-022-00312-w

[11] N. Convery, N. Gadegaard, 30 years of microfluidics, Micro Nano Eng. 2 (2019) 76-91. https://doi.org/10.1016/j.mne.2019.01.003

[12] C. Jia, B. Li, J. Pan, F. Zhu, X. Bai, X. Liu, B. Lin, Y. Chen, Y. Lu, Braiding three-dimensional paper-based microfluidic devices, Analysis Sensing 4 (2024) e202300065. https://doi.org/10.1002/anse.202300065

[13] M. Liang, G. Zhang, J. Song, M. Tan, W. Su, Paper-based microfluidic chips for food hazard factor detection: Fabrication, modification, and application, Foods 12 (2023) 4107. https://doi.org/10.3390/foods12224107

[14] M.Z. Hua, X. Lu, Development of a microfluidic paper-based immunoassay for rapid detection of allergic protein in foods, ACS Sens. 5 (2020) 4048-4056. https://doi.org/10.1021/acssensors.0c02044

[15] L. Ma, A. Nilghaz, J.R. Choi, X. Liu, X. Lu, Rapid detection of clenbuterol in milk using microfluidic paper-based ELISA, Food Chem. 246 (2018) 437-441. https://doi.org/10.1016/j.foodchem.2017.12.022

[16] T. Ozer, C. McMahon, C.S. Henry, Advances in paper-based analytical devices, Annu. Rev. Anal. Chem. 13 (2020) 85-109. https://doi.org/10.1146/annurev-anchem-061318-114845

[17] M. Naseri, G.P. Simon, W. Batchelor, Development of a paper-based microfluidic system for a continuous high-flow-rate fluid manipulation, Anal. Chem. 92 (2020) 7307-7316. https://doi.org/10.1021/acs.analchem.0c01003

[18] X. Weng, G. Gaur, S. Neethirajan, Rapid detection of food allergens by microfluidics ELISA-based optical sensor, Biosensors 6 (2016) 24. https://doi.org/10.3390/bios6020024

[19] O. Skurtys, J.M. Aguilera, Applications of microfluidic devices in food engineering, Food Biophys. 3 (2008) 1-15. https://doi.org/10.1007/s11483-007-9043-6

[20] S.M. Scott, Z. Ali, Fabrication methods for microfluidic devices: An overview, Micromachines 12 (2021) 319. https://doi.org/10.3390/mi12030319

[21] M.K. Morsy, O.M. Morsy, E.M. Abd-Elaaty, R. Elsabagh, Development and validation of rapid colorimetric detection of nitrite concentration in meat products on a polydimethylsiloxane (PDMS) microfluidic device, Food Anal. Methods 15 (2022) 552-564. https://doi.org/10.1007/s12161-021-02139-5

[22] C.C. Liu, C.H. Ko, L.M. Fu, Y.L. Jhou, Light-shading reaction microfluidic PMMA/paper detection system for detection of cyclamate concentration in foods, Food Chem. 400 (2023) 134063. https://doi.org/10.1016/j.foodchem.2022.134063

[23] R. Su, F. Wang, M.C. McAlpine, 3D printed microfluidics: Advances in strategies, integration, and applications, Lab Chip 23 (2023) 1279-1299. https://doi.org/10.1039/D2LC01177H

Materials Research Forum LLC
https://doi.org/10.21741/9781644904213-3

[24] S.A. Papatheodorou, D. Houhoula, S. Magoulas, A.G. Tsantes, E. Tsakali, S. Akkermans, J. Impe, P. Halvatsiotis, A.E. Tsantes, Development of a 3d microfluidic analytical device for the detection of pathogenic bacteria in commercial food samples with loop-mediated isothermal amplification, Acta Microbiol. Hell. 69 (2024) 41-49. https://doi.org/10.3390/amh69010006

[25] X. Cheng, C.C. Wang, A. Liu, Y. Liu, J. Dong, Z. Wang, W. Wei, S. Liu, Simultaneous detection of foodborne pathogenic bacteria in milk by fluorescence immunoassay, Spectrochim. Acta, Part A 285 (2023) 121830. https://doi.org/10.1016/j.saa.2022.121830

[26] H. Ito, M. Kaneko, On-chip cell manipulation and applications to deformability measurements, Robomech J. 7 (2020) 3. https://doi.org/10.1186/s40648-020-0154-x

[27] H.C.S. Hong, L. Wu, X. Yan, S. Wang, Developments and applications of microfluidic pumps and valves, Highl. Sci. Eng. Technol. 37 (2023) 218-227. https://doi.org/10.54097/hset.v37i.6077

[28] Y.T. Atalay, S. Vermeir, D. Witters, N. Vergauwe, B. Verbruggen, P. Verboven, B.M. Nicolaï, J. Lammertyn, Microfluidic analytical systems for food analysis, Trends Food Sci. Technol. 22 (2011) 386-404. https://doi.org/10.1016/j.tifs.2011.05.001

[29] Q.R. Guo, L.L. Zhang, J.F. Liu, Z. Li, J.J. Li, W.M. Zhou, H. Wang, J.Q. Li, D.Y. Liu, X.Y. Yu, J.Y. Zhang, Multifunctional microfluidic chip for cancer diagnosis and treatment, Nanotheranostics 5 (2021) 73-89. https://doi.org/10.7150/ntno.49614

[30] G. Cai, L. Xue, H. Zhang, J. Lin, A review on micromixers, Micromachines 8 (2017) 274. https://doi.org/10.3390/mi8090274

[31] I. Podunavac, M. Djocos, M. Vejin, S. Birgermajer, Z. Pavlovic, S. Kojic, B. Petrovic, V. Radonic, 3D-printed microfluidic chip for real-time glucose monitoring in liquid analytes, Micromachines 14 (2023) 503. https://doi.org/10.3390/mi14030503

[32] Y. Zhang, Y. Chen, J. Huang, Y. Liu, J. Peng, S. Chen, K. Song, X. Ouyang, H. Cheng, X. Wang, Skin-interfaced microfluidic devices with one-opening chambers and hydrophobic valves for sweat collection and analysis, Lab Chip 20 (2020) 2635-2645. https://doi.org/10.1039/D0LC00400F

[33] R. Bosma, J. Devasagayam, A. Singh, C.M. Collier, Microchip capillary electrophoresis dairy device using fluorescence spectroscopy for detection of ciprofloxacin in milk samples, Sci. Rep. 10 (2020) 13548. https://doi.org/10.1038/s41598-020-70566-1

[34] W. Yu, Y. Chen, Z. Wang, L. Qiao, R. Xie, J. Zhang, S. Bian, H. Li, Y. Zhang, A. Chen, Multiple authentications of high-value milk by centrifugal microfluidic chip-based real-time fluorescent LAMP, Food Chem. 351 (2021) 129348. https://doi.org/10.1016/j.foodchem.2021.129348

[35] Z. Xie, H. Pu, D.W. Sun, Computer simulation of submicron fluid flows in microfluidic chips and their applications in food analysis, Compr. Rev. Food Sci. Food Saf. 20 (2021) 3818-3837. https://doi.org/10.1111/1541-4337.12766

[36] B. Song, J. Yu, Y. Sun, Q. Wang, S. Xu, Y. Jia, S. Lin, Y. Zhang, C. Wang, Y. Zhang, X. Zhang, Microfluidics for the rapid detection of Escherichia coli O157:H7 using antibody-coated microspheres, Bioengineered 12 (2021) 392-401. https://doi.org/10.1080/21655979.2020.1870805

[37] Z. Gu, M.L. Wu, B.Y. Yan, H.F. Wang, C. Kong, Integrated digital microfluidic platform for colorimetric sensing of nitrite, ACS Omega 5 (2020) 11196-11201. https://doi.org/10.1021/acsomega.0c01274

[38] Z. Liu, S. Wang, K. Wang, J. Tong, Z. Zhao, X. Liu, Y. Liu, Digital microfluidic-based fluorescence methods for the automated determination of copper ions in wine, Microchim. Acta 192 (2025) 157. https://doi.org/10.1007/s00604-025-07019-x

[39] R. Eswar, C.H. Brodie, C.M. Collier, Lock-in amplified fluorescence spectroscopy in a digital microfluidic configuration for antibiotic detection of ciprofloxacin in milk, J. Microelectromech. Syst. 31 (2022) 927-934. https://doi.org/10.1109/JMEMS.2022.3199167

[40] H. Gao, C. Yan, W. Wu, J. Li, Application of microfluidic chip technology in food safety sensing, Sensors 20 (2020) 1792. https://doi.org/10.3390/s20061792

[41] S. Kumar, J.B. Kaushal, H.P. Lee, Sustainable sensing with paper microfluidics: applications in health, environment, and food safety, Biosensors 14 (2024) 300. https://doi.org/10.3390/bios14060300

[42] E. Trofimchuk, A. Nilghaz, S. Sun, X. Lu, Determination of norfloxacin residues in foods by exploiting the coffee-ring effect and paper-based microfluidics device coupling with smartphone-based detection, J. Food Sci. 85 (2020) 736-743. https://doi.org/10.1111/1750-3841.15039

[43] N. Yang, L.L. Xie, C. Pan, M.F. Yuan, Z.H. Tao, H.P. Mao, A novel on-chip solution enabling rapid analysis of melamine and chloramphenicol in milk by smartphones, J. Food Process Eng. 42 (2019) e12976. https://doi.org/10.1111/jfpe.12976

[44] O. Milark, M. Buttkewitz, E. Agócs, B. Legutko, B. Bergmann, J. Bahnemann, A. Heisterkamp, M.L. Torres-Mapa, Design and fabrication of 3D-printed lab-on-a-chip devices for fiber-based optical chromatography and sorting, Adv. Photonics Res. 5 (2024) 2400011. https://doi.org/10.1002/adpr.202400011

[45] Z. Zhang, X. Ma, M. Jia, B. Li, J. Rong, X. Yang, Deposition of CdTe quantum dots on microfluidic paper chips for rapid fluorescence detection of pesticide 2,4-D, Analyst 144 (2019) 1282-1291. https://doi.org/10.1039/C8AN02051E

[46] M.S. Chiriacò, I. Parlangeli, F. Sirsi, P. Poltronieri, E. Primiceri, Impedance sensing platform for detection of the food pathogen Listeria monocytogenes, Electronics 7 (2018) 347. https://doi.org/10.3390/electronics7120347

[47] B. Ang, T. Jirapanjawat, K.P. Tay, D. Ashtiani, C. Greening, K.L. Tuck, A. Neild, V.J. Cadarso, Rapid Concentration and detection of bacteria in milk using a microfluidic surface acoustic wave activated nanosieve, ACS Sens. 9 (2024) 3105-3114. https://doi.org/10.1021/acssensors.4c00291

[48] Y. Wang, W. Qi, L. Wang, J. Lin, Y. Liu, Magnetic bead chain-based continuous-flow dna extraction for microfluidic PCR detection of Salmonella, Micromachines 12 (2021) 384. https://doi.org/10.3390/mi12040384

[49] L. Tang, R. Chen, J. Liao, Y. Fan, R. Meng, L. Fu, L. Jia, L-Tryptophan/Levofloxacin-based ratiometric fluorescence biosensor for Escherichia coli by a portable device, Anal. Lett. (2025) 1-16. https://doi.org/10.1080/00032719.2025.2483323

Materials Research Foundations 191 (2026) 72-95 https://doi.org/10.21741/9781644904213-3

[50] X. Jing, H. Wang, X. Huang, Z. Chen, J. Zhu, X. Wang, Digital image colorimetry detection of carbaryl in food samples based on liquid phase microextraction coupled with a microfluidic thread-based analytical device, Food Chem. 337 (2021) 127971. https://doi.org/10.1016/j.foodchem.2020.127971

[51] S. Asgari, G. Wu, S.A. Aghvami, Y. Zhang, M. Lin, Optimisation using the finite element method of a filter-based microfluidic SERS sensor for detection of multiple pesticides in strawberry, Food Addit. Contam.,:Part A 38 (2021) 646-658. https://doi.org/10.1080/19440049.2021.1881624

[52] A.M. Alam, M. Kamruzzaman, T.D. Dang, S.H. Lee, Y.H. Kim, G.M. Kim, Enzymeless determination of total sugar by luminol-tetrachloroaurate chemiluminescence on chip to analyze food samples, Anal. Bioanal. Chem. 404 (2012) 3165-3173. https://doi.org/10.1007/s00216-012-6429-1

[53] T.K. Dhiman, G.B.V.S. Lakshmi, A. Roychoudhury, S.K. Jha, P.R. Solanki, Ceria-nanoparticles-based microfluidic nanobiochip electrochemical sensor for the detection of ochratoxin-A, ChemistrySelect 4 (2019) 4867-4873. https://doi.org/10.1002/slct.201803752

[54] L.A. Pradela-Filho, D.A.G. Araújo, R.M. Takeuchi, A.L. Santos, C.S. Henry, Thermoplastic electrodes as a new electrochemical platform coupled to microfluidic devices for tryptamine determination, Anal. Chim. Acta 1147, (2021) 116-123. https://doi.org/10.1016/j.aca.2020.12.059

[55] H.S. Magar, R.Y.A. Hassan, A. Mulchandani, Electrochemical impedance spectroscopy (EIS): Principles, construction, and biosensing applications, Sensors 21 (2021) 6578. https://doi.org/10.3390/s21196578

[56] M. Carminati, Advances in high-resolution microscale impedance sensors, J. Sens. (2017) 7638389. https://doi.org/10.1155/2017/7638389

[57] H. Qi, J. Zhang, X. Zhang, L. Lv, T. Wang, E. Huang, J. Wu, L. Zheng, A disposable microfluidic aptasensor for one-step and real-time detection of sub-femtomolar-level aflatoxin B1 in food, Food Res. Int. 197 (2024) 115144. https://doi.org/10.1016/j.foodres.2024.115144

[58] N. Alanazi, M. Almutairi, A.N. Alodhayb, A Review of Quartz crystal microbalance for chemical and biological sensing applications, Sens Imaging 24 (2023) 10. https://doi.org/10.1007/s11220-023-00413-w

[59] F. Bonyadi, M. Kavruk, S. Ucak, B. Cetin, G. Bayramoglu, A.D. Dursun, Y. Arica, V.C. Ozalp, Real-time biosensing bacteria and virus with quartz crystal microbalance: Recent advances, opportunities, and challenges, Crit. Rev. Anal. Chem. (2023) 2888-2899. https://doi.org/10.1080/10408347.2023.2211164

[60] D. Mandal, S. Banerjee, Surface acoustic wave (SAW) sensors: Physics, materials, and applications, Sensors 22 (2022) 820. https://doi.org/10.3390/s22030820

[61] G. Cai, Z. Yang, Y.C. Chen, Y. Huang, L. Liang, S. Feng, J. Zhao, Magnetic bead manipulation in microfluidic chips for biological application, Cyborg Bionic Syst. 4 (2023) 0023. https://doi.org/10.34133/cbsystems.0023

[62] M. Mujika, S. Arana, E. Castaño, M. Tijero, R. Vilares, J.M. Ruano-López, A. Cruz, L. Sainz, J. Berganza, Magnetoresistive immunosensor for the detection of Escherichia coli

O157:H7 including a microfluidic network, Biosens. Bioelectron. 24 (2009) 1253-1258. https://doi.org/10.1016/j.bios.2008.07.024

[63] M. Zhang, J. Liao, X. Kong, Q. Yu, M. Zhang, A.X. Wang, Ultra-sensitive, rapid and on-site sensing harmful ingredients used in aquaculture with magnetic fluid SERS, Biosensors (2022) 169. https://doi.org/10.3390/bios12030169

[64] B. Pang, K. Fu, Y. Liu, X. Ding, J. Hu, W. Wu, K. Xu, X. Song, J. Wang, Y. Mu, C. Zhao, J. Li, Development of a self-priming PDMS/paper hybrid microfluidic chip using mixed-dye-loaded loop-mediated isothermal amplification assay for multiplex foodborne pathogens detection, Anal. Chim. Acta 1040 (2018) 81-89. https://doi.org/10.1016/j.aca.2018.07.024

[65] J. Wu, H. Yuan, Y. Yang, P. Yang, X. Yan, Y. Mu, Q. Jin, P. Yang, W. Gao, A comb-shaped microfluidic aptasensor for rapid and sensitive on-site simultaneous detection of aflatoxin B1 and deoxynivalenol, Food Chem. 473 (2025) 143072. https://doi.org/10.1016/j.foodchem.2025.143072

[66] M. Pan, X. Han, S. Chen, J. Yang, Y. Wang, H. Li, S. Wang, Paper-based microfluidic device for selective detection of peanut allergen Ara h1 applying black phosphorus-Au nanocomposites for signal amplification, Talanta 267 (2024) 125188. https://doi.org/10.1016/j.talanta.2023.125188

[67] Y. Deng, Q. Li, Y. Zhou, J. Qian, Fully inkjet printing preparation of a carbon dots multichannel microfluidic paper-based sensor and its application in food additive detection, ACS Appl. Mater. Interfaces 13 (2021) 57084-57091. https://doi.org/10.1021/acsami.1c14435

[68] H.P. Kim, J.H. Park, M.G. Kim, Y. Seok, Lab-on-paper for molecular testing with USB-powered isothermal amplification and fluidic control, Microchim. Acta 192 (2025) 90. https://doi.org/10.1007/s00604-024-06889-x

[69] Z. Jin, G. Ding, G. Li, G. Yang, Y. Han, N. Hao, J. Deng, Y. Zhang, W. Zhang, W. Li, Rapid detection of foodborne bacterial pathogens using visual high-throughput microfluidic chip, J. Chem. Technol. Biotechnol. 95 (2020) 1460-1466. https://doi.org/10.1002/jctb.6331

[70] M. Shalaby, V. Busin, X. Yan, S. Cengiz, M.C. Adiguzel, J.M. Cooper, T. Forde, J. Reboud, Sample-to-answer microfluidic device towards the point-of-need detection of Staphylococcus aureus enterotoxin genes in ruminant milk, Lab Chip 25 (2025) 524-535. https://doi.org/10.1039/D4LC00907J

[71] A. Teixeira, J.L. Paris, F. Roumani, L. Diéguez, M. Prado, B. Espiña, S. Abalde-Cela, A. Garrido-Maestu, L. Rodriguez-Lorenzo, Multifuntional gold nanoparticles for the SERS detection of pathogens combined with a LAMP-in-microdroplets approach, Materials 13 (2020) 1934. https://doi.org/10.3390/ma13081934

[72] Y. Chen, Y. Hu, X. Lu, Polyethersulfone-based microfluidic device integrated with DNA extraction on paper and recombinase polymerase amplification for the detection of Salmonella enterica, ACS Sens. 8 (2023) 2331-2339. https://doi.org/10.1021/acssensors.3c00387

[73] Y. Jin, J. Wang, Z. Wang, P. Xiong, J. Cheng, T. Xu, An integrated microfluidic biosensing system based on a versatile valve and recombinase polymerase amplification for rapid and sensitive detection of Salmonella typhimurium, Biosensors 13 (2023) 790. https://doi.org/10.3390/bios13080790

[74] L.D. Renner, J. Zan, L.I. Hu, M. Martinez, P.J. Resto, A.C. Siegel, C. Torres, S.B. Hall, T.R. Slezak, T.H. Nguyen, D.B. Weibel, Detection of ESKAPE bacterial pathogens at the point of care using isothermal DNA-based assays in a portable degas-actuated microfluidic diagnostic assay platform, Appl. Environ. Microbiol. 83 (2017) e02449-16. https://doi.org/10.1128/AEM.02449-16

[75] Y. Lu, M.Z. Hua, Y. Luo, X. Lu, Q. Liu, Hybrid paper/PDMS microfluidic device integrated with RNA extraction and recombinase polymerase amplification for detection of norovirus in foods, Appl. Environ. Microbiol. 90 (2024) e01208-24. https://doi.org/10.1128/aem.01208-24

[76] C. Zhang, X. Lv, X. Han, Y. Man, Y. Saeed, H. Qing, Y. Deng, Whole-cell based aptamer selection for selective capture of microorganisms using microfluidic devices, Anal. Methods 7 (2015) 6339-6345. https://doi.org/10.1039/C5AY01016K

[77] P. Nolan, S. Auer, A. Spehar, M. Oplatowska-Stachowiak, K. Campbell, Evaluation of mass sensitive micro-array biosensors for their feasibility in multiplex detection of low molecular weight toxins using mycotoxins as model compounds, Talanta 222 (2021) 121521. https://doi.org/10.1016/j.talanta.2020.121521

[78] X. Xiang, M. Song, X. Xu, J. Lu, Y. Chen, S. Chen, Y. He, Y. Shang, Microfluidic biosensor integrated with signal transduction and enhancement mechanism for ultrasensitive noncompetitive assay of multiple mycotoxins, Anal. Chem. 95 (2023) 7993-8001. https://doi.org/10.1021/acs.analchem.3c00813

[79] M.Z. Hua, S. Li, M.S. Roopesh, X. Lu, Development of a microfluidic device to enrich and detect zearalenone in food using quantum dot-embedded molecularly imprinted polymers, Lab Chip 24 (2024) 2700-2711. https://doi.org/10.1039/D4LC00193A

[80] I.A. Rather, W.Y. Koh, W.K. Paek, J. Lim, The sources of chemical contaminants in food and their health implications, Front. Pharmacol. 8 (2017) 830. https://doi.org/10.3389/fphar.2017.00830

[81] K.S. Deepak, S.K. Dubey, S. Goel, A. Javed, A portable hand-held microfluidic colorimetric device for the detection of organophosphorus pesticides, IEEE Sens J. (2024). https://doi.org/10.1109/JSEN.2024.3499873

[82] H. Pang, J. Xie, X. Meng, R. Sun, J. Chen, C. Guo, T. Zhou, Portable organophosphorus pesticide detection device based on controllable microfluidic and luminol composite nanofibers, J. Food Eng. 364 (2024) 111810. https://doi.org/10.1016/j.jfoodeng.2023.111810

[83] X. Yan, Z. Zhang, R. Zhang, T. Yang, G. Hao, L. Yuan, X. Yang, Rapid detection of dimethoate in soybean samples by microfluidic paper chips based on oil-soluble cdse quantum dots, Foods 10 (2021) 2810. https://doi.org/10.3390/foods10112810

[84] W. Peng, Z. Xu, C. Yi, Y. Zhang, Q. Liao, Silver coated PS microsphere array SERS microfluidic chip for pesticide detection, Heliyon 10 (2024). https://doi.org/10.1016/j.heliyon.2024.e33647

[85] Y. Yang, C. Tong, R. Zhou, Z. Qin, J. Xu, C. Liao, S. Zhang, S. Shi, Y. Guo, Hinge-like paper-based dual-channel enzyme-free ratiometric fluorescent microfluidic platform for simultaneous visual detection of carbaryl and glyphosate, Food Chem. 431 (2024) 137127. https://doi.org/10.1016/j.foodchem.2023.137127

[86] M.K. Filippidou, A.I. Kanaris, E. Aslanidis, A. Rapesi, D. Tsounidi, S. Ntouskas, E. Skotadis, G. Tsekenis, D. Tsoukalas, A. Tserepi, S. Chatzandroulis, Integrated plastic microfluidic device for heavy metal ion detection, Micromachines 14 (2023) 1595. https://doi.org/10.3390/mi14081595

[87] W.-H. Huang, V.-P. Mai, R.-Y. Wu, K.-L. Yeh, R.-J. Yang, A microfluidic aptamer-based sensor for detection of mercury(II) and lead(II) ions in water, Micromachines 12 (2021) 1283. https://doi.org/10.3390/mi12111283

[88] M. Yuan, C. Li, Y. Zheng, H. Cao, T. Ye, X. Wu, L. Hao, F. Yin, J. Yu, F. Xu, A portable multi-channel fluorescent paper-based microfluidic chip based on smartphone imaging for simultaneous detection of four heavy metals, Talanta 266 (2024) 125112. https://doi.org/10.1016/j.talanta.2023.125112

[89] M. Taghizadeh-Behbahani, M. Shamsipur, B. Hemmateenejad, Detection and discrimination of antibiotics in food samples using a microfluidic paper-based optical tongue, Talanta 241 (2022) 123242. https://doi.org/10.1016/j.talanta.2022.123242

[90] Á. Écija-Arenas, V. Román-Pizarro, J.M. Fernández-Romero, Usefulness of hybrid magnetoliposomes for aminoglycoside antibiotic residues determination in food using an integrated microfluidic system with fluorometric detection, J. Agric. Food Chem. 69 (2021) 6888-6896. https://doi.org/10.1021/acs.jafc.1c01571

[91] J.B. Uppin, G. Vaidya, G.M.N. Mamdapur, Understanding food allergies and allergens: A comprehensive guide to diagnosis, treatment, and management, in: J.S. Yaradoddi, B.S. Meti, S.B. Madgulkar, D. Agsar, (Eds.), Frontiers in Food Biotechnology, Springer Nature, Singapore, 2024, pp. 143-161. https://doi.org/10.1007/978-981-97-3261-6_9

[92] R.L. Sayers, L.A. Gethings, V. Lee, A. Balasundaram, P.E. Johnson, J.A. Marsh, A. Wallace, H. Brown, A. Rogers, J.I. Langridge, E.N.C. Mills, Microfluidic separation coupled to mass spectrometry for quantification of peanut allergens in a complex food matrix, J. Proteome Res. 17 (2018) 647-655. https://doi.org/10.1021/acs.jproteome.7b00714

[93] X. Weng, S. Neethirajan, A microfluidic biosensor using graphene oxide and aptamer-functionalized quantum dots for peanut allergen detection, Biosens. Bioelectron. 85 (2016) 649-656. https://doi.org/10.1016/j.bios.2016.05.072

[94] X. Chen, D. Zhang, Q. Liu, S. Liu, H. Li, Z. Li, Enzyme-linked immunosorbent assay-based microarray on a chip for bioaerosol sensing: Toward sensitive and multiplexed profiling of foodborne allergens, Anal. Chem. 95 (2023) 7354-7362. https://doi.org/10.1021/acs.analchem.3c00591

[95] T.A. Baldo, C. A. Proença, F. S. Felix, T.A. Freitas, S.K. Sakata, L. Angnes, R.C. Faria, Disposable electrochemical microfluidic device for ultrasensitive detection of egg allergen in wine samples, Talanta 232 (2021) 122447. https://doi.org/10.1016/j.talanta.2021.122447

[96] C. Polonschii, M. Rosu-Hamzescu, S. David, A. Oloumi, V.D. Ursu, M. Szardenings, K. Kern, A.E.E. Salhi, E. Gheorghiu, Point-of-care personalized rapid diagnosis of allergies using peptide epitopes and SPR multiplexed detection, Sens. Actuators, B Chem. 418 (2024) 136359. https://doi.org/10.1016/j.snb.2024.136359

[97] F. Balogh-Hartmann, C. Páger, A. Bufa, I.M. Horváth, Z. Verzár, T. Marosvölgyi, L. Makszin, Microfluidic analysis for the determination of protein content in different types of plant-based drinks, Molecules 28 (2023) 6684. https://doi.org/10.3390/molecules28186684

[98] A.G. Bozkurt, Determination of ascorbic acid (vitamin C) with paper based colorimetric method, AKU J. Sci. Eng. 22 (2022) 597-606. https://doi.org/10.35414/akufemubid.1015952

[99] N. Bagheri, H.A.J.A. Lawati, J. Hassanzadeh, I.A. Lawati, Novel amino-functionalized magnetic metal-organic framework/layered double hydroxide adsorbent for microfluidic solid phase extraction: Application for vitamin D3, Talanta 256 (2023) 124272. https://doi.org/10.1016/j.talanta.2023.124272

[100] S. Demirbaş, M. Şen, Developing a Paper-based sensor for cholesterol measurement and processing images using ImageJ, Int. Conf. Appl. Eng. Nat. Sci. 1 (2023) 544-546. https://doi.org/10.59287/icaens.1053

[101] M.G. Pereira, A. Machado, A. Leite, M. Rangel, A. Bordalo, A.O.S.S. Rangel, R.B.R. Mesquita, Microfluidic paper-based device incorporated with silica nanoparticles for iodide quantification in marine source dietary supplements, Sensors 24 (2024) 1024. https://doi.org/10.3390/s24031024

[102] X. Tian, J. Li, K. Wang, S. Fei, X. Zhang, C. Wu, M. Tan, W. Su, Microfluidic fabrication of core-shell fucoxanthin nanofibers with improved environmental stability for reducing lipid accumulation in vitro, Food Chem. 442 (2024) 138474. https://doi.org/10.1016/j.foodchem.2024.138474

[103] D. Yu, M. Guo, M. Tan, W. Su, Lipid-lowering and antioxidant effects of self-assembled astaxanthin-anthocyanin nanoparticles on high-fat Caenorhabditis elegans, Foods 13 (2024) 514. https://doi.org/10.3390/foods13040514

[104] T. Kasputis, K.E. Hosmer, Y. He, J. Chen, Ensuring food safety: Microfluidic-based approaches for the detection of food contaminants, Anal. Sci. Adv. 5 (2024) e2400003. https://doi.org/10.1002/ansa.202400003

[105] Y. Lu, J. Zhang, X. Lu, Q. Liu, Isothermal nucleic acid amplification based microfluidic "lab-on-a-chip" for the detection of pathogenic bacteria and viruses in agri-foods, Trends Food Sci. Technol. 148 (2024) 104482. https://doi.org/10.1016/j.tifs.2024.104482

[106] T.N.D. Trinh, K.T.L. Trinh, N.Y. Lee, Microfluidic advances in food safety control, Food Res. Int. 176 (2024) 113799. https://doi.org/10.1016/j.foodres.2023.113799

[107] C. Xiao, G. Ross, M.W.F. Nielen, J. Eriksson, G.I. Salentijn, W.C. Mak, A portable smartphone-based imaging surface plasmon resonance biosensor for allergen detection in plant-based milks, Talanta 257 (2023) 124366. https://doi.org/10.1016/j.talanta.2023.124366

Materials Research Foundations 191 (2026) 96-127
https://doi.org/10.21741/9781644904213-4

Chapter 4

Beam Electron Lithography for Microfluid Devices Fabrication

M. Rizwan[1*], A. Ahmed[2], A. Ayub[3], S. Hanif[1], A. Munir[1], Alishba[1], M. Perveen[1]

[1]Department of Physics, University of the Punjab, Lahore, Pakistan

[2]Department of Physics, University of Education, Lahore, Pakistan

[3]Quantum Technology & Nanoscience, Scuola Normale Superiore di Pisa, Pisa, Italy

*rizwan.sps@pu.edu.pk

Abstract

Electron Beam Lithography (EBL) is a high-resolution patterning method that is mostly used in nanofabrication for semiconductor devices, photonics, and bioengineering applications. Unlike optical lithography, EBL offers superior resolution by utilizing focused electron beams to define intricate nanoscale features. This chapter explores the fundamental principles of EBL, its working mechanisms, resistance systems, and key processing parameters. The advantages and limitations of EBL are discussed, along with its role in modern nanotechnology. Future advancements in the field, including improvements in throughput and hybrid lithographic approaches, are also examined.

Keywords

Electron Beam Lithography (EBL), Nanofabrication, High-Resolution Patterning, Electron-Material Interactions, Maskless Lithography, Hybrid Lithographic Techniques

Contents

Materials Research Foundations 191 (2026) 96-127 https://doi.org/10.21741/9781644904213-4

Materials Research Foundations 191 (2026) 96-127 https://doi.org/10.21741/9781644904213-4

1. Introduction

In lithography, the pattern is transferred from one medium to another. Many beams of particles are used in this technique. An electron beam is used as a source because of its high diffraction resolution, and it can transfer the pattern whose size is in nanometers [1]. Electron beam resist is applied to the substrate in order to capture the image pattern that is to be transferred [2]. By using computers, we can control the position of the beam of electrons, which directly transfers the real image on the substrate. EBL was developed later after the scanning electron microscope [3]. Other lithography methods are widely used because they provide better resolution by using other types of radiation, like ion beam, electron beam, and ultraviolet radiation. Numerous studies have been conducted on these methods to bring lithography to the nanoscale [4].

1.1 Overview of Microfluidic Devices

Microfluidics is also known as micro-hydrodynamics. It is the branch of science and technology that controls small amounts of fluids by using small channels whose size is of the order of ten to hundreds of micrometers. It is possible to integrate, automate, and combine many processes by manipulating tiny quantities of fluid [5]. The materials through which microfluidic devices are made are categorized into five groups. The first group consists of Polydimethylsiloxane (PDMS), elastomers, and polymers [6]. Thermoplastics are included in the second group. The third group contains a thermosetting polymer processed by photopolymerization. In the fourth group, the paper is used for a Point of Care (POC) system and a disposable system. In the fifth group, hydrogels are used to study living [7-9] Cells. In microfluidic devices, the flow generation system is categorized into two major groups: Active systems and Passive flow generation systems [10]. Active pumping methods are complex but can be used due to continuous flow control. These systems consist of electric, magnetic, or thermal forces, solenoid valves, pumps, micropumps, and stepper motors. The active pressure-driven system depends on fluid flow through valves, electromagnetic pumps, or syringe pumps [11]. Active pumping system mechanisms show a microfluidic flow regulatory system, high efficiency, robustness, and full automation. By using different pumps and valves in the active system, fluid flow can be precisely controlled. In passive mechanisms, the liquid is moved through the chip channel via surface tension, capillary action, and gravity without the need for an external power source. It means that these systems have a very low cost and are portable for microfluidic applications. Depending on the forces generating the flow, passive flow control techniques can be classified as gravity-induced, capillary action, osmosis, vacuum suction, surface tension, and pressure-driven [12].

The Reynolds numbers, viscosity, density, and velocity gradient are factors that define microfluidic flow [13]. By changing these variables, the user can regulate the fluid flow rate and direction. Additional flow control can be achieved by the physical properties of the liquid and the geometry of the channel. Simple microchannels have laminar flow and a tiny Reynolds number, usually less than 1, due to the liquid's incompressibility; as a result, turbulent flow is not seen [13].

Compared to bigger systems, these systems can integrate and combine many processes by manipulating small quantities of fluid. Compared to larger systems, these systems are easier to use, have low fluid flow, and offer greater control over spatio-temporal dynamics. High levels of efficiency and reproducibility are offered by these systems [14].

2. Importance of High-Resolution Patterning in Microfluidic

High-resolution patterning is essential for the development of microfluidic technology, which enables us to control fluid flow precisely, better biosensing capabilities, and reduce the complexity of the lab-on-a-chip (LOC) systems. High-resolution patterning in microfluidic devices can be obtained from different methods, such as photolithography, three-dimensional printing, soft lithography, electrodynamics, etc.

Microfluidic devices are microscopic systems that maintain small amounts of liquid in fields like biology, chemistry, environmental science, etc. These devices are used for diagnosis, chemical

Materials Research Foundations 191 (2026) 96-127 https://doi.org/10.21741/9781644904213-4

testing, and creation of the LOC system. Microfluidic devices are manufactured in a clean room. Digital Micromirror Device (DMD) based on maskless lithography shows well-defined generation pattern capabilities and fast processing speed, which is why it is mostly used because it requires an initial low cost [15]. By using step-and-repeat projection lithography, a high-resolution pattern is created [16]. Soft lithography combines with printing and forms new methods that enable the fabrication of high-resolution patterns [17]. Electron beam lithography also has features of high-resolution patterning. In this technique, direct writing and projection printing are used. In projection printing, the high-precision lens system is used [18]. We introduce a systematic method through which microfluidic devices can be manufactured over a wide range. Initially, by using image-based feedback, we can control the flow in maskless lithography, the speed of UV light, and the projection pattern precisely. Whatever the material, this technique provides device thickness, light uniformity, UV power, and the type of device [19].

The devices that use microfabrication and microfluidics that enable the culture and manage the brain-related tissues are called brain-on-a-chip devices [20]. The organ-on-a-chip (OOC) method is a new development in the manufacturing of microfluidic devices. These devices have high-resolution patterning that enables us to study human organs and their activities properly. OOC is a multichannel, three-dimensional, microfluidic cell culture integrated circuit that represents the mechanisms and activities of a whole organ system [21]. New development in organ chips helps us to study the complex pathophysiology of human vital infections [22].

3. Role of Electron Beam Lithography in Microfluidic Fabrication

EBL has high potential to produce sub-micro patterns [23]. The gap between high-resolution patterning and high output can be filled by EBL. The basis of EBL is the electron beam that is scanned on the surface. This beam is coated with a layer that is sensitive to electrons. A chemical reaction is induced in the resist when energy is transferred by a beam of electrons, which changes its solubility. To derive chemical processes, energy per unit area is required. The final resolution of the manufacturing process in EBL is limited by scattered electrons. As the electrons pass through the resist layer, the scattering event causes the electrons to broaden. The width of the main beam of an electron can be as small as less than 1 nm. This widening restricts the minimum size of manufactured structures and makes the resolution reliant on the resistance of the film thickness. Furthermore, some electrons from the beam of electrons may be scattered in the substrate or resist at sharp angles and have a horizontal travel distance of around 30 μm [24]. The manufacturing of microstructure by Electron Beam Lithography can be done when different substrates are coated with polymers that are sensitive to electrons. In EBL, the epoxy-based resist is known as SU-8. It is sensitive to electron irradiation, while it was first created for UV-photolithography [25]. It can be used on substrates of many different thicknesses, from over 300 μm to sub-micrometer thickness. We can make microfluidics channels by negative tone resist SU-8, which is highly dependent on the energy of electrons that show penetration properties. In this method, there are two steps in which EBL is used. In the first step, when SU-8 is disclosed to the substrate, supporting structures are formed. In the second step, elevated structures are formed when the subjected parameters are changed accordingly. This process creates the microfluidic channel with an exact arrangement over several mm [26].

Materials Research Foundations 191 (2026) 96-127 | https://doi.org/10.21741/9781644904213-4

4. Comparison with Other Lithography Techniques

The process of creating three-dimensional patterns of the image onto the substrate is called lithography. It is a Greek word that means to write something on stone. Silicon wafer can also be used as a stone in this technique. Positive and negative masks are the patterns that transfer onto substrates via lithography. Lithography has other subcategories: extreme UV lithography and X-ray lithography, ion beam lithography, electron beam lithography, soft lithography, photolithography, colloidal lithography, nanoimprint lithography, and many more [27].

Photolithography creates the pattern of the image onto the photoresist when radiation falls near the Ultraviolet (UV) visible part of the electromagnetic spectrum. Glass or silicon can be used as a substrate. The resolution of Photolithography is of the order of 3 to 5 micrometers. The system in which radiation falls consists of a condenser lens and a source that provides uniform lighting to the mask. In this method, a mercury-rare gas discharge lamp is used as a source. Instead of a discharge lamp, when we use the laser as a source then it provides high output to the photoresist. The measurement of the phase relationship between emitted wavefronts is called spatial coherence. A laser amplifies the radiation that falls on the photoresist and can produce coherent radiation. This process increases resolution. Photolithography with the projection technique is widely used for the manufacturing of Integrated Circuits (IC). There are limited applications of this technique because of wafer damage, the mask, small size, and difficulty in arranging. These limitations are overcome by manufacturing such tools whose resolution is below 0.25 micrometers [28].

Soft lithography is a method for quickly prototyping a wide range of devices and structures at the micro- and nanoscale on flat and flexible surfaces. This method is not too expensive. Different techniques of soft lithography are listed here: nano-transfer printing, micro-transfer molding, decal transfer lithography, phase-shifting edge lithography, and so on [29]. An elastomer stamp is an essential component of soft lithography that transfers the pattern from the master mold to the substrate. This stamp is made when a precursor liquid is spread out on a master that has comparable structures imprinted on its surface. An elastomer stamp can be made from silicon-based rubber or cross-linked Polydimethylsiloxane (PDMS) [30]. The combination of self-assembled monolayers (SAMs) with microcontact printing to create a patterned surface is made possible by soft lithography. Elastomer enables the micropatterned surfaces to make comparable contact with the surfaces across wide regions. Since most masters are hard, using an elastomer makes it easier to separate the duplicate from the master [31].

Imprint lithography is a new technique that offers resolution of less than 10 nm. It is an effective method and has nanoscale features [32]. The first step is the imprint process, in which a mold is pressed into thin resist that is spread out on the substrate, and then the mold is removed. By the imprint process, a thickness contrast pattern is produced in the resist. During step two, any remaining resist in the compressed area is removed via anisotropic etching (e.g., RIE). During this step, heat is applied at a temperature that is greater than the resist transition temperature. The thermoplastic resin turns into a viscous liquid at that temperature, which allows it to flow and easily deform into the mold's shape. When the temperature increases, the viscosity of the resist decreases. In imprint lithography, no energetic beam is used. The resolution of lithography is not constrained by backscattering from a substrate, interference and scattering in a resist, and wave diffraction. Additionally, imprint lithography differs from stamping with a monolayer of self-assembled molecules in essential ways [33]. The imprinting fluid is known as a polymeric substance, ink, resist, or monomer. The mask is a known template or mold [34].

Materials Research Foundations 191 (2026) 96-127
https://doi.org/10.21741/9781644904213-4

In the microfabrication process, electron beam lithography uses a concentrated electron beam that produces small patterns on a substrate with a resolution as low as 10 nm by using different techniques. EBL creates perfect microchannels and complex designs, which are important for the maintenance of fluid flow at the microscale in microfluidic devices. It is time-consuming and more costly than other techniques. But EBL is useful for the fabrication of microfluidic devices because it has high-resolution features.

5. Fundamentals of Electron Beam Lithography

EBL is widely used in nanofabrication for surficial patterns at the nanoscale. EBL uses a focused electron beam to write patterns on a surface. The technique of nanofabrication is crucial in fabricating the advanced materials and devices for vast applications, including the fields of microelectronics, photonics, and nanotechnology [35]. EBL is used in semiconductor manufacturing and nanotechnology to precisely produce complex nanostructures at the sub-10-nanometer scale. In photolithography, UV light is used to transfer the pattern on a substrate. Electron beams directly fall on electron-sensitive resist. This technique provides unmatched flexibility in the design and prototyping of nanoscale devices by doing away with the need for a physical mask. The procedure starts with a thin coating of resist material, such as hydrogen silsesquioxane (HSQ) or polymethyl methacrylate (PMMA), applied to a substrate, usually a silicon wafer. Electromagnetic lenses and deflection devices are used to guide the electron beam in a specific way when it passes through the surface. A development step reveals the chemical changes brought about by electron and resist interaction, which might be either crosslinking for negative resists or chain scission for positive resists. To provide a patterned template for etching, deposition, or lift-off procedures later on, the exposed portions are either dissolved or hardened [36,37].

The following are the featured and fundamental principles of electron beam lithography:

5.1 Electron source

EBL is based on an accelerated beam of electrons. The process of their high-energy acceleration is accomplished with an internally built electron gun that generates them. This beam of high-energy electrons is set to achieve the required patterns at the nanometer scale [37].

5.2 Sensitivity of Resist Material

The surface of the substrate in EBL is coated with specific materials that are highly sensitive to electron exposure. These materials are known as resists, e.g., polymethylmethacrylate (PMMA) and hydrogen silsesquioxane (HSQ). These resists are chemically amplified.
There are two main types of resistance employed:

5.2.1 Positive Resists

When the beam of electrons is exposed to these materials, they become soluble in the developer. These materials become soluble in the developer, which allows the exposed areas to undergo fine patterns. Polymethyl methacrylate (PMMA) was the first resist that was mostly used as a positive resist and has a low glass transition temperature of 114 °C. PMMA is available as a powder that dissolves in a solvent with the appropriate concentration, like anisole or chlorobenzene. To create a thin coating, first, a resist liquid is placed on the substrate and then spun quickly. The next step

Materials Research Foundations 191 (2026) 96-127 https://doi.org/10.21741/9781644904213-4

is soft bake processing, in which a hot plate is used to bake out the casting solution at temperatures between 130 and 170 degrees Celsius. The concluding resist thickness is affected by PMMA concentration and spin speed [37].

5.2.2 Negative Resists

The materials that exhibit cross-linking on exposure to an electron beam and become insoluble by illustrating final patterns in the exposed areas. Although negative resists typically exhibit less bias, they struggle with bridging between features and scum and swelling during development. Commonly used negative e-beam resist comprises a partly chloromethylated polystyrene (CMS), also referred to as Shipley advanced lithography (SAL) product line. Additionally, when PMMA is subjected to a dose that is an order of magnitude larger, it may show negative tone. CMS has a modest resolution at a modest sensitivity, whereas COP has a high sensitivity [37].

5.3 Generation of Patterns

The cycle of pattern generation is accomplished with the precision and controlled beams of electrons. Some specific computer-aided design (CAD) software is used for patterns of interest, which are converted into a beam by the EBL system. This is a unique and direct capability of EBL that enables distinct flexibility of patterns. A common EBL operating environment, 50 keV, has an electron de Broglie wavelength of less than 10 pm, or 0.01 nm, which is significantly smaller than an ordinary atomic size. E-beam widths of approximately 1 nm are ideal. However, this limit is greatly weakened by beam-material or scattering interaction [38].

In a solid resist, small-angle forward scattering of electrons leads to beam widening. Some electrons undergo large-angle scattering at the initial point after passing through the resist into the substrate, which causes backscattering. This happens when a beam of electrons turns back through a resist that is far from the desired exposure. This effect is called electron beam proximity and results in more exposure in the resist. The deceleration of primary electrons results in the emission of secondary electrons. In the sub-micron region, a population of fast secondary electrons arises, which is primarily responsible for resist exposure and contributes significantly to the proximity effect [38].

5.4 Exposure and Development Process

When an electron beam falls on the substrate that is coated with resist, this process changes its chemical properties. Following exposure, development uses a chemical bath to strip away the soluble or, for other resists, the insoluble portions.

5.5 High-Resolution Capabilities

Known for its very high resolution, EBL features as fine as a few nanometers. EBL is an invaluable tool for applications requiring intricate nanoscale patterns due to its ability to surpass the constraints of conventional optical lithography [38].

6. Key Components

An electron source, two or more lenses, electron beam deflectors, and a sample stage are the components of EBL. These components have different functions [38].

Materials Research Foundations 191 (2026) 96-127 https://doi.org/10.21741/9781644904213-4

6.1 Electron Sources

Electron emission from a conducting material occurs either through thermal excitation, where heating provides electrons with sufficient energy to escape (thermionic emission), or through strong electric fields that enable electrons to overcome the work function barrier (field emission).There are three main characteristics of a virtual source: brightness, distribution of energy that is released by emitted electrons (in electron volts), and size of the source.

The thermal field emission source is used in EBL and various electron microscopes [39]. It combines heating from a thermal source with a sharp tungsten needle from the field emission source. The tip can function steadily for months at a time since it runs at 1800 K, which reduces its susceptibility to the environmental gases. Although the term "thermal field emitter" is frequently used, the term "Schottky emitter" is more appropriate because the electrons are excited by heat and are able to cross the work function barrier.

Its brightness is almost equal to that of its cold field emission source, its energy spread is low, and the virtual source is small. The work function barrier can be reduced by coating zirconium oxide on tungsten. Material that evaporates from the tip of the electron cannon is continuously refilled by a heated reservoir of zirconium oxide [40].

6.2 Electron Lenses

Either magnetic or electrostatic forces can concentrate electrons. There are discrepancies, even though electron lenses and optical lenses operate similarly in theory. Electron lenses are used for convergence, not divergence, with a few exceptions. Additionally, in terms of aberrations, electron lenses are not as good as optical lenses. Because of the poor quality of electron lenses, convergence angles and field size are restricted.

EBL depends on two forms of aberration: one is spherical, and the other is chromatic aberration. In chromatic aberrations, electrons with different energies focus on different image planes. When the lens has spherical aberrations, then its outer zone focuses more intensely than its interior zone. Although this reduces the beam current significantly, both types of aberrations can be minimized by decreasing the convergence angle, thereby confining electrons to the lens center [39].

6.3 Electron Beam Deflectors

By using deflection, the beam of electrons is moved across the surface of the sample. It can be done electrostatically or magnetically, just like with lenses. When plates or coils are placed, the field is positioned perpendicularly to the optical axis. The beam diameter deteriorates when the beam is deflected off-axis due to extra aberrations introduced, and X and Y deviations from linearity rise with increasing deflection. The deflection size that can be employed is constrained by these effects [41].

Similar to lenses, electrostatic deflection causes more distortions than magnetic deflection. To further minimize deflection aberrations, two matched coils in double magnetic deflection are occasionally employed. Although the frequency response is restricted by magnetic deflection inductance, electrostatic deflection can reach significantly greater speeds. Additionally, the speed of magnetic deflection may be further limited by eddy currents created by the magnetic fields. In order to avoid the fields interacting with conducting metal components, deflection mechanisms are

Materials Research Foundations 191 (2026) 96-127 https://doi.org/10.21741/9781644904213-4

often positioned inside the final lens. To reduce eddy currents, the final lens is typically insulated by ferrite [41].

Certain tools employ two different deflection systems: long-range deflection is magnetic, and high-speed, short-range deflection is electrostatic. In both cases, the field size of the tool is constrained by the deflection of the system's aberrations; some tools improve the maximum field size by dynamically correcting the deflection, focus, and stigmator, but doing so comes at the expense of complexity.

6.4 Sample Stage

During nanoscale patterning, the sample stage is a precise part that holds and places the substrate, such as glass or a silicon wafer. With nanometer-scale resolution, it moves the sample in three dimensions to guarantee precise placement. Piezoelectric or high-precision mechanical actuators are frequently used, and laser interferometry is occasionally used to provide real-time feedback. To avoid pattern distortion, the stage must run smoothly in a high vacuum environment with as few vibrations as possible. For alignment corrections, some higher stages further provide tilt and rotation adjustments. Pattern stitching and multi-layer overlay alignment are made possible by the stage's integration with the EBL system's software, which is essential for intricate production. Stages can be mobile (for writing across a vast area), stationary (with beam deflection performing the job), or hybrid (combining both ways), depending on the design. Resolution, throughput, and the dependability of producing complex nanostructures are all directly impacted by its accuracy [41].

7. Electron Sample Interactions and Resolution Limits

Even though electron beam lithography equipment can create incredibly small probes, the process becomes more complicated when the electrons encounter the workpiece. The initial beam width tends to be widened by the numerous small-angle scattering events (forward scattering) that the electrons encounter as they pass through the resist. Occasionally, electrons penetrating the resist and entering the substrate undergo large-angle scattering. The proximity effect occurs when electrons are scattered from the surroundings due to backscattering, which alters the quantity of the pattern. Secondary electrons, also known as low-voltage electrons, are created as a result of the electrons' constant slowing down throughout this process [42].

7.1 Forward Scattering

As electrons traverse the resist, small-angle scattering can make the beam noticeably wider at the bottom than at the top. The thinnest resist and the maximum accelerating voltage are used to reduce forward scattering.

Forward scattering effects can sometimes be advantageous, although it is usually advisable to avoid them wherever feasible. For instance, by varying the development period, it could be feasible to customize the resist sidewall angle in thick resist. The resist alters its slope from positive to negative as time advances. This is particularly advantageous for pattern transfer via liftoff [43].

Materials Research Foundations 191 (2026) 96-127 https://doi.org/10.21741/9781644904213-4

7.2 Backscattering

As the electrons continue to enter the substrate through the resist, many electrons experience significant angle scattering. Additional resist exposure might result from these electrons returning through the resist at a considerable distance from the incoming beam. When electrons cover the distance in a bulk material before dissipating all energy, depending on initial electron energy and the type of substrate, it is known as the electron range. The portion of backscattering electrons depends on the type of material through which the substrate is formed and is mostly independent of beam energy [43].

7.3 Secondary Electrons

A large portion of the initial electrons' energy is released as they slow down, known as secondary electrons, whose energy ranges from 2 to 50 eV. Most of the actual resist exposure procedures are their responsibility. Their contribution to the proximity effect is less because of their limited resistivity, which is only a few nanometers. Rather, the end effect may be seen as an increment in its width of about 10 nm. This largely explains the low resolution of about 20 nm, which was observed in the highest resolution of the electron beam, as well as contributing to the bias observed in the positive resist system [43].

7.4 Resolution Limits

The resolution attained for "isolated" nanostructures, those in which the feature size is much smaller than the pitch between adjacent features, is described as:

We now look at a different element of resolution, which is the lowest possible period of a given size rather than the smallest size possible with EBL. A thick collection of nanostructures. In fact, for certain uses, such as high-density data storage, in contrast to the actual size of the nanostructure, the density of the nanostructure is the more important factor. Another issue with the fabrication of dense arrays is extremely tiny dimensions, which is the small resist feature for development. The smallest pitch currently achieved for pillar arrays using the whole procedure is 30 nm. The pillar's diameter doesn't show any discernible change throughout a dosage range of 7.6 to 8.4 fC. It is important to note that following exposure and growth [44].

Our whole manufacturing process, which combines lift-off, reactive ion etching (RIE), and very high-resolution EBL of SiO_2 pillars, is used to produce effective molds for nanoimprint lithography (NIL). These molds have undergone testing for imprinting experiments, and arrays with x down to a pitch of 50 nm showed satisfactory reproduction. NIL has emerged as a novel patterning method with low cost and great productivity that can achieve sub-10 nm resolution. Quantum Magnetic disks (QMDs) consist of bits in a single magnetic domain surrounded by non-magnetic material. The traditional storage density constraint of continuous magnetic medium should be addressed by this novel idea [45].

8. Advantages and Limitations of Microfluidic Applications

The capacity of EBL to attain ultra-high resolution is one of its many benefits. It is ideal for developing nanoscale channels that need to have an extremely precise shape because of its characteristics that extend less than 10 nm. The method is also very adaptable; EBL transfers the pattern onto the substrate, which can be modified easily via computer control. For research and

Materials Research Foundations 191 (2026) 96-127 https://doi.org/10.21741/9781644904213-4

development, where iterations are frequent, EBL is a strong fit. EBL is further expanded due to its versatility, which allows it to be employed with a variety of substrates and materials [46].

EBL is not without its drawbacks, though. Its limited throughput is a major shortcoming. EBL is significantly slower than parallel methods like photolithography because it is a consecutive process. Because of this, EBL is less suited for mass manufacturing, particularly when the surface area is large. Another difficulty is the proximity effect, which can lower pattern fidelity by causing unintentional exposure in nearby areas because electrons are dispersed in the resist and substrate.

8.1 Advantages in Microfluidic Applications

Electron beam lithography (EBL) is an advanced research technique because it provides a number of significant benefits for the manufacturing of microfluidic devices. Its unparalleled resolution and accuracy, which allow for the fabrication of nanoscale features down to a few nanometers, are among its most important advantages. This ability is essential for creating high-performance microfluidic parts that need precise dimensional control, such as plasmonic sensors, nanochannels, and Loc systems.

Furthermore, EBL is a maskless method, which means that costly photomasks are not required and that design iterations may be completed quickly. This is very useful for prototyping and customized microfluidic systems. Additionally, the technique creates high aspect ratio structures and smooth sidewalls, which improve fluidic performance by reducing undesired turbulence or adsorption in microchannels. Additionally, EBL offers versatility in material selection for a broad range of substrates, such as silicon, glass, and polymers, as well as different resist materials, such as PMMA and HSQ.

Because of its direct-write capacity, complex and unusual geometries that are hard or impossible to do with traditional lithography techniques, including gradient patterns or fractal designs, may be fabricated [35].

5.8.2 Limitations in Microfluidic Applications

EBL has several drawbacks that prevent it from being widely used in microfluidics, despite its benefits. The major disadvantage of electron beam writing is its poor throughput, which results from its serial nature. This makes it impractical for mass manufacturing and far slower than photolithography or nanoimprinting. The technique also involves high operational costs, requiring expensive equipment, high-vacuum environments, and specialized expertise, which can be prohibitive for many research labs. The EBL system's stage limits the substrate size, which makes it difficult to produce large-scale microfluidic devices and usually limits creation to tiny regions (a few cm^2) [47].

Fine features may be distorted by proximity effects brought on by electron scattering, deviating from the original design. Additionally, some microfluidic applications may require extra post-processing procedures due to the lack of chemical or mechanical resilience of the resist materials frequently employed in EBL, such as PMMA. Lastly, in order to produce functioning microfluidic devices, EBL frequently calls for additional methods like etching or bonding, which lengthens and complicates the fabrication process overall. Despite EBL's superior accuracy and adaptability, these drawbacks make it more appropriate for specialist research and prototyping than large-scale production [48].

Materials Research Foundations 191 (2026) 96-127 https://doi.org/10.21741/9781644904213-4

9. Materials and Resists for EBL-Based Microfluidics

9.1 Positive versus negative resists (PMMA, HSQ, and ZEP-520A)

The lithography process with electron beam contains two kinds of resists, known as positive and negative. Positive ones are easy to remove and provide high resolution, ideal for metal deposition, lift-off masks, and wet etching. A popular positive resist, PMMA, has an excellent roughness-to-sensitivity ratio. Varying molecular weights of PMMA can be used in a number of applications of fabrication in EBLs [49]. A negative tone resist with extremely excellent etch resistance, resolution, and stability in microscope examination of electron is hydrogen silsesquioxane HSQ [50].

However, negative tone resist prevents cross-linking that is chemically inactive, and in substrate dry etching, creates a mask of excellent quality [51]. Furthermore, processed negative resists turn into a long-lasting structure, for the Nano print lithography techniques, which makes the direct fabrication more suitable [52]. The PMMA, which is used instead of utilizing PMMA, which has a similar or greater contrast, is now obtained using HSQ resist [53]. The reason for this is that exposed HSQ essentially turns into silicon dioxide, which is harder than the resist made by an organic polymer.

The resist adhesion, either the resist ZEP-520 having methyl ethyl ketone and polystyrene resist for electron beam lithography developer, was found to be significantly improved by a thin layer of polymer on the substrate [54,55]. In metallic or ceramic substrates, the electron beam lithography using HSQ, the resist adhesion can be improved using trimethylsilane by modifying its surface [56]. The establishment of the SML acceptability can be obtained for electron beam lithography resist by comparing the SML and PMMA [57]. ZEP's high sensitivity may be attributed to the existence of a CL inside a compound, while SML and PMMA do not include this group, which is a reason why the two substances have a lower sensitivity compared to ZEP in electron exposure [58].

9.2 Substrate Considerations (Si, Glass, Quartz, Polymers)

Substrate consideration is an important factor for electron beam lithography for microfluidics. Due to numerous possible uses, including NEMS/MEMS sensors and biomedical engineering, the creation of micro and nanostructures on quartz substrate has generated interest and attention from researchers in recent years [59]. The creation of nanostructures on quartz will create several new opportunities due to its exceptional UV transmission, high temperature tolerance, and hardness. The X-ray diffractive optical elements are the application of this structure, microfluidics, and light trapping structures [60].

Particularly in three-dimensional batteries, biomedical engineering, and the integration of integrated circuits with 3D integrated circuits, numerous layers are involved. The fabrication procedure is challenging for insulating materials with structures. For quartz etching, processes like cryogenic and pseudo Bosch, which are responsible for etching silicon, are not so good [61]. On OTI glass, the critical dose was of the order of magnitude less than ZEP and high-resolution resists like PMMA and HSQ [62]. But the CD is bigger than UV1116.To show dense and complex geometry, sub-wavelength metallic features could be patterned on the OTI substrate of glass over 0,25cm area with quick write times when combined with high-energy EBL.

Materials Research Foundations 191 (2026) 96-127 https://doi.org/10.21741/9781644904213-4

This work presents an appealing path toward the use of high energy in the large area fabrication of nanostructures and transmissive optical elements [63,64]. As of right now, a lot of fabrication processes, including lithography, etching, and transfer, are needed for their incorporation into electronic circuits [65]. A lot of work has gone into creating cleaner alternative fabrication techniques, as an alternative to optical lithography and electron beam [66,67]. Directly patterning the substrate has been one strategy to accomplish selective growth in the targeted regions [68, 69]. Photolithography, O2, plasma, and a particular chemical surface treatment were used in the instance of MoS2 on SIO/Si before chemical vapor deposition CVD growth [70]. The end product of this process can be a single-layer MoS2 with the required shapes.

9.3 Multi-Layer Resist Systems for Complex Structures

The main benefit of the EBL is that it has higher resolution than the optical technique [71,72]. The EBL can produce a new three-dimensional pattern in many resist layers in one exposure. The application of Electron beam lithography in 3D patterning of several resist layers is restricted to polymer resist with varying permeability, and with the help of the soft polymer resist material, electron energy deposition is nearly the same [73-76]. In EBL, the electron energy deposition profile across these multilayers would change, so we can add a metal layer to the resist. The rate of elastic scattering and sine both increase with the target density. The metallic layer in a multilayer resist with an intermediate metal layer will experience a greater loss of electron energy [77].

The multilayer resist can have almost the same thickness and resolution as a one-layer resist. The glass substrate with a photo-lithographically printed OTI is the primary step in the process of device fabrication. The substrate can be covered with a vertical PMMA/AI/ PMMA stack film of three layers to create the vertical transistor template [78-80]. For the multilayer, the projected line width resolution and electron energy deposition profile of a 30 keV electron beam Lithography.

10. Electron Beam Lithography Process for Microfluidic Device Fabrication

10.1 Pre-Processing: Substrate Preparation and Cleaning

Successful electron beam lithography requires a well-prepared substrate. By cleaning the substrate, impurities that can impede the process are eliminated. Chemical baths and plasma cleaning are common cleaning techniques used to get rid of organic particles and residues. For more complex uses, washing the substrate and then applying a silicon dioxide barrier layer can improve the substrate's capability with the resist.

A mild bake at 130 to 170 degrees Celsius eliminates solvents and enhances substrate resistance to adhesion. When preparing the substrate, it is essential to keep the area dust-free. The tiny parents via made by electron beam lithography may have flaws due to dust particles. HEPA filters should be used in a cleanroom for operators to reduce airborne particulates. Wearing clean room clothing and using anti-static equipment significantly lowers the chance of contamination [81].

10.2 Eelectron Beam Lithography Patterning Steps: Exposure, Development, and Etching

Exposure: A unique electron gun, the centerpiece of EBL, is at the center of EBL. With the use of lenses and aperture, this cannon creates an electron beam that is then concentrated into a very small area. Next, a computer-controlled pattern is used to scan the electron beam across the resist

Materials Research Foundations 191 (2026) 96-127 https://doi.org/10.21741/9781644904213-4

control substrate. Usually, specialized software is used in advance to construct this pattern. The resist's chemical Characteristics in the exposed regions are changed because of the way the resist and beam interact during the scanning [82].

Development: The pattern created by EBL is revealed by the development process. The developer solution is used to dip the substrate during an exposure. Depending upon which kind of resist is utilized, this solution dissolves the part of the resist that is exposed or not. The exposed areas dissolve and are eliminated in the case of positive resist, such as PMMA. Conversely, unexposed portions dissolve when exposed to the negative resist, like HSQ. Common developers are TMAH-based solutions for negative resist and MIBK: IP mixes for PMMA.

Etching: it is the last stage in the electron beam lithography, which involves transferring the created resist onto the substrate. We have two primary ways used:

Dry Etching: Methods such as plasma and reactive ion etching employ ionized gases to remove material from the substrate's exposed area alone.

Wet Etching: material in the exposed substrate regions is dissolved using chemical solutions. After etching is finished, the final pattern structure remains on the substrate, and the residual resist is eliminated with the proper solvents. Now the structure is produced for a number of uses in photonics and microelectronics [83,84].

10.3 Critical Parameters: Dose Control, Proximity Effect Correction, and Line Edge Roughness

Dose Control: A greater dosage results in more electrons in each shot, which means that the exposure is averaged across a large number of electrons. A major practical and basic issue in electron beam lithography is dosage fluctuation from shot to shot or between pixels. Optical system drift, charge-related dose fluctuations, and electron source instability. For increasing the quality pattern and flow rate beyond the shot noise limit, real-time feedback dose control of electrons would be a significant advancement [85] in e-beam-based pattern technology [86,87].

Proximity effect Correction: In EBL, the proximity effect is a phenomenon whereby the resist and the substrate are affected by the primary electron beam, causing the exposure dosage distribution [88]. Thus, the produced pattern is broader than the scanned pattern. The result is a non-zero dosage applied to the resist outside the scanned pattern; both overexposure and underexposure arise from the proximity effect [89].

Line Edge Roughness: The resist growth process, distribution of spices of the chemicals in the resist, including the photo acid generators PAG, and other stochastically changing phenomena are the causes of LER [90]. The increase in fluctuation can cause an increase in LER and reduced contrast of exposure. Therefore, in order to keep improving the clarity of features and density of patterns the EBL could attain, LER must be kept to a minimum. A lot of previous attempts to lower the LER used an empirical or trial-and-error methodology [85,91].

10.4 Post Processing: Hard Masking, Lift-off Techniques, surface treatments

Hard Masking: Currently, the most advanced method for creating masks is resist patterning with electron beam lithography. Being metal oxides, the resists are attractive options for both EUB and electron beam lithography because of their great etch resistance, which is comparable to hard mask materials [92]. Thus, two or three organic layer systems with lift-off or an inorganic hard mask

Materials Research Foundations 191 (2026) 96-127 https://doi.org/10.21741/9781644904213-4

interlayer for hard mask deposition following organic resist patterning can be investigated. All of these strategies have a drawback of adding complexity to the process, despite the inorganic hard masks' advantageous introduction [93].

Lift-off Techniques: A lift-off technique involves applying, exposing, and developing a resist layer upon a substrate for creating a mask. Then the required substance is deposited, and the undesirable substance is removed by removing that mask, leaving the substrate with a desired structure. Lift off is preferred in situations when the material to be etched does not adhere to the photoresist mask correctly [94]. An appropriate electron beam resist, like PMMA, can be exposed to an electron beam to define high resolution. Additionally, two PMMA layers of varying molecular weight upon MMA, beneath PMMA, a metallic layer, or two layers of metal can be used to create a reentrant profile for lift-off technology [95].

Surface Treatments: It is possible to create homogeneous layers of electron-sensitive materials on uneven surfaces by spraying and evaporating polymeric and non-polymeric materials; moreover, self-assembled monolayers or polymer brushes may be formed on these surfaces with the help of chemical bonding. Furthermore, by using a concentrated e-beam to expose layers of anisole on organic ice, 3D nanostructures may be created on uneven surfaces [96].

11. Integration of EBL with Microfluidic Fabrication Methods

11.1 Hybrid Lithography: Combining EBL with UV Lithography and Soft Lithography

In semiconductor manufacturing, ultraviolet lithography (UVL) has been a cornerstone technology for decades. It enables compact and dense chip features, thereby enhancing performance and functionality. UVL also remains valuable for high-volume production due to its cost-effective equipment, scalability, and mature infrastructure. However, with continued advances in the industry, there is a growing shift toward extreme ultraviolet (EUV) lithography, which employs shorter wavelengths to achieve finer features and higher levels of integration.

In Soft UV-NIL, a compliant PDMS stamp is employed, which alleviates constraints tied to substrate nonuniformity and bow. The method is well-suited to producing gold nanostructures for biomolecular detection. When specific geometries or high pattern densities risk elastomeric mold collapse or wettability complications, Hard UV-NIL can be substituted [2].

11.2 Pattern Transfer Techniques (Reactive Ion Etching, Wet Etching)

Plasma based dry etching processes offer higher accuracy in replicating device patterns than wet etching processes, but further advances may depend on a better understanding of gas phase phenomena and plasma surface interactions [3]. Anisotropic dry etching by a number of different techniques is widely employed in III–V compound semiconductor technology for pattern transfer, device isolation, mesa formation, grating fabrication, and via hole etching [4].

11.3 Integration with MEMS/NEMS Devices

Silicon photonic devices represent a well-established class of on-chip photonic technologies, fabricated from materials such as single-crystal silicon, polycrystalline silicon, silicon oxide, and silicon nitride. These platforms have already demonstrated strong capability in modulating and confining infrared (IR) light. Owing to the widespread use of silicon and silica in MEMS/NEMS fabrication, photonic devices can be monolithically integrated with MEMS/NEMS actuators, allowing seamless combination of their manufacturing processes. This integration enables a wide

Materials Research Foundations 191 (2026) 96-127 https://doi.org/10.21741/9781644904213-4

range of applications, including tunable optical modulators, couplers, filters, routers, and switches, through mechanical displacement or deformation of photonic structures [5].

12. Applications OF Electron Beam Lithography in Microfluidic Devices

12.1 Single Cell Analysis Platforms

Quantitative phenotypes are best captured with single-cell methods, which have become foundational to contemporary microbiology. By probing living bacteria one cell at a time, these strategies resolve genotype–phenotype coupling that bulk assays often blur. With fluorescent probes, microfluidic platforms, and time-lapse microscopy, researchers can run in vivo measurements over hundreds of cells and thousands of divisions, delivering sharp insight into mechanisms that drive antimicrobial resistance [97].

12.2 Lab-on-a-Chip (LOC) Systems

A new age of possibilities for micro-scale sample manipulation and detection has been brought about by the combination of optical circuits with microfluidic Loc devices. On-chip optical components reduce dependence on costly laboratory photonic equipment that has hampered microfluidic progress while increasing control and analytical capabilities. Notably, as inexpensive, portable, and extremely sensitive systems, in-situ LoC devices for biochemical applications like environmental monitoring and diagnostics could be very beneficial. However, a number of difficulties still exist because of how difficult it is to integrate photonics with microfabricated devices [98].

12.3 Nano-pore and Bio-sensing Application

A transducer and a bio-recognizer work together to quantitatively and selectively detect an analyte in a biosensor. A biological event is brought on by the bio-recognizers' particular binding or reaction with an analyte. A transducer then transforms the biological event into a diagnostic signal, like an optical signal [99].

12.4 Drug Delivery and Organ-on-a-Chip Technologies

The field has proposed multi-organ microfluidic devices, known as body- or human-on-a-chip, that replicate interactions among different tissues. Such devices allow continuous observation of linked pharmacokinetics (ADME: absorption, distribution, metabolism, excretion) for various administration paths, with resulting datasets supporting model-based predictions of drug efficacy [100].

13. Challenges and Future Perspectives

13.1 Scalability and Throughput Issues in Electron Beam Lithography

Alternative lithographic techniques must be developed since the boundaries of classic optical lithography have been stretched by the need for ever-smaller feature sizes in Integrated Circuits (ICs). With its high-resolution patterning capabilities, Electron Beam Lithography (EBL) is widely acknowledged as a viable next-generation lithography solution. Scalability and throughput are still

Materials Research Foundations 191 (2026) 96-127
https://doi.org/10.21741/9781644904213-4

major issues that need to be resolved, though, if EBL is to be a practical high-volume manufacturing (HVM) option [101].

13.2 Challenges in Scalability and Throughput

The expensive equipment costs of alternatives draw attention to the need for scalable solutions, which in turn drives research to increase efficiency without sacrificing resolution. Beam blurring causes slow single-angle EBL. While corrections add optical complexity, higher beam currents increase throughput but lower resolution. It's still difficult to balance resolution, current, and throughput [101].

5.13.3 MEMS-Based Multi-Beam Electron Beam Lithography

MEMS-based emitter arrays are used in Multi-Beam Electron Beam Lithography (MEBL) to increase throughput while preserving the resolution and scalability of EBL. With near-term improvements, a multi-column system may produce 10–60 wafers per hour at 50 nm precision. Scalable, high-throughput, maskless lithography with accurate emitter control is made possible by MEMS integration in EBL. MEMS-based EBL is perfect for next-generation lithography because precision motion stages further increase viability [101].

5.13.4 Character Projection for Enhanced Throughput

Character projection in EBL exposes preset patterns on a stencil, increasing throughput. Efficiency is increased by optimizing character distribution and selection; a multi-column system can achieve a speedup of up to 1782×. By reducing exposure events, character projection speeds up EBL and is perfect for repeating patterns in logic circuits and memory. Its combination with multi-beam EBL improves resolution and speed [102]. 1D gridded layouts are used in the semiconductor industry for smaller chips, increased efficiency, and better scaling. While line cutting requires high-resolution lithography, line creation employs a variety of lithographic processes. With sub-10 nm widths, EBL is the most economical option for high-resolution line cutting. It makes scalable, high-volume production possible when paired with multibeam technology and complementary EBL (CEBL) [101].

13.5 Future Perspectives

Multi-beam EBL, character projection, and optical lithography integration are being investigated by researchers to improve EBL scalability and throughput. Sub-50 nm lithography can produce 60 wafers per hour, while boutique manufacturing can produce 1–10 wafers per hour due to developments in MEMS and smart materials [101].

13.5.1 Cost-Effective Alternatives and Emerging Technologies

High-resolution, reasonably priced lithography is becoming more and more necessary as semiconductor technology develops. In order to supplement or replace EBL in high-volume manufacturing, this chapter examines substitutes and new technologies [102].

13.5.2 Nanoimprint Lithography (NIL)

By mechanically deforming resist rather than employing photons or electrons, Nanoimprint Lithography (NIL) makes it possible to perform high-throughput, inexpensive nanoscale

Materials Research Foundations 191 (2026) 96-127 https://doi.org/10.21741/9781644904213-4

patterning. It provides excellent precision for applications in biotechnology, photonics, semiconductors, and nanofluidics [103]. NIL expands its applications by enabling high-resolution patterning in polymers. Roll-to-roll imprinting and material advancements increase throughput, making NIL a potent substitute for EBL in a variety of applications [102].

13.5.3 Extreme Ultraviolet Lithography (EUVL)

Using plasma sources, reflecting masks, and reduction imaging, Extreme Ultraviolet Lithography (EUVL) at 13 nm wavelength gets the better of DUV's limitations for sub-100 nm features in modern semiconductor fabrication [104]. Despite its advancements since the 1980s, EUVL still has issues with defect-free masks, power output, and resist performance. Improvements in resist and tin-based plasma sources have made it possible to achieve 2.7 nm line edge roughness and 25 nm resolution. Although improvements in optical lithography are necessary for EUVL adoption, mask fabrication and high prices continue to be obstacles. With room for expansion, it is scheduled for full-scale production at 22 nm and pilot production at 32 nm.

13.5.4 Evolution of Lithographic Techniques and Future Challenges

In order to accommodate decreasing IC feature sizes, lithography has changed from visible to deep UV sources. Although nodes down to 22 nm are possible with 193 nm immersion lithography, optical restrictions become noticeable below 20 nm [103].There are issues with throughput, cost, and imaging with next-generation lithography (NGL) techniques such as EUVL, X-ray, and charged particle lithography. EUVL needs improvements for high-volume manufacturing since it has issues with source power, mask defects, and resist performance. Since 1988, EUVL has been developed by the EUCLIDES consortium, which is led by ASML and others. Adoption depends on ongoing research and development to advance optical lithography. Since 1988, EUVL has been developed by the EUCLIDES consortium, which is led by ASML and others. Adoption depends on ongoing research and development to advance optical lithography [103].

13.5.5 Directed Self-Assembly (DSA) as an Alternative to Electron Beam Lithography

Unlike traditional EBL, Directed Self-Assembly (DSA) offers better feature control by using block copolymers to self-organize into nanoscale structures guided by chemical patterns, improving resolution and throughput.

14. Future Trends: AI-Assisted Lithography, Ultrafast Beam Control, and Quantum Lithography

Sub-10 nm resolution is made possible by EBL, although HVM has issues with scalability and throughput. DSA, AI-assisted lithography, ultrafast beam control, and quantum lithography are examples of innovations that seek to increase efficiency without sacrificing accuracy.

14.1 AI-Assisted Lithography for Improved Throughput

Deep learning improves simulations, while AI and ML speed up lithographic optimization. Methods such as Litho-Aware Data Augmentation (LADA) create artificial masks to increase precision. AI enhances process control and mask design, increasing overlay correction by 35% and defect identification by 50%. Large-scale fabrication calculation times are reduced from days to hours thanks to GPU-based AI, which speeds up lithography modeling [105].

Materials Research Foundations 191 (2026) 96-127 https://doi.org/10.21741/9781644904213-4

14.2 Ultrafast Beam Control for Enhanced EBL Throughput

Throughput is limited by EBL's serial writing; however, switching can occur in 1–1.5 ps with less energy consumption thanks to ultrafast beam control using perovskite-based vortex micro lasers [106]. Phased array antennas (PAAs) from optical communications could improve EBL by increasing wafer processing speed, beam steering, and precision [107,108]. By enabling ultra-fast beam management, microwave photonics improves EBL scalability, cuts down on processing time, and boosts competitiveness [109].

14.3 Quantum Lithography for Sub-Rayleigh Patterning

To achieve ultra-fine interference fringes, quantum lithography overcomes the classical diffraction limit through the use of entanglement and two-photon diffraction. [110]Since patterning resolution can be doubled using quantum lithography beyond classical bounds, practical application is still difficult [111]. The development of specialized light sources and materials is essential to quantum lithography [112]. As new materials and light sources are developed, quantum lithography offers sub-5 nm patterning, which could surpass EBL [113-115].

15. Future Directions in Microfluidic Research

15.1 Impact of Electron Beam Lithography on Microfluidic Device Innovation

EBL makes high-resolution microfluidic fabrication possible, which is more precise than conventional techniques and is necessary for Loc devices, diagnostics, and cooling systems [116].

15.2 Fabrication Methods for Micro and Nanochannels

15.2.1 Photolithography

For microfluidic fabrication, photolithography is one of the most popular techniques, especially in the semiconductor sector. A light-sensitive resist layer that has been previously applied to a substrate is covered with a design that is transferred from a lithographic mask. Although photolithography is a useful technique for producing metal nanoparticles and microfluidic devices, its scalability and resolution are limited [117].

15.2.2 Soft Lithography

Another popular method for creating microfluidics is soft lithography. As a stamp mold, it uses polydimethylsiloxane (PDMS) to form accurate micro- and nanochannels. In biological applications, including stem cell development and cell capture, this technique is frequently employed. Soft lithography, on the other hand, has a poor throughput and can be expensive and time-consuming. Reconfigurable molds are a recent breakthrough that provides more design flexibility for microfluidic systems [118-121]. Focused Ion Beam (FIB) Milling

FIB milling is a highly specialized method that makes it possible to precisely fabricate structures at the micro- and nanoscales. It works by hitting a target substance with a concentrated beam of ions, usually gallium, enabling accurate material removal. For the creation of nanopores and nanochannels, which are crucial for molecular analysis, biological sensing, and LOC devices, this technique is especially helpful [122-123].

Materials Research Foundations 191 (2026) 96-127 https://doi.org/10.21741/9781644904213-4

15.2.3 Etching Techniques (Wet and Dry Etching)

Etching is essential to microfluidic fabrication because it makes it possible to precisely remove material to design structures and channels. Although wet etching, which uses chemical solutions, is less expensive, its isotropic nature causes undercutting. For high aspect-ratio structures, dry etching—including deep reactive ion etching, or DRIE—offers greater control over feature size and is frequently employed. However, dry etching's cost and intricacy prevent it from being widely used [60,124].

15.2.4 3D Printing for Microfluidics

By making it possible to produce complex microfluidic devices quickly and affordably, 3D printing has completely changed the field of microfluidic fabrication. High-precision creation of intricate micro- and nanochannels is made possible by technologies including inkjet printing, digital light processing, fused deposition modelling, and stereolithography. It is anticipated that 3D printing will become increasingly important in next-generation microfluidic devices as the technology develops [125-128].

5.15.2.5 EBL's Role in Microfluidic Device Innovation

The capacity of EBL to reach sub-10 nm resolution has made it a revolutionary tool in microfluidic creation. Masks are not needed for EBL, in contrast to standard lithographic techniques, which increases design flexibility and accuracy. For the fabrication of nanoscale features in microfluidic devices, such as high-resolution biosensors, customized Loc platforms, and nanopores for DNA sequencing, this benefit makes it very helpful [116]. One of EBL's main advantages is its capacity to produce complex and highly customizable microfluidic structures, which are crucial for biomedical research and diagnostics. The method has been widely utilized to create microfluidic chips for point-of-care testing (POCT), which allows for quick monitoring and illness identification. The accuracy and effectiveness of diagnostic applications can be improved by EBL-fabricated devices that combine microfluidics with lateral flow tests and nucleic acid amplification [129].

5.16 Future Directions in Microfluidic Research

5.16.1 Liquid Biopsies and Cancer Diagnostics

Applying microfluidics to liquid biopsy applications, which provide a non-invasive way to find cancer biomarkers, has drawn a lot of interest. By detecting circulating tumour cells (CTCs), exosomes, and other biomarkers in real time, liquid biopsies offer important information about how a disease develops and how well a treatment works. Liquid biopsy assays are a potent tool for early cancer diagnosis and tailored treatment because EBL-fabricated microfluidic devices increase their sensitivity and precision [130-134].

5.16.2 Circulating Tumour Cells (CTCs) and Exosomes

CTCs are important in the spread of cancer, and their identification offers useful data for planning treatment and staging the disease. By effectively capturing and analyzing CTCs from patient blood samples, microfluidic devices designed with EBL allow for real-time tumor progression monitoring [135-138]. Similarly, EBL-fabricated nonporous membranes can be used to isolate

Materials Research Foundations 191 (2026) 96-127 https://doi.org/10.21741/9781644904213-4

exosomes, which are extracellular vesicles implicated in the development of cancer. These developments open the door to better methods for diagnosing and treating cancer [140-142].

Although blood-based diagnostics are still the gold standard for illness monitoring, other body fluids, like urine, provide a practical and non-invasive substitute. A number of indicators from the systemic circulation are found in urine, which makes it a useful tool for cancer screening. By offering high-precision biomarker analysis and allowing for time-dependent illness progression tracking, EBL-enabled microfluidic devices have the potential to improve urine-based diagnostics [143-144].

Conclusion

The use of EBL in high-volume manufacturing is constrained by scalability and throughput. Promising innovations include character projection, MEBL, and 1D layouts. As semiconductor technology develops, NIL and EUVL provide high-resolution, reasonably priced substitutes for EBL. DSA enhances lithography by using block copolymers to self-organize into nanoscale structures, improving resolution and line edge roughness over conventional EBL. Future semiconductor production will be shaped by the development of other technologies as DSA and AI advance. For biosensing and diagnostics, EBL enables customizable microfluidic devices with high resolution. With AI and 3D printing, its integration is set to increase biomedical innovation. Electron Beam Lithography remains an indispensable tool in the field of nanotechnology, enabling the fabrication of complex structures with unprecedented precision. While EBL provides superior resolution and flexibility, challenges such as slow processing speeds and high costs limit its widespread adoption for mass production. Advances in multi-beam systems, resist development, and integration with other lithographic techniques hold promise for improving its efficiency. As semiconductor technology moves towards sub-10 nm regimes, EBL is expected to play a crucial role in pushing the boundaries of nanoscale fabrication.

References

[1] S.K. Ghandhi, VLSI Fabrication Principles: Silicon and Gallium Arsenide, John Wiley & Sons, 1994, pp.161-432.

[2] J.N. Helbert, Handbook of VLSI Microlithography, William Andrew. Norwich, NY, 2001, 1-71.

[3] K.C.A. Smith, C.W. Oatley, The scanning electron microscope and its fields of application, Br. J. Appl. Phys. 6 (1955) 391. https://doi.org/10.1088/0508-3443/6/11/304

[4] I. Brodie, J.J. Muray, The Physics of Micro/nano-Fabrication, Springer Sci. Bus. Media. NY, 2013, pp. 1-49.

[5] J. El-Ali, P.K. Sorger, K.F. Jensen, Cells on chips, Nature 442 (2006) 403-411. https://doi.org/10.1038/nature05063

[6] C. Y. Jiang, L. Dong, J.-K. Zhao, X. Hu, C. Shen, Y. Qiao, X. Zhang, Y. Wang, R.F. Ismagilov, S.-J. Liu, High-throughput single-cell cultivation on microfluidic streak plates, Appl. Environ. Microbiol. 82 (2016) 2210-2218. https://doi.org/10.1128/AEM.03588-15

[7] K. Giri, C.-W. Tsao, Recent advances in thermoplastic microfluidic bonding, Micromachines. 13 (2022) 486-491. https://doi.org/10.3390/mi13030486

[8] S. Kumar, P. Bhushan, A.K. Agarwal, S. Bhattacharya, A historical perspective on paper microfluidic based point-of-care diagnostics, Paper Microfluidics: Theory Appl. 1 (2019) 1-5. https://doi.org/10.1007/978-981-15-0489-1_1

[9] C.B. Goy, R.E. Chaile, R.E. Madrid, Microfluidics and hydrogel: A powerful combination, React. Funct. Polym. 145 (2019) 104314-104317. https://doi.org/10.1016/j.reactfunctpolym.2019.104314

[10] W. Han, X. Chen, A review on microdroplet generation in microfluidics, J. Braz. Soc. Mech. Sci. Eng. 43 (2021) 247-250. https://doi.org/10.1007/s40430-021-02971-0

[11] Z.Z. Chong, S.H. Tan, A.M. Gañán-Calvo, S.B. Tor, N.H. Loh, N.-T. Nguyen, Active droplet generation in microfluidics, Lab on a Chip. 16 (2016) 35-58. https://doi.org/10.1039/C5LC01012H

[12] A.P. Iakovlev, A.S. Erofeev, P.V. Gorelkin, Novel pumping methods for microfluidic devices: a comprehensive review, Biosensors. 12 (2022) 956-965. https://doi.org/10.3390/bios12110956

[13] H.A. Stone, Introduction to fluid dynamics for microfluidic flows, CMOS Biotechnol. 1 (2007) 5-30. https://doi.org/10.1007/978-0-387-68913-5_2

[14] S.O. Catarino, R.O. Rodrigues, D. Pinho, J.M. Miranda, G. Minas, R. Lima, Blood cells separation and sorting techniques of passive microfluidic devices: From fabrication to applications, Micromachines. 10 (2019) 593-610. https://doi.org/10.3390/mi10090593

[15] B. Yang, J. Zhou, Q. Chen, L. Lei, K. Wen, Fabrication of hexagonal compound eye microlens array using DMD-based lithography with dose modulation, Opt. Express. 26 (2018) 28927-28937. https://doi.org/10.1364/OE.26.028927

[16] G.L.T. Chiu, J.M. Shaw, Optical lithography: Introduction, IBM J. Res. Dev. 41 (1997) 3-6. https://doi.org/10.1147/rd.411.0003

[17] B. Michel, A. Bernard, A. Bietsch, E. Delamarche, M. Geissler, D. Juncker, H. Kind, J.-P. Renault, H. Rothuizen, H. Schmid, Printing meets lithography: Soft approaches to high-resolution patterning, IBM J. Res. Dev. 45 (2001) 697-719. https://doi.org/10.1147/rd.455.0697

[18] S.D. Berger, J.M. Gibson, New approach to projection-electron lithography with demonstrated 0.1 μm linewidth, Appl. Phys. Lett. 57 (1990) 153-155. https://doi.org/10.1063/1.103969

[19] A.K. Miri, D. Nieto, L. Iglesias, H. Goodarzi Hosseinabadi, S. Maharjan, G.U. Ruiz-Esparza, P. Khoshakhlagh, A. Manbachi, M.R. Dokmeci, S. Chen, Microfluidics-enabled multimaterial maskless stereolithographic bioprinting, Adv. Mater. 30 (2018) 1800242-1800249. https://doi.org/10.1002/adma.201800242

[20] S.N. Bhatia, D.E. Ingber, Microfluidic organs-on-chips, Nat. Biotechnol. 32 (2014) 760-772. https://doi.org/10.1038/nbt.2989

[21] B. Zhang, A. Korolj, B.F.L. Lai, M. Radisic, Advances in organ-on-a-chip engineering, Nat. Rev. Mater. 3 (2018) 257-278. https://doi.org/10.1038/s41578-018-0034-7

[22] H. Tang, Y. Abouleila, L. Si, A.M. Ortega-Prieto, C.L. Mummery, D.E. Ingber, A. Mashaghi, Human organs-on-chips for virology, Trends Microbiol. 28 (2020) 934-946. https://doi.org/10.1016/j.tim.2020.06.005

[23] B. Bilenberg, S. Jacobsen, M.S. Schmidt, L.H.D. Skjolding, P. Shi, P. Bøggild, J.O. Tegenfeldt, A. Kristensen, High resolution 100 kV electron beam lithography in SU-8, Microelectron. Eng. 83 (2006) 1609-1612. https://doi.org/10.1016/j.mee.2006.01.142

[24] D.A. Czaplewski, L.E. Ocola, Measurement of backscattered 100 keV electrons on a solid substrate, Appl. Phys. Lett. 99 (2011) 1-10. https://doi.org/10.1063/1.3660712

[25] H. Lorenz, M. Despont, N. Fahrni, N. LaBianca, P. Renaud, P. Vettiger, SU-8: A low-cost negative resist for MEMS, J. Micromech. Microeng. 7 (1997) 121-125. https://doi.org/10.1088/0960-1317/7/3/010

[26] D.M. Koller, N. Galler, H. Ditlbacher, A. Hohenau, A. Leitner, F.R. Aussenegg, J.R. Kren, Direct fabrication of micro/nano fluidic channels by electron beam lithography, Microelectron. Eng. 86 (2009) 1314-1316. https://doi.org/10.1016/j.mee.2008.11.025

[27] C. Mack, Fundamental principles of optical lithography: The science of microfabrication, John Wiley & Sons. England, 2008, pp. 1-37. https://doi.org/10.1002/9780470723876

[28] B.W. Smith, Optics for photolithography, Marcel Dekker. NY, 1998, pp. 1-137.

[29] Q. Xu, R.M. Rioux, M.D. Dickey, G.M. Whitesides, Nanoskiving: A new method to produce arrays of nanostructures, Acc. Chem. Res. 41 (2008) 1566-1577. https://doi.org/10.1021/ar700194y

[30] T.W. Odom, J.C. Love, D.B. Wolfe, K.E. Paul, G.M. Whitesides, Improved pattern transfer in soft lithography using composite stamps, Langmuir. 18 (2002) 5314-5320. https://doi.org/10.1021/la020169l

[31] Y. Xia, J.J. McClelland, R. Gupta, D. Qin, X.-M. Zhao, L.L. Sohn, R.J. Celotta, G.M. Whitesides, Replica molding using polymeric materials: A practical step toward nanomanufacturing, Adv. Mater. 9 (1997) 147-149. https://doi.org/10.1002/adma.19970090211

[32] M.D. Austin, H. Ge, W. Wu, M. Li, Z. Yu, D. Wasserman, S.A. Lyon, S.Y. Chou, Fabrication of 5 nm linewidth and 14 nm pitch features by nanoimprint lithography, Appl. Phys. Lett. 84 (2004) 5299-5301. https://doi.org/10.1063/1.1766071

[33] A. Kumar, G.M. Whitesides, Magnetization process observation of a CoNiCr film in the field of a weakly excited objective lens in transmission electron microscope, Appl. Phys. Lett. 63 (2002) 1993-1996. https://doi.org/10.1063/1.110625

[34] A. Kumar, G.M. Whitesides, Features of gold having micrometer to centimeter dimensions can be formed through a combination of stamping with an elastomeric stamp and an alkanethiol "ink" followed by chemical etching, Appl. Phys. Lett. 63 (1993) 2002-2004. https://doi.org/10.1063/1.110628

[35] K. Stokes, K. Clark, D. Odetade, M. Hardy, P. Goldberg Oppenheimer, Advances in lithographic techniques for precision nanostructure fabrication in biomedical applications, Digit. Diagn. Nanomed. 18 (2023) 153-157. https://doi.org/10.1186/s11671-023-03938-x

[36] E. Ogunnowo, E.L.E.M. Ogu, P. Egbumokei, I. Dienagha, W. Digitemie, Conceptual model for topology optimization in mechanical engineering to enhance structural efficiency and material utilization, Iconic Res. Eng. J. 7 (2024) 2456-8880.

[37] R.J. Pease, Electron beam lithography, CRC Crit. Rev. Solid State Mater. Sci. 22 (1981) 265-290. https://doi.org/10.1080/00107518108231531

[38] S. So, J. Yun, B. Ko, D. Lee, M. Kim, J. Noh, J. Rho, Radiative cooling for energy sustainability: From fundamentals to fabrication methods toward commercialization, Adv. Sci. 11 (2024) 2305067-2305068. https://doi.org/10.1002/advs.202305067

[39] P. Hawkes, E. Kasper, Principles of electron optics, Academic Press, NY, 1989, pp. 1211-1220.

[40] H. Boersch, Experimentelle Bestimmung der Energieverteilung in thermisch ausgelösten Elektronenstrahlen, Z. Phys. 139 (1954) 115-146. https://doi.org/10.1007/BF01375256

[41] S. Rishton, H. Schmid, D. Kern, H. Luhn, T. Chang, G. Sai-Halasz, M. Wordeman, E. Ganin, M. Polcari, Lithography for ultrashort channel silicon field effect transistor circuits, J. Vac. Sci. Technol. B Microelectron. Nanometer Struct. Process. Meas. Phenom. 6 (1988) 140-145. https://doi.org/10.1116/1.584032

[42] M.A. Mohammad, M. Muhammad, S.K. Dew, M. Stepanova, Fundamentals of electron beam exposure and development, Nanofabrication: Tech. Principles. 1 (2011) 11-41. https://doi.org/10.1007/978-3-7091-0424-8_2

[43] G. Rius, Electron beam lithography for nanofabrication, Univ. Autònoma Barcelona. 1 (2008) 17-39.

[44] C. Vieu, F. Carcenac, A. Pepin, Y. Chen, M. Mejias, A. Lebib, L. Manin-Ferlazzo, L. Couraud, H. Launois, Electron beam lithography: Resolution limits and applications, Appl. Surf. Sci. 164 (2000) 111-117. https://doi.org/10.1016/S0169-4332(00)00352-4

[45] A.N. Broers, A. Hoole, J.M. Ryan, Electron beam lithography-resolution limits, Microelectron. Eng. 32 (1996) 131-142. https://doi.org/10.1016/0167-9317(95)00368-1

[46] V.R. Manfrinato, J. Wen, L. Zhang, Y. Yang, R.G. Hobbs, B. Baker, D. Su, D. Zakharov, N.J. Zaluzec, D.J. Miller, Determining the resolution limits of electron-beam lithography: Direct measurement of the point-spread function, Nano Lett. 14 (2014) 4406-4412. https://doi.org/10.1021/nl5013773

[47] D. Xia, J. Yan, S. Hou, Fabrication of nanofluidic biochips with nanochannels for applications in DNA analysis, Small 8 (2012) 2787-2801. https://doi.org/10.1002/smll.201200240

[48] S. Kumar, S. Kumar, M.A. Ali, P. Anand, V.V. Agrawal, R. John, S. Maji, B.D. Malhotra, Microfluidic-integrated biosensors: Prospects for point-of-care diagnostics, Biotechnol. J. 8 (2013) 1267-1279. https://doi.org/10.1002/biot.201200386

[49] S. Yasin, D. Hasko, H. Ahmed, Comparison of MIBK/IPA and water/IPA as PMMA developers for electron beam nanolithography, Microelectron. Eng. 61 (2002) 745-753. https://doi.org/10.1016/S0167-9317(02)00468-9

[50] A.S. Gangnaik, Y.M. Georgiev, J.D. Holmes, New generation electron beam resists: A review, Chem. Mater. 29 (2017) 1898-1917. https://doi.org/10.1021/acs.chemmater.6b03483

[51] M. Blideran, M. Häffner, B.-E. Schuster, C. Raisch, H. Weigand, M. Fleischer, H. Peisert, T. Chassé, D. Kern, Improving etch selectivity and stability of novolak based negative resists by fluorine plasma treatment, Microelectron. Eng. 86 (2009) 769-772. https://doi.org/10.1016/j.mee.2008.12.057

[52] G. Das, C. Liberale, F. De Angelis, M.L. Coluccio, E. Di Fabrizio, P. Candeloro, Manipulation and nanostructuring for biological applications, Nanomedicine. CRC Press, NY, 2012, pp. 353-481.

[53] J.K. Yang, B. Cord, H. Duan, K.K. Berggren, J. Klingfus, S.-W. Nam, K.-B. Kim, M.J. Rooks, Understanding of hydrogen silsesquioxane electron resist for sub-5-nm-half-pitch lithography, J. Vac. Sci. Technol. B Microelectron. Nanometer Struct. Process. Meas. Phenom. 27 (2009) 2622-2627. https://doi.org/10.1116/1.3253652

[54] S. Ma, C. Con, M. Yavuz, B. Cui, Polystyrene negative resist for high-resolution electron beam lithography, Nanoscale Res. Lett. 6 (2011) 1-6. https://doi.org/10.1186/1556-276X-6-446

[55] B. Shokouhi, J. Zhang, B. Cui, Very high sensitivity ZEP resist using MEK:MIBK developer, Nano Lett. 6 (2011) 992-994. https://doi.org/10.1049/mnl.2011.0548

[56] Z. Zhang, H. Duan, Y. Wu, W. Zhou, C. Liu, Y. Tang, H. Li, Improving the adhesion of hydrogen silsesquioxane (HSQ) onto various substrates for electron-beam lithography by surface chemical modification, Microelectron. Eng. 128 (2014) 59-65. https://doi.org/10.1016/j.mee.2014.05.027

[57] A.S. Gangnaik, Electron beam lithography assisted high-resolution pattern generation, PhD Thesis, University College Cork, 2015, pp. 119-132.

[58] A. Gangnaik, Y.M. Georgiev, J.D. Holmes, Correlation of lithographic performance of the electron beam resists SML and ZEP with their chemical structure, J. Vac. Sci. Technol. B. 33 (2015) 041601-041602. https://doi.org/10.1116/1.4926387

[59] G. Das, C. Liberale, F. De Angelis, M.L. Coluccio, E. Di Fabrizio, P. Candeloro, The investigation of microstructures fabrication on quartz substrate employing electron beam lithography (EBL) and ICP-RIE process, Adv. Mater. Res. 980 (2014) 69-73. https://doi.org/10.4028/www.scientific.net/AMR.980.69

[60] M. Guo, X. Chou, J. Mu, B. Liu, J. Xiong, Fabrication of micro-trench structures with high aspect ratio based on DRIE process for MEMS device applications, Microsyst. Technol. 19 (2013) 1097-1103. https://doi.org/10.1007/s00542-012-1720-9

[61] K. Mohamed, M.M. Alkaisi, The fabrication of high aspect ratio, Updates Adv. Lithogr. 1 (2013) 211-215.

[62] G. Wiederrecht, Handbook of nanofabrication, Academic Press, Netherland, 2010, pp. 1-10.

[63] T. Takigawa, K. Kawabuchi, M. Yoshimi, Y. Kato, High voltage electron beam lithography, Microelectron. Eng. 1 (1983) 121-142. https://doi.org/10.1016/0167-9317(83)90025-4

[64] C.S. Wu, Y. Makiuchi, C. Chen, High-energy electron beam lithography for nanoscale fabrication, Lithogr. 13 (2010) 241-247.

[65] Y. Chen, X.L. Gong, J.G. Gai, Progress and challenges in transfer of large-area graphene films, Adv. Sci. 3 (2016) 1500343-1500350. https://doi.org/10.1002/advs.201500343

[66] L. Ju, B. Geng, J. Horng, C. Girit, M. Martin, Z. Hao, H.A. Bechtel, X. Liang, A. Zettl, Y.R. Shen, Graphene plasmonics for tunable terahertz metamaterials, Nat. Nanotechnol. 6 (2011) 630-634. https://doi.org/10.1038/nnano.2011.146

[67] Y. Chen, N. Zhang, Y.F. Li, Y.G. Bi, Y.Y. Yue, J. Feng, H.B. Sun, Microscale-patterned graphene electrodes for organic light-emitting devices by a simple patterning strategy, Adv. Opt. Mater. 6 (2018) 1701348-1701350. https://doi.org/10.1002/adom.201701348

[68] K. Kim, Y. Zhao, H. Jang, S. Lee, J. Kim, K.-S. Kim, B.H. Hong, Large-scale pattern growth of graphene films for stretchable transparent electrodes, Nature. 457 (2009) 706-710. https://doi.org/10.1038/nature07719

[69] Z. Liu, L. Ma, G. Shi, W. Zhou, Y. Gong, S. Lei, X. Yang, J. Zhang, J. Yu, K.P. Hackenberg, In-plane heterostructures of graphene and hexagonal boron nitride with controlled domain sizes, Nat. Nanotechnol. 8 (2013) 119-124. https://doi.org/10.1038/nnano.2012.256

[70] Y. Guo, P.-C. Shen, C. Su, A.-Y. Lu, M. Hempel, Y. Han, Q. Ji, Y. Lin, E. Shi, E. McVay, Additive manufacturing of patterned 2D semiconductor through recyclable masked growth, Proc. Natl. Acad. Sci. U.S.A. 116 (2019) 3437-3442. https://doi.org/10.1073/pnas.1816197116

[71] D.F. Kyser, Spatial resolution limits in electron beam nanolithography, J. Vac. Sci. Technol. B Microelectron. Process. Phenom. 1 (1983) 1391-1397. https://doi.org/10.1116/1.582706

[72] H.G. Craighead, R. Howard, L. Jackel, P. Mankiewich, 10-nm linewidth electron beam lithography on GaAs, Appl. Phys. Lett. 42 (1983) 38-40. https://doi.org/10.1063/1.93757

[73] R.E. Howard, E.L. Hu, L.D. Jackel, Multilevel resist for lithography below 100 nm, IEEE Trans. Electron Devices. 28 (1981) 1378-1381. https://doi.org/10.1109/T-ED.1981.20617

[74] H. Yang, A. Jin, Q. Luo, J. Li, C. Gu, Z. Cui, Electron beam lithography of HSQ/PMMA bilayer resists for negative tone lift-off process, Microelectron. Eng. 85 (2008) 814-817. https://doi.org/10.1016/j.mee.2008.01.006

[75] M. Li, L. Chen, S.Y. Chou, Direct three-dimensional patterning using nanoimprint lithography, Appl. Phys. Lett. 78 (2001) 3322-3324. https://doi.org/10.1063/1.1375006

[76] S. Kim, B. Lim, H. Lee, D.-H. Shin, S. Kim, H. Park, J. Rhee, Sub-100 nm T-gate fabrication using a positive resist ZEP520/P (MMA-MAA)/PMMA trilayer by double exposure at 50 kV e-beam lithography, Mater. Sci. Semicond. Process. 7 (2004) 7-11. https://doi.org/10.1016/j.mssp.2003.12.003

[77] D. Newbury, Electron beam-specimen interactions in the analytical electron microscope, Principles Anal. Electron Microsc. 1 (1986) 1-27. https://doi.org/10.1007/978-1-4899-2037-9_1

[78] S.y. Fujimoto, K.-i. Nakayama, M. Yokoyama, Fabrication of a vertical-type organic transistor with a planar metal base, Appl. Phys. Lett. 87 (2005) 1-10. https://doi.org/10.1063/1.2061866

[79] Y.C. Chao, M.-C. Ku, W.-W. Tsai, H.-W. Zan, H.-F. Meng, H.-K. Tsai, S.-F. Horng, Polymer space-charge-limited transistor as a solid-state vacuum tube triode, Appl. Phys. Lett. 97 (2010) 1-10. https://doi.org/10.1063/1.3513334

[80] Y. Watanabe, H. Iechi, K. Kudo, Characteristics of organic inverters using vertical- and lateral-type organic transistors, Thin Solid Films 516 (2008) 2731-2734. https://doi.org/10.1016/j.tsf.2007.04.130

[81] J. Nabity, L.A. Compbell, M. Zhu, W. Zhou, E-beam nanolithography integrated with scanning electron microscope, Scanning Microsc. Nanotechnol.: Tech. Appl. 1 (2006) 120-151. https://doi.org/10.1007/978-0-387-39620-0_5

[82] A. Mukhopadhyay, S. Maji, S. Kayal, N.H. Huilgol, P.A. Kumar, Estimation of the interfacial Dzyaloshinskii-Moriya interaction in all-metallic multilayer with perpendicular magnetic anisotropy, J. Phys. D Appl. Phys. 57 (2024) 225001. https://doi.org/10.1088/1361-6463/ad2aae

[83] A.A. Tseng, K. Chen, C.D. Chen, K.J. Ma, Electron beam lithography in nanoscale fabrication: Recent development, IEEE Trans. Electron. Packag. Manuf. 26 (2003) 141-149. https://doi.org/10.1109/TEPM.2003.817714

[84] M.S. Saifullah, N. Tiwale, R. Ganesan, Review of metal-containing resists in electron beam lithography: Perspectives for extreme ultraviolet patterning, J. Micro/Nanolithogr. MEMS MOEMS. 21 (2022) 041402-041402. https://doi.org/10.1117/1.JMM.21.4.041402

[85] J. Bolten, T. Wahlbrink, N. Koo, H. Kurz, S. Stammberger, U. Hofmann, N. Ünal, Improved CD control and line edge roughness in E-beam lithography through combining proximity effect correction with gray scale techniques, Microelectron. Eng. 87 (2010) 1041-1043. https://doi.org/10.1016/j.mee.2009.11.097

[86] Y. Y. Keathley, S.A. Maloney, J.T. Hastings, Real-time dose control for electron-beam lithography, Nanotechnology. 32 (2020) 095302. https://doi.org/10.1088/1361-6528/abcaca

[87] M.N. Hasan, S.-Y. Lee, B.-S. Ahn, J. Choi, J.-S. Park, Shape and dose control for proximity effect correction on massively parallel electron-beam systems, J. Vac. Sci. Technol. B. 38 (2020) 1-10. https://doi.org/10.1116/6.0000556

[88] L. Ren, B. Chen, Proximity effect in electron beam lithography, Proc. 7th Int. Conf. Solid-State Integr. Circuits Technol. 12 (2004) 1271-1275. https://doi.org/10.1116/1.568515

[89] T. Chang, Proximity effect in electron-beam lithography, J. Vac. Sci. Technol. 12 (1975) 1271-1275. https://doi.org/10.1116/1.568515

[90] X. Zhao, S.-Y. Lee, J. Choi, S.-H. Lee, I.-K. Shin, C.-U. Jeon, Minimization of line edge roughness and critical dimension error in electron-beam lithography, J. Vac. Sci. Technol. B. 32 (2014) 1-10. https://doi.org/10.1116/1.4878317

[91] J. Bolten, T. Wahlbrink, M. Schmidt, H.D. Gottlob, H. Kurz, Implementation of electron beam grey scale lithography and proximity effect correction for silicon nanowire device fabrication, Microelectron. Eng. 88 (2011) 1910-1912. https://doi.org/10.1016/j.mee.2010.12.047

[92] X. Thrun, K.-H. Choi, M. Freitag, A. Grenville, M. Gutsch, C. Hohle, J.K. Stowers, J.W. Bartha, Evaluation of direct patternable inorganic spin-on hard mask materials using electron beam lithography, Microelectron. Eng. 98 (2012) 226-229. https://doi.org/10.1016/j.mee.2012.07.017

[93] G. Grenci, E. Zanchetta, A. Pozzato, G. Della Giustina, G. Brusatin, M. Tormen, High resolution spin-on electron beam lithography resist with exceptional dry etching resistance, Appl. Mater. Today. 1 (2015) 13-19. https://doi.org/10.1016/j.apmt.2015.06.003

[94] H. Northfield, O. Krupin, R.N. Tait, P. Berini, Tri-layer contact photolithography process for high-resolution lift-off, Microelectron. Eng. 241 (2021) 111545. https://doi.org/10.1016/j.mee.2021.111545

[95] S.A. Maier, Plasmonics: fundamentals and applications, Springer. NY, 1 (2007) 110-120. https://doi.org/10.1007/0-387-37825-1

[96] C. Zhu, H. Ekinci, A. Pan, B. Cui, X. Zhu, Electron beam lithography on nonplanar and irregular surfaces, Microsyst. Nanoeng. 10 (2024) 52-57. https://doi.org/10.1038/s41378-024-00682-9

[97] M.S. Vincent, S. Uphoff, Bacterial phenotypic heterogeneity in DNA repair and mutagenesis, Biochem. Soc. Trans. 48 (2020) 451-462. https://doi.org/10.1042/BST20190364

[98] D. Psaltis, S.R. Quake, C. Yang, Developing optofluidic technology through the fusion of microfluidics and optics, Nature. 442 (2006) 381-386. https://doi.org/10.1038/nature05060

[99] S.K. Metkar, K. Girigoswami, Diagnostic biosensors in medicine-a review, Biotechnol. Appl. Biochem. 17 (2019) 271-283. https://doi.org/10.1016/j.bcab.2018.11.029

[100] A. Sin, K.C. Chin, M.F. Jamil, Y. Kostov, G. Rao, M.L. Shuler, The design and fabrication of three-chamber microscale cell culture analog devices with integrated dissolved oxygen sensors, Biotechnol. Prog. 20 (2004) 338-345. https://doi.org/10.1021/bp034077d

[101] J.G. Chase, B.W. Smith, Overview of modern lithography techniques and a MEMS-based approach to high throughput rate electron beam lithography, J. Intell. Mater. Syst. Struct. 12 (2001) 807-817. https://doi.org/10.1177/104538901400438109

[102] L.J. Guo, Nanoimprint lithography: methods and material requirements, Adv. Mater. 19 (2007) 495-513. https://doi.org/10.1002/adma.200600882

[103] B. Wu, A. Kumar, Extreme ultraviolet lithography: A review, J. Vac. Sci. Technol. B Microelectron. Nanometer Struct. Process. Meas. Phenom., 25 (2007) 1743-1761. https://doi.org/10.1116/1.2794048

[104] R.H. Stulen, D.W. Sweeney, Extreme ultraviolet lithography, IEEE J. Quantum Electron. 35 (2002) 694-699. https://doi.org/10.1109/3.760315

[105] A.A.R. Tonoy, M. Ahmed, M.R. Khan, Precision mechanical systems in semiconductor lithography equipment design and development, Am. J. Adv. Technol. Eng. Solutions. 1 (2025) 71-97. https://doi.org/10.63125/j6tn8727

[106] C. Huang, C. Zhang, S. Xiao, Y. Wang, Y. Fan, Y. Liu, N. Zhang, G. Qu, H. Ji, J. Han, Ultrafast control of vortex microlasers, Sci. 367 (2020) 1018-1021. https://doi.org/10.1126/science.aba4597

[107] R. Bonjour, S.A. Gebrewold, D. Hillerkuss, C. Hafner, J. Leuthold, Continuously tunable true-time delays with ultra-low settling time, Opt. Express. 23 (2015) 6952-6964. https://doi.org/10.1364/OE.23.006952

[108] R. Bonjour, M. Singleton, P. Leuchtmann, J. Leuthold, Comparison of steering angle and bandwidth for various phased array antenna concepts, Opt. Commun. 373 (2016) 35-43. https://doi.org/10.1016/j.optcom.2015.10.032

[109] R. Bonjour, M. Singleton, S.A. Gebrewold, Y. Salamin, F.C. Abrecht, B. Baeuerle, A. Josten, P. Leuchtmann, C. Hafner, J. Leuthold, Ultra-fast millimeter wave beam steering, IEEE J. Quantum Electron. 52 (2015) 1-8. https://doi.org/10.1109/JQE.2015.2509242

[110] A.N. Boto, P. Kok, D.S. Abrams, S.L. Braunstein, C.P. Williams, J.P. Dowling, Quantum interferometric optical lithography: Exploiting entanglement to beat the diffraction limit, Phys. Rev. Lett. 85 (2000) 2733. https://doi.org/10.1103/PhysRevLett.85.2733

[111] A. Einstein, B. Podolsky, N. Rosen, Einstein-Podolsky-Rosen, Phys. Rev. 47 (1935) 777-781. https://doi.org/10.1103/PhysRev.47.777

[112] M. D'Angelo, M.V. Chekhova, Y. Shih, Two-photon diffraction and quantum lithography, Phys. Rev. Lett. 87 (2001) 013602-0136024. https://doi.org/10.1103/PhysRevLett.87.013602

[113] M. Scully, Proceedings of the Conference on Effects of Atomic Coherence and Interference in Quantum Optics, Crested Butte, Colorado, 1993, IOP.

[114] R.W. Boyd, J.P. Dowling, Quantum lithography: Status of the field, Quantum Inf. Process. 11 (2012) 891-901. https://doi.org/10.1007/s11128-011-0253-y

[115] M. Liu, H. Yang, B. Khailany, H. Ren, An adversarial active sampling-based data augmentation framework for AI-assisted lithography modeling, Proc. 2023 IEEE/ACM Int. Conf. Comput.-Aided Des. (ICCAD). 1 (2023) 1-9. https://doi.org/10.1109/ICCAD57390.2023.10323949

[116] K. Karimi, A. Fardoost, N. Mhatre, J. Rajan, D. Boisvert, M. Javanmard, A thorough review of emerging technologies in micro- and nanochannel fabrication: Limitations, applications, and comparison, Micromachines. 15 (2024) 1274. https://doi.org/10.3390/mi15101274

[117] F. Hubenthal, Noble metal nanoparticles: Synthesis and optical properties, (2011). https://doi.org/10.1016/B978-0-12-374396-1.00034-9

[118] Z. Geng, X. Zhang, Z. Fan, X. Lv, H. Chen, A route to terahertz metamaterial biosensor integrated with microfluidics for liver cancer biomarker testing in early stage, Sci. Rep. 7 (2017) 16378. https://doi.org/10.1038/s41598-017-16762-y

[119] M.A. Rose, J.J. Bowen, S.A. Morin, Emergent soft lithographic tools for the fabrication of functional polymeric microstructures, ChemPhysChem. 20 (2019) 909-925. https://doi.org/10.1002/cphc.201801140

[120] H. Kavand, H. van Lintel, S. Bakhshi Sichani, S. Bonakdar, H. Kavand, J. Koohsorkhi, P. Renaud, Cell-imprint surface modification by contact photolithography-based approaches: direct-cell photolithography and optical soft lithography using PDMS cell imprints, ACS Appl. Mater. Interfaces. 11 (2019) 10559-10566. https://doi.org/10.1021/acsami.9b00523

[121] A.D. Sedeh, F. Karimzadeh, M. Kharaziha, A high-performance single-electrode triboelectric nanogenerator based on polydimethylsiloxane surface modified using zinc oxide nanotubes: Fabrication and simulation, Sustain. Energy Technol. Assess. 56 (2023) 103058-103061. https://doi.org/10.1016/j.seta.2023.103058

[122] L. Harriott, A second generation focused ion beam micromachining system, Proc. SPIE, (1987). https://doi.org/10.1117/12.940370

[123] L.D. Menard, J.M. Ramsey, Fabrication of sub-5 nm nanochannels in insulating substrates using focused ion beam milling, Nano Lett. 11 (2011) 512-517. https://doi.org/10.1021/nl103369g

[124] S. Iqbal, X. Guo, Y. Yi, X.Y. Zhang, T. Zhang, Recent frontiers in anisotropic wet chemical etching techniques for precision surface modification of industrial-grade bulk crystalline silicon wafers: enhancing performance in photovoltaics and MEMS devices, Silicon. 16 (2024) 4139-4158. https://doi.org/10.1007/s12633-024-02963-4

[125] H.B. Mamo, M. Adamiak, A. Kunwar, 3D printed biomedical devices and their applications: a review on state-of-the-art technologies, existing challenges, and future perspectives, J. Mech. Behav. Biomed. Mater. 143 (2023) 105930. https://doi.org/10.1016/j.jmbbm.2023.105930

[126] Y. Bozkurt, E. Karayel, 3D printing technology; methods, biomedical applications, future opportunities and trends, J. Mater. Res. Technol. 14 (2021) 1430-1450. https://doi.org/10.1016/j.jmrt.2021.07.050

[127] K. Pathak, R. Saikia, A. Das, D. Das, M.A. Islam, P. Pramanik, A. Parasar, P.P. Borthakur, P. Sarmah, M. Saikia, 3D printing in biomedicine: Advancing personalized care through additive manufacturing, Explor. Med. 4 (2023) 1135-1167. https://doi.org/10.37349/emed.2023.00200

[128] P. Prabhakar, R.K. Sen, N. Dwivedi, R. Khan, P.R. Solanki, A.K. Srivastava, C. Dhand, 3D-printed microfluidics and potential biomedical applications, Front. Nanotechnol. 3 (2021) 609355-609360. https://doi.org/10.3389/fnano.2021.609355

[129] S. Sachdeva, R.W. Davis, A.K. Saha, Microfluidic point-of-care testing: Commercial landscape and future directions, Front. Bioeng. Biotechnol. 8 (2021) 602659-6022663. https://doi.org/10.3389/fbioe.2020.602659

[130] C. Alix-Panabières, Perspective: The future of liquid biopsy, Nature 579 (2020) S9-S9. https://doi.org/10.1038/d41586-020-00844-5

[131] J. Kaiser, 'Liquid biopsy' for cancer promises early detection, Science (News). 359 (2018) 258-259. https://doi.org/10.1126/science.359.6373.259

[132] J. Marrugo-Ramírez, M. Mir, J. Samitier, Blood-based cancer biomarkers in liquid biopsy: A promising non-invasive alternative to tissue biopsy, Int. J. Mol. Sci. 19 (2018) 2870-2877. https://doi.org/10.3390/ijms19102877

[133] C. Alix-Panabières, K. Pantel, Liquid biopsy: From discovery to clinical application, Cancer Discov. 11 (2021) 858-873. https://doi.org/10.1158/2159-8290.CD-20-1311

[134] M. Ignatiadis, G.W. Sledge, S.S. Jeffrey, Liquid biopsy enters the clinic-implementation issues and future challenges, Nat. Rev. Clin. Oncol. 18 (2021) 297-312. https://doi.org/10.1038/s41571-020-00457-x

[135] A.H. Fischer, Circulating tumor cells: seeing is believing, Arch. Pathol. Lab. Med. 133 (2009) 1367-1369. https://doi.org/10.5858/133.9.1367

[136] G.T. Budd, M. Cristofanilli, M.J. Ellis, A. Stopeck, E. Borden, M.C. Miller, J. Matera, M. Repollet, G.V. Doyle, L.W. Terstappen, Circulating tumor cells versus imaging-predicting overall survival in metastatic breast cancer, Clin. Cancer Res. 12 (2006) 6403-6409. https://doi.org/10.1158/1078-0432.CCR-05-1769

[137] G. Cristofanilli, M. Budd, A.S. Ellis, J. Matera, M.C. Miller, J.M. Reuben, G.V. Doyle, W.J. Allard, L.W.M.M. Terstappen, D.F. Hayes, Circulating tumor cells, N. Engl. J. Med. 351 (2004) 781-791. https://doi.org/10.1056/NEJMoa040766

[138] D.F. Hayes, M. Cristofanilli, G.T. Budd, M.J. Ellis, A. Stopeck, M.C. Miller, J. Matera, W.J. Allard, G.V. Doyle, L.W. Terstappen, Circulating tumor cells at each follow-up time point during therapy of metastatic breast cancer patients predict progression-free and overall survival, Clin. Cancer Res. 12 (2006) 4218-4224. https://doi.org/10.1158/1078-0432.CCR-05-2821

[139] E.A. Punnoose, S.K. Atwal, J.M. Spoerke, H. Savage, A. Pandita, R.-F. Yeh, A. Pirzkall, B.M. Fine, L.C. Amler, D.S. Chen, Molecular biomarker analyses using circulating tumor cells, PLoS One. 5 (2010) e12517-e12518. https://doi.org/10.1371/journal.pone.0012517

[140] E. Cocucci, G. Racchetti, J. Meldolesi, Shedding microvesicles: artefacts no more, Trends Cell Biol. 19 (2009) 43-51. https://doi.org/10.1016/j.tcb.2008.11.003

[141] R. Vaidyanathan, R.H. Soon, P. Zhang, K. Jiang, C.T. Lim, Cancer diagnosis: from tumor to liquid biopsy and beyond, Lab on a Chip. 19 (2019) 11-34. https://doi.org/10.1039/C8LC00684A

[142] Y. Sun, T.A. Haglund, A.J. Rogers, A.F. Ghanim, P. Sethu, Microfluidics technologies for blood-based cancer liquid biopsies, Anal. Chim. Acta. 1012 (2018) 10-29. https://doi.org/10.1016/j.aca.2017.12.050

[143] C. Chinello, V. L'imperio, M. Stella, A.J. Smith, G. Bovo, A. Grasso, M. Grasso, F. Raimondo, M. Pitto, F. Pagni, The proteomic landscape of renal tumors, Expert Rev. Proteomics. 13 (2016) 1103-1120. https://doi.org/10.1080/14789450.2016.1248415

[144] F.M. Bosisio, A. Antoranz, Y. van Herck, M.M. Bolognesi, L. Marcelis, C. Chinello, J. Wouters, F. Magni, L. Alexopoulos, M. Stas, Functional heterogeneity of lymphocytic patterns in primary melanoma dissected through single-cell multiplexing, Life. 9 (2020) e53008-e53009. https://doi.org/10.7554/eLife.53008

Materials Research Foundations 191 (2026) 128-179 https://doi.org/10.21741/9781644904213-5

Chapter 5

Microfluidic Devices for Water Quality Monitoring: A Cross-Bibliometric Analysis of Scientific and Technological Trends

N. Outili[1]*, A. Kadri[1], A H. Meniai[1]

[1]Laboratory of environmental process engineering (LIPE). Process engineering faculty. Constantine3 Salah Boubnider. Algeria

*nawel.outili@univ-constantine3.dz

Abstract

Microfluidics, at the intersection of miniaturization and analytical innovation, is revolutionizing sectors such as health, environment, and energy. This chapter combines a comprehensive literature review with a bibliometric analysis using the Lens database to explore advances in materials, fabrication methods, and applications, with a focus on water monitoring. Patent data emphasize functional integration and miniaturization, while academic research advances analytical principles and detection methods. Beyond mapping the field, the originality of this work lies in the integrated analysis of scientific publications and patents, enabling a structured identification and distinction between research gaps and technological gaps. Paper-based microfluidics (µPADs) emerge as a key innovation for portable, low-cost, in situ water analysis. Despite technical challenges, this synthesis provides strategic insights into emerging trends and future directions for microfluidic technologies.

Keywords

Micofluidic Device, Water Quality Monitoring, Bibliometric Analysis, Technology Trends, Research Innovations

Contents

Materials Research Foundations 191 (2026) 128-179 https://doi.org/10.21741/9781644904213-5

1. Introduction

Access to water quality is a global health and environmental issue. In response to the growing threats of water pollution, climate change, and sustainability, there is an urgent need for innovative solutions for water quality analysis and monitoring towards real-time, on-site and more efficient devices, providing a cost-effective, portable, and accurate alternative to traditional laboratory equipment [1]. Faced with the growing challenges of water quality monitoring, microfluidic devices are emerging as a promising technological solution for real-time contaminant analysis [2,3]. Among these devices, paper-based microfluidic sensors stand out for their low cost, portability, and ease of use, making them particularly suited to resource-poor environments and in situ applications [4].

Materials Research Foundations 191 (2026) 128-179 https://doi.org/10.21741/9781644904213-5

This cross-disciplinary dimension of water monitoring directly aligns with several Sustainable Development Goals (SDGs), including: Good health and well-being, by limiting exposure to unsafe water; Clean water and sanitation, by facilitating access to quality water and regular monitoring; Responsible consumption and production, by optimizing water use in industry and agriculture; Combating climate change, by detecting alterations linked to extreme events; and Aquatic and terrestrial life, by protecting ecosystems from contamination [5,6].

Microfluidics, a multidisciplinary field at the intersection of biology, physics, chemistry, and engineering, studies and implements flows in microchannels on the order of a micron [7-9]. It leverages advances in microelectronics for its fabrication and specific physical phenomena at small scales [10]. Inspired by nature and its natural capillary networks, this biomimicry has opened new avenues of innovation for more efficient, greener systems that are closer to natural principles [11,12].

Microfluidics has evolved over the decades from a theoretical concept to a key technology for various industrial, medical, and environmental applications [13]. Its physics is primarily described by the Navier-Stokes equation, which simplifies to the Poiseuille equation for laminar flows at low Reynolds numbers [7,14,15]. For porous media such as paper, Darcy's law and the Washburn equation (for imbibition) are used, with more advanced models such as the Richards equation and numerical simulations for a more precise description [16,17].

Among the most commonly monitored analytes for water monitoring are pH [18], turbidity, temperature, nutrients (nitrates, nitrites, phosphates, ammonium) [19], heavy metals (lead, cadmium, mercury) [4,20], organic and inorganic compounds [21], pesticides, herbicides [22], and microbiological indicators (coliforms, *E. coli*, etc.) [8]. Microfluidic devices for monitoring water quality, through these parameters, rely on a diversity of materials (paper, polymers, composites) [8,15], manufacturing technologies (printing, laser cutting, molding, screen printing) [23] and detection methods (colorimetry, electrochemistry, fluorescence, biosensors) [18,20,24,25]. These devices integrate innovative approaches allowing miniaturization, rapid analysis, low cost and field use. However, these technologies still have limitations in terms of sensitivity, selectivity, stability, integration of analysis steps, and adaptability to real environments [26]. Current research trends and perspectives aim to overcome these obstacles by integrating new detection methods, developing more efficient and durable materials, improving portability and exploiting artificial intelligence or IoT for more reliable and automated real-time monitoring [27-29].

Scientific research and technological innovation in this field are experiencing rapid growth, as evidenced by the growing number of academic publications and patents filed in recent years [30]. Advances in materials, manufacturing methods, and integration strategies with digital technologies (e.g., smartphone-assisted detection, data analysis using artificial intelligence) have improved the sensitivity, selectivity, and robustness of these devices [27,28,30]. However, despite these advances, several gaps remain, particularly in terms of sensor stability, standardization of analytical methods, automation of devices, and environmental assessment of these technologies [31].

In this context, this chapter offers a bibliometric analysis of scientific and technological trends in the field of paper-based microfluidic devices applied to water monitoring. The objective is to identify, through a dual approach—based on scientific databases and patent databases—the dominant research areas, recent technological developments, and unexplored innovation opportunities.

Materials Research Foundations 191 (2026) 128-179 https://doi.org/10.21741/9781644904213-5

This approach not only tracks the evolution of publications in each subfield, but also identifies major players, influential journals, emerging trends, and research gaps. Although several studies have analyzed bibliometric trends in microfluidics in general [32,33], few studies have combined scientific and technological analysis in the specific field of microfluidic devices for water monitoring, which lends an original and relevant character to the approach proposed in this chapter.

The value of this dual approach lies in the complementarity between academic knowledge and industrial innovation. By combining these analyses, this chapter will highlight not only the state of the art for the past two decades, but also the gaps and untapped opportunities in the field of paper-based microfluidic devices for water monitoring.

The Lens.org platform was used to conduct the bibliometric analysis due to its open, transparent, and interdisciplinary nature, as well as its unique ability to cross-reference data from scientific publications and patents [34]. It provides broad access to structured and reliable metadata (titles, authors, affiliations, dates, research fields, etc.) and offers extraction and visualization tools for in-depth statistical and bibliometric analysis [35].

A single search equation was used to query both academic publications and patents, using relevant keywords. Date filters (initially focusing on the last 20 years, then refined to the last decade) were applied, as well as specific search fields tailored to each document type (e.g., "title", "abstract" and "claims"). This bibliometric approach makes it possible to map the evolution of the field of microfluidics applied to water monitoring, to identify key players (institutions, inventors, authors, countries) and to highlight the dominant research axes, technological trends and emerging subjects. Visualisation tools were used to map and highlight key findings using the Lens analysis tools, VOSviewer and Excel [36,37].

For an in-depth analysis, results targeted for the last decade including the most relevant patent documents and academic publications were selected according to criteria of relevance, innovation, and representativeness. These documents were analyzed in detail according to several dimensions: manufacturing methods, detection methods, types of water analyzed, and innovative contributions. The objective was to identify the key trends in research and innovation in this field.

The combined analysis of scientific publications and patents provided a perspective on the differences between academic research and industrial innovation, by identifying technologies studied but still underexploited on the market, underexplored research avenues, particularly through quantitative and qualitative analysis of areas underrepresented in the literature and patents, the opportunities for innovation and technology transfer, by highlighting sectors where the transition from laboratory to real-world application is still limited and expected future developments, by identifying areas of rapid growth and emerging technologies with the greatest potential for impact.

This chapter will therefore be a valuable tool for researchers wishing to better focus their work on issues with high industrial impact and for companies seeking to identify emerging trends in the field of paper-based microfluidic sensors for water monitoring.

2. Fundamentals and Evolution of Microfluidics

To better understand the development and potential of microfluidic devices for water monitoring, it is essential to begin by exploring their origins and the fundamental physical principles upon

Materials Research Foundations 191 (2026) 128-179 https://doi.org/10.21741/9781644904213-5

which they are based. This section provides a historical overview of the emergence of microfluidics, followed by the fundamental scientific and technological principles that govern their design and operation, as well as their evolution over time and their main applications and advantages.

2.1 Definition and origins

Microfluidics combines miniaturization, precision, and innovation for the control and manipulation of fluids in channels ranging from 1 to 500 μm in diameter (the thickness of a human hair is approximately 100 μm) [29]. It mimics nature and its highly efficient fluidic networks (Fig. 1), such as plant vascular systems, where the capillary network in plant leaves is between 100 and 10 nm, or the blood vascular network, whose capillaries are on the order of micrometers [30].

Researchers such as Whitesides et al. [38] have explored microfluidic devices inspired by biology, particularly in terms of fluid manipulation on soft polymer surfaces such as PDMS. Dawson et al. have studied vascular-inspired structures for fluid distribution control [39]. Lee et al. have proposed biomimetic devices using networked channels to mimic the properties of blood microcirculation in tissues [40]. Hossain et al. have reviewed new microfluidic techniques that enable the creation of bio-inspired microparticles with tunable properties through microscopic control [41]. This characteristic of liquid flow in the infinitesimally small channels allows microfluidics to create lab-on-a-chip devices, which are miniaturized laboratories allowing very rapid analyses with a minimum of reagents.

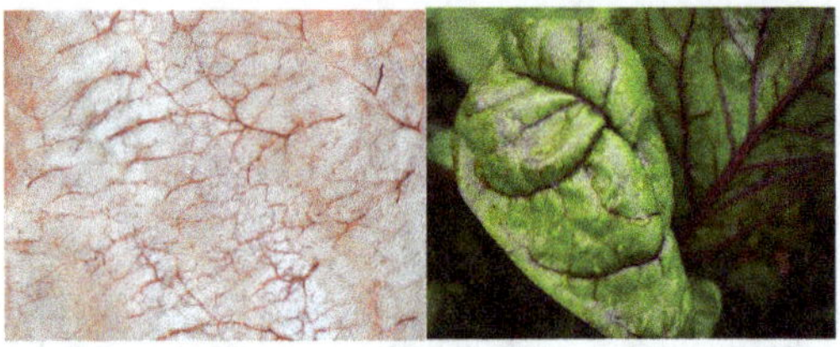

Figure 1. Microfluidics and bio-mimitic of the capillary network of the skin and a green leaf (from https://www.istockphoto.com)

2.2 Microfluidics principles

At micrometric dimensions, the flow laws remain those of classical fluidics and can be studied using the Navier-Stokes equation in the case of incompressible liquids, whose inertial terms can be neglected and viscous forces are dominant. The flow regime is predominantly laminar, characterized by a low Reynolds number (Re < 1) [7].

The Navier-Stokes equation (Eq. 1) in the case of an incompressible Newtonian fluid is as follows [42,43]:

$$\rho\left(\frac{\partial u}{\partial t} + u\nabla u\right) = -\nabla p + \eta\Delta u + f \quad (1)$$

Materials Research Foundations 191 (2026) 128-179
https://doi.org/10.21741/9781644904213-5

In steady state and neglecting the inertia terms (convective terms u·∇u and acceleration term (∂u/∂t), the Navier-Stokes equation simplifies to (Eq. 2) as follows:

$$\nabla p=\eta\Delta u \tag{2}$$

The resolution of this equation depends on the channel geometry. The Poiseuille equation is a particular solution for a stationary, incompressible, and laminar flow in a cylindrical tube of radius *R* with constant pressure gradient along the *z* axis and "no-slip" boundary conditions. It describes a parabolic velocity profile (Eq. 3) and a linear behavior of the pressure drop Δp (Eq. 5) applied along a channel of length L, and the resulting flow rate Q (m^3/s) described by (Eq. 4) [44]:

$$u_z(r) = \frac{\Delta p\, R^2}{4\eta L}\left(1-\left(\frac{r}{R}\right)^2\right) \tag{3}$$

$$Q = \frac{\pi\, R^4}{8\eta L}\,\Delta p \tag{4}$$

Therefore, the pressure drop is expressed as:

$$\Delta p = \frac{8\eta L}{\pi\, R^4}\, Q \tag{5}$$

This is considered as a disadvantage or limitation of microfluidics as the pressure drop increases linearly with the decrease in the channel diameter as shown in (Fig. 2).

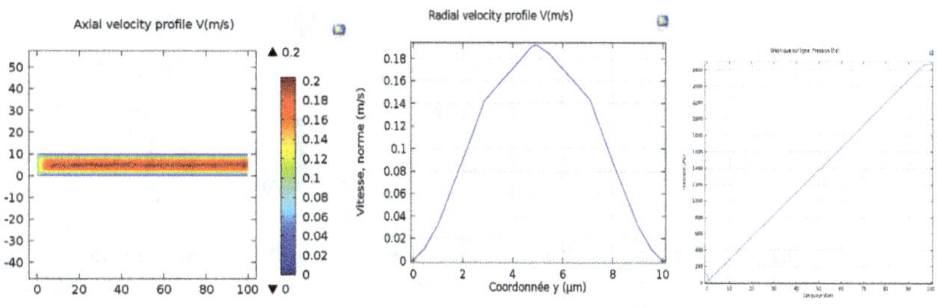

(a) (b) (c)

Figure 2. Laminar flow regine in a microchanal

(Comsol Multiphysics simulation of a microchanel of 10 µm diameter and 100 µm in length, a water flow with 100 mm/s at the inlet. (a) fluid flow is laminar with Reynolds number of 9.21. (b)The velocity distribution at a cross section is parabolic in shape with the maximum velocity at the center and (c) pressure drop profile is linear). Produced by the authors

Regarding the mass transfer in these laminar flows, the phenomenon is characterized by the Peclet number (Pe), which is the ratio between convective transfer and diffusion transfer. For the diameter d of the channel considered as the characteristic length of the flow, (Eq. 6) gives the Peclet number [43]:

$$Pe = \frac{Diffusion\ time}{Convection\ time} = \frac{d^2/D_{iff}}{d/u} = \frac{d\ V}{D_{iff}} \tag{6}$$

With D_{iff} : diffusion coefficient (m²/s), V : mean velocity (m/s)

If Pe > 1, the flow is convective and the mass is mainly transported by the flow.

If Pe < 1, diffusion dominates, allowing mixing by diffusion, with a very short diffusion time as shown by Eq.7 (very rapid mixing):

$$t_{diff} = \frac{d^2}{D_{iff}} \tag{7}$$

For a microchannel of the order of 10 μm and a liquid flow with a diffusion coefficient of the order of 10^{-9} m^2/s, the diffusion time would be on the order of 1 second.

This fast diffusion is a key advantage in microfluidics, where the micrometric scale enables rapid mixing by diffusion [43], despite the generally laminar flow conditions as simulated in (Fig. 3).

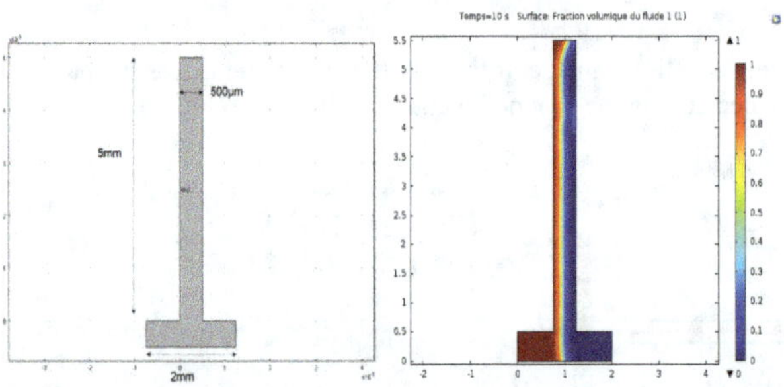

Figure 3. Mixing by diffusion in a micochannel (Comsol Multiphysics simulation carried out by the author)

In microfluidics, specific geometric structures are often needed to accelerate mixing, reaction, mass and heat transfer, such as flow focusing, elbows, droplet generation, T or Y junctions [16]. Flow focusing (Fig. 4) is a technique where a dispersed fluid is injected into a continuous fluid via a cross or T-shaped geometry, allowing the formation of fine jets or droplets [45,46]. The capillary number (Eq. 8) is a dimensionless number that represents the relative effect of viscous forces to surface tension forces and determines the fluid regimes [47]:

Materials Research Foundations 191 (2026) 128-179
https://doi.org/10.21741/9781644904213-5

$$Ca = \frac{\mu V}{\gamma} \tag{8}$$

Ca≪1 → **Surface tension dominates**: fluid tends to form droplets (dripping regime)

Ca≫1 → **Viscous forces dominate**: fluid forms continuous streams or jets (jetting regime)

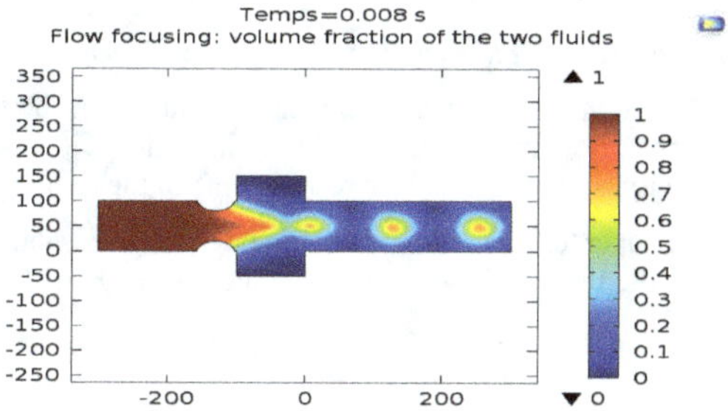

Figure 4. Generating water drops in oil by flow focusing (Comsol Multiphysics simulation carried out by the author)

In the specific case of paper-based microfluidic devices (μPADs), which are porous media, spontaneous fluid transport relies on capillary forces without the need for external pumps. The description of fluid motion in paper (porous media) typically uses equations such as Darcy's law and the continuity equation. Darcy's law relates velocity (or volume flow rate) to pressure (Eq.9-10) and the properties of the porous medium and fluid [16,17].

$$V = \frac{K}{\mu} \nabla P \tag{9}$$

V : mean volume velocity of the fluid (m/s), **K** : permeability of the porous medium (m²), **μ**: fluid dynamic viscosity (Pa·s), **∇P**: Pressure gradient (Pa/m)

$$Q = \frac{K\,A}{\mu} \frac{\Delta P}{L} \tag{10}$$

With *A* the cross-sectional area and *L* its length. Darcy's law is suitable for laminar flows in porous media. It plays a fundamental role in paper microfluidics, as it allows modeling and predicting the capillary flow of liquids through the porous matrix of the paper, without the use of external pumps. (Fig. 5).

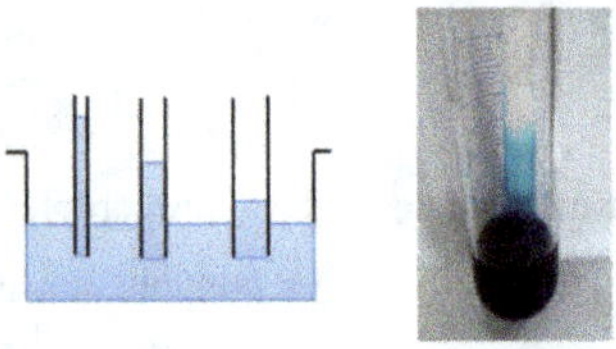

Figure 5. Flow capillary in tubes and in paper

Chemical or biological reactions of on paper depend on the contact time between the fluid and the deposited reagents. Darcy's law coupled with Washburn's law, makes it possible to estimate these times to optimize sensitivity, selectivity, and the reading time of the results [17].

In a porous material like paper, capillary flow can be modeled by Washburn's law, derived from Darcy's law, it describes the penetration distance or imbibition L (Eq. 11) as a function of time [44] :

$$L(t) = \sqrt{\frac{K\,\gamma\,\cos\theta}{2\,\mu\,\varphi}\,t} \tag{11}$$

With : L(t) the length traveled by the liquid (m), K : permeability of the paper (m^2), γ : surface tension of the liquid (N/m), θ : contact angle (liquid/paper), μ : dynamic viscosity (Pa·s), φ : paper porosity (pore volume fraction), t : time (s)

2.3 History and evolution of microfluidics

The history of microfluidics has been marked by significant successive phases of innovation, from the first fundamental studies in the 1990s to emerging technologies, particularly for environmental, biotechnological, and medical solutions. This evolution is illustrated by (Fig. 6), which represents the evolution of the number of scientific documents by year.

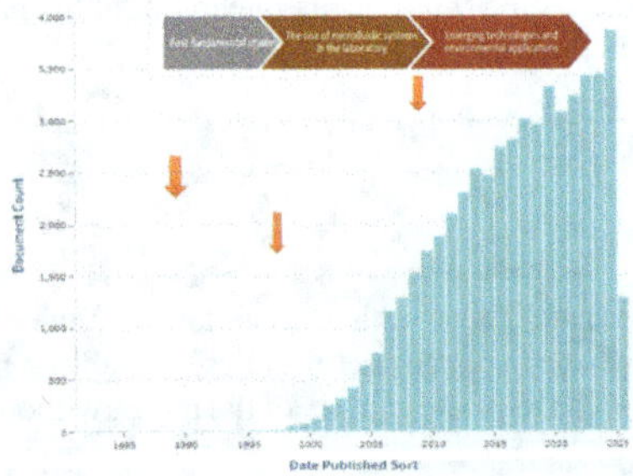

Figure 6. Evolution of the number of scientific publications over time (Produced by the author using the Lens.org database research in Title field using the query: (microfluidic OR microfluidics) : 52721 academic documents. NB : the results were refined excluding unknown documents)

Materials Research Foundations 191 (2026) 128-179 https://doi.org/10.21741/9781644904213-5

This evolution can be presented in three main phases: First, around the 1990s the first fundamental studies introduced the idea of miniaturization of analytical devices for applications in chemistry and biology and established the bases of modern microfluidic systems by demonstrating the advantages of reducing the devices' sizes [48]. Then, the 2000s were marked by the rise of microfluidic systems in the laboratory [49] and the improvement of devices and chips. This allowed applications for biological analyses, as well as faster and less expensive tests [50], thanks to the fabrication of silicone microfluidic chips. And finally, during the phase of emerging technologies and environmental applications around the 2010s, the evolution towards 3D printing allowed researchers to design microfluidic devices at lower cost and with greater flexibility [51,52]. The emergence of paper-based microfluidics (μPADs), notably with the start of work by Whitesides and his team [53] which has opened up new possibilities for applications in resource-limited laboratories. Also, there is strong growth in the use of microfluidics for environmental monitoring, particularly for the detection of pollutants in water [54-56].

The map based on text data analysis using VOSviewer of the most relevant results of the scholarly research in Lens.org is presented in (Fig. 7). The network visualization map (Fig. 7.a) revealed four major thematic clusters after manual cleaning of generic terms (e.g., experiment, result, output…). The first cluster, in red, centers around terms such as pump, silicon, glass, micropump, integration, and geometry, indicating a strong focus on the design and integration of microfluidic components and materials. The second cluster, in yellow, including microfluidic systems, cost, feasibility, miniaturization, reflecting the drive toward scalable and low-cost system development. The third cluster, in green, revolves around analytical applications with terms like separation, electrophoresis, sensitivity, peptide, protein, antibody, highlighting the importance of bioanalytical and diagnostic uses. The fourth cluster, in blue, is structured around PDMS, surface modification, and micromixer, pointing to efforts in material innovation and fluid manipulation. Together, these clusters reflect the diversity of research efforts in microfluidics, from fundamental component design to analytical applications and material advancements. On the other hand, the overlay visualization map (Fig. 7.b) highlights the temporal dynamics of research topics. The cluster centered around PDMS appears as the most recent, with emerging terms such as PMMA, DNA, and protein, indicating a growing interest in advanced materials and bioanalytical applications. This is followed by the cluster involving separation, geometry, and related concepts, reflecting ongoing innovation in fluid manipulation and structural optimization. These observations suggest a shift in research focus toward more application-oriented and biomolecular detection themes in microfluidic device development.

a- Network visualization map *b- Overlay visualization map*

Figure 7. VOSviewer text map analysis based on Lens research results

Materials Research Foundations 191 (2026) 128-179 https://doi.org/10.21741/9781644904213-5

2.4 Advantages and applications of microfluidics

Microfluidics is transforming research and applications across disciplines thanks to several fundamental advantages: time reduction, massive parallel analyses, use of small quantities, high safety and control, miniaturization and portability. These advantages have a major impact on disciplines that affect human health, process efficiency and environmental monitoring and sustainability (Fig. 8), namely:

- ***Medicine and Biomedical Engineering:*** through rapid diagnostics and point-of-care testing. Viral tests, pathogen detection (viruses, bacteria) [41,56]. Rapid blood tests for multiple infectious diseases are possible through multi-microdrop systems. Microfluidics could significantly accelerate and reduce the costs of molecule testing. It allows the creation of new, more efficient molecules and drug delivery systems and improves pharmacokinetics and pharmacodynamics [57].

-***Chemical Engineering***: through process intensification, the synthesis of new materials and molecules and the formulation of complex fluids, the study of complex fluid flows in confined environments, the control of reaction conditions [58], the study of rapid, exothermic or explosive reactions thanks to the control of contact times and heat exchange capacities, which allows access to kinetic data that are difficult to obtain otherwise. Also, the possibility of scaling up high value-added products using parallel circuits or droplet systems performing thousands to millions of simultaneous reactions [59-61].

-***Agriculture and Environmental Monitoring***: through early detection of plant diseases caused by bacteria, fungi and viruses and by analyzing agricultural fertilizers and nutrients. Also, the detection and quantification of pollutants in water, soil and air; and integrated biosensors for environmental monitoring [31,56].

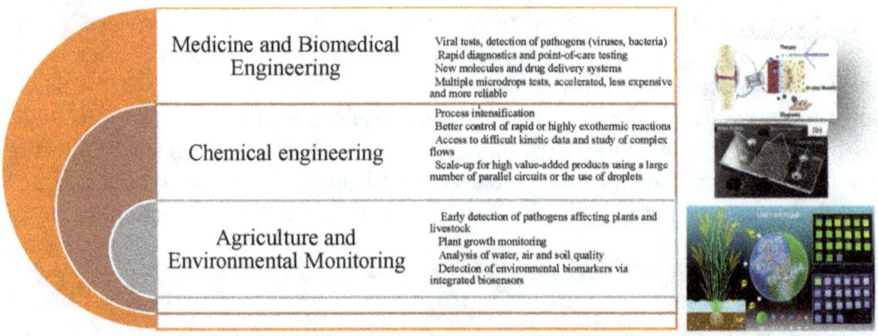

Figure 8. Main applications of microfluidics (produced by the author using images from [57,61,56])

3. A Microfluidic devices for water quality monitoring

Water quality monitoring, an emerging application of microfluidics, is not solely an environmental or analytical issue; it is emerging as a central challenge at the intersection of several strategic areas, including public health, the environment, agriculture, food security, and even climate change resilience [62]. By enabling the early detection of chemical, biological, or physical contaminants,

Materials Research Foundations 191 (2026) 128-179 https://doi.org/10.21741/9781644904213-5

water monitoring plays a key role in preventing waterborne diseases, protecting aquatic ecosystems, sustainably managing agricultural resources, and reducing the risk of industrial pollution [28].

In this context, microfluidic devices for water monitoring are emerging as technological tools with high impact potential. Their portability, low cost, and adaptability to constrained environments make them particularly suited to rural and limited resource countries, and decentralized monitoring systems [28,63]. They can also be integrated into smart environmental monitoring networks such as IoT and connected sensors, strengthening early warning and integrated water resource management capabilities [18,25].

This interest, placing water monitoring at the heart of sustainability, justifies particular attention to scientific and technological trends related to these devices through a cross-bibliometric analysis of scientific publications and patents in the field of microfluidic devices for water analysis. This approach is justified by the fact that, for technical devices, patents are particularly representative of concrete technical solutions currently under development and of industrialization prospects while academic publication content is mainly basic research findings [64,65].

Patent bibliometrics is thus increasingly used to complement traditional studies based on academic publications, particularly in fields where research is oriented towards technological innovation [65].

The bibliometric results of scientific and patent publications will be grouped into three main parts: (1) the temporal evolution of production, (2) the identification of key actors, and (3) the analysis of research and innovation themes in this domain [66,67].

4. Scientometric and Technological Landscape

To identify scientific and industrial trends in microfluidic devices applied to water monitoring, a bibliometric analysis was performed by combining academic publications and patents. This approach allows identifying time evolution of publications and patents, dominant research areas and trends, key contributors, including authors, inventors, countries, academic institutions, and companies. Also, this approach allows identifying the gaps between academic research and industrial innovation based on a cross-analysis of scientific research trends (academic publications) and technological trends (patents) (Fig. 9), as follows:

-Technologies widely studied in the laboratory but rarely patented or commercialized help identify the technology transfer opportunities.

-Patented technologies that are rarely covered in the scientific literature highlight underexplored topics requiring further academic research.

-Emerging innovations in industry that have not yet been well documented by research indicate the prospects for scientific development.

Materials Research Foundations 191 (2026) 128-179 https://doi.org/10.21741/9781644904213-5

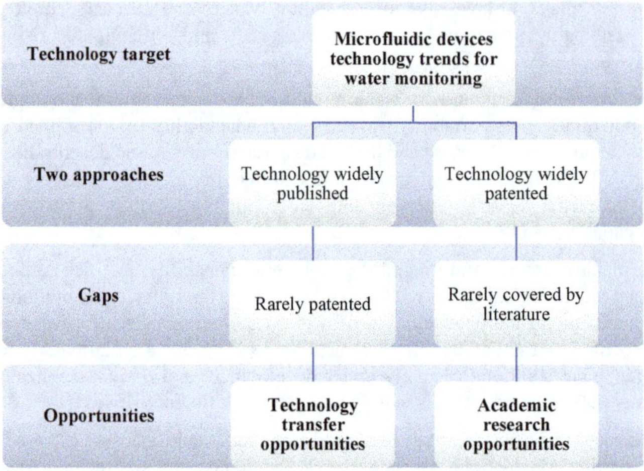

Figure 9. Academic research and industrial innovations approach

4.1 The bibliometric methodology

A bibliometric study, whether in scientific or technological databases, follows common steps ranging from the choice of databases, the selection of relevant keywords, the application of appropriate filters according to the intended objective and the desired refinement of the relevance of the results. This refinement is done by applying filters to key fields common to both documents types as well as specific fields to each type (Table 1): publication date, authors/inventors, fields/ technical fields classifications, affiliations/applicants [68,69].

Table 1. Comparaison key fields for academic publication and patent publication

Scientific paper	Patent
Title	Title
Abstract	Abstract
Authors	Inventors
Publication date	Filing/Publication date
Affiliation (university/research lab)	Applicants (enteprise/lab)
Fields/Journal	Classification (IPC/CPC)
/	Claims (innovation)
Citations	Citations

4.2 Data collection and cleaning

To collect and analyse the obtained results, these main steps were followed:

- In this work, Lens.org was used as a database for both academic and patent documents ~~and~~ using the same search query. This database is a free and open-access platform that integrates data from open sources such as CORE, CrossRef, PubMed, OpenAlex, and Unpaywall, ensuring extensive coverage of the scientific literature [70]. Several studies

Materials Research Foundations 191 (2026) 128-179 https://doi.org/10.21741/9781644904213-5

have demonstrated the effectiveness of Lens.org in bibliometric analyses: For instance, a comparative study analyzed the PubMed, Scopus, Web of Science, and Lens.org databases, highlighting the latter's advantages in terms of coverage and accessibility [34]. Also, other bibliometric studies have used Lens.org in combination with VOSviewer to map research trends, demonstrating its ability to integrate scientific and patent data [35,71].

- To define the relevant keywords and the relevant search query, we defined two key concepts : « microfluidic devices », and « water quality monitoring ». The Boolean ***AND*** was used to connect the main concepts and the Boolean ***OR*** to connect the synonyms of the same concept. Synonyms are the most likely terms to define a key concept and are chosen to cover both scientific terms for academic publications and more generic terms for patents, given that the same search query will be used to limit disparities in the results without being too complicated or long:

  microfluidic* ***AND*** (device* ***OR*** sensor* ***OR*** material* ***OR*** equipment* ***OR*** biosensor*) ***AND*** ("water quality") ***AND*** (monitoring ***OR*** measurement ***OR*** Analysis)
  The wildcard * was used to encompass grammatical variants (singular/plural) and quotation marks (" ") to search for an exact sequence of words in the same order, without any variation or extra words in between.
- First, the period of the study: The bibliometric study was conducted for a period of 20 years to cover the validity period of a patent and starting from 01/01/2004.
- Data cleaning: Once data are collected, it often requires manual cleaning before analysis. This involves removing possible duplicates, unknown documents, irrelevant and off-topic results, affiliations under different transcriptions, etc.
- Filtering results for more relevant results: date filter (last twenty and ten years) and document field filter (title, abstract for academic documents and adding claims for patents documents). This choice is justified by the fact that patent abstracts are not always indexed or available and are also often very technical and shorter than those of scientific documents [72,73].
- The results of this last analysis were used to analyze the trends in this field through a small and relevant number of patent or academic documents. The search query used for this case:

  title:(microfluidic*) **AND** ((device*) **OR** (sensor*) **OR** (material*) **OR** (equipment*) **OR** (biosensor*)) **AND** (("water quality") **AND** (monitoring **OR** (measurement **OR** Analysis))) **AND** Claims: (microfluidic*) **AND** ((device*) **OR** (sensor*) **OR** (material*) **OR** (equipment*) **OR** (biosensor*)) **AND** (("water quality") **AND** (monitoring **OR** (measurement **OR** Analysis))) **AND** abstract: (microfluidic*) **AND** ((device*) **OR** (sensor*) **OR** (material*) **OR** (equipment*) **OR** (biosensor*)) **AND** (("water quality") **AND** (monitoring **OR** (measurement **OR** Analysis)))

- Analysis and visualization: for this purpose we used three tools: Lens analysis tool [70], VOSViewer for text mining functionality to visualize co-occurrence networks of important terms extracted from a corpus of research results [74,75,37] and Excel for cross-tabulations [36].

Materials Research Foundations 191 (2026) 128-179 https://doi.org/10.21741/9781644904213-5

4.3 Bibliometic results and analysis

Applying the above-mentioned search equation in the Lens.org database for both academic and patent documents yielded ~~to~~ the number of results and citations shown in (Fig. 10). In this case, unknown academic documents were not considered and patent documents were restricted to simple patent families.

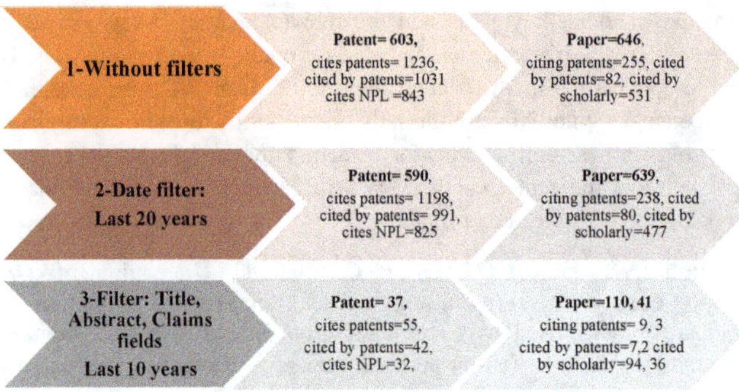

Figure 10. Search Results and citations for three cases search methodology

Initially, the search was performed globally and without applying filters and yielded 646 academic documents and 603 patents. This result indicates an almost balanced dynamic between scientific production and technological innovation, which explains that the field of microfluidic devices for water monitoring is highly oriented towards innovation and generates both publications and patents. This dynamic has been maintained even over the last twenty years, and the results indicate that the growth of this field has indeed taken place during this period as shown in (Fig. 11).

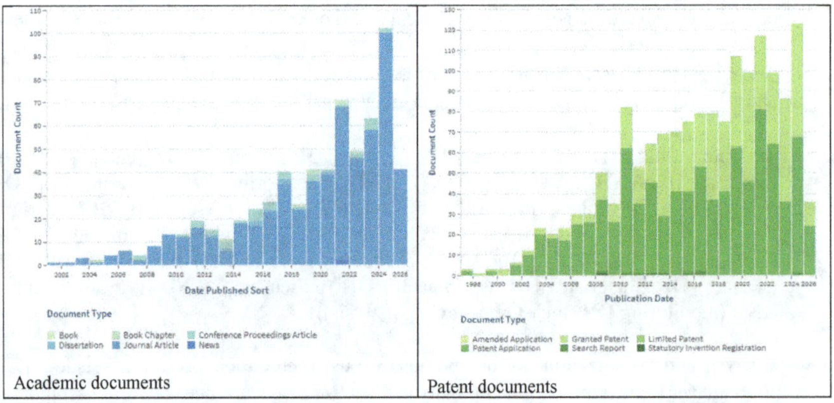

Figure 11. Academic and Patent documents publication over time (without any filter)

Comparative analysis of citation rates (citations/document) shows that patents are, on average, more cited than academic publications in the field of microfluidic devices for water monitoring (Fig. 12). They are also more cited by other patents. In contrast, academic documents and other non-patent sources (NPLs) are mostly cited by patents rather than by scientific articles. Finally, there has been a notable decline over the last ten years in patent citations in academic publications. This result reflects that the field has a strong dynamic of technological innovation, where new inventions are directly based on previous inventions. Also, this result shows that academic research strongly serves as a source of knowledge for technological developments. On the other hand, this highlights a lack of integration between basic and applied research. The decline in patent citations in academic publications over the last ten years could indicate a refocusing of academic research on more fundamental work that is less related to patented inventions [76].

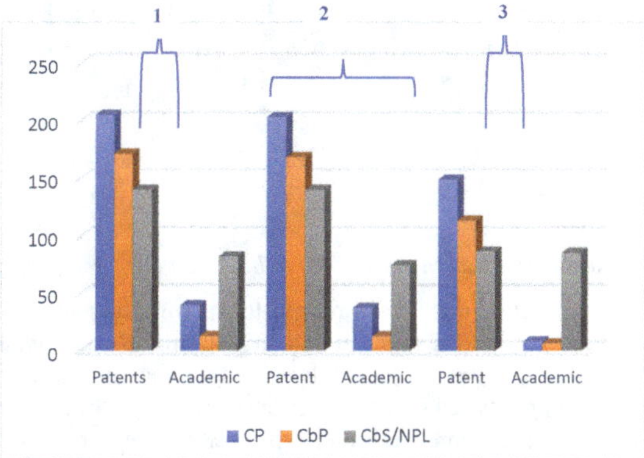

Figure 12. Analysis of citations rates for patents and academics documents resulted from the Lens search (CP : cites patents , CbP : cited by patents , CbS/NPL : cited by scholarly, 1,2 and 3 refer to the three cases of Lens search as presented in Fig. 10)

4.3.1 Dynamics of scientific and technological production

Analyzing the results obtained without filtering provides a global view of the subject, but only a filtered analysis allows significant and exploitable trends to be identified. This refines the quality and relevance of the strategic interpretation of the technological and scientific landscape.

Once the scientific publications have been extracted and filtered, according to defined criteria, a quantitative analysis is conducted, including the number of publications over the years, their geographic distribution, the most influential authors and institutions, as well as the leading journals and conferences. Furthermore, a qualitative analysis is conducted through the identification of key thematic trends (materials, manufacturing methods, types of analyses performed) and methodological advances [64,74].

To understand when and how the field of water monitoring using microfluidic devices has evolved, we examine the evolution over the past 20 years, with the aim of highlighting the emergence, intensification, or slowdown of research and innovation in the field.

Materials Research Foundations 191 (2026) 128-179 https://doi.org/10.21741/9781644904213-5

To better compare technological and scientific developments, we first extracted the evolution of each, as previously described, and then created a cross-tabulation to visualize the comparison presented in (Fig. 13).

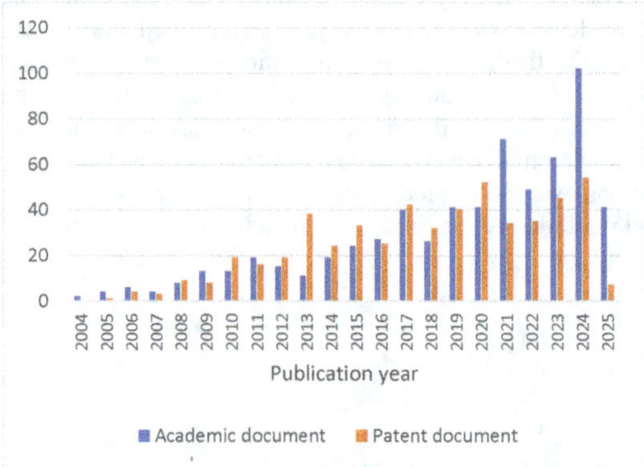

Figure 13. Academic and Patent documents publication over time (last two decades)

Until 2020, the number of patent documents related to microfluidics for water monitoring was higher than or comparable to scientific publications. In 2013, there was even a very marked peak in patent documents. This can be explained by the fact that microfluidics applied to the environment (particularly paper-based systems) was entering a phase of technological innovation, with industrial actors and institutions seeking to protect device concepts and manufacturing methods. After 2020, more laboratories around the world began working on environmental microfluidics. The COVID-19 pandemic also stimulated the use of simple, local, and rapid technologies [77,78], leading to more publications on inexpensive and adaptable systems, such as paper-based microfluidic devices.

4.3.2 Contributors and stakeholders

To identify the most contributors and leading countries in this field, we used the analysis of Lens results to identify the main contributors, both scientific (authors, institutions) and industrial (inventors, companies) [79].

First, the cross-tabulation for academic document country affiliations and patent document priority judisdictions is presented in (Fig. 14). The analysis shows four leading countries that stand out in the field of microfluidic devices for water quality monitoring, with the United States leading. China, India, and the United Kingdom have a higher volume of scientific publications than patents, reflecting sustained academic momentum. In contrast, the United States stands out with a number of patents approximately four times higher than its scientific publications, representing more than two-thirds of all patents filed in this field over the past two decades. This dominance is explained by a particularly mature innovation ecosystem, where advances in microfluidics are quickly converted into patented technologies, driven by partnerships between universities and industry partners, and by a strong focus on the valorization and commercialization of research results applied to the environment and public health.

Materials Research Foundations 191 (2026) 128-179	https://doi.org/10.21741/9781644904213-5

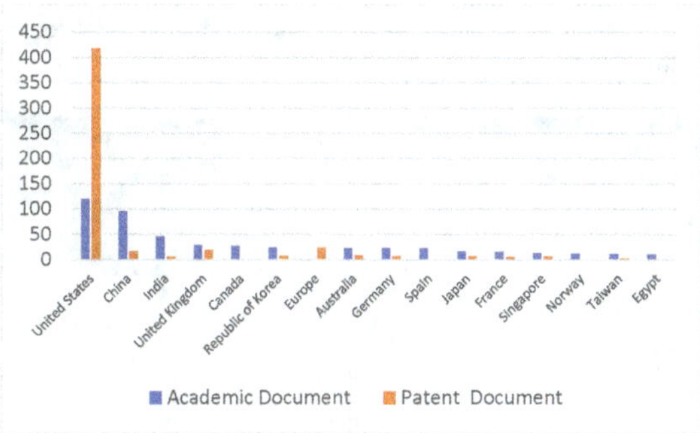

Figure 14. Country affiliations and priority jurisdictions for academic and patent documents for the last two decades

The affiliation-applicant analysis reveals that, as expected, scientific publications are predominantly produced by universities and research institutes. However, the ranking of patent filers is dominated by certain institutions, notably the University of California, which stands out for its significantly higher number of patents than publications in this field. MIT follows a similar pattern. In contrast, universities in the top five for scientific publications do not appear in the ranking of the main patent filers (Fig. 15).

This lack of alignment between academic output and intellectual property protection suggests that few universities, even among the most productive in research, have an active or systematic patenting strategy. Conversely, institutions like the University of California have a well-established technology transfer policy, with efficient patenting offices, enabling them to transform research results into protected assets in applied fields such as environmental microfluidics.

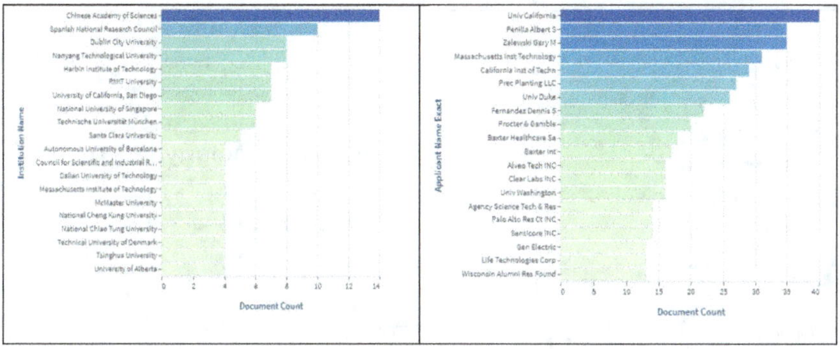

Figure 15. Affiliations and applicants for academic and patent documents for the last two decades

This trend is also confirmed by the analysis of authors and inventors: none of the most prolific authors in the Top 20 of scientific publications are among the top 20 inventors in patents (Fig. 16),

Materials Research Foundations 191 (2026) 128-179 https://doi.org/10.21741/9781644904213-5

which reflects a marked separation between academic production and technology transfer and innovation in this field.

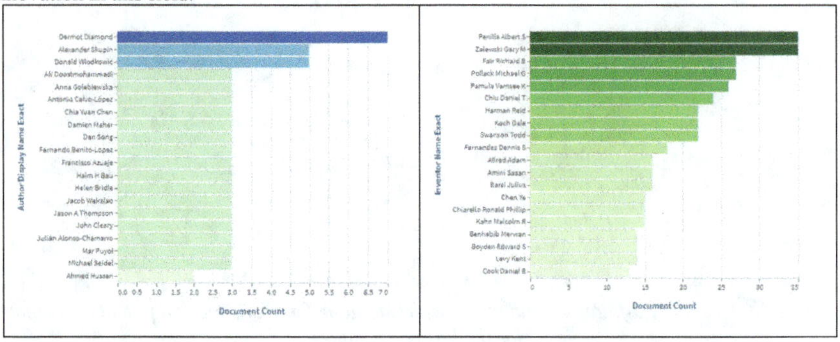

Figure 16. Auteurs/Inventors for academic and patent documents for the last two decades

4.3.3 Thematics and technical orientations

To understand the dominant themes and areas of application, we need to analyze the most frequent sub-themes, the co-occurrences of keywords, the IPC/CPC technological classifications for patents, or the most active journals [67].

The analysis of the themes between scientific publications and patents is carried out separately due to the nature of the data analyzed in both cases. Academic publications, identified via the word cloud generated from Lens analysis, focus on fundamental research fields, and specialized journals (e.g., environmental monitoring and assessment, sensors, microkemika, nanotechnology, microfluidics, chemistry, biology, engineering, …) focus on specialized fields (Fig. 17).

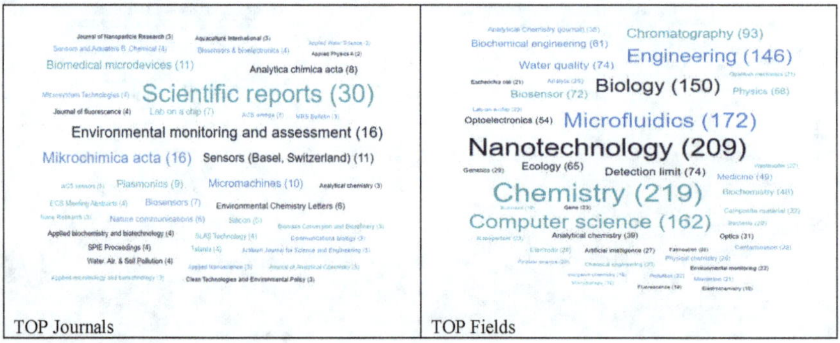

Figure 17. Word Cloud from Lens for Top Journals and Top Fields for academic documents for the last two decades

On the other hand, patents, analyzed through relevant IPC codes and their descriptions (Table 2), reveal more applied themes oriented toward technological innovation as illustrated by the word

Materials Research Foundations 191 (2026) 128-179 https://doi.org/10.21741/9781644904213-5

cloud of (Fig. 18). The dominant terms in patents mainly concern detection methods, microfluidic devices, analysis methods, as well as specific application areas (e.g., colorimetric, optical analysis, electrochemical, fluorescence, lab-on-a-chip, enzymatic, automatic analysis,...).

Table 2. IPC patent classification description for patent documents for the last two decades

IPC Classification	Desciption	Patents
B01L3/00	General **physical** or **chemical** methods or **devices**, **laboratory apparatus**	195
G01N33/18	**Mesearing, testing, analyzing water**	130
G01N33/543	Use of **immunoassay** for **diagnosis**	95
G01N21/64	**Optical analysis, colorimetric methods**	92
C12Q1/68	Measuring or testing processes involving **enzymes or microorganisms**	78
G01N21/78	**colorimetry**, chemical reactions creating **color**	67
G01N35/00	Automatic analysis — e.g., **lab-on-chip, auto-samplers**	61
A61B5/00	Measuring **body parameters**	60
G01N15/14	Investigating materials using wave, ray or **particle radiation** Especially flow **cytometry or fluorescence** in cells	58
G01N33/53	**biological/chemical** analysis of **blood** or similar	58
G01N35/10	**Automated** analysis using **flow injection** techniques	53
C12M1/34	Apparatus for **biochemistry/microbiology** With **integrated measuring devices**	49
G01N27/447	**Electrochemical** analysis, **Biosensors** using **enzyme/electrode**	45
G01N33/487	**Diagnosis** related to **hematopoietic** (blood-forming) tissues	44
G01N21/25	**Optical analysis using fluorescence** (e.g., **biomarkers** tagged with **fluorophores**)	41

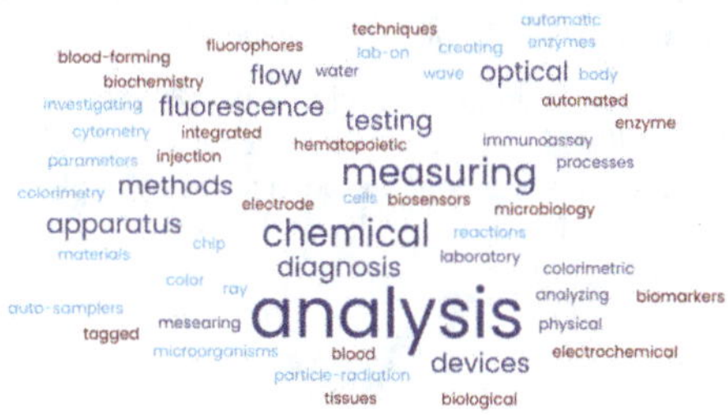

Figure 18. Word cloud generated by the author using the IPC description (after data cleaning) and the online wordcloud generator: freewordcloudgenerator.com

Materials Research Foundations 191 (2026) 128-179 https://doi.org/10.21741/9781644904213-5

Indeed, academic articles aim to advance scientific knowledge, while patents aim to protect an innovative technical solution. Thus, if the major themes are common (chemical, water, biological, ...), the content orientations diverge: conceptual and exploratory in the academic publications, technical, operational and specific in the patents.

5. Microfluidic Devices for Water Monitoring: an in-depth analysis

After identifying the main academic and industrial developments through bibliometric analyses of the most relevant academic publications and patents, it is now essential to deepen the operational technologies currently applied to microfluidics-based water monitoring over the last decade.

5.1 Technological and Academic development of the last decade

To obtain fewer but more relevant results, the search was refined by limiting the period to the last decade and using filters on the document fields. Given that the explored fields influence the volume of results -patents include the "claims" field, which is absent from scientific publications- it was necessary to restrict the in-depth analysis to the ten most relevant results for each document type, manually excluding "review-type articles", to ensure a more balanced and representative comparison.

Relevant results were summarized in Tables 3 and 4 to compare the approaches in terms of monitored water and type of analytes, materials used, manufacturing and detection methods, detection devices, and innovative contributions. This analysis makes it possible to present ofand describe the produced devices, to identify key trends in research and innovation in microfluidic for water quality monitoring.

5.1.1 Technological development-Patent documents analysis

In general, the patented microfluidic devices for water monitoring concern different types of water, different analytes and used different detection methods as presented in Table 3. Patents place more emphasis on the technical and functional aspects of the devices as presented in patent US10232382B2 [80] where a single-walled carbon nanotube (SWNT)-based sensor was described. Also, a paper-based layer with a test cell divided into two regions with different contact angles to regulate liquid flow in EP 4552740 A1 [81], and a device with two main components : premixing chip and anlysis one in US 2022/0032297 A1 [82].

As a technical document, the patent aims to provide a sufficiently detailed description of the invention to enable a person skilled in the art to reproduce it. This includes technical diagrams, precise claims and structured descriptions, often accompanied by explicit figures illustrating the components, arrangement, or operation of the device as the example illustrated in (Fig. 19).

Materials Research Foundations 191 (2026) 128-179 https://doi.org/10.21741/9781644904213-5

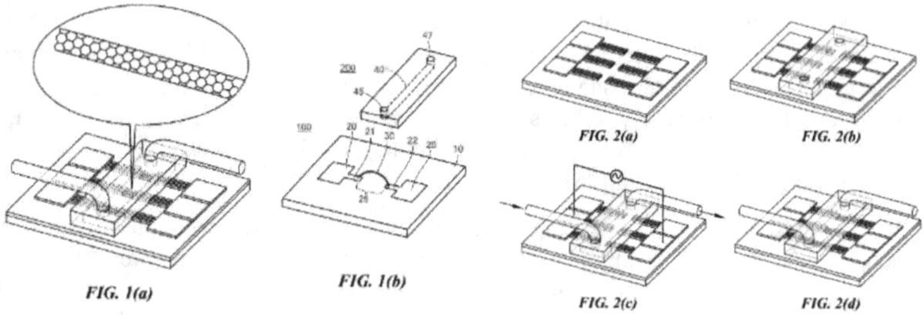

Figure 19. Schematic representation of a microfluidic device « Wireless SWNT sensor integrated with microfluidic system for various liquid sensing applications » from the patent US10232382 B2 [80]

On the other hand, patents of (Table 3) prestent innovations in microfluidic integration such as the integration of separate sample and reagent in US 2022/0362766 A1[83], the optical detection integrated with pumps and data acquisition electronics for high multiplexing and real-time autonomous detection in US 11209412 B2 [84], a fully passive system integrating multiple reactant and separation tanks in US 9410894 B2 [85] and a fully integrated portable system with on-board sampling, reagent storage, mixing, sensing, data processing, cleaning, and data transmission in US 2018/0128779 A1 [86].

The US 2024/0125699 A1 [87] presents an innovative bubble brightening structure that enhances the optical signal. The use of new materials and detection methods is also present in patent documents such as the use of CVD for multiplex detection in a µPAD platform in US 2019/0118175 A1 [88] and the use of a sustainable material for a wood-based microfluidic cassette for nitrate detection in WO 2025/042580 A2 [89].

However, the overarching objective remains the continued improvement of portability, autonomy, lightness, rapid analysis, and multiplexed detection, in response to the growing demands of in-situ water quality monitoring.

5.1.2 Academic development-Scholarly documents analysis

In parallel, academic publications (Table 4) focus mainly on detection mechanisms and analytical methods: colorimetric as presented in the works of Gillian et al. [90], Xiaolu Xiong et al. [91], DaSilva et al. [18], electrochemical as a detection in a multifluidic senor of Dai et al. [92], biolumeniscence for rapid screening of *E.Coli* [93], fluorescence as used in a magnetic Separation (µFMS) platform for water monitoring by Castillo et al. [2]. Il also emphasize analytical validation and performance evaluation across different types of water: drinking, industrial, wastewater and different analytes: heavy metals, bacteria, nutriments, COD, *E.Coli*, pesticides, pH, … (Table 4). A clear trend towards the use of µPADs and the exploitation of the simplicity, low cost and modularity of microfluidic devices is notable in academic work [91, 18, 94, 95], in particular to extend their applications to the monitoring of various parameters such as pesticides [22] or *E.Coli* [95].

Several results from academic research are presented as promising scientific innovations, such as the use of "stacked electrospun nanofibers" in a 3D µPADs as proposed by Chen et al. [94]. and the use of magnetic tagging for bacteria separation as used by Castillo et al [2]. Other examples include the integration of Immunomagnetic separation (IMS), IFAST (Immiscible Filtration Assisted by Surface Tension), and ATP biolumeniscence assay in a single microfluidic device as proposed by Bangkot et al. [93] as well as the application of the Ion Concentration Polarization (ICP) on µPADs used by Perera et al. [95].

However, the in-depth analysis revealed that these advances presented as new in the academic literature, actually correspond to technologies already patented (prioity dates before 2014). Their novelty lies in the adaptation of these technologies within µPADs architecture and/or their application in water quality monitoring. (Fig. 20) illustrates some previously patented technologies that have been re-exploited in new applications, underscoring the ability of researchers to repurpose existing inventions to adress emerging challenges to developing new µPADs for water quality monitoring [96,97]. Reusing previously patented technologies is a common and beneficial practice that helps accelerate innovation by building on well-established technological foundations to explore new applications and contexts.

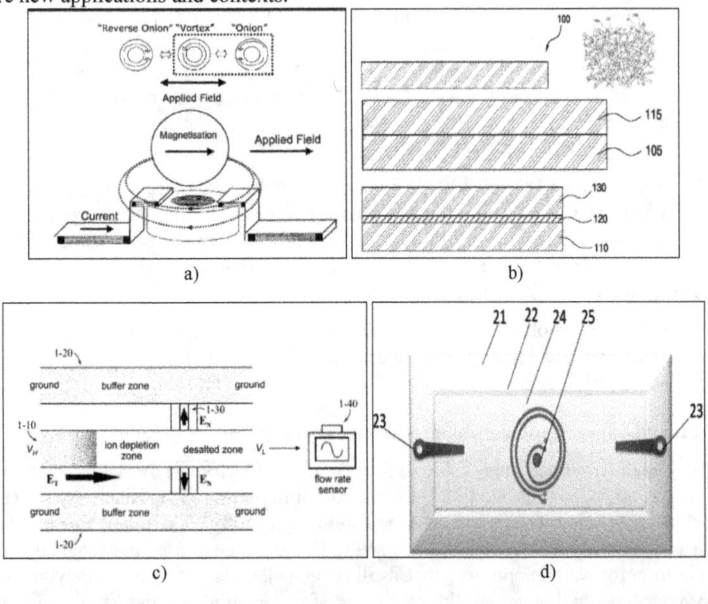

Figure 20. Examples of Patented technologies exploited in advances academic micofluidic devices for water monitoring (a-Insulation sheet, from WO2015026116A1 [98], b-Magnetic tagging techniques US8283916B2 [99] , c) Amplified electrokinetic fluid pumping switching and desalting, US8834696B2 [100], d) Trace cell capture system based on microfluidic and immunomagnetic separation dual strategies, CN109486653B [101].

Materials Research Foundations 191 (2026) 128-179 https://doi.org/10.21741/9781644904213-5

This observation underscores the importance of cross-analyzing scientific literature and patent databases to assess the true degree of originality, technological maturity, and freedom to exploit innovations for research studies.

This analytical approach, rarely undertaken systematically, highlights a potential gap between academic research and protected technological innovation, as well as a form of complementarity between the two spheres. While academic research explores new concepts, patents reflect concrete applications and commercialization intentions.

Finally, a common trend in both types of documents clearly emerges: the development of compact, integrated, in-situ, rapid, and precise analytical systems, meeting the growing demands of water monitoring.

Table 3. The most relevant document patents for microfluidic devices for water monitoring for the last decade

Applicant/ inventors	**Title**	**Wter type and analytes**	**Fabication/material**	**Method of detection**	**Illustration**	**Key innovation**	**Patent document Number/ Date/ Ref**
Univ Northeastern/ Wang Ming , Liu Yu , Dokmeci Mehmet Remzi	Wireless SWNT sensor integrated with microfluidic system for various liquid sensing applications	**General water** quality monitoring for **Various chemical agents**	Single-walled carbon nanotube (SWNT)-based sensors, conductive microelectrodes	**Electrical detection** via dielectrophoretic assembly across microelectrodes in a microfluidic channel	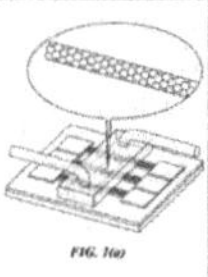	A **multiplexed nanosensor** platform with functionalized SWNTs; **autonomous and wireless monitoring**; robust integration of delicate nanotube sensors into microfluidic systems.	US 10232382 B2 Mar 19, **2019** [80]
Univ South Australia/ Priest Craig Ian , Elmas Sait , Pospisilova Aneta , Sekulska Aneta Anna	Microfluidic Sensor for Continuous or Semi-Continuous Monitoring of Quality of an Aqueous Solution	Any fluid, including **drinking water, wastewater, or industrial water** for pH (extendable to other parameters)	Microfluidic channels fabricated on a solid substrate (likely polymeric or glass); includes optical window for detection.	**Optical detection of colorimetric** response via a pH indicator after microfluidic mixing; measured through an optical reading window.	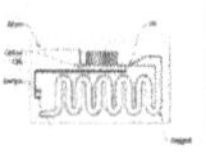	**Integration** of separate sample and reagent (pH indicator) channels with **a mixing unit** and optical readout, enabling compact, real-time pH measurement.	US 2022/0362766 A1 Nov 17, **2022** [83]
Univ Texas Tech System/Kim Jungkyu , Devadhasan Jasmine Pramila , Howse Ryan	Chemically patterned microfluidic paper-based analytical device (c-µpad) for multiplex analyte detection	**water** (real-world water quality testing) **for Heavy metals** (multiplex detection)	Microfluidic paper-based analytical device (**µPAD**) fabricated via chemical vapor deposition (CVD); uses hydrophobic barriers, substrate layer, and immobilized reagents.	**Colorimetric detection** via chromogenic reagents that change color upon contact with the target analytes	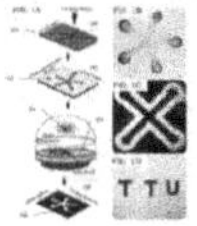	**A µPAD platform using CVD for multiplex detection**; multilayer capability allowing the multiplex detection of heavy metals; low-cost and field-usable	US 2019/0118175 A1 Apr 25, **2019** [88]
Univ Maryland/ Rao Govind , Kumar Sanjeev , Srinivasan Venkatesh , Ahuja Preety , Kadolkar Revati	Wood microfluidic engineering: development of highly stable polymer nanocomposit	**Nitrate in water samples**	Microfluidic cassette made of wood or polymer - Sensor includes a graphite-based electrode coated with Pani/C nanocomposite	**Electrochemical** impedimetric detection using a sensor based on a polyaniline/carbon (Pani/C) nanocomposite functionalized		**- Use of a wood-based microfluidic cassette** for nitrate detection - Impedimetric sensor offering high specificity to nitrate - Wide temperature operability (5°C–	WO 2025/042580 A2 Feb 27, **2025** [102]

	e electrochemical nitrate sensor		- Metal complexes (e.g., Ni(II))	with nitrate-binding metal ion complexes		65°C), **reproducibility,** and commercial-grade performance	
Univ California/ Hasty Jeff , Baumgart Leo Alexander , Cookson Scott , Ferry Michael , Graham Garrett , Huerta Ramon , Johnson Ryan , Tsimring Lev	Microbial microfluidic biosensor	Suitable for **drinking water, surface water, wastewater, and groundwater. For Heavy metals** (e.g., arsenic, lead, mercury, cadmium, copper, chromium, zinc, cobalt), ammonia, and malathion (a pesticide)	Microfluidic chip with parallel trapping chambers. Uses genetically engineered microorganisms (bacteria, microalgae, fungi, etc.). Compatible with robotic spotting tools.	**Biosensors. Optical detection (fluorescence/luminescence)**, integrated with pumps and data acquisition electronics.	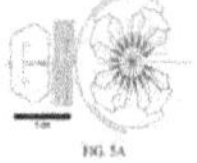 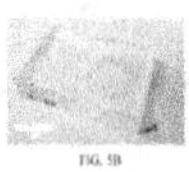	**High multiplexing** (many analytes at once). **Real-time autonomous detection**. Field-deployable and long-term storable system. Can store lyophilized (freeze-dried) cultures. Supports up to 4,000 strains	US 11209412 B2 Dec 28, **2021** [84]
China Petroleum & Chem Corp , Sinopec Res Institute of Safety Engineering Co Ltd/ Jiang Huiyun , Jin Yan , Wang Shiqiang , Feng Junjie , Zhu Liang , Xiao Anshan , Sun Bing , An Fei , Wang Haozhi	Microfluidic Paper-Based Chip, Microfluidic Test System, Liquid Test Method, and Use	General liquid samples (intended for **water, environment, food, or medical testing**)	- Paper-based layer with a test cell divided into two regions with different contact angles - hydrophobic and hydrophilic treatments - cover layer, bottom layer, vent hole, diffusion channel, chromogenic reagent	Chromatic migration-based **colorimetric detection** on a **microfluidic paper-based** chip with controlled contact angle gradients	  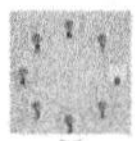	- **Regulates liquid flow** by creating regions with different contact angles to drive chroma toward the center - **Increases chromaticity, sensitivity, and repeatability** - Improves color uniformity and detection limit -includes control unit (for T°, humidity, vacuum)	EP 4552740 A1 May 14, **2025** [81]
Chambre De Commerce Et D'industrie De Paris/ Angelescu Dan Eugen , Freyermuth Pierre	Microfluidic device for analyzing a pressurized fluid	**pressurized water in pipelines, including drinking water distribution networks**. Adaptable to **various analytes (eg. Chlorine)** depending on the reactant used	Microfluidic chip located outside the pipe, connected via a sampling channel with a protruding tip inserted into the pipe. Reactant tank includes: A water compartment, A reactant compartment, A separator (piston, flexible membrane, or bag).	Analyzer is an **optical cell**, various configurations described: With filtered or broadband light sources, optical cavity, monochromators, optical fibers, or liquid waveguides.	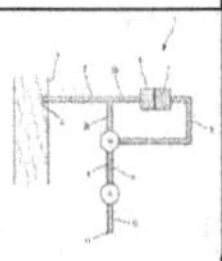	**Fully passive system: operates without any pumps**—flow is driven by existing pipe pressure and ambient pressure at outlet. Suitable for in situ, real-time monitoring in water infrastructure. Modular design allows integration of multiple tanks (e.g., for different reagents or cleaning agents). Low-energy, maintenance-friendly setup .	US 9410894 B2 Aug 9, **2016** [85]
Haimann Richard Alan , Li Bo/ Haimann Richard Alan , Li Bo	Microfluidics method for detecting chemicals in water in near real time	Likely applicable to **environmental water, industrial effluents, or natural waters** For Total	Microfluidics device with integrated: Electrically actuated peristaltic pumps and valves, Ultrasonic transducers (for mixing and cleaning),	**Electrochemical** stripping voltammetry using multi-electrode setup, Pattern recognition / **machine learning**		**Fully integrated portable system: On-board sampling, reagent storage, mixing, sensing, data processing, cleaning, and data transmission**,	US 2018/0128779 A1 May 10, **2018** [86]

Materials Research Foundations 191 (2026) 128-179 https://doi.org/10.21741/9781644904213-5

		recoverable **metals and additional parameters: pH, temperature, electrical conductivity** (EC)	Electrodes, Sensors: pH, EC, temperature, Central processing unit (CPU), computer code, Transceiver for wireless data transmission, Battery, USB, and power management circuits	**(ANN)** on remote servers to correlate signal to concentration,		Autonomous operation via embedded software, Self-calibrating and self-cleaning features	
Klearia / Doungmene Floriant , Gautrin Lisa , Magnier Aurélie , Nanteuil Clémen	Microfluidic Method for Analyzing Metals	**Especially for drinking or environmental water for Any metal trace element**, Primary analyte: Arsenic (both **As(III) and As(V)**)	Two main components: Premixing chip: Combines sample and reagents (Nitric acid (HNO_3 and L-cysteine) Analysis chip: Performs detection with electrodes Reservoirs: For reagents and standard solutions	**Electrochemical detection** (voltammetry or similar). Uses at least 3 electrodes: Gold working electrode, Platinum reference electrode and Platinum counter-electrode	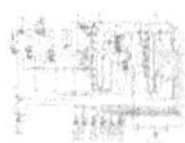	On-chip speciation of arsenic. **Accurate detection via standard addition method**. Reusable thanks to electrode cleaning and reconditioning. Designed for ultra-trace detection (ppb level)	US 2022/0032297 A1 Feb 3, **2022** [82]
Univ South China Tech/ Li Zongtao , Wang Hong , Xing Yongheng , Li Jiasheng , Li Jiexin , Lu Zhou , Wu Junhao	Microfluidic Ion Detection Chip Having Bubble Brightening Structure, and Detection Method Thereof	**environmental water** analysis, particularly focusing on water quality impacted by **toxic ion contamination, for Heavy metal ions**	Substrate and upper plate: Made from PMMA, PDMS, or quartz. UV adhesive, plasma bonding, and hot pressing. Glass cover plates: Serve for optical detection. Excitation light source (UV lamp or blue LED)	**Optical detection using a photoelectric detector**, Uses gas-generated microbubbles to increase the optical signal (bubble brightening effect)	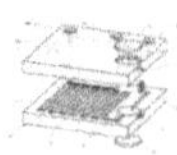	Novelty lies in the **"bubble brightening structure" that enhance the optical signal of ion detection**. Integrated gas and liquid microfluidics to amplify signal without increasing reagent consumption. Modular and scalable design for compact, low-cost, and field-deployable water testing	US 2024/0125699 A1 Apr 18, **2024** [87]

Materials Research Foundations 191 (2026) 128-179 https://doi.org/10.21741/9781644904213-5

Table 4. The most relevant document patents for microfluidic devices for water monitoring for the last decade

Affiliation/ Authors	Title	Water type and analytes	Fabication/material	Method of detection	Illustration	Key innovation	document Type/Date/ Ref
Dublin City University. School of Chemical Sciences/ Gillian Duffy et al.	Development and optimisation of colourimetric microfluidic sensors for water quality monitoring	**Fresh water, Wastewater effluent, Agricultural and environmental waters for Phosphate (PO_4^{3-}) and Cr(III)/Cr(VI)**	Microfluidic discs (likely thermoplastics such as PMMA or PC). Motors (for disc rotation – centrifugal platform). Optical components (LEDs, photodiodes for absorbance). Syringe pumps (for autonomous chip-based version). Electronics board and power supply	**Colourimetric detection**: Detection via absorbance measurement (optical sensing).		**Autonomous** with on-board data storage and periodic sampling. Use of **Centrifugal microfluidics for automated mixing**/analysis. Cost-effectiveness Regulatory monitoring needs	Dissertation Sep 1, **2017** [90]
Centre for Quantum Physics, Key Laboratory of Advanced Optoelectronic Quantum Architecture and Measurement (MOE), …./Xiaolu Xiong et al.	Simultaneous Cross-type Detection of Water Quality Indexes via a Smartphone-App Integrated Microfluidic Paper-Based Platform.	**Industrial wastewater and river water. pH: 5–9** **Nitrite**: 2.3–90 ppm (LOD: 0.4 ppm) **Cu(II):** 3.8–400 ppm (LOD: 1.9 ppm) **Ni(II):** 2.9–1000 ppm (LOD: 2.9 ppm) **Fe(III):** 2.8–500 ppm (LOD: 1.1 ppm)	Wax-patterned paper µPAD. Optimization of shape, baking time, reagent amount/ratios for better color uniformity and intensity	**Colorimetric detection via µPAD** (microfluidic paper-based analytical device) **and Smartphone** app for image analysis and concentration readout	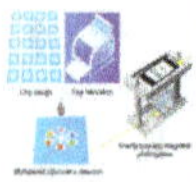	**Simultaneous multi-analyte detection** (metal ions, nitrite, and pH) **Integration of µPAD with a smartphone** app for real-time, on-site quantification Low-cost, portable solution	Journal article Nov 23, **2022** [91]
Institute of Microchemical Technology, Japan, …/ Hidekatsu Tazawa et al.	Development of microfluidic devices for on-site water quality testing using glass molding process.	**Drinking water and environmental water** (industrial contamination context) for Residual chlorine in the range of 0.1–1 ppm	Glass microfluidic device (50 mm diameter, 300 µm channel depth). Glass molding method. Diamond-like carbon (DLC) coating for durability	Microfluidic chemical sensing N,N-diethyl-p-phenylenediamine (DPD) **colorimetric method**	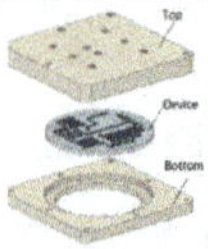	**a robust, weather-resistant glass microfluidic device .** the flow rate was unchanged after flow testing water samples for 10 days	Journal article Apr 27, **2023** [23]
Federal University of São Carlos, Brazil , …/Vinicius A O P da Silva et al.	Microfluidic paper-based device integrated with smartphone for point-of-use colorimetric monitoring of water quality index	Distribution water samples (**tap or potable water**). **Water hardness** (Ca^{2+}, Mg^{2+}): 20–560 mg/L, LOD: 0.083 mg/L **Catechol** (total phenols): 0.20–16 mg/L, LOD: 0.124 mg/L	µPAD (microfluidic paper-based device). 3D-printed support (polylactic acid filament)	**Colorimetric analysis using smartphone** camera	A reproduire avec boite et telephone 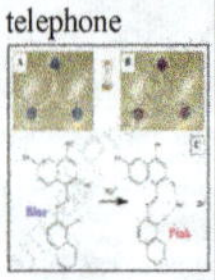	**Integration of µPAD and smartphone in a 3D-printed holder. Multi-parameter detection** (in a portable, low-cost format Simple, eco-friendly	Journal article June, **2020** [18]

Materials Research Foundations 191 (2026) 128-179 https://doi.org/10.21741/9781644904213-5

		pH: Range 4.7–12, LOD: 0.262					
amueli School f Engineering, Jniversity of Salifornia,… / Hamsa N Gowda t al.	Development of a proof-of-concept microfluidic portable pathogen analysis system for water quality monitoring.	**Contaminated water sources in vulnerable/developing areas**. **pathogen analysis** system that can detect bacteria in water (Enterococcus faecalis)	Centrifugal microfluidic disc (CD) Add-on bacterial concentration cup with absorbent polymer beads Integrated in a portable CD Driver system	Digital droplet loop-mediated isothermal amplification (ddLAMP). **On-disc integration of lysis, extraction, mixing, and amplification**	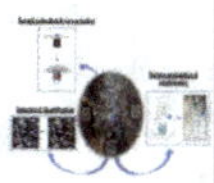	**Microfluidic disc integrates a complete digital DNA amplification assay.** Sample-to-answer results achieved within one hour in a single step. On disc bacterial concentration through integration with adsorption beads.	Dec 21, **2021** [103]
Shanjiang Jormal Jniversity / Xiaojun Chen et al.	A Flexible Method for Nanofiber-based 3D Microfluidic Device Fabrication for Water Quality Monitoring.	**Drinking water Iron ions (Fe^{3+})** Concentration range tested: 0.001–0.09 mol/L	3D multilayer nanofiber-based µPADs. Electrospun nanofiber membranes Electrostatic wax printing for hydrophobic barriers	**Colorimetric reaction** using $K_4Fe(CN)_6$ with $FeCl_3$ Image analysis via grayscale intensity (Adobe Photoshop)	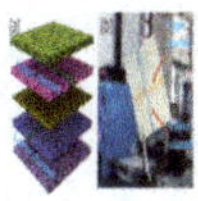	First use of **stacked electrospun nanofibers in 3D µPADs.** **Electrostatic printing** Flexible and adhesive-free device assembly	Mar 6, **2020** [94]
Jniversity of lorida/Keisha Y Castillo-Torres t al.	A High-Throughput Microfluidic Magnetic Separation (µFMS) Platform for Water Quality Monitoring.	**Drinking water and recreational water for Escherichia coli** (E. coli) and potentially other bacterial targets (with ~94% capture efficiency for tagged bacteria)	High-throughput microfluidic magnetic separation (µFMS) device. Uses 1.5 µm magnetic microdiscs to tag bacteria. External magnet array (not integrated)	**Magnetic separation of tagged bacteria.** Provides imaging-ready substrates for fluorescent microscopy	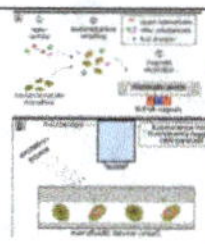	a promising proof-of-concept for rapid, high-volume **bacterial detection in water using magnetic tagging and microfluidic separation,** with a future aim of automated fluorescence-based quantification.	Dec 22, 2019 [2]
Jniversity of Hull, UK, …/ Pongkot Igamsom et al.	A Microfluidic Device for Rapid Screening of E. coli 0157:H7 Based on IFAST and ATP Bioluminescence Assay for Water Analysis	**Wastewater effluent for Escherichia coli** (10^4 CFU/mL in wastewater (without pre-concentration)	Microfluidic device integrating: Immunomagnetic separation (IMS) IFAST (Immiscible Filtration Assisted by Surface Tension) -based ATP assay detection zone Photomultiplier tube (PMT) for bioluminescence readout	**Immunomagnetic capture + ATP-based bioluminescence assay** **Bioluminescence signal** measured using a PMT. Fully integrated on-chip sample processing and detection	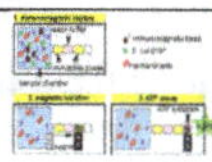	**First reported system integrating IMS, IFAST, and ATP assay in a single microfluidic device.** No off-chip pre-concentration required. Rapid, user-friendly, low-cost, and point-of-need compatible	18 Aug **2017** [93]
Nanyang Technological Jniversity, Singapoure/ A T Perera et al	Rapid pre-concentration of Escherichia coli in a microfluidic paper-based device using ion concentration polarization	**Coastal water (high salinity) and Freshwater** (with adjusted ion concentration). **Escherichia coli** (E. coli)	µPAD (microfluidic paper-based analytical device), Contains: Nafion® membrane Built-in micro wire electrodes (for voltage application)	**Ion Concentration Polarization (ICP) to pre-concentrate E. coli.** **Followed by fluorescence detection**	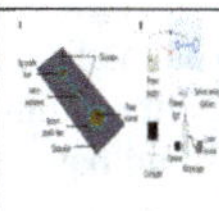	**First demonstration of ICP on a µPAD for E. coli pre-concentration** Rapid, inexpensive, and suitable for high-salinity environments	Nov 10, **2019** [95]

Ningbo University Chinese Academy of Sciences/ Dai Jinying et al.	Microfluidic sensor integrated with nanochannel liquid conjunct Ag/AgCl reference electrode for trace Pb(II) measurement.	**Various water samples (Drinking water, Tap water, Lake water, River water) Lead ions (Pb^{2+})**, (LOD): 0.13 mg/L	Integrated electrochemical microsensor with: "Glass–silicon–glass" sandwich structure Nanochannel-based Ag/AgCl reference electrode (RE) Silicon-glass double-sided bonding for detection chamber	**Electrochemical detection** (current signal from Pb^{2+} oxidation) Amperometric measurement with reference/working electrode system	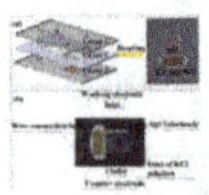	**Integrated, miniaturized, and stable sensor for trace Pb^{2+} detection High selectivity, sensitivity, and reproducibility**. Stable signal over time	Apr 14, **202** [92]
Ningbo University/Jiawen Yin et al.	A batch microfabrication of a microfluidic electrochemical sensor for rapid chemical oxygen demand measurement.	**Environmental waters, Chemical Oxygen Demand (COD)**. Interval : 4.17–200 mg/L, LOD: 2.05 mg/L	MEMS-based microfabrication Electrode materials: Ag/AgCl thin film PbO_2 working electrode (oxidation of organic matter) Pt counter electrode Silicon substrate with Pt–Si–Pt contact structure	**Electrochemical oxidation in a microfluidic detection** chamber using Potassium Acid Phthalate (PAP) as standard organic matter	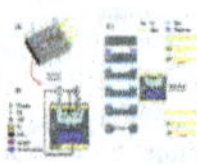	compact, fast, and scalable electrochemical sensor for real-time COD detection, with low-cost, high-throughput devices.	an 26, **2021** [104]
iochemistry Department, Faculty of Science, Ege University, Turkey / Hichem Moulahoum	Dual Chromatic Laser-Printed Microfluidic Paper-Based Analytical Device (µPAD) for the Detection of Atrazine in Water.	**Environmental water. Pesticide: Atrazine**. (LOD): 3.5 µM with AgNPs 10.9 µM with AuNPs	Laser printing (LP) used to create µPADs. Signal molecules: (AgNPs) and (AuNPs)	**Colorimetric reaction with visual change analyzed by a smartphone** camera	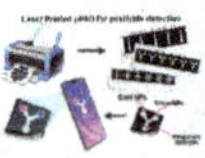	**laser printing for fabricating µPAD biosensors Dual-signal detection (AgNPs and AuNPs) enhances reliability Good repeatability, reproducibility, and selectivity** even in presence of interferents (e.g., NaCl, pH)	Oct 28, **202** [22]

5.2 Applications in water monitoring

Water quality is a global health, environmental, and economic priority. Access to clean and safe water is a major challenge for public health, food security, ecosystem protection, and sustainable development. To assess water quality, it is essential to monitor a range of physical, chemical, and microbiological indicators depending on the nature of water and in accordance with environmental and health regulations such as the limits imposed by the WHO and FAO [105,106]. The development of rapid, portable, and low-cost analysis technologies such as microfluidic devices addresses the need to make this monitoring more accessible, in real time and decentralized, particularly in areas with limited resources or in contexts of intensive environmental monitoring.

To further support the thematic analysis – particularly with regard to the materials used, detection methods, application contexts, limitations, and development prospects – it is necessary to draw on additional publications and patents. This will strengthen the observed trends, fill in certain gaps, and broaden the discussion across the field.

The field of microfluidics in general, and in particular with the rise of paper-based devices (µPADs), opens the way to the analysis and detection of a wide variety of elements and analytes.

Materials Research Foundations 191 (2026) 128-179 https://doi.org/10.21741/9781644904213-5

The applications of microfluidic devices for water monitoring can be grouped into two main classes of analytes and elements which are at the heart of trends and perspectives:

1. ***Physico-chemical parameters***: Microfluidics is being actively developed for monitoring these water quality parameters, and key analytes include:

- Heavy metals [88,87],such as iron (Fe) [91, 94], copper (Cu, Cu(II)) [91], mercury (Hg, Hg(II), Hg^{2+}) [107], lead (Pb, Pb(II), Pb^{2+}) [92], silver (Ag) [108], chromium (Cr) [90], arsenic (As) [82], cadmium (Cd, Cd(II)), and nickel (Ni, Ni(II)) [91].
- Nitrite [91], nitrate [89], ammonium [84], phosphates [90].
- Water hardness [18,109] .
- COD [104,110,111]
- Halides [23,85].
- Pesticides [22,84], herbicides.
- pH [18,83,86,91], temperature [86], electrical conductivity [86]
- Other chemicals such as surfactants [112], phenols [18], etc

2. Microbiological parameters: Microfluidic devices, and especially μPADs, are considered to have immense potential for the analysis and detection of microbiological parameters of water quality including, the detection and quantification of pathogenic microorganisms [103] or contamination indicators such as Escherichia coli [2,93,95] and total coliforms [113].

5.3 Detection and monitoring methods

Water quality monitoring relies on various analytical methods depending on the type of targeted analyte. These methods can sometimes be combined on a single chip (lab-on-a-chip) for multiplex detection systems, and often, they are combined with methods of extraction or concentration of the analyte before its analysis. The main methods for detecting and monitoring water quality based on microfluidics are:

5.3.1 Colorimetric Detection

This is the most common and simplest method. It relies on the production of a color or a change in color intensity in the presence of the target analyte (Fig. 21). It is easy to use, and results can often be read visually or using additional equipment such as a smartphone for quantitative measurements. Color intensity is generally proportional to the analyte concentration [114]. This method has been widely used in both academic work and patents, particularly using μPADs (Tables 3 and 4), it has various applications in water monitoring, for the detection of pH as presented in the patent document US 2022/0362766 A1[83] and the publication of Xiong et al. [91], heavy metals, for the detection of which it has been the most used method such as in [90] and [91], nitrite as used in [115] and nitrate as used in [91]. It was also used for the detection of pathogens such as *E.Coli* [116] or chemical parameters such as COD [117].

Materials Research Foundations 191 (2026) 128-179
https://doi.org/10.21741/9781644904213-5

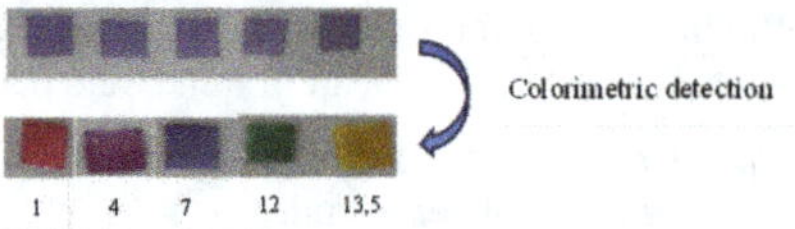

Figure 21. Color change of cabbage-impregnated paper (anthocyanins) depending on water pH. (Image produced by the authors)

5.3.2 Electrochemical Detection

This method involves electrochemical reaction signals measured using electrodes integrated into the microfluidic device. It can provide more stable and quantifiable signals than colorimetry, and can achieve lower detection limits and better sensitivity/selectivity for some applications [33]. Common techniques include amperometry, voltammetry, potentiometry, electrochemical impedance, and capacitance [24]. Two or three electrode systems can be used (Fig. 22), and different materials are employed, such as carbon or metals (silver, gold, copper). This method has been used in microfluidic devices for water monitoring for the detection of temperature, pH, and electrical conductivity of several types of water, as presented in the patent document US 2018/0128779 A1 [86], the detection of heavy metals in the patent document US 2022/0032297 A1[82] and the scientific publication of Dai et al. [92]. It was also used for COD detection by Jiaw et al. [104], for phenol [118], ethanol [119], for multianalyte halide detection as used by Cuartero et al. [120], and nitrate as presented in the patent document WO 2025/042580 A2.

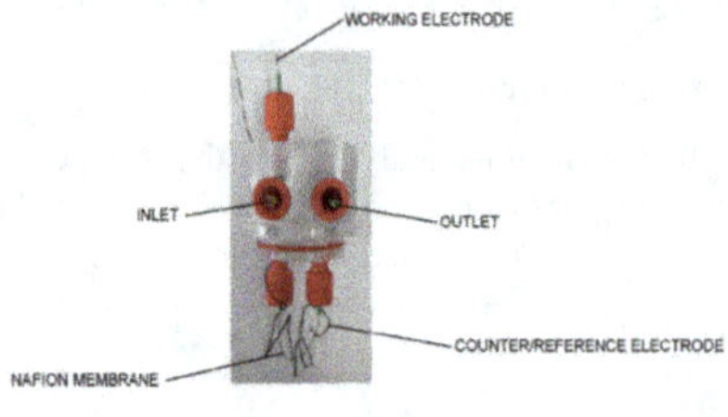

Figure 22. Microfluidic Nafion-based electrochemical cell [120]

5.3.3 Fluorescence Detection

This method uses fluorescent markers that emit light upon excitation, with the amount of light emitted being related to the concentration of the analyte [20] (Fig. 23). It can offer superior sensitivity to colorimetry for the detection of low concentrations of analytes. Its applications in water monitoring have included, for example, the detection of heavy metals and pesticides as presented in the patent US 11209412 B2 [84]. It was also used by Perera et al. [95] for *E.Coli* with a Nafion membrane in a μPAD device and by Lefevre et al. [121] for a herbicide detection using a biosensor.

Materials Research Foundations 191 (2026) 128-179 https://doi.org/10.21741/9781644904213-5

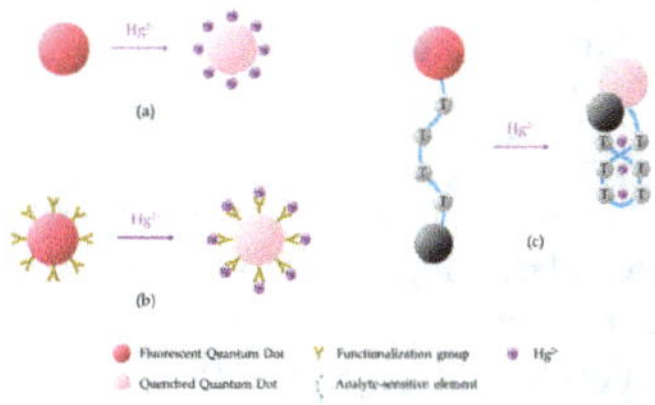

Figure 23. Sensing mechanisms based on fluorescent quantum dots. [20]

5.3.4 Detection by Chemiluminescence

For this method of detection, light is generated through a chemical reaction (or a biological one in the case of bioluminescence) without requiring an external light source [25] (Fig. 24). The technique is sensitive and efficient. The absence of an external light source reduces background noise compared to fluorescence. It has been applied in water monitoring for the detection of heavy metals by chemiluminescence-for example Cu(II) by Tyrell et al. [122], Co(II) by Chen et al. [123], Cr(III) [124] by Shang et al. and for microbiological contamination by bioluminescence, as used by Bongkot et al. for *E.Coli* detection without pre-concentration [93].

5.3.5 Electrochemiluminescence Detection

Electrochemiluminescence combines the principles of chemiluminescence and electrochemistry to generate light. A light signal is produced as a result of an electrochemical reaction [25] (Fig. 24). The method offers high sensitivity and selectivity and has been successfully used in microfluidic devices for water monitoring for the detection of Trenbolone by Song et al. [125] and the detection of dibutylaminoethanol (DBAE) and nicotinamide adenine dinucleotide (NADH) by [126] as examples of applications.

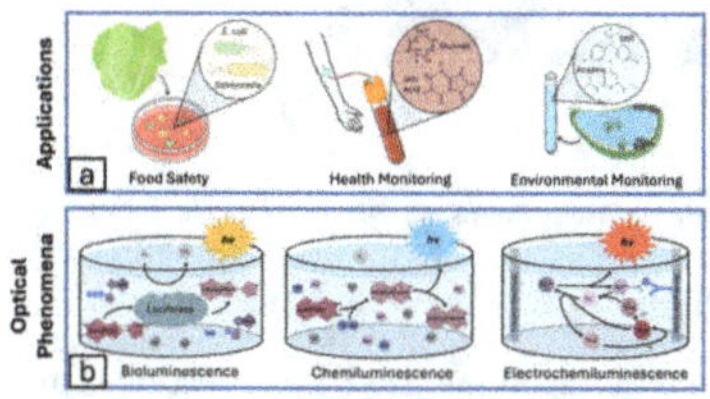

Figure 24. Chemiluminescence sensors: (a) application of analytical chemiluminescence in food safety, health monitoring, and environmental monitoring; (b) optical phenomena including bioluminescence, chemiluminescence, and electrochemiluminescence. [25]

5.3.6 Other Detection Methods

Other optical methods have been used for water monitoring, such as light scattering, used for direct optical detection of pollutants in water [127], or Surface-Enhanced Raman Spectroscopy (SERS),

Materials Research Foundations 191 (2026) 128-179 https://doi.org/10.21741/9781644904213-5

used for real-time sensing of toxic substances in water [128] and for the detection of glyphosate in drinking water [129] as application examples.

5.4 Manufacturing materials

The materials used to manufacture microfluidic devices are constantly evolving. Initially, devices were often made from silicon or glass in scientific publications [23,48,63,92] and patents (US 2022/0362766 A1) [83]. These materials, while useful, often require sophisticated equipment for processing and are expensive. Polymers, particularly PolyDiMethylSiloxane (PDMS) (Fig. 25), are also commonly used materials [123], they are valued for their transparency, ease of shaping, biocompatibility, and elasticity, which allows the integration of flexible elements. Poly(Methyl Methacrylate) (PMMA) [90,130] and other thermoplastics are also used in microfluidic devices [28].

However, a significant breakthrough has occurred with the use of paper as a substrate for microfluidic devices (Fig. 25), giving rise to paper-based microfluidic devices (μPADs) in scientific publications [116,124] and patent documents US 2019/0118175 A1 [88], EP 4552740 A1 [81]. Paper is considered the most widely used material for water monitoring as presented in (Table 4). Paper also has a high surface area that can improve detection limits for colorimetric methods and has the ability to store reagents. Commonly used paper types include chromatography paper, filter paper [18], and nitrocellulose membranes, as well as other porous substrates such as nanofibers [94], fiberglass membranes and fabrics are also used [130,131]. Carbone nanotubes and wood-based cassettes for microfluidic devices were also used as reported in patent documents US 10232382 B2 [80] and WO 2025/042580 A2 [102].

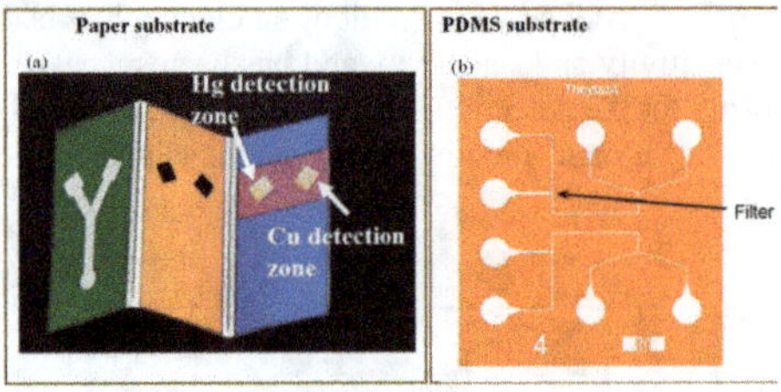

Figure 25. Paper and PDMS materials as microfluidic devices for water monitoring [63]

5.5 Devices and fabrications

Microfabrication is the manufacturing method that produces micrometer-scale structures on materials such as silicon, glass, polymers, or metals, using processes inspired by the semiconductor industry [30,104,110]. Manufacturing methods vary considerably depending on the material and the intended application. For traditional materials such as silicon, etching is used to form the channels [48], followed by assembly with a substrate to close the network. For polymers such as PDMS, molding is a common method [23], often performed using a mold obtained by photolithography [30] (Fig. 26). Hot embossing is a manufacturing method for thermoplastics that allows the high-volume production by printing a raised pattern onto the heated polymer.

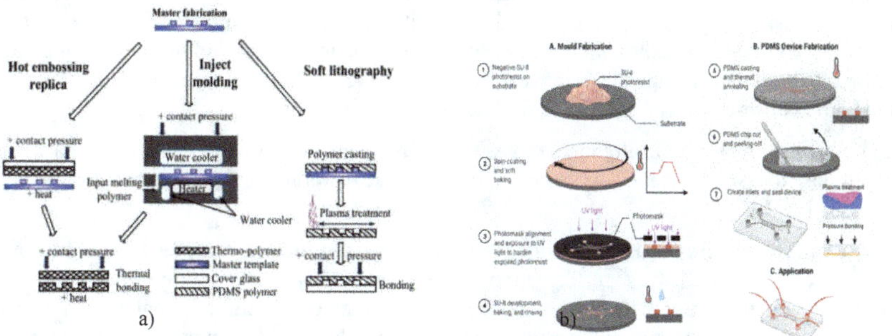

Figure 26. Representation of the fundamental microfabrication methods: a) Hot embossing and soft lithography b) molding methods for microfluidics fabrication. [30]

3D printing, also known as additive manufacturing, is a layer-by-layer manufacturing technology that creates three-dimensional objects from a digital model [128,132]. It is increasingly used to manufacture microfluidic devices for water quality monitoring [52]. It is a fast, flexible, and cost-effective alternative to traditional cleanroom microfabrication methods (Fig. 27). Rapid prototyping (RPT) technologies offer an efficient and flexible solution for the rapid manufacturing of microfluidic devices [51]. Using RPT molds and PDMS allows the production of high-resolution devices in a few hours, thus facilitating the complete design–manufacturing–testing–modification cycle in one day [133]. This approach is particularly advantageous for water monitoring, accelerating the development of customized sensors and support platforms [18].

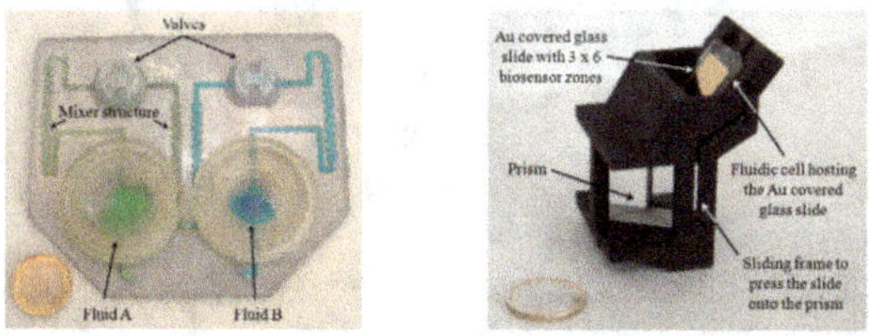

Figure 27. Representation of a) 3D-printed microfluidic devices for fluid mixing, b) prism holder platform. [51]

For paper-based devices (μPADs), the key manufacturing principle is to structure the paper by defining hydrophilic regions for fluid transport (the microchannels) and hydrophobic regions that act as barriers. Several techniques are used for this purpose (Fig. 28):

Photolithography: This is one of the first methods used for μPADs. It involves immersing the paper in a photosensitive resin, exposing it to UV light through a mask, and then washing it to remove the uncured resin, leaving the hydrophobic barriers [30,134]. This method achieves good precision but requires specialized and expensive equipment, and sometimes additional steps such as plasma treatment for aqueous solutions.

Materials Research Foundations 191 (2026) 128-179 https://doi.org/10.21741/9781644904213-5

Wax Printing: This is a very popular method for µPADs due to its simplicity, low cost, and potential for mass production [91]. A solid wax pattern is printed onto paper (using several methods, as presented in Fig. 28), then the paper is heated (in an oven, on hot plate, or through hot lamination) to melt the wax and allow it to diffuse through the paper, creating hydrophobic barriers [30,114]. It is considered very promising for high-volume processing.

Inkjet printing: Enables the automated deposition of reagents or hydrophobic materials (such as PDMS in ink form). It offers good control over the deposition volume and enables precise geometries [135,136].

Laser Cutting/Patterning/Writing: Allows channels to be defined by removing paper material or modifying its properties as performed by Nie et al. [137] using a mini-type CO_2 laser cutting/engraving machine. It is a non-contact method that can minimize contamination and achieve high precision, comparable to photolithography, while using less expensive equipment. It can enable mass production.

Plasma Treatment: Can be applied to hydrophilize areas on initially hydrophobized paper [138].

Simple physical methods: Include using craft cutting, scissors, hole punches, or embossing to create the patterns.

Several approaches are used to create 3D paper devices, which allow for complex fluid networks and the separation of reaction/detection zones, such as ***stacking layers*** of structured paper and adhesives [139,140] or using ***origami*** (Fig. 25) to fold 2D structured sheets of paper to create compact 3D structures [141].

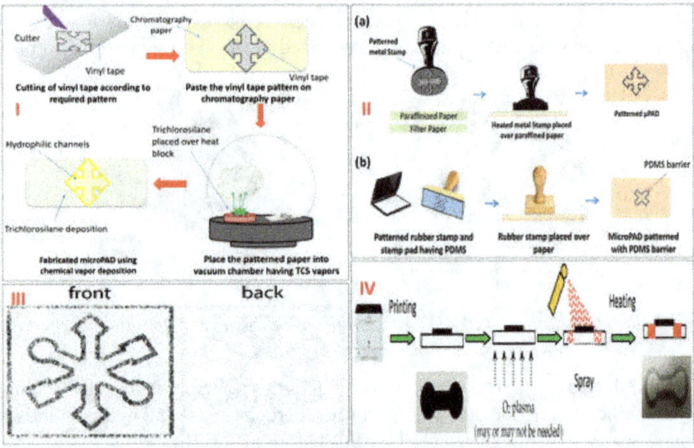

Figure 28. Schematic representation of several µPADs fabrication methods [142]

5.6 Integration of new technologies

The trend is toward integrating microfluidic devices (particularly µPADs) with electronic, optical, mobile communication technologies, and advanced signal processing methods [31,80,143]. These

Materials Research Foundations 191 (2026) 128-179
https://doi.org/10.21741/9781644904213-5

combinations aim to transform simple analytical devices into portable, inexpensive tools that are more sensitive, more versatile, potentially connected, and autonomous, and that require less expertise for their use, thus bringing them closer to "lab-on-a-chip" applications [144] for point-of-care, environmental monitoring, and potentially the IoT [8].

Electronic integration allows for the addition of power, control, display, and analysis functions directly onto the microfluidic device, such as the use of USB-powered μPADs [145]. Paper-based microfluidic fuel cells are being explored to power next-generation lateral flow devices [146].

The integration with Smartphones and Mobile devices allows for the reading and analyzing of results [18]. They can be used as signal readers for colorimetric detection by performing image analysis. Modern smartphones, with their digital cameras and processing capabilities, can replace more expensive laboratory equipment such as spectrophotometers or fluorometers [114].

To detect low-concentration analytes, increasing sensitivity is crucial, often through analyte preconcentration or signal amplification. Microfluidics allows the integration of preconcentration techniques such as ion concentration polarization (ICP) [95]. These techniques, sometimes powered by portable sources such as USB, can significantly improve the limit of detection (LOD). The use of nanoparticles (gold, silver) [147,148] is also a common method for signal amplification, particularly in colorimetric detections.

Interfacing paper devices with electronics and communication capabilities (smartphones, USB or wireless connectivity) makes these sensors suitable for Internet of Things (IoT) applications for distributed and real-time monitoring. The aim is to enable connected analysis and monitoring, potentially without requiring significant user expertise [149] (Fig. 29).

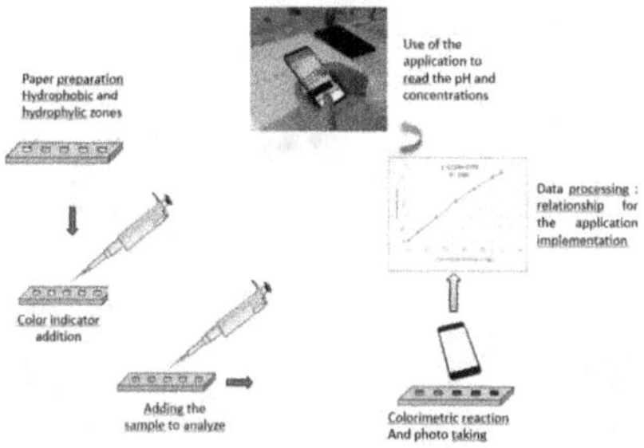

Figure 29. Water quality analysis with smartphone application integrated paper micofluidic device (image produced by the authors)

5.7 Advantages of microfluidic devices for water monitoring

In addition to the general advantages of microfluidics mentioned above, the main advantage of paper-based microfluidic analytical devices (μPADs) lies in their simplicity, low cost, and compatibility with field detection. These characteristics make them a promising and accessible solution for large-scale water quality monitoring, particularly in resource-limited environments. Significant advantages include [1,114,136,146]:

Materials Research Foundations 191 (2026) 128-179
https://doi.org/10.21741/9781644904213-5

• Low cost: Paper is a very affordable and widely available substrate, and the associated manufacturing techniques (such as wax printing) are also economical and allow for rapid and inexpensive mass production.

• Energy-free fluid transport: The capillary action of the cellulose fibers in paper allows samples and reagents to be transported through defined channels without the need for external pumps or a power supply, which is essential for portability and field use.

• Ease of use: The devices are designed to be simple to handle, often with direct visual interpretation of the results.

• Portability and light weight: The paper is thin and lightweight, making the devices easy to transport.

• Disposability and biodegradability: Paper-based devices are often designed for single use and can be easily incinerated or biodegraded, simplifying the safe disposal of biological waste.

• Biocompatibility: Cellulose paper is compatible with biological samples and allows for the immobilization of biomolecules.

• Reagent storage ability: Reagents can be stored in a dried form within the paper's fibrous matrix, extending their shelf life and eliminating the need for the user to handle liquid solutions.

• Compatibility with various detection methods: While colorimetric detection is common because it is visible to the naked eye, paper-based devices can integrate electrochemical, chemiluminescent, fluorescent, and other methods.

• Multiplexing potential: The paper's structure allows for the creation of multiple reaction zones for simultaneous testing of multiple analytes.

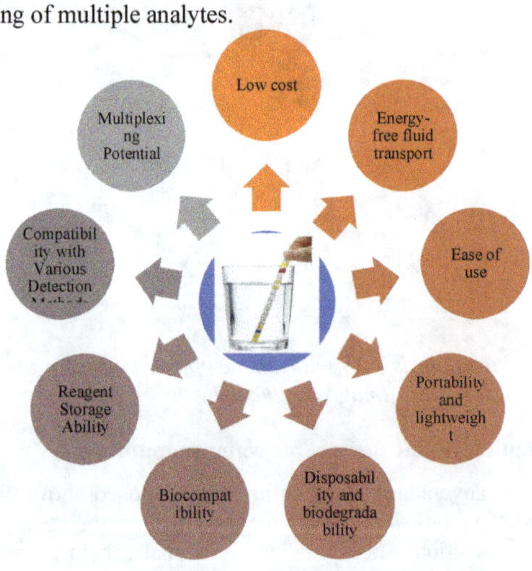

Figure 30. Avantages of microfluidic devices for water monitoring

Materials Research Foundations 191 (2026) 128-179
https://doi.org/10.21741/9781644904213-5

5.8 Limitations

The main challenges faced by microfluidic devices, whether conventional or paper-based, include the control of the diffusion-driven mixing. Conventional devices are limited by their high cost and the need for external equipment, while µPADs, despite their cost and portability advantages, face challenges related to flow speed and control, sensitivity/LOD, reproducibility, and robustness [26,109].

1- Flow management and control :

- Relatively slow flow rate: Passive capillary transport in paper can be slow, potentially taking hours for long distances (> 5 cm).
- Evaporation and drying: Long flow times can cause liquid evaporation and device drying.
- Less precise flow control: Although passive and simple, capillary transport can exhibit non-uniformities and backflow. Active methods (often used in conventional devices) offer greater control over flow. Moreover, control of flow rates, mixing, and sample-reagent interaction times is not yet fully understood because these devices are relatively new.

2- Sensitivity and Limit of Detection (LOD): µPADs often exhibit higher limits of detection (LOD) compared to established methods or traditional laboratory tests. This is partly attributed to detection methods (such as colorimetry) that are compatible with paper [26]. The increasing demand for sensitivity presents a challenge for the current format of µPAD tests, particularly those using colorimetric detection [150].

3- Reproducibility: µPADs can have poor reproducibility compared to traditional laboratory tests [30].

4- Mechanical Robustness: they lack the mechanical strength of traditional materials such as glass or silicon [23].

5- Manufacturing Complexity for Patterning: Although paper is inexpensive, creating hydrophobic barriers to define channels (e.g., by wax printing or photolithography) can be challenging for large-scale manufacturing. Also, high-resolution patterns and the fabrication 3D structures by layering are difficult to achieve [31].

5.9 Trends and prospects

-The rise of "Lab-on-a-chip systems": Microfluidics promises the imminent arrival of complete analysis systems that can analyze a sample from a single droplet and extract information about its contents. These systems are seen as technological tools capable of implementing revolutionary lab-on-chip plateforms [144].

-Growing and diversified applications: Microfluidics is experiencing considerable growth in many industrial fields such as food, biotechnology, medical and chemical analysis. Its application naturally extends to the life sciences [3,143].

-Paper-based microfluidics (µPADs) as a major trend: Paper-based microfluidic devices (µPADs) are particularly highlighted in the litterature as a technology with high potential. They have emerged as an attractive platform for chemical and biological detection [88].

Materials Research Foundations 191 (2026) 128-179 https://doi.org/10.21741/9781644904213-5

-Improvement of the colorimetric method: The use of nanoparticles (gold, silver) is common to enhance the colorimetric signal. The "coffee-ring" effect can be exploited to amplify the colorimetric signal and improve the limit of detection (LOD) [26].

Developments and prospects:

1. **Sensitivity Improvement (LOD):** Techniques such as sample preconcentration (e.g., using ion concentration polarization - ICP) are being explored to lower detection limits [95].

2. **Advanced Fabrication Techniques:** The development of simpler, faster, and less expensive fabrication methods that enable mass production is essential. Improving pattern resolution (to create smaller channels and denser devices) is a significant challenge. Techniques such as wax printing, inkjet printing, photolithography, paper die-cutting, plasma processing, etc., are being used and continuously refined [30].

3. **3D Patterning:** The fabrication of three-dimensional devices, often by assembling paper layers, is a trend aimed at enabling multi-step analyses and improve sample dispersion. Understanding and predicting 3D flow behavior is the subject of several experimental and modeling research projects [51,52].

4. **Flow Control:** Better control of flow velocity, mixing, and interaction times is being sought. Active approaches, sometimes with miniaturized components, are being studied to improve mixing uniformity and efficiency [149].

5. **New Materials:** Exploring porous materials as alternatives to standard cellulose paper is an avenue for expanding applications and improving performance. Using flexible and inexpensive materials with greater control over their structure is one potential area of research [28].

6. **Functional Integration:** Integrating different analytical steps (preparation, reaction, detection) and different technologies (electronics, optics) into the same device is also a research avenue [80].

7. **Robustness and Reproducibility:** Improving mechanical robustness and the reproducibility of results compared to laboratory methods is necessary [26].

8. Biomolecule Immobilization: The ability to stably and efficiently immobilize delicate biomolecules (antibodies, enzymes, DNA) on microfluidic supports is an ongoing challenge [31,77] .

9. **Manufacturing and Commercialization:** Continued efforts are needed to develop simple, reproducible, and inexpensive mass manufacturing methods, which is crucial for the large-scale commercialization of devices designed to detect these diverse analytes, particularly for field applications [31].

Finally, (Table 5) compares the main limitations with the associated trends/perspectives to better synthesize, clarify, and leverage this analysis.

Conclusion

This chapter used a targeted bibliometric approach to highlight the developments, trends, and prospects in microfluidics, its materials, manufacturing methods, and applications, particularly for the purpose of water quality monitoring. The quantitative analysis identified the most active topics, the most widely used technologies, and the most influential players, while the qualitative review provided an in-depth discussion of major advances and current limitations.

In summary, the general trend in microfluidics is toward the miniaturization of analytical systems, creating portable and efficient "labs-on-a-chip." In this context, paper-based microfluidics (μPADs) are emerging as a key technology due to their economic and practical advantages for on-site analysis. Future research aims to overcome current limitations of microfluidic devices by improving their sensitivity, reproducibility, fluid control, structural complexity, and manufacturing techniques, to facilitate the widespread adoption and commercialization of these systems of water quality monitoring.

The prospects of microfluidics involve expanding the range of detectable analytes through increasingly powerful, portable, and cost-effective devices, making chemical, biological, and environmental analyses accessible both at the point of care and in resource-limited settings.

The results of this bibliometric analysis can thus guide future research by highlighting not only past and current trends, but also gaps that need to be filled and potential collaborations to be established. Finally, this analysis methodology, reproducible in other sub-domains, demonstrates the value of integrating bibliometric tools to guide and optimize scientific research efforts, as well as incorporating patent analysis to assess innovation and exploitation opportunities, particularly for researchers and project leaders wishing to ensure the relevance, novelty, and valorization of their scientific results.

References

[1] M.I.G.S. Almeida, B.M. Jayawardane, S.D. Kolev, I.D. McKelvie, Developments of microfluidic paper-based analytical devices (μPADs) for water analysis: A review, Talanta 177 (2018) 176-190. https://doi.org/10.1016/j.talanta.2017.08.072

[2] K.Y. Castillo-Torres, E.S. McLamore, D.P. Arnold, A high-throughput microfluidic magnetic separation (μFMS) platform for water quality monitoring, Micromachines 11 (2019) 16. https://doi.org/10.3390/mi11010016

[3] A.W. Martinez, S.T. Phillips, E. Carrilho, S.W. Thomas, H. Sindi, G.M. Whitesides, Simple telemedicine for developing regions: Camera phones and paper-based microfluidic devices for real-time, off-site diagnosis, Anal. Chem. 80 (2008) 3699-3707. https://doi.org/10.1021/ac800112r

[4] Y. Lin, D. Gritsenko, S. Feng, Y. C. Teh, X. Lu, J. Xu, Detection of heavy metal by paper-based microfluidics, Biosens. Bioelectron. 83 (2016) 256-266. https://doi.org/10.1016/j.bios.2016.04.061

[5] D. V. Chapman, T. Sullivan, The role of water quality monitoring in the sustainable use of ambient waters, One Earth. 5 (2022) 132-137. https://doi.org/10.1016/j.oneear.2022.01.008

[6] J. Alcamo, Water quality and its interlinkages with the Sustainable Development Goals, Curr. Opin. Environ. Sustain. 36 (2019) 126-140. https://doi.org/10.1016/j.cosust.2018.11.005

[7] D.F. Fletcher, B.S. Haynes, J. Aubin, C. Xuereb, Modeling of microfluidic devices, in: V. Hessel, A. Renken, J.C. Schouten, J. Yoshida (Eds.), Micro Process Engineering, First ed., Wiley, 2009, pp. 117-144. https://doi.org/10.1002/9783527631445.ch5

[8] B.A. AlMashrea, A.M. Almehdi, S. Damiati, Simple microfluidic devices for in situ detection of water contamination: A state-of-art review, Front. Bioeng. Biotechnol. 12 (2024) 1355768. https://doi.org/10.3389/fbioe.2024.1355768

[9] S.K. Mitra, S. Chakraborty (Eds.), Microfluidics and Nanofluidics Handbook: Fabrication, Implementation, and Applications, CRC Press, Boca Raton, FL, 2012.

[10] F.A. Perdigones, A. Luque, J.M. Quero, Correspondence between electronics and fluids in MEMS: Designing microfluidic systems using electronics, IEEE Ind. Electron. Mag. 8 (2014) 6-17. https://doi.org/10.1109/MIE.2014.2318062

[11] B. Sebastian, P.S. Dittrich, Microfluidics to mimic blood flow in health and disease, Annu. Rev. Fluid Mech. 50 (2018) 483-504. https://doi.org/10.1146/annurev-fluid-010816-060246

[12] G. Buchberger, et al., Bio-inspired microfluidic devices for passive, directional liquid transport: Model-based adaption for different materials, Procedia Eng. 120 (2015) 106-111. https://doi.org/10.1016/j.proeng.2015.08.576

[13] S. Kumar, P. Bhushan, A.K. Agarwal, S. Bhattacharya, A historical perspective on paper microfluidic based point-of-care diagnostics, in: S. Bhattacharya, S. Kumar, A.K. Agarwal (Eds.), Paper Microfluidics, in Advanced Functional Materials and Sensors, Springer, Singapore, 2019, pp. 1-5. https://doi.org/10.1007/978-981-15-0489-1_1

[14] R.B. Channon, M.P. Nguyen, C.S. Henry, D.S. Dandy, Multilayered microfluidic paper-based devices: Characterization, modeling, and perspectives, Anal. Chem. 91 (2019) 8966-8972. https://doi.org/10.1021/acs.analchem.9b01112

[15] BioProcessing Research Institute of Australia (BioPRIA), Australian Pulp and Paper Institute (APPI), Department of Chemical Engineering, Monash University, G. Garnier, W. L. Then, et BioProcessing Research Institute of Australia (BioPRIA), Australian Pulp and Paper Institute (APPI), Department of Chemical Engineering, Monash University, Review: Paper Microfluidics: Applications and Perspectives », in Trans. of the XVth Fund. Res. Symp. Cambridge, 2013, S. J. I' Anson, Éd., Fundamental Research Committee (FRC), Manchester, 2013, p. 541-583. https://doi.org/10.15376/frc.2013.2.541

[16] A. Venkateshwarlu, R.P. Bharti, Effects of capillary number and flow rates on the hydrodynamics of droplet generation in two-phase cross-flow microfluidic systems, J. Taiwan Inst. Chem. Eng. 129 (2021) 64-79. https://doi.org/10.1016/j.jtice.2021.07.045

[17] E. Moreno, P. Kumar, R.O. Adansi, A. Rodriguez, V. Kumar, Exploring capillary flow in paper-based microfluidics: Unraveling mechanics and applications, in: Proc. Volume 2: Computational Fluid Dynamics (CFDTC); Micro and Nano Fluid Dynamics (MNFDTC); Flow Visualization, American Society of Mechanical Engineers, Anaheim, CA, USA, July 2024, pp. V002T06A004. https://doi.org/10.1115/FEDSM2024-131425

[18] V.A.O.P. Da Silva, et al., Microfluidic paper-based device integrated with smartphone for point-of-use colorimetric monitoring of water quality index, Measurement 164 (2020) 108085. https://doi.org/10.1016/j.measurement.2020.108085

[19] A.M. Nightingale, et al., A droplet microfluidic-based sensor for simultaneous in situ monitoring of nitrate and nitrite in natural waters, Environ. Sci. Technol. 53 (2019) 9677-9685. https://doi.org/10.1021/acs.est.9b01032

[20] N. De Acha, C. Elosúa, J. Corres, F. Arregui, Fluorescent sensors for the detection of heavy metal ions in aqueous media, Sensors 19 (2019) 599. https://doi.org/10.3390/s19030599

[21] R. Wang, X. Wang, Sensing of inorganic ions in microfluidic devices, Sens. Actuators B Chem. 329 (2021) 129171. https://doi.org/10.1016/j.snb.2020.129171

[22] H. Moulahoum, Dual chromatic laser-printed microfluidic paper-based analytical device (μPAD) for the detection of atrazine in water, ACS Omega 8 (2023) 41194-41203. https://doi.org/10.1021/acsomega.3c04387

[23] H. Tazawa, et al., Development of microfluidic devices for on-site water quality testing using glass molding process, Anal. Sci. 39 (2023) 1269-1277. https://doi.org/10.1007/s44211-023-00335-3

[24] J. Baranwal, B. Barse, G. Gatto, G. Broncova, A. Kumar, Electrochemical sensors and their applications: A review, Chemosensors 10 (2022) 363. https://doi.org/10.3390/chemosensors10090363

[25] R. Abbasi, X. Hu, A. Zhang, I. Dummer, S. Wachsmann-Hogiu, Optical image sensors for smart analytical chemiluminescence biosensors, Bioengineering 11 (2024) 912. https://doi.org/10.3390/bioengineering11090912

[26] R. Ghosh, V. Vaishampayan, A. Mahapatra, R. Malhotra, S. Balasubramanian, A. Kapoor, Enhancement of limit of detection by inducing coffee-ring effect in water quality monitoring microfluidic paper-based devices, Desalination Water Treat. 156 (2019) 316-322. https://doi.org/10.5004/dwt.2019.23715

[27] Y. Zhang, J. Li, Y. Zhou, X. Zhang, X. Liu, Artificial intelligence-based microfluidic platform for detecting contaminants in water: A review, Sensors 24 (2024) 4350. https://doi.org/10.3390/s24134350

[28] J. Saez, R. Catalan-Carrio, R.M. Owens, L. Basabe-Desmonts, F. Benito-Lopez, Microfluidics and materials for smart water monitoring: A review, Anal. Chim. Acta 1186 (2021) 338392. https://doi.org/10.1016/j.aca.2021.338392

[29] S. Yazhini, B.S. Sreeja, Comprehensive overview of microfluidic flow sensors: Principles, materials, and fabrication, Microsyst. Technol. 31 (2025) 1259-1288. https://doi.org/10.1007/s00542-024-05787-9

[30] A.-G. Niculescu, C. Chircov, A. C. Bîrcă, et A. M. Grumezescu, Fabrication and Applications of Microfluidic Devices: A Review, Int. J. Mol. Sci. 22 (2021) 2011. https://doi.org/10.3390/ijms22042011

[31] P. Aryal, C. Hefner, B. Martinez, C.S. Henry, Microfluidics in environmental analysis: Advancements, challenges, and future prospects for rapid and efficient monitoring, Lab Chip 24 (2024) 1175-1206. https://doi.org/10.1039/D3LC00871A

[32] K. Fan, et al., Visualization and analysis of mapping knowledge domain of fluid flow related to microfluidic chip, ACS Omega 9 (2024) 22801-22818. https://doi.org/10.1021/acsomega.4c00966

Materials Research Foundations 191 (2026) 128-179 https://doi.org/10.21741/9781644904213-5

[33] X. Zhang, Y. Zhang, W. Guo, S. Jiao, X. Liu, The global research trend in electrochemical microfluidic technology: A bibliometric review, Chemosensors 11 (2023) 14. https://doi.org/10.3390/chemosensors11010014

[34] A. Tarazi, Comparative analysis of the bibliographic data sources using PubMed, Scopus, Web of Science, and Lens: Comparative analysis of bibliographic data sources, High Yield Med. Rev. 2 (2024) 1. https://doi.org/10.59707/hymrUNHW4628

[35] O.A. Jefferson, et al., Mapping CRISPR-Cas9 public and commercial innovation using The Lens institutional toolkit, Transgenic Res. 30 (2021) 585-599. https://doi.org/10.1007/s11248-021-00237-y

[36] H. Mokhtari, S. Barkhan, D. Haseli, M.K. Saberi, A bibliometric analysis and visualization of the Journal of Documentation: 1945-2018, J. Doc. 77 (2020) 69-92. https://doi.org/10.1108/JD-08-2019-0165

[37] VOSviewer - Visualizing scientific landscapes, VOSviewer (2025). https://www.vosviewer.com/(accessed 24 May 2025).

[38] G.M. Whitesides, E. Ostuni, S. Takayama, X. Jiang, D.E. Ingber, Soft lithography in biology and biochemistry, Annu. Rev. Biomed. Eng. 3 (2001) 335-373. https://doi.org/10.1146/annurev.bioeng.3.1.335

[39] J.G. Dawson, et al., Bioinspired vascular structures via 3D printing and suspended microfluidics, in: Proc. 2017 IEEE 30th Int. Conf. Micro Electro Mechanical Systems (MEMS), IEEE, (2017) 426-429. https://doi.org/10.1109/MEMSYS.2017.7863433

[40] J.B. Lee, et al., Microchannel network hydrogel induced ischemic blood perfusion connection, Nat. Commun. 11 (2020) 615. https://doi.org/10.1038/s41467-020-14480-0

[41] A. Hossain, P. Rajput, Z. Li, A.V. Soldatov, E.A. Mukhanova, C.Y. Cárdenas Rodriguez, Engineering bioinspired microfluidics; biomimetic self-healing/cleaning coating designs and unique advanced materials, Chem. Eng. J. 489 (2024) 151336.. https://doi.org/10.1016/j.cej.2024.151336

[42] S. Mohammadi, et al., Novel concept of washing for microfluidic paper-based analytical devices based on capillary force of paper substrates, Anal. Bioanal. Chem. 408 (2016) 7559-7563. https://doi.org/10.1007/s00216-016-9853-9

[43] T.M. Squires, S.R. Quake, Microfluidics: Fluid physics at the nanoliter scale, Rev. Mod. Phys. 77 (2005) 977-1026. https://doi.org/10.1103/RevModPhys.77.977

[44] Y. Wang, et al., Spontaneous Imbibition in Paper-Based Microfluidic Devices: Experiments and Numerical Simulations, Langmuir 38 (2022) 2677-2685. https://doi.org/10.1021/acs.langmuir.1c03403

[45] N. Hassan, A. Stocco, A. Abou-Hassan, Droplet Liquid/Liquid Interfaces Generated in a Microfluidic Device for Assembling Janus Inorganic Nanohybrids, J. Phys. Chem. C 119 (2015) 10758-10765. https://doi.org/10.1021/acs.jpcc.5b02527

[46] M.S.N. Oliveira, F.T. Pinho, M.A. Alves, Divergent streamlines and free vortices in Newtonian fluid flows in microfluidic flow-focusing devices, J. Fluid Mech. 711 (2012) 171-191. https://doi.org/10.1017/jfm.2012.386

[47] E. Elizalde, R. Urteaga, C.L.A. Berli, Precise capillary flow for paper-based viscometry, Microfluid. Nanofluidics 20 (2016) 135. https://doi.org/10.1007/s10404-016-1800-8

[48] A. Manz, Y. Miyahara, J. Miura, Y. Watanabe, H. Miyagi, K. Sato, Design of an open-tubular column liquid chromatograph using silicon chip technology, Sens. Actuators B Chem. 1 (1990) 249-255. https://doi.org/10.1016/0925-4005(90)80210-Q

[49] J.C. McDonald, et al., Fabrication of microfluidic systems in poly(dimethylsiloxane), Electrophoresis 21 (2000) 27-40. https://doi.org/10.1002/(SICI)1522-2683(20000101)21:1<27::AID-ELPS27>3.0.CO;2-C

[50] C. Hansen, S.R. Quake, Microfluidics in structural biology: smaller, faster… better, Curr. Opin. Struct. Biol. 13 (2003) 538-544.. https://doi.org/10.1016/j.sbi.2003.09.010

[51] A. Bonyár, H. Sántha, B. Ring, M. Varga, J. Gábor Kovács, G. Harsányi, 3D Rapid Prototyping Technology (RPT) as a powerful tool in microfluidic development, Procedia Eng. 5 (2010) 291-294.. https://doi.org/10.1016/j.proeng.2010.09.105

[52] P.J. Kitson, M.H. Rosnes, V. Sans, V. Dragone, L. Cronin, Configurable 3D-Printed millifluidic and microfluidic 'lab on a chip' reactionware devices, Lab. Chip 12 (2012) 3267-3271. https://doi.org/10.1039/c2lc40761b

[53] G.M. Whitesides, The origins and the future of microfluidics, Nature 442 (2006) 368-373.. https://doi.org/10.1038/nature05058

[54] A.W. Martinez, S.T. Phillips, G.M. Whitesides, E. Carrilho, Diagnostics for the Developing World: Microfluidic Paper-Based Analytical Devices, Anal. Chem. 82 (2010) 3-10. https://doi.org/10.1021/ac9013989

[55] A.W. Martinez, S.T. Phillips, G.M. Whitesides, Three-dimensional microfluidic devices fabricated in layered paper and tape, Proc. Natl. Acad. Sci. 105 (2008) 19606-19611. https://doi.org/10.1073/pnas.0810903105

[56] S. Kumar, J.B. Kaushal, H.P. Lee, Sustainable Sensing with Paper Microfluidics: Applications in Health, Environment, and Food Safety, Biosensors 14 (2024) 300. https://doi.org/10.3390/bios14060300

[57] Z. Zou, X. Luo, Z. Chen, Y.S. Zhang, C. Wen, Emerging microfluidics-enabled platforms for osteoarthritis management: from benchtop to bedside, Theranostics 12 (2022) 891-909. https://doi.org/10.7150/thno.62685

[58] J.P. McMullen, K.F. Jensen, Rapid Determination of Reaction Kinetics with an Automated Microfluidic System, Org. Process Res. Dev. 15 (2011) 398-407. https://doi.org/10.1021/op100300p

[59] H. Cong, N. Zhang, Perspectives in translating microfluidic devices from laboratory prototyping into scale-up production, Biomicrofluidics 16 (2022) 021301. https://doi.org/10.1063/5.0079045

[60] L. Ma, et al., Droplet Microfluidic Devices: Working Principles, Fabrication Methods, and Scale-Up Applications, Small Methods 8 (2024) 2301406.. https://doi.org/10.1002/smtd.202301406

[61] J. Yue, Green process intensification using microreactor technology for the synthesis of biobased chemicals and fuels, Chem. Eng. Process. - Process Intensif. 177 (2022) 109002. https://doi.org/10.1016/j.cep.2022.109002

[62] J. Cleary, D. Maher, D. Diamond, Development and Deployment of a Microfluidic Platform for Water Quality Monitoring, in: S.C. Mukhopadhyay, A. Mason (Eds.), Smart Sensors for Real-Time Water Quality Monitoring, Springer, Berlin, Heidelberg, 2013, pp. 125-148. https://doi.org/10.1007/978-3-642-37006-9_6

[63] S.A. Jaywant, K.M. Arif, A comprehensive review of microfluidic water quality monitoring sensors, Sensors 19 (2019) 4781. https://doi.org/10.3390/s19214781

[64] F. Narin, K.S. Hamilton, Bibliometric performance measures, Scientometrics 36 (1996) 293-310. https://doi.org/10.1007/BF02129596

[65] [1] E. Bakthavachalam, A scientometric analysis of international collaboration and growth of literature at the macro level, Malaysian J. Libr. Inf. Sci. 20 (2015) 1-17.

[66] K. Börner, L. Dall'Asta, W. Ke, A. Vespignani, Studying the emerging global brain: Analyzing and visualizing the impact of co-authorship teams, Complexity 10 (2005) 57-67. https://doi.org/10.1002/cplx.20078

[67] K. Börner, C. Chen, K.W. Boyack, Visualizing knowledge domains, Annu. Rev. Inf. Sci. Technol. 37 (2003) 179-255. https://doi.org/10.1002/aris.1440370106

[68] M. Alnoukari, A. Hanano, Integration of business intelligence with corporate strategic management, J. Intell. Stud. Bus. 7 (2017) 5-16. https://doi.org/10.37380/jisib.v7i2.235

[69] S. Adams, The text, the full text and nothing but the text: Part 1 - Standards for creating textual information in patent documents and general search implications, World Pat. Inf. 32 (2010) 22-29. https://doi.org/10.1016/j.wpi.2009.06.001

[70] S. McReynolds, Research Guides: Lens Guide: How to Use Lens.org. https://researchguides.library.syr.edu/lensguide/use, (accessed 24 May 2025).

[71] Bibliometric Analysis on Digital Payment ... | Open Research Europe. https://open-research-europe.ec.europa.eu/articles/4-191, (accessed 24 May 2025).

[72] Organisation for Economic Co-operation and Development, OECD Patent Statistics Manual. OECD, 2009.

[73] A. Mogoutov, B. Kahane, Data search strategy for science and technology emergence: A scalable and evolutionary query for nanotechnology tracking, Res. Policy 36 (2007) 893-903. https://doi.org/10.1016/j.respol.2007.02.005

[74] N.I. Mohd Ishar, M.H. Mior Harun, A. Hanif, N.A. Mustapha, R. Kusa, J. Duda, Bibliometric analysis on digital payment using Lens.org and VOSviewer: A comparison of research between Malaysia and Poland, Open Res. Eur. 4 (2024) 191. https://doi.org/10.12688/openreseurope.18248.1

[75] J.T. McAllister, L. Lennertz, Z. Atencio Mojica, Mapping a discipline: A guide to using VOSviewer for bibliometric and visual analysis, Sci. Technol. Libr. 41 (2022) 319-348. https://doi.org/10.1080/0194262X.2021.1991547

[76] F. Narin, K.S. Hamilton, D. Olivastro, The increasing linkage between U.S. technology and public science, Res. Policy 26 (1997) 317-330. https://doi.org/10.1016/S0048-7333(97)00013-9

[77] M. Tayyab, M.A. Sami, H. Raji, S. Mushnoori, M. Javanmard, Potential microfluidic devices for COVID-19 antibody detection at point-of-care (POC): A review, IEEE Sens. J. 21 (2021) 4007-4017. https://doi.org/10.1109/JSEN.2020.3034892

[78] J.A. Berkenbrock, R. Grecco-Machado, S. Achenbach, Microfluidic devices for the detection of viruses: aspects of emergency fabrication during the COVID-19 pandemic and other outbreaks, Proc. R. Soc. A Math. Phys. Eng. Sci. 476 (2020) 20200398. https://doi.org/10.1098/rspa.2020.0398

[79] S. Arora, Knowledge flows and social capital: a network perspective on rural innovation, Maastricht University, 2009.

[80] M. Wang, Y. Liu, and M. R. Dokmeci, Wireless SWNT sensor integrated with microfluidic system for various liquid sensing applications, US10232382B2, 19 Mar. 2019. https://patents.google.com/patent/US10232382B2/en?oq=US10232382+B2 (accessed 24 May 2025).

[81] H. Jiang et al., Microfluidic paper-based chip, microfluidic test system, liquid test method, and use, EP4552740A1, 14 May 2025. https://patents.google.com/patent/EP4552740A1/en?oq=EP+4552740+A1 (accessed 24 May 2025).

[82] F. Doungmene, L. Gautrin, A. Magnier, and C. Nanteuil, Microfluidic method for analyzing metals, US20220032297A1,3Feb.2022.https://patents.google.com/patent/US20220032297A1/en?oq=US+2022%2f0032297+A1 (accessed 24 May 2025).

[83] C. I. Priest, S. Elmas, A. Pospisilova, and A. A. Sekulska, Microfluidic sensor for continuous or semi-continuous monitoring of quality of an aqueous solution, US20220362766A1, 17 Nov. 2022. https://patents.google.com/patent/US20220362766A1/en?oq=US+2022%2f0362766+A1 (accessed 24 May 2025).

[84] J. Hasty et al., Microbial microfluidic biosensor, US11209412B2, 28 Dec. 2021. https://patents.google.com/patent/US11209412B2/en?oq=US+11209412+B2 (accessed 24 May 2025).

[85] D. E. Angelescu and P. Freyermuth, Microfluidic device for analyzing a pressurized fluid, US9410894B2, 9 Aug. 2016. Available online: https://patents.google.com/patent/US9410894B2/en?oq=US+9410894+B2 (accessed 24 May 2025).

[86] R. A. Haimann and B. Li, Microfluidics method for detecting chemicals in water in near real time, US20180128779A1,10May2018.https://patents.google.com/patent/US20180128779A1/en?oq=US+2018%2f0128779+A1 (accessed 24 May 2025).

Materials Research Foundations 191 (2026) 128-179 https://doi.org/10.21741/9781644904213-5

[87] Z. Li et al., Microfluidic ion detection chip having bubble brightening structure, and detection method thereof,US20240125699A1,18Apr.2024.https://patents.google.com/patent/US20240125699A1/en?oq=US+2024%2f0125699+A1 (accessed 24 May 2025).

[88] J. Kim, J. P. Devadhasan, and R. Howse, Chemically patterned microfluidic paper-based analytical device (C-µPAD) for multiplex analyte detectionUS20190118175A1,25Apr.2019. https://patents.google.com/patent/US20190118175A1/en?oq=US+2019%2f0118175+A1 (accessed 24 May 2025).

[89] G. Rao, S. Kumar, V. Srinivasan, P. Ahuja, and R. Kadolkar, Wood microfluidic engineering: development of highly stable polymer nanocomposite electrochemical nitrate sensor, WO2025042580A2, 27 Feb. 2025. https://patents.google.com/patent/WO2025042580A2/en?oq=WO+2025%2f042580+A2 (accessed 24 May 2025).

[90] G. Duffy, Development and optimisation of colourimetric microfluidic sensors for water quality monitoring, Ph.D. thesis, Dublin City University, 2017.

[91] X. Xiong, Y. Zhang, H. Li, J. Wang, L. Liu, Simultaneous cross-type detection of water quality indexes via a smartphone-app integrated microfluidic paper-based platform, ACS Omega 7 (2022) 44338-44345. https://doi.org/10.1021/acsomega.2c05938

[92] J. Dai, W. Gao, J. Yin, L. Liang, J. Zou, Q. Jin, Microfluidic sensor integrated with nanochannel liquid conjunct Ag/AgCl reference electrode for trace Pb(II) measurement, Anal. Chim. Acta 1164 (2021) 338511. https://doi.org/10.1016/j.aca.2021.338511

[93] B. Ngamsom, S. Choi, M. Sakamoto, A microfluidic device for rapid screening of E. coli O157:H7 based on IFAST and ATP bioluminescence assay for water analysis, Chem. Eur. J. 23 (2017) 12754-12757. https://doi.org/10.1002/chem.201703487

[94] X. Chen, D. Mo, M. Gong, A flexible method for nanofiber-based 3D microfluidic device fabrication for water quality monitoring, Micromachines 11 (2020) 276. https://doi.org/10.3390/mi11030276

[95] A.T.K. Perera, S. Pudasaini, S.S.U. Ahmed, D.-T. Phan, Y. Liu, C. Yang, Rapid pre-concentration of Escherichia coli in a microfluidic paper-based device using ion concentration polarization, Electrophoresis 41 (2020) 867-874. https://doi.org/10.1002/elps.201900303

[96] Using design strategies from microfluidic device patents to support idea generation. ResearchGate.https://www.researchgate.net/publication/325769493_Using_design_strategies_from_microfluidic_device_patents_to_support_idea_generation, 2018 (accessed 25 May 2025).

[97] R. and Markets, Microfluidic Technologies for Diagnostic Applications Patent Landscape Report 2024-2032: Patent Surge with 70 New Entries Filed, Highlighting Advances in Portable and Automated Diagnostics. GlobeNewswire News Room, 21 Apr. 2025. https://www.globenewswire.com/news-release/2025/04/21/3064482/28124/en/Microfluidic-Technologies-for-Diagnostic-Applications-Patent-Landscape-Report-2024-2032-Patent-Surge-with-70-New-Entries-Filed-Highlighting-Advances-in-Portable-and-Automated-Diagn.html.(accessed 25 May 2025).

Materials Research Foundations 191 (2026) 128-179 https://doi.org/10.21741/9781644904213-5

[98] S.-J. Hwang, Insulation sheet, manufacturing method therefor, and insulation panel using same, WO2015026116A1, 26 Feb. 2015. Available online: https://patents.google.com/patent/WO2015026116A1/en?oq=WO2015026116A1. (accessed 25 May 2025).

[99] J. Llandro et al., Magnetic tagging techniques, US8283916B2, 9 Oct. 2012. Available online: https://patents.google.com/patent/US8283916B2/en?oq=US8283916B2 (accessed 25 May 2025).

[100] S. J. Kim, J. Han, Amplified electrokinetic fluid pumping switching and desalting, US8834696B2, 16 Sept. 2014. Available online: https://patents.google.com/patent/US8834696B2/en?oq=US8834696B2 (accessed 25 May 2025).

[101] P. Zhang, H. Han, Trace cell capture system based on micro-fluidic and immunomagnetic separation dual strategies, CN109486653B, 22 Mar. 2022. Available online: https://patents.google.com/patent/CN109486653B/en?oq=CN109486653B (accessed 25 May 2025).

[102] G. Rao, S. Kumar, V. Srinivasan, P. Ahuja, R. Kadolkar, Wood microfluidic engineering: development of highly stable polymer nanocomposite electrochemical nitrate sensor, WO2025042580A2, 27 Feb. 2025. Available online: https://patents.google.com/patent/WO2025042580A2/en?oq=WO+2025%2f042580+A2 (accessed 26 May 2025).

[103] H.N. Gowda, S. Bhat, P. Kulkarni, A. Narayan, Development of a proof-of-concept microfluidic portable pathogen analysis system for water quality monitoring, Sci. Total Environ. 813 (2022) 152556. https://doi.org/10.1016/j.scitotenv.2021.152556

[104] J. Yin, L. Liang, J. Dai, Q. Jin, A batch microfabrication of a microfluidic electrochemical sensor for rapid chemical oxygen demand measurement, Analyst 146 (2021) 1956-1964. https://doi.org/10.1039/D0AN02133D

[105] World Health Organization, Guidelines for Drinking-water Quality: Fourth Edition Incorporating the First and Second Addenda.https://www.who.int/publications/i/item/9789240045064, 2022 (accessed 25 May 2025).

[106] Food and Agriculture Organization of the United Nations, Water Quality for Agriculture. https://www.fao.org/4/t0234e/t0234e00.htm, 1994 (accessed 25 May 2025).

[107] W.-H. Huang, V.-P. Mai, R.-Y. Wu, K.-L. Yeh, R.-J. Yang, A microfluidic aptamer-based sensor for detection of mercury(II) and lead(II) ions in water, Micromachines 12 (2021) 11283. https://doi.org/10.20944/preprints202109.0056.v1

[108] S. Wu, M. Wu, G. Wang, T.-H. Chen, Visual quantitation of silver contamination in fresh water via accumulative length of microparticles in capillary-driven microfluidic devices, Talanta 235 (2021) 122707. https://doi.org/10.1016/j.talanta.2021.122707

[109] O.D. Oyewunmi, S.H. Safiabadi-Tali, S. Jahanshahi-Anbuhi, Dual-modal assay kit for the qualitative and quantitative determination of the total water hardness using a permanent

marker fabricated microfluidic paper-based analytical device, Chemosensors 8 (2020) 97. https://doi.org/10.3390/chemosensors8040097

[110] J. Yin, L. Liang, J. Dai, Q. Jin, A batch microfabrication of a microfluidic electrochemical sensor for rapid chemical oxygen demand measurement, Analyst 146 (2021) 1956-1964. https://doi.org/10.1039/D0AN02133D

[111] S. Xie, H. Wu, X. Yang, Microfluidic electrochemical sensor for online detection of chemical oxygen demand based on AuNPs/Au electrodes, IEEE Sens. J. 24 (2024) 38584-38588. https://doi.org/10.1109/JSEN.2024.3438289

[112] B. Riechers, F. Maes, E. Akoury, B. Semin, P. Gruner, J.-C. Baret, Surfactant adsorption kinetics in microfluidics, Proc. Natl. Acad. Sci. U.S.A. 113 (2016) 11465-11470. https://doi.org/10.1073/pnas.1604307113

[113] N.S.K. Gunda, S.K. Mitra, Rapid water quality monitoring for microbial contamination, Electrochem. Soc. Interface 25 (2016) 73. https://doi.org/10.1149/2.F06164if

[114] O. Zouggar, C.E. Hamailia, N. Pr. Outili, Conception d'un support papier microfluidique pour l'analyse de l'eau.http://localhost:8080/xmlui/handle/123456789/5944, 2024 (accessed 25 May 2025).

[115] D. Cogan, P. Cleary, E. Phelan, R. Hughes, Development of a low-cost microfluidic sensor for the direct determination of nitrate using chromotropic acid in natural waters, Anal. Methods 7 (2015) 5396-5405. https://doi.org/10.1039/C5AY01357G

[116] J.-Y. Kim, M.-K. Yeo, A fabricated microfluidic paper-based analytical device (μPAD) for in situ rapid colorimetric detection of microorganisms in environmental water samples, Mol. Cell. Toxicol. 12 (2016) 101-109. https://doi.org/10.1007/s13273-016-0013-2

[117] H. Jiang, X. Zhang, L. Wang, Y. Liu, A disposable multiplexed chip for the simultaneous quantification of key parameters in water quality monitoring, ACS Sens. 5 (2020) 3013-3018. https://doi.org/10.1021/acssensors.0c00775

[118] F.R. Caetano, R. Paixão, A. Wong, C. Delerue-Matos, A. Sotomayor, Combination of electrochemical biosensor and textile threads: A microfluidic device for phenol determination in tap water, Biosens. Bioelectron. 99 (2018) 382-388. https://doi.org/10.1016/j.bios.2017.07.070

[119] A. Leo, A.G. Monteduro, S. Rizzato, A. Milone, G. Maruccio, Miniaturized sensors for detection of ethanol in water based on electrical impedance spectroscopy and resonant perturbation method-A comparative study, Sensors 22 (2022) 2742. https://doi.org/10.3390/s22072742

[120] M. Cuartero, G.A. Crespo, M. Ghahraman Afshar, E. Bakker, Exhaustive thin-layer cyclic voltammetry for absolute multianalyte halide detection, Anal. Chem. 86 (2014) 11387-11395. https://doi.org/10.1021/ac503344f

[121] F. Lefèvre, A. Chalifour, L. Yu, V. Chodavarapu, P. Juneau, R. Izquierdo, Algal fluorescence sensor integrated into a microfluidic chip for water pollutant detection, Lab Chip 12 (2012) 787-793. https://doi.org/10.1039/C2LC20998E

[122] É. Tyrrell, K. Leonard, P. McCormack, Development of a micro-fluidic manifold for copper monitoring utilising chemiluminescence detection, Lab Chip 4 (2004) 384-390. https://doi.org/10.1039/b400805g

[123] X. Chen, C. Liu, Z. Xu, Y. Pan, J. Liu, L. Du, An effective PDMS microfluidic chip for chemiluminescence detection of cobalt(II) in water, Microsyst. Technol. 19 (2013) 99-103. https://doi.org/10.1007/s00542-012-1551-8

[124] Q. Shang, P. Zhang, H. Li, R. Liu, C. Zhang, A flow chemiluminescence paper-based microfluidic device for detection of chromium(III) in water, J. Innov. Opt. Health Sci. 12 (2019) 1950016. https://doi.org/10.1142/S1793545819500160

[125] X. Song, Y. Zhang, H. Zhao, W. Zhou, Y. Zhao, A portable microfluidic-based electrochemiluminescence sensor for trace detection of trenbolone in natural water, Anal. Chem. 94 (2022) 12531-12537. https://doi.org/10.1021/acs.analchem.2c02780

[126] J.L. Delaney, C.F. Hogan, J. Tian, W. Shen, Electrogenerated chemiluminescence detection in paper-based microfluidic sensors, Anal. Chem. 83 (2011) 1300-1306. https://doi.org/10.1021/ac102392t

[127] R. Lanfranco, J. Saez, E. Di Nicolò, F. Benito-Lopez, M. Buscaglia, Phantom membrane microfluidic cross-flow filtration device for the direct optical detection of water pollutants, Sens. Actuators B Chem. 257 (2018) 924-930. https://doi.org/10.1016/j.snb.2017.11.024

[128] S. Bai, D. Serien, A. Hu, K. Sugioka, 3D microfluidic surface-enhanced Raman spectroscopy (SERS) chips fabricated by all-femtosecond-laser-processing for real-time sensing of toxic substances, Adv. Funct. Mater. 28 (2018) 1706262. https://doi.org/10.1002/adfm.201706262

[129] G. Emonds-Alt, P. Giraud, F. Renaud, Development and validation of an integrated microfluidic device with an in-line surface enhanced Raman spectroscopy (SERS) detection of glyphosate in drinking water, Talanta 249 (2022) 123640. https://doi.org/10.1016/j.talanta.2022.123640

[130] G. Chen, J. Guo, X. Xu, Fabrication of a fiberglass-packed channel in a microchip for flow injection analysis, Microchim. Acta 159 (2007) 191-199. https://doi.org/10.1007/s00604-007-0750-y

[131] T. Zhang, A.M. Ratajczak, H. Chen, J.A. Terrell, C. Chen, A step forward for smart clothes-Fabric-based microfluidic sensors for wearable health monitoring, ACS Sens. 7 (2022) 3857-3866. https://doi.org/10.1021/acssensors.2c01827

[132] G. Ballacchino, F. Moreira, R. Garcia, Manufacturing of 3D-printed microfluidic devices for the synthesis of drug-loaded liposomal formulations, Int. J. Mol. Sci. 22 (2021) 8064. https://doi.org/10.3390/ijms22158064

[133] Y. Chen, L. Booth, C. Schaper, B.T. Khuri-Yakub, K.C. Saraswat, 3D modeling of rapid thermal processors for design optimization of a new flexible RTP system, in: Proc. IEEE Int. Electron Devices Meet., IEEE, 1994, pp. 545-548.

[134] H. Asano, Y. Shiraishi, Development of paper-based microfluidic analytical device for iron assay using photomask printed with 3D printer for fabrication of hydrophilic and hydrophobic

zones on paper by photolithography, Anal. Chim. Acta 883 (2015) 55-60. https://doi.org/10.1016/j.aca.2015.04.014

[135] K. Maejima, S. Tomikawa, K. Suzuki, D. Citterio, Inkjet printing: an integrated and green chemical approach to microfluidic paper-based analytical devices, RSC Adv. 3 (2013) 9258-9263. https://doi.org/10.1039/c3ra40828k

[136] W. Lee, F. Gomez, Experimental analysis of fabrication parameters in the development of microfluidic paper-based analytical devices (μPADs), Micromachines 8 (2017) 99. https://doi.org/10.3390/mi8040099

[137] J. Nie, Y. Liang, Y. Zhang, S. Le, D. Li, S. Zhang, One-step patterning of hollow microstructures in paper by laser cutting to create microfluidic analytical devices, Analyst 138 (2012) 671-676. https://doi.org/10.1039/C2AN36219H

[138] X. Li, J. Tian, T. Nguyen, W. Shen, Paper-based microfluidic devices by plasma treatment, Anal. Chem. 80 (2008) 9131-9134. https://doi.org/10.1021/ac801729t

[139] M.M. Thuo, S. Martinez, C. Sweeney, J. Barber, G.M. Whitesides, Fabrication of low-cost paper-based microfluidic devices by embossing or cut-and-stack methods, Chem. Mater. 26 (2014) 4230-4237. https://doi.org/10.1021/cm501596s

[140] D. Das, A. Dsouza, N. Kaur, S. Soni, B.J. Toley, Paper stacks for uniform rehydration of dried reagents in paper microfluidic devices, Sci. Rep. 9 (2019) 15755. https://doi.org/10.1038/s41598-019-52202-9

[141] D. Gregucci, R. Berton, A. Pezzato, A. Fiorati, M. Selva, A. Perosa, An origami colorimetric paper-based sensor for sustainable on-site and instrument-free analysis of nitrite, Sens. Diagn. 4 (2025) 239-246. https://doi.org/10.1039/D4SD00308J

[142] R. Kumari, A. Singh, U.P. Azad, P. Chandra, Insights into the fabrication and electrochemical aspects of paper microfluidics-based biosensor module, Biosensors 13 (2023) 891. https://doi.org/10.3390/bios13090891

[143] L. Busa, S. Mohammadi, M. Maeki, A. Ishida, H. Tani, M. Tokeshi, Advances in microfluidic paper-based analytical devices for food and water analysis, Micromachines 7 (2016) 86. https://doi.org/10.3390/mi7050086

[144] R. Pol, F. Céspedes, D. Gabriel, M. Baeza, Microfluidic lab-on-a-chip platforms for environmental monitoring, TrAC Trends Anal. Chem. 95 (2017) 62-68. https://doi.org/10.1016/j.trac.2017.08.001

[145] F. Schaumburg, P.A. Kler, C.S. Carrell, C.L.A. Berli, C.S. Henry, USB powered microfluidic paper-based analytical devices, Electrophoresis 41 (2020) 562-569. https://doi.org/10.1002/elps.201900273

[146] J.H. Cho, Y. Gao, S. Choi, A portable, single-use, paper-based microbial fuel cell sensor for rapid, on-site water quality monitoring, Sensors 19 (2019) 5452. https://doi.org/10.3390/s19245452

[147] G.-H. Chen, W.-Y. Chen, Y.-C. Yen, C.-W. Wang, H.-T. Chang, C.-F. Chen, Detection of mercury(II) ions using colorimetric gold nanoparticles on paper-based analytical devices, Anal. Chem. 86 (2014) 6843-6849. https://doi.org/10.1021/ac5008688

[148] A. Saadati, F. Farshchi, M. Hasanzadeh, Y. Liu, F. Seidi, Colorimetric and naked-eye detection of arsenic(III) using a paper-based microfluidic device decorated with silver nanoparticles, RSC Adv. 12 (2022) 21836-21850. https://doi.org/10.1039/D2RA02820D

[149] W. Park, S. Choe, M. Kim, Microscale flow control and droplet generation using Arduino-based pneumatically controlled microfluidic device, Biosensors 14 (2024) 469. https://doi.org/10.3390/bios14100469

[150] B. Giri, A perspective on the sensitivity of paper-analytical devices for bioanalysis, J. Appl. Bioanal. 2 (2016) 6-9. https://doi.org/10.17145/jab.16.002

Microfluidics: Synthesis, Device Fabrication and Applications | Materials Research Forum LLC
Materials Research Foundations 191 (2026) 180-223 | https://doi.org/10.21741/9781644904213-6

Chapter 6

Advanced Nanomaterials for Microfluidic Biosensors

Tabina Amin[1], Muhammad Ravish Khan[1], Muhammad Zeeshan Ali Janjua[1], Usman Ahmad[1], Fatima Masood[1], Midhat Areeba[1], Shiza Noor[1], Faheem K. Butt[2], Ghulam Nabi[3], Sadia Z. Bajwa[1]*

[1]National Institute for Biotechnology and Genetic Engineering College, Pakistan Institute of Engineering and Applied Sciences (NIBGE-C, PIEAS), Faisalabad 38000, Pakistan

[2]Department of Physics, Division of Science and Technology, University of Education, Lahore 54770, Pakistan

[3]University of Gujrat, Gujrat 51700, Pakistan

*sadya2002pk@yahoo.co.uk

Abstract

This chapter explores how advanced nanomaterials enhance performance of microfluidic biosensors by improving sensitivity, specificity, and speed of diagnostics. Nanomaterials such as gold nanoparticles, carbon nanotubes, graphene, quantum dots, and nanocomposites amplify signal strength and can enable multiplexed detection. Detection methods include colorimetric, chemiluminescent, magnetic, photothermal, distance-based, and paper-based assays. It covers microscale transport, biomolecule detection, optical and electrochemical transduction, and signal processing. The chapter also highlights real-world applications in early disease diagnosis, environmental and food safety monitoring, and drug development, emphasizing the role of nanomaterials in advancing microfluidic biosensing technologies.

Keywords

Microfluidic Biosensors, Nanomaterials, Biosensor Functionalization, Surface Modification, Nanomaterial Integration, Point-of-Care Diagnostics, Detection Mechanism, Biomedical Applications

Contents

Materials Research Foundations 191 (2026) 180-223 https://doi.org/10.21741/9781644904213-6

Materials Research Foundations 191 (2026) 180-223 https://doi.org/10.21741/9781644904213-6

Materials Research Foundations 191 (2026) 180-223 https://doi.org/10.21741/9781644904213-6

1. Introduction

Microfluidic biosensors, a revolutionary technological approach, are set to transform the landscape of diagnostic services, healthcare management, environmental surveillance, and point-of-care (POC) applications. These systems, which integrate fluid management methods at the micrometer level with biological detection capabilities, have the potential to identify specific substances such as pathogens, toxins, or biomarkers. Their compact size, higher processing speed, fewer testing requirements, and field-based testing capability make them a promising tool for rapid diagnostic assessments and real-time monitoring, instilling optimism about the future of diagnostics [1-4].

The emergence of microfluidic biosensors addresses the pressing need for better, more cost-effective, and portable diagnostic equipment. The urgency of this need has been further underscored by the COVID-19 pandemic, highlighting the importance of rapid and reliable diagnostic tools. Microfluidic systems have risen to this challenge, demonstrating their success in detecting SARS-CoV-2. PCR-based and antibody-based microfluidic sensors have shown promising results, reinforcing the importance of the audience's work in this field. It also demonstrates the two key applications of microfluidic biosensors: biomarker detection for diseases and assessing human exposure to water pollutants for personalized medical care. With compact devices that provide quick results, real-time microfluidic biosensing technology can be used in POC facilities. Biosensing has significant utility in health protection and monitoring chronic diseases, for example, cancer, diabetes mellitus, and cardiovascular diseases [5].

Nanomaterial incorporation in these systems has become a promising solution. Their distinctive nanoscale properties significantly advance such biosensing platforms' performance. These materials feature a greater surface area and are feasibly engineered with a high degree of precision; thus, they constitute an ideal choice for optimizing the sensitivity and specificity of the biosensor. For instance, the metallic nanoparticles of gold and silver enhance the biosensor's activity based on the localised surface plasmon resonance principle. On the other hand, carbon-containing nanoparticles, like graphene, produce optimal molecular recognition and signal transduction. Therefore, nanomaterials contribute to increasing the detection limits, speed, and accuracy of microfluidic biosensors, making them indispensable in sectors requiring high precision, such as early disease detection, environmental monitoring, and point-of-care diagnostics.

Nanomaterials offer several distinct advantages when applied to microfluidic systems, enhancing their overall capabilities and broadening the scope of their potential applications. One significant benefit is their ability to amplify the signal generated during biosensing, thereby enabling the detection of low-concentration biomarkers that conventional materials cannot detect. The high surface area of nanomaterials ensures effective functionalization with biomolecular probes, which enhances specificity in detecting biosensors and reduces interference from non-target substances. Additionally, nanomaterials are crucial in miniaturizing microfluidic devices, making them more compact, less expensive, and suitable for real-time point-of-care diagnostics. With miniaturization achieved, there is no loss of performance, quite the opposite: nanomaterials increase the sensitivity

Materials Research Foundations 191 (2026) 180-223 https://doi.org/10.21741/9781644904213-6

and accuracy of detection. Consequently, miniaturized fluidic tools powered by nanomaterials are poised to revolutionize diagnostic capabilities with speed, reliability, and portability across various fields, including medical diagnostics, environmental monitoring, and food safety [6].

In this regard, microfluidic biosensors with nanomaterials represent a groundbreaking method for developing more effective, sensitive, and portable diagnostic equipment that can overcome current limitations and pave the way for further advancements in overall health and environmental monitoring.

2. Fundamentals of Microfluidic Biosensing

2.1 Basic Principle of Microfluidics

Microfluidic science deals with systems that work at micro- to pico-litre sizes, where surface tension and viscosity govern how fluids move. In the case of the microfluidic systems, the most common is laminar flow, characterized by the smooth, parallel motion of fluid layers. Microfluidic devices consist of microchannels, microvalves, micropumps, and microsensors; the microchannels primarily govern fluid flow. The movement of fluids within these devices is controlled by pressure control, while soft lithography is usually part of the fabrication process. One significant advantage of the microfluidic systems is their ability to safely transport biological substrates and reagents, permitting several processes in biological experimentation; sample handling, mixing, reaction, and analysis can occur on a single chip.

2.2 Working Mechanisms of Microfluidic Biosensing

A microfluidic biosensor contains three basic components: biological recognition elements, transducers, and output signal processing systems. The target molecule engages with the sensor's biological sensing unit, followed by the transducer converting this reaction to a detectable output signal, as shown in Figure 1. After signal transduction, the processed information reveals quantitative or qualitative measurements about the detected analyte.

2.2.1 Biological Recognition Elements

A microfluidic biosensor implements a biological recognition element as one of its three main components. The mechanisms through which biosensors detect targets depend on specific biological interactions that determine their sensitivity features and specific recognition power. For example, nucleic acid-based sensors achieve binding through DNA or RNA probes, which recognise complementary target sequences.

2.2.2 Transducers

The biological reaction must pass through the transducer to become a detectable signal. Two primary transduction mechanisms are used within microfluidic biosensors: Optical and Electrochemical transduction.

Numerous biosensors commonly use **optical transduction**. The recognition process between the analyte and element in optical biosensors causes light properties to change, including absorption, fluorescence, and surface plasmon resonance effects. In **electrochemical transduction,** an electrochemical biosensor measures changes in electrical characteristics (current, voltage, impedance) upon analyte-recognition element binding [7].

Materials Research Foundations 191 (2026) 180-223 https://doi.org/10.21741/9781644904213-6

2.2.3 Output Signal Processing

The output signal must be detected and processed after the interaction between the analyte and the biological recognition element is transduced. Amplification, filtering, or conversion to digital data for display or further analysis is carried out in signal processing. For example, photodetectors are usually used for detection in optical biosensors and as potentiometers in electrochemical systems [8].

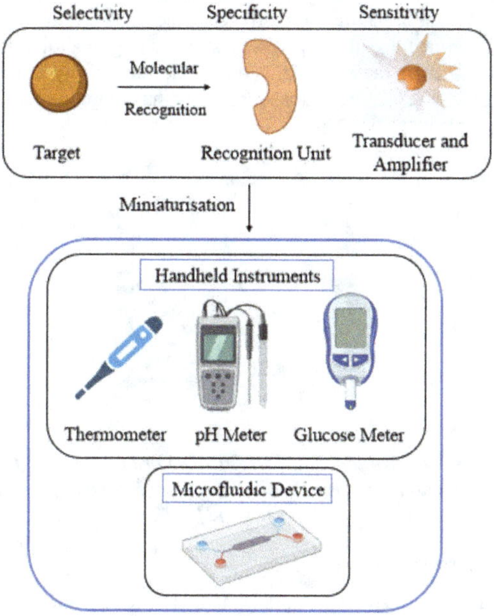

Figure 1 Mechanism of Microfluidic Biosensing: Schematic illustration of how target analytes bind with specific recognition elements and produce detectable changes, such as electrochemical, optical, or colourimetric. Handheld tools, such as thermometers, pH meters, and glucose meters, are being miniaturised into smaller, integrated devices that analyse samples more easily.

2.3 Types of Microfluidic Sensors

Sensor developers choose between many microfluidic detectors based on the application requirements, such as detection mechanism, measurement sensitivity, and portability. There are three major types of microfluidic biosensors: microfluidics on paper, lateral flow assays (LFAS), and microarrays. These technologies have unique features and applications and are preferred for specific diagnostic tasks.

2.3.1 Paper-Based Microfluidics

Materials Research Foundations 191 (2026) 180-223 https://doi.org/10.21741/9781644904213-6

The microfluidics involved in paper-based microfluidics is a microfluidic subset where fluid flows are realised by porous materials, such as paper, to integrate sensors and fluid transport. The transport of fluid through porous paper material occurs through capillary action. Two detection methods exist for paper-based microfluidic devices: optical, which represents colourimetric changes as shown in Figure 2, and electrochemical detection, which shows changes in current resulting from target interactions [9].

Specifically, these devices are incredibly beneficial for point-of-care (POC) applications because they are less expensive, can be fabricated easily, are practical for quick testing, and are portable. They have been developed to identify diseases, including malaria, HIV, and Zika, using colorimetric analysis. Its main drawback is its diminished sensitivity since it needs large reagent and sample volumes; however, detection sensitivity has improved by using wax patterns, nanomaterials, and graphene [10].

Figure 2 Paper-based microfluidic devices (μPADs) are used for performing inexpensive bioassays in remote locations, and results can be exchanged with off-site technicians. Adapted from [3]

2.3.2 Lateral Flow Assays (LFAs)

Lateral flow assays (LFAs) are an immunoassay method to detect signals through specific antigen-antibody interactions. A routine LFA device enables samples to flow through membranes from its entry point before completing lateral movement across the strip, as shown in Figure 3. When the target analyte interacts with the reagents, the signal is detected by direct visual observation or by a reader device to measure signal intensity. Signals are amplified through gold nanoparticles and quantum dots applications, which substantially increase the sensitivity and reliability of LFAs.

LFAs are portable devices that are well-suited for POC diagnostic applications. It is used in pregnancy tests and to detect infectious diseases, including HIV and COVID-19. The technology of LFAs has progressed through the development of multi-pathogen or multi-biomarker detection capabilities within a unified testing platform [11,12].

Materials Research Foundations 191 (2026) 180-223 https://doi.org/10.21741/9781644904213-6

2.3.3 Microarrays

Microarrays function as high-performance microfluidic biosensors, simultaneously detecting diverse biomarkers (DNA, RNA, proteins) to investigate genomic and proteomic studies. The probes (DNA, RNA, or antibodies) are embedded into a solid foundation to seek target molecules, including complementary DNA or proteins. The detection of this interaction uses fluorescence or chemiluminescence. Microarrays decrease handler requirements and enhance accuracy when implemented in microfluidic devices that automate fluid operations. Improvements in microfluidic integration make the microarray more viable for clinical diagnostics due to cost reductions in production.

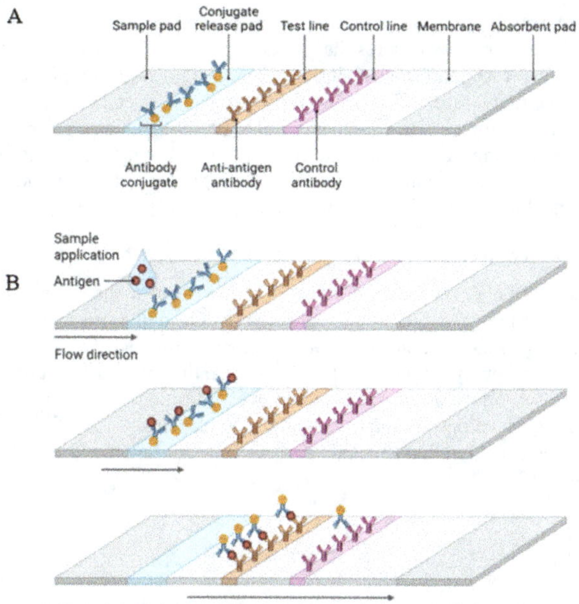

Figure 3 Lateral Flow Assay: (A) Structure of a complete lateral flow strip, consisting of key components such as the sample pad, conjugate pad, test line, control line, nitrocellulose membrane, and absorbent pad. (B) Schematic of the detection process showing sample application, antigen–antibody interaction, and signal development at the test and control lines.

2.4 Key Performance Parameters in Microfluidic Biosensing

Microfluidic biosensors rely on three key parameters to perform optimally: sensitivity, specificity, and detection limit.

Materials Research Foundations 191 (2026) 180-223 | https://doi.org/10.21741/9781644904213-6

2.4.1 Sensitivity

Biosensors must be sensitive to detect small concentrations of analytes, especially for targets such as pathogens and cancer biomarkers, which often occur in femtomolar to attomolar concentrations. The sensitivity of a sensor is affected by the molecular interaction surface area, reagent concentration, and signal conversion method. However, these sensitivity levels can be boosted by using nanomaterials that increase the area of interaction and thus help enhance the signal. Moreover, signal amplification through ELISA and PCR-based procedures enhances the detection threshold of analytical systems so they can identify lower quantities of materials [13].

2.4.2 Specificity

A biosensor must have high specificity for its target analyte to stay separate from all other substances. This high specificity ensures that the device delivers precise and correct results. The selection of biological recognition elements, including aptamers, DNA probes, and antibodies, determines the specificity level of microfluidic biosensors. Specificity can be improved by optimizing the recognition element-analyte binding interactions. Furthermore, surface functionalization, blocking agents, and selective coatings enhance microfluidic biosensor selectivity through non-specific binding reduction. Microfluidic systems provide control over the local environment, which reduces background interferences [14].

2.4.3 Limit of Detection (LOD)

A sensor can detect analyte concentrations at very low levels. The lowest level that can be measured accurately is known as the limit of detection or LOD of the sensor. This precise detection is vital for diagnosing diseases at an early stage and for environmental screening. The sensitivity of a sensor improves when the LOD decreases. To reduce the value of Limit of Detection (LOD) in microfluidic biosensors, it is possible to achieve a desirable sensitivity level through the optimisation of the channel design, improvement of the reagents, usage of sensitive methods for transduction, and preconcentration methods such as microfiltration and electrokinetic trapping [15].

Materials Research Foundations 191 (2026) 180-223 https://doi.org/10.21741/9781644904213-6

3. Advanced Nanomaterials for Microfluidic Biosensing

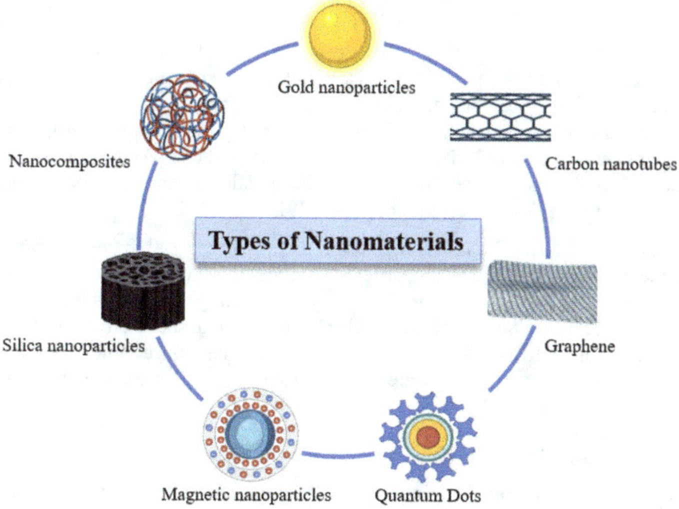

Figure 4 Different types of advanced nanomaterials used in microfluidic biosensing

3.1 Gold Nanoparticles (AuNPs) in Microfluidic Biosensing

The use of gold nanoparticles (AuNPs) in microfluidic biosensing assays has been significant due to their optical chemistries and biological properties. One of the essential properties of gold nanoparticles is that they exhibit the SPR due to the electrons' oscillating behavior on the outer perimeter of the nanoparticle in response to the incidence of light. This property results in discrete and tunable optical signals, which makes gold nanoparticles very susceptible to environmental changes, such as changes in the interaction between target biomolecules and receptor sites [16].

In microfluidic biosensing, gold nanoparticles (AuNPs) play a pivotal role as a result of their specific photoactive, electronic, and catalytic behaviors. One of the most prevalent applications is in colorimetric detection. Dispersed AuNPs typically exhibit a ruby-red color resulting from their LSPR behavior. However, when they aggregate, often triggered by specific binding interactions including DNA strand pairing or immunological interactions, the plasmonic coupling effect induces a visible color change from red to blue or purple. This shift can be readily observed by the naked eye or quantified using integrated optical detectors within the microfluidic platform. This enables rapid, low-cost, and equipment-free diagnostics, especially for point-of-care applications [17].

Beyond simple colorimetric sensing, AuNPs are extensively used in surface-enhanced Raman scattering (SERS)-based biosensing. When analyte molecules are adsorbed near the surface of AuNPs, particularly those arranged into hot spots such as nanoparticle aggregates or nanogaps, the local electromagnetic fields are significantly amplified. This enhancement enables the detection of trace levels of biomolecules (even down to single-molecule levels), making SERS-AuNP

Materials Research Foundations 191 (2026) 180-223 https://doi.org/10.21741/9781644904213-6

platforms ideal for ultrasensitive applications such as early disease diagnostics and environmental pollutant monitoring [18].

In addition to optical techniques, AuNPs are crucial in electrochemical biosensors within microfluidic environments. Their high surface area and excellent conductivity facilitate the efficient binding of biological molecules like enzymes, antibodies, or DNA probes, and facilitate rapid electron transfer between the analyte and the electrode. This property substantially increases signal intensity, enhances sensitivity and lowers the LOD of electroanalytical assays. Moreover, their biocompatibility and ease of surface functionalization make them highly adaptable for constructing complex and multiplexed biosensors [19].

Recent advances have also explored hybrid nanostructures, where AuNPs are combined with other materials like graphene, magnetic nanoparticles, or silica shells to further enhance performance in microfluidic biosensors. These composite systems offer synergistic properties, such as improved stability, multi-modal sensing capabilities, and enhanced analyte specificity. Additionally, AuNPs have been integrated into **lab-on-a-chip** platforms with automated fluid handling and signal readouts, bringing us closer to real-time, miniaturized diagnostics for clinical and field-based settings [20].

3.2 Carbon Nanotubes (CNTs) in Microfluidic Biosensing

Carbon nanotubes (CNTs) are also one of the nanomaterial classes that are very useful for microfluidic biosensing due to their highly remarkable electrical, mechanical, and chemical properties. Sealing graphene sheets, CNTs have excellent electrical conductivity and can convey electrons effectively. The large surface area permits the attachment of many biomolecules, including enzymes, antibodies, and DNA probes, which are needed to identify specific targets. In addition, mechanically strong and flexible, CNTs are stable inside microfluidic channels where a fluid flow could otherwise disturb weaker structures. Furthermore, CNTs are highly robust and resistant to degradation, which makes them amenable to incorporate into biological microfluidic devices intended for continuous monitoring or highly hostile environments. Their flexibility will enable them to embed them in flexible microfluidic chips manufactured from PDMS (polymethylsiloxane), for example, leaving the ground to wearable or implantable biosensors [21-23].

CNTs are excellent transducers in microfluidic biosensors due to their exceptional charge transfer capability, greater surface area, and stable chemical structure. Their ability to promote rapid electron transfer makes them most suitable for electrochemical sensing applications. When used with microelectrodes in a microfluidic channel, CNTs can transduce biomolecular recognition events such as antigen-antibody binding or nucleic acid hybridization, giving rise to quantifiable changes in the current, resistance, or impedance. Such signal changes are directly proportional to the presence and concentration of target analytes and allow in-situ quantitative measurement in a small, low-volume system with real-time monitoring capability [24].

A significant strength of CNTs is their ease of chemical functionalization with a diverse array of recognition elements like enzymes/antibodies/ aptamers/ DNA probes. This surface modification confers selectivity on the biosensor, thus enabling it to distinguish between specific analytes even if they are part of a complex biological matrix. This can be seen in that glucose oxidase enzyme bound on carbon nanotubes allows detection of glucose with high sensitivity in biological samples from blood. At the same time, functionalized CNTs have been successfully used to detect neurodegenerative disease markers, cardiac events, and even bacteria or virus strain differentiation.

Another sophisticated application concerns field-effect transistors, also known as FET biosensors, in which carbon nanotubes are used as semiconducting transport material. In these devices, the adhesion of a charged biomolecule close to the CNT surface disturbs the local electrostatic milieu, which controls the altered carrier density that changes the current between the source and drain terminals. This mechanism enables label-free and real-time detection of analytes at ultra-sensitive levels, typically femtomolar levels. CNT-FETs are especially useful for low-abundance biomarkers that can prove helpful as markers for early-stage disease diagnostics [25].

New advances in nanofabrication have also allowed the incorporation of CNT networks or arrays into multiplexed microfluidic platforms, thus enabling simultaneous analysis of multiple analytes using a single sample. Moreover, hybrid nanomaterials (such as CNTs with or without gold nanoparticles, graphene, conductive polymers, etc.) are under investigation to improve the functionality and versatility of biosensors. These composite systems may demonstrate synergistic effects such as enhanced biocompatibility, better signal amplification, and improved mechanical stability [26].

Prospecting, integrating CNT-based biosensors with wireless data transmission, smartphone readouts, and AI-guided analytics, is leading to the development of innovative diagnostic systems. Such platforms promise great things for continuous health monitoring, personalized medicine, and rapid disease detection.

3.3 Graphene in Microfluidic Biosensing

Graphene is one layer of carbon atoms bonded in a two-dimensional honeycomb lattice. It has extraordinary characteristics, e.g., outstanding electrical conductivity, structural integrity, durability, flexibility, and greater area-to-volume ratio. Graphene is in demand in microfluidic biosensing because it can present a highly sensitive platform for detecting biological molecules in small fluid volumes. The large surface area makes it possible to immobilise various biomolecules, such as DNA, antibodies, or enzymes, at substantial sites, enabling graphene to serve as a functional interface between biological targets and electronic readout systems [27].

Graphene is a versatile sensing medium in microfluidic biosensors that can transduce biological recognition events into measurable electrical or optical events. With an atomically thin structure, high conducting properties, and a high surface-to-volume ratio, graphene's ability to react to surface-bound molecular events on a surface makes it suitable for the implementation of microfluidic platforms where accurate detection within confined volumes is vital [28].

The main application of graphene in this field is electrochemical biosensing. Microelectrodes within the microfluidic device are where graphene is deposited or printed in such systems. The high electron mobility and superior chemical stability of its high quality make it easier to transfer electrons between immobilized biomolecules (enzymes, antibodies, or DNA strands) to the electrode surface. This leads to improved signal intensity and response times versus the conventional materials. For instance, graphene-modified electrodes have been used to detect glucose, the cancer biomarker, viral RNA, and heavy metals with outstanding sensitivity and specificity in serum or saliva [29,30].

Materials Research Foundations 191 (2026) 180-223
https://doi.org/10.21741/9781644904213-6

Apart from the electrochemical approaches, graphene also supports the development of graphene-based field-effect transistor (GFET) biosensors. Graphene is the conductive channel between the source and drain in GFET configurations embedded with microfluidic devices on the source-drain electrode interface. Upon binding to the functionalized graphene surface at modified sites by target biomolecules, a displacement of the local electric fields occurs. This regulates the graphene's conductivity, providing an immediate, label-free electrical signal. Due to this direct signal transduction mechanism, GFET biosensors can perform real-time detection of very low analyte levels, sometimes at an attomolar level, which indeed makes them ideal for early disease diagnosis and trace contaminant analysis [31,32].

An emerging area of interest is optical biosensing with graphene, including surface plasmon resonance (SPR) and fluorescence quenching. The highlight interaction and the phenomenon of light quenching in graphene have been utilized to make optical sensors more sensitive. For example, microfluidic chips have been made from graphene oxide, as well as reduced graphene oxide, for label-free optical detection of DNA hybridization and protein interactions [33].

Over the last several years, progress in scalable manufacturing methods, including chemical vapor deposition (CVD) and inkjet printing, has enabled low-cost and reproducible production of graphene-based sensing elements. Graphene-based biosensors are moving quickly towards next-generation compact, disposable, user-friendly, and point-of-care diagnostic systems, which will be used with microfluidic automation and smartphone-based data capture functionalities [34].

3.4 Quantum Dots (QDs) in Microfluidic Biosensing

QDs are quantum dots and tiny-sized semiconductor nanoparticles (2-10 nm) with unique optical properties. When excited by light, quantum dots emit fluorescence at defined wavelengths, which depend on their size and composition. Such size-tunable fluorescence, coupled with its high brightness, photostability, and broad absorption with narrow emission spectra, makes quantum dots excellent materials for biosensing, particularly in microfluidic platforms where small volumes of samples require highly sensitive detection [35,36].

In microfluidic biosensing, quantum dots (QDs) are extensively adopted for optical detection because of their specific size-tunable photoluminescence, broad excitation spectra, and narrow, tunable emission windows. The semiconductor nanocrystals involve conjugating high specificity biological recognition elements to target analytes (antibodies, aptamers, peptides, nucleic acids). Once bound to the target biomolecule, the quantum dot-labelled probe generates a fluorescent signal that is individually detectable and quantifiable in the confines of the microfluidic device, this being done either off or on-chip photo-detectors or external imaging systems [37].

A specific, handy function of QDs applicable in multiplexed biosensing. As the QDs of various sizes and compositions produce different colors of fluorescence upon excitation with the exact source of illumination, several biomarkers can be determined concurrently using a single test by labelling each of the targets with quantum dots of different colors. This multiplexing ability is exciting in diagnosing multifactorial illnesses, for example, cancer, autoimmune diseases, and infections. Accurately quantifying multiple biomolecular markers is vital for prognosis and treatment plans [38].

Besides fluorescence-based detection, QDs are increasingly applied in FRET (Förster Resonance Energy Transfer) biosensing configurations. With appropriate energy acceptors, QDs can become

Materials Research Foundations 191 (2026) 180-223 https://doi.org/10.21741/9781644904213-6

effective energy donors to enable proximity sensing methods. This arrangement exposes the QD sensors to detect molecular interactions and conformational changes, increasing the utility of QD-based sensors beyond the simple detection of presence or absence to dynamic biochemical monitoring [39].

Recent innovations also involve the use of quantum dot barcodes for high-throughput diagnostics, and integration of QDs in point-of-care microfluidic diagnostic devices utilizing smartphone-based fluorescence imaging for spot-checking analysis. These have even driven quantum dot-enabled microfluidic biosensors towards practical implementation on account of point-of-care testing, environmental monitoring, and personalized medicine [40].

3.5 Magnetic Nanoparticles (MNPs) in Microfluidic Biosensing

Magnetic nanoparticles (MNPs), which are usually based on iron oxide (Fe_3O_4 or γ–Fe_2O_3), fulfill an indispensable role in microfluidic biosensing due to their magnetic responsiveness, biocompatibility, and easy surface functionalization. These particles can be controlled externally with magnetic paths, and the ability to control particles within these small fluid volumes is beneficial in microfluidic applications. In biosensing, MNPs are usually covered with biological recognition molecules, including antibodies, aptamers, or enzymes. The Nature of these functionalized MNPs is to be channeled into microfluidic channels to capture the target analytes from intricate biological samples such as nature's blood, saliva, or urine [41].

In microfluidic devices (on which tiny volumes of liquid are operated in precisely designed microchannels), magnetic nanoparticles (MNPs) provide unmatched possibilities for the control, isolation, and detection of biological targets. Thanks to their nanoscale size and superparamagnetic behavior, MNPs can be moved inside the chip using external magnetic field excitation without leaving residual magnetization that may lead to unwanted aggregation. This makes them ideal in dynamic biosensing applications in a constrained microfluidic environment [42].

MNPs' surfaces are generally functionalized with biological recognition units, antibodies, DNA/RNA probes, aptamers, etc. This enables specific binding with target analytes, which may be pathogens (e.g., E. coli, SARS-CoV-2), cancer biomarkers (e.g., PSA, HER2), toxins, or environmental pollutants. Once target molecules are trapped, however, a magnetic environment is imposed to control or immobilize the MNP-analyte assemblies in some areas of the microfluidic chip. This allows for fast enrichment and pre-concentration of the analyte, significantly impacting detection sensitivity via an improved local concentration in the sensing area. However, beyond sample preparation and target enrichment, MNPs actively participate in signal generation and transduction on different biosensing fronts. MNPs form detectable magnetic labels in magneto-resistive sensors. Their binding to a sensor surface changes the local magnetic field, which is transduced to an electrical signal [43].

In electrochemical biosensors, MNPs are the most commonly employed carriers of either enzymatic or redox-active tags. These deposits at an electrode surface within a microfluidic device catalyse electron transfer reactions, which affect observable changes in current or potential.

With the ongoing integration of magnetism, nanotechnologies, and microfluidics, the MNP-based biosensing systems are rapidly developing toward a highly integrated, automated, and ultra-sensitive diagnostic system most suitable for individual medicine, environmental control, and rapid outbreak response.

Materials Research Foundations 191 (2026) 180-223 | https://doi.org/10.21741/9781644904213-6

3.6 Silica Nanoparticles (SiNPs) in Microfluidic Biosensing

Silica nanoparticles (SiNPs) are a family of highly versatile nanomaterials used in microfluidic biosensing owing to excellent chemical stability, ease of surface modification, large surface area, and biocompatibility. These nanoparticles usually possess a porous or solid silica structure containing different chemical groups for functionalizing and immobilizing antibodies, nucleic acids, or enzymes. The surface of SiNPs is readily modified using silane chemistry, and consequently, one can design very specific and stable biosensors [44].

Silicon nanoparticles have one of the most prominent applications in optical biosensing via fluorescently labelled silica nanoparticles. These nanoparticles can be doped (or conjugated) with fluorescent dyes or quantum dots to generate bright emission signals to recognize the target. Fluorescent SiNPs display improved photostability, more intense signals, and light-induced bleaching resistance compared to free fluorophores; They are handy for long-term or in situ monitoring in flowing microenvironments [45].

SiNPs also have a critical role in multiplex signaling. It is possible to prepare a panel of probes with different emission spectra by labelling distinct populations of SiNPs with various fluorescent dyes or reporter molecules. When incorporated with microfluidic biosensors, this enables the simultaneous detection of multiple analytes, which is particularly important for analyses made in complex platforms, including profiling cancer biomarkers, infectious agents, or metabolic panels [46].

In the foreseeable future, new stimuli-responsive SiNPs, responsive to pH, temperature, enzymes, or light, are making possible the development of innovative, microfluidic biosensors that adaptively sense and feedback-control therapy. Complemented by recent developments in smartphone-based fluorescence detection and microfluidic automation, SiNP-activated biosensors are poised to become essential to the next generation's POC testing and environmental monitoring.

3.7 Nanocomposites in Microfluidic Biosensing

Nanocomposites play a critical role in enhancing the performance of microfluidic biosensors. These hybrid materials are synthesized by combining matrixing substances like polymers, metals, or ceramics with materials such as CNT, graphene sheets, and metal NPs. Their surface properties, electrical conductivity, and compatibility with biological materials enhance the biosensors' sensitivity, selectivity, and reliability. Hence, when integrated into microfluidic biosensors, it improves signal transduction. For instance, combining gold nanoparticles with Carbon Nanotubes or graphene significantly increases the electrochemical sensitivity, which in turn allows the monitoring of biomarkers often present in minute concentrations [47].

Nanocomposites also immobilize biomolecules with higher binding efficiency and specificity, which is especially suitable for complex biological environments, where high specificity is essential to differentiate different biomolecules. Furthermore, nanocomposites display remarkable synergistic effects. When different nanoparticles are mixed, the produced nanocomposite typically has properties superior to those of the individual components. For example, magnetic nanoparticles in conjunction with polymers or silica give magnetically responsive nanocomposites that are easier to handle externally using magnets for target separation or pre-concentration. Not only do these nanocomposites increase the sensitivity of this biosensor, but they are also able to facilitate the preparation of the sample within a microfluidic device since magnetic forces can be applied to

Materials Research Foundations 191 (2026) 180-223 https://doi.org/10.21741/9781644904213-6

remove unbound particles/contaminants, thereby minimizing the requirement for complex fluidic elements [48,49].

As research progresses, these innovative materials will allow us to achieve more complex and responsive sensing, further broadening the possibilities for dynamic and interactive biosensing systems.

Table 1 Advanced nanoparticles applied in microfluidic biosensing platforms

Nanomaterial	Key Role in Microfluidic Biosensing
Gold NPs	Colorimetric detection, signal amplification
Carbon Nanotubes	Electrical signal generation, high sensitivity
Graphene	Ultra-sensitive electrical sensing, FET biosensors
Quantum Dots	Fluorescent labelling, optical detection
Magnetic NPs	Target capture, separation, and pre-concentration
Silica NPs	Dye/drug carriers, enzyme-based detection
MOFs/Nanocomposites	Multifunctional sensing, tunable platforms

4. Functionalization and Surface Modification of Nanomaterials

Integrating nanomaterials into microfluidic biosensors by introducing nanoparticle-based catalysts increases their ability and flexibility in detecting biological and chemical targets. However, the nature of these materials requires modification to allow appropriate interaction with biological molecules, thus improving sensor performance. This process, known as functionalization, refers to tailoring the surface features of nanomaterials so that the reactivity, selectivity, and biocompatibility are increased. Surface modification further sharpens these interactions so that the nanomaterials have a stable and reproducible response in their operational state in sensor mode. These changes are critical in enabling reliable biosensing capability [50].

Integrating nanomaterials into biosensing applications requires the preservation of their biomolecules in a proper state, such as proteins and enzymes, for them to be functional. Nevertheless, the preparation process can cause denaturation of such biomolecules, reducing their biological activity. State-of-the-art techniques combined with biomimetic mineralization are used to preserve the biological function of proteins and functionalized nanoparticles. On one hand, these methods provide stability; on the other, they facilitate the interaction between the biomolecule and the nanomaterial. Functionalized magnetic nanoparticles (MNPs), for example, can also be applied in different biological applications with an incredibly high surface area, separability of MNPs using a magnetic field, and better stability, selectivity, and specificity. Additionally, functionalized nanomaterials are used for electrode modification by droplet film casting, jet printing, hand coating, electroplating, and electrostatic spraying to enable sensitive and selective detection [51].

The surface modification of covalent and non-covalent strategies addresses such concerns as toxicity and absorption. Crosslinkers, including amino silanes and thio-carboxylic acids, are frequently used to introduce functional groups for biological molecules' attachment to the nanoparticle's surface. Non-covalent strategies, such as weak interactions like charge-based forces

Materials Research Foundations 191 (2026) 180-223 https://doi.org/10.21741/9781644904213-6

and H-bonding, are often utilized to modify metallic and silica nanoparticles.. These techniques facilitate attaching bioreceptors to the altered nanomaterial surfaces, a critical process of biosensor function. The following step of bioreceptor immobilization is critically essential for the successful development of nanomaterials capable of specific recognition and adsorption of target analytes, allowing sensitive and selective analysis [52].

4.1 Bioreceptor Immobilization Strategies

Nanomaterials offer a large biocompatible surface area for immobilizing bioreceptors. The immobilization strategies use bioreceptors like antibodies, enzymes, DNA, and aptamers. These bioreceptors are used to develop efficient microfluidic biosensors to enhance sensitivity and performance in various applications [53,54].

4.1.1 Antibody Immobilization Strategy

Antibodies are primary tools applied in the immunoassay because of their binding affinity and capability to target ligands to probes for analytes in complex sample matrices. Due to the great affinity associated with specific antigens, they are commonly employed in biosensors. Attaching them to a solid surface or a nanoparticle can be key towards developing sensors for analytes of interest. Anticipated performance relies on the approach used for antibody immobilization, which manages area coating, allows precise placement of antibodies to capture the target compound, controls the extent of random interaction, and affects the stability period of the sensor. Electrochemical immunosensing platforms appeal because they are sensitive, straightforward devices, and easy to incorporate into microchannel networks. Microfluidic immunosensors are designed to measure target molecules from ecological samples, biofluids, or drug formulations. For instance, CNT-based electrodes were used to bind elevated antibody count onto biosensors, due to increased surface area. Nevertheless, antibody immobilization needs careful consideration to preserve their functionality and orientation [55].

The growing need for precise and rapid disease diagnosis and the emergence of personalized medicine have made DNA-based electrochemical biosensors highly popular in the past few decades. Genosensors involve electrochemical methods such as amperometry, potentiometry, impedimetry, and nanomaterials like carbon nanotubes, graphene, and metal nanoparticles. Their unique hybridization specificity, fast response times, and ease of miniaturization make them stand out.

This technology frequently relies on the adhesion of a nucleic acid probe to signal-transducing interfaces, which can be light-based, current- or voltage-based, or gravimetric. This may be in the form of a probe or the transducer itself. The DNA probe is selective when recognizing the corresponding target DNA sequence using base-pairing, which produces a measurable signal. The particular ssDNA probe is bound to a surface, and its complementary DNA strand is converted to read-out responses. Peptide nucleic acid, or PNA, is much more precise and selective than DNA for detecting missing biomolecules. In terms of structure, it is similar to DNA. Still, unlike DNA, which has a fully negatively charged sugar-phosphate backbone, PNAs are neutral, made up of repeating unit N-(2-aminoethyl) glycine linked by peptide linkages [56].

The techniques used for DNA immobilization include electrochemical detection of hybridization on transducers, covalent bonding using nanomaterial layers, and avidin/streptavidin to facilitate effective tethering of biotinylated DNA to streptavidin-coated surfaces.

Materials Research Foundations 191 (2026) 180-223
https://doi.org/10.21741/9781644904213-6

4.1.2 Aptamer Immobilization Strategy

Aptamers are short sequences of DNA, RNA, or proteins created artificially in the laboratory to firmly and accurately attach to a specific target, such as a tiny compound, a peptide, DNA/RNA, or a complete cell [57]. The creation of nucleic-acid-based aptamers follows a particular methodology as outlined in Figure 5.

SELEX incorporates several iterative stages, including selection, elution, and amplification, aiming to discover aptamers with high affinity for a specific target molecule. A target molecule is incubated with a library of random nucleic acids to initiate the SELEX process. The unattached sequences are removed before the bound nucleotide sequences are released and multiplied. They are the starting library in the subsequent SELEX cycles. Usually, five to sixteen cycles of SELEX are carried out, each incorporating increasingly rigorous selection procedures. Then the remaining sequences are analyzed and sequenced, from which those that are highly specific and selective towards the analyte are obtained and dubbed possible aptamers [58,59].

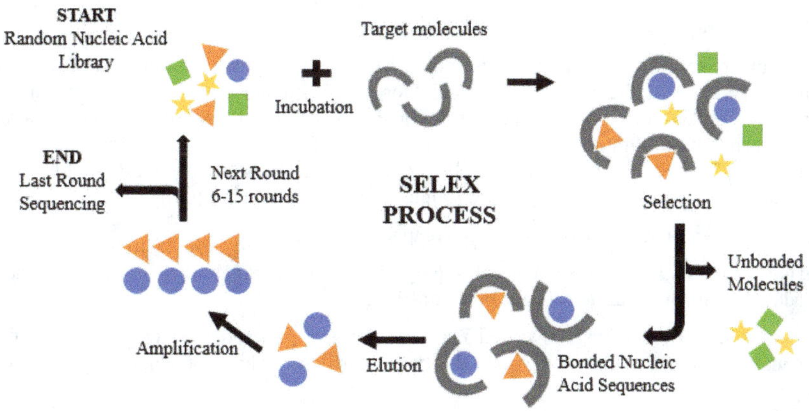

Figure 5 Schematic overview of the SELEX Process to identify aptamers that bind to the target molecule

4.2 Surface Chemistry and Biocompatibility

4.2.1 Surface Chemistry

The material's surface chemistry determines how nanomaterials interact with living systems to a large extent. Modifying the surface charge, hydrophilicity, and functionality content allows for alteration of nanomaterials to enhance their biocompatibility, reduce toxicity, and achieve targeted actions on biomolecules. Surface charge impacts cellular uptake, nanoparticle stability, and interactions with the biological environment. Surface hydrophobicity influences cell adhesion and, by extension, biosensor performance. Microfluidic methods make controlling size, charge, and surface chemistry possible to fabricate dual-ligand systems to optimize these properties [60-62].

Materials Research Foundations 191 (2026) 180-223 https://doi.org/10.21741/9781644904213-6

The surface chemistry of common nanomaterials is as below:

Gold Nanoparticles (AuNPs): AuNPs can be fabricated into different structures, such as nanospheres, nanorods, nanoshells, and nanoclusters. Their surfaces can easily undergo conjugation with various biomolecules and thiol group-terminated grafts via thiol-Au bonds.

Carbon Nanomaterials: Carbon-based nanostructures, for example, carbon nanotubes and oxides of graphene, possess remarkable mechanical as well as electrical properties. Oxidation can attach reactive groups to the edge surfaces [63].

Silica Nanoparticles (SiO2): Silica nanoparticles possess customizable physical, chemical, optical, and mechanical characteristics. Their surfaces can be easily functionalized for cellular imaging and biosensing [64].

Quantum Dots (QDs): Quantum dots have a core constituent made from group II-VI or III-V atoms encapsulated in a shell mostly of zinc sulfide. Capping agents improve their dispersibility in water and enable biomolecule conjugation [65].

4.2.2 Biocompatibility

Biocompatibility is the capacity to achieve a desired objective without causing harmful local or general reactions in the host. Microfluidic biosensors and biocompatibility ensure that the nanomaterials do not cause cytotoxicity, immunological reactions, or any other harmful phenomena that can limit the functionality and efficacy of the biosensors or worsen the patient's condition. The definition of biocompatibility varies from one host to another, depending on the characteristics of the biomaterials used to synthesize the nanomaterials [66].

There are different methods to improve biocompatibility, such as:

Surface Functionalization: Biomolecules, polymers, and even inorganic elements like silica can be added to nanomaterials to enhance biocompatibility.

Passivation: By preventing unwanted reactions, surface passivation extends the life of the nanomaterial's surface, improving its stability and shelf life while maintaining optoelectronic properties [67].

Ligand Exchange: Substituting existing ligands on the surface of the nanomaterial with biocompatible ones can reduce toxicity and improve biocompatibility.

4.3 Signal Enhancement Strategies

Combining plasmonic effects and catalytic amplification in nanomaterials' design has significantly enhanced imaging, sensing, and therapeutic potential outcomes. The distinctive properties of nanomaterials are applied to intensify the strength of the signal, thereby improving the possibility of detecting and distinguishing target substances.

4.3.1 Plasmonic Effects in Signal Enhancement

Enhancing optical imaging systems is particularly important because it improves resolution. Electric fields and optical signals are further augmented, which is facilitated by the enhancement brought about by the optical excitation of metal nanoparticles, i.e., the plasmonic effects. The underlying phenomena involve Localized Surface Plasmon Resonance, whose behavior is dictated by the nanoparticles' shape, size, composition, and environmental factors. When plasmonic

Materials Research Foundations 191 (2026) 180-223
https://doi.org/10.21741/9781644904213-6

nanoparticles are subjected to resonant excitation, it enhances light-matter interaction owing to LSPR, which results in a striking field boost in the vicinity of their surfaces, as shown in Figure 6 [68].

The studies of plasmon resonance (SPR) based nano biosensors are widespread and have been recognized and constructed with a prism, grating, visual fibers, couplers, and coupled with the optical waveguide for the analysis of biochemical and biotic indicators.

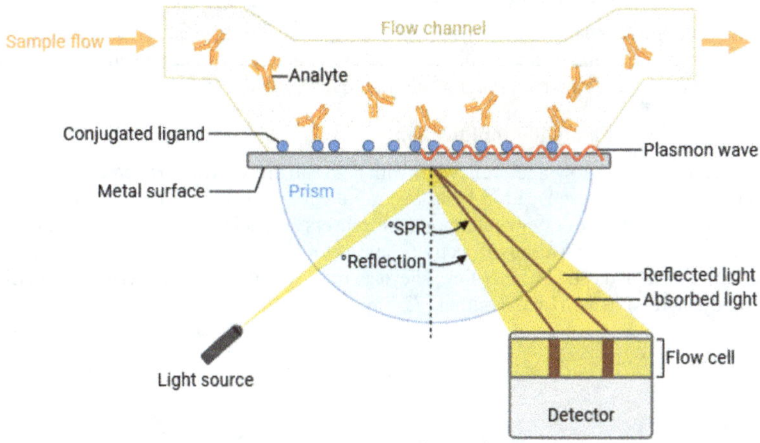

Figure 6 Schematic Principle of Surface Plasmon Resonance (SPR): Analytes bind to immobilized ligands and cause changes in the refractive index, which is detected as a shift in the SPR signal and allows analysis of binding kinetics

4.3.1.1 Metal Enhanced Fluorescence (MEF)

Metal-enhanced fluorescence, often abbreviated as MEF, occurs when the proximity of metallic layers or nanoparticles modifies the fluorophores' excitation and emission rates, photostability, radiation patterns, and quantum yield. Surface plasmon polaritons, which are created at the boundary that separates metals from dielectrics during illumination of incident light at the metal plasmon resonance peak, lead to these changes. The formation of surface plasmons increases the local electromagnetic field, raising the fluorophore's excitation rate. Silver and gold are often used in MEF because of their high concentration of free electrons and plasmon resonance [69].

4.3.1.2 Surface-Enhanced Raman Scattering

Surface-Enhanced Raman Scattering, also called SERS, is a process which improves the scattering of Raman signals of molecules glued to the rough borders of metals or metal nanostructures, as shown in Figure 7. With carefully chosen modulation, excitation light frequency, and LSPR associated with the nanomaterial, the enhancement can be improved (up to 10 orders of magnitude) for SERS active molecules with rough surfaces. Tags known as SERS, which have composite structures and possess Raman-active molecules and nanostructures with known Raman-active

Materials Research Foundations 191 (2026) 180-223 https://doi.org/10.21741/9781644904213-6

groups, are the detection probes. Quantifying the concentration of the particles or analytes is done by measuring the peak intensity of the Raman active molecules. The principal mechanisms of enhancement in SERS are electrolyte anodization and chemical enhancement. Metallic nanoparticles such as silver or gold are used in SERS nano biosensors to create an intense localized electromagnetic field when struck with light, which enormously increases the Raman signal of molecules adsorbed on their surface [70-72].

4.3.2 Catalytic Amplification in Signal Enhancement

The catalytic amplification methodology utilizes a catalyst to amplify biosensing outputs, typically by incorporating enzymatic or nanozyme approaches. This approach significantly improves the sensitivity, accuracy, and spatiotemporal resolution in immunosensors.

4.3.2.1 Nanozymes in Catalytic Amplification

Nanozymes are nano-sized, synthetic enzymes with high and diverse responsiveness, which allows them to be economically produced in bulk, making them inexpensive. Electrochemical enzyme biosensors are preferred over enzyme-based biosensors, especially in biological diagnostics, because of their exceptional catalytic activity and stability. Like other enzymes, they exhibit oxidase, peroxidase, and catalase activities. They can remarkably improve paper-based biosensors' sensitivity by increasing their signal output [73-75].

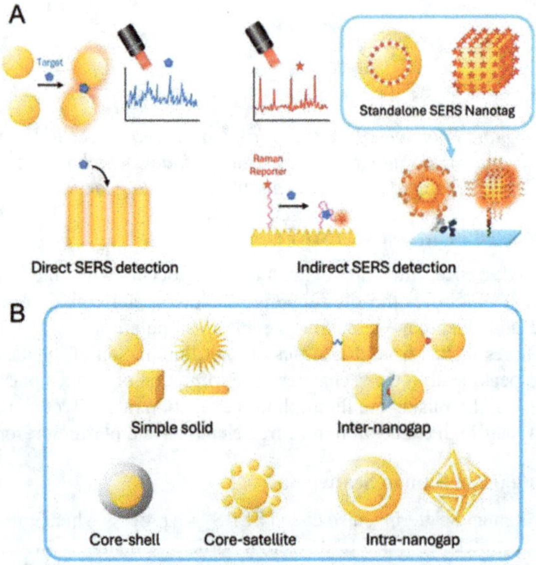

*Figure 7 **(A)** Direct and Indirect SERS detection strategies. **(B)** Nanostructures used in standalone SERS nanoprobes. Adapted from ADDIN EN.CITE [2], licensed under CC BY 4.0 (https://creativecommons.org/licenses/by/4.0/). No changes were made.*

Materials Research Foundations 191 (2026) 180-223 https://doi.org/10.21741/9781644904213-6

4.3.2.2 Single-Atom Catalysts (SACs)

Single-atom catalysts (SACs) have promising activity, selectivity, and durability. They have been investigated for their role as signal relays in biosensing applications. SAC-supported biosensors with signal amplification have shown reasonable sensitivity and selectivity [76,77].

5. Integration of Nanomaterials in Microfluidic Devices

Nanomaterials have many valuable properties that have provided us with many benefits in medicine, medical technology, agriculture, textiles, and the food industry. Some remarkable properties of the nanomaterials that distinguish them from bulk materials are their size, design, and surface area, which provides many atoms for reactivity. On the other hand, microfluidic devices are gaining attention in science due to their remarkable design, which includes the formation of precise microchannels with a regulated flow of liquid [78,79].

When nanomaterials are integrated into microfluidic devices, efficiency and performance increase due to the fine regulation of the movement of liquid within the microchannels. The tiny size of the nanomaterials increases the sensitivity and stability of the microfluidic device, because they can flow through the microchannels easily, and help catalysis according to need. Different types of nanomaterials can be integrated into microfluidic devices according to the application of the microfluidic device. The restricted fluid flow in the microfluidic channels and the properties of nanotechnology make them very efficient in detecting different biomolecules, such as protein, DNA, and other biomolecules [80-82].

5.1 Techniques of Integrating Nanomaterials into Microfluidic Devices

There are different techniques for integrating nanomaterials into microfluidic devices. Some of them are mentioned below.

5.1.1 Ex-situ Technique

In this technique, the nanomaterials are manufactured separately according to the specific properties needed and then integrated into the microfluidic devices by chemical or other means. For example, porous polymers are fabricated outside the microfluidic chips and inserted into the thermoplastic microchannel [83,84].

5.1.2 In situ Technique

In this, the nanomaterials are synthesized by various methods inside the microfluidic devices. This technique is exact and cost-effective. For instance, ZnO particles are synthesized by the hydrothermal method, in which the size and structure of the nanoparticles are adjusted by the use of heat in the microchannels [85-87].

5.2 Fabrication of Microfluidic Device

Microfluidic devices have a highly interactive structure consisting of microchannels, as shown in Figure 8. Pressure is a driving force that moves the liquid in these microchannels. Due to the microfluidic device's small structure and the fluid's precise control, they can be used in sample separation, bioanalysis, and biosensors. Sometimes, they also have pumps and mixers, which help flow liquids, mix reagents, and perform other crucial functions. Nanomaterials are integrated into

Materials Research Foundations 191 (2026) 180-223 https://doi.org/10.21741/9781644904213-6

the microchannels of these microfluidic devices by chemical or other methods. There is also a closed channel system of microfluidics and an open channel system [88].

Usually, silicon and glass are used to fabricate microfluidic devices because silicon can act as a semiconductor, and glass can remain stable at high temperatures. Other materials used for fabrication include polymers such as polydimethylsiloxane, polymethacrylate, and polycarbonate, which are more flexible and cheaper than glass or silicon [89-92].

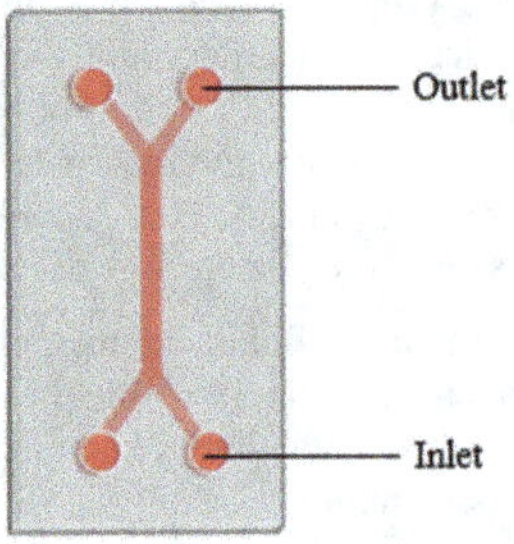

Figure 8 Basic design of a microfluidic device featuring an inlet for sample introduction, a microchannel, and an outlet to collect samples or remove waste.

5.2 Fabrication Techniques for Microfluidic Devices

5.2.1 Lithography-based Techniques

This technique makes precise microchannels in microfluidic devices and includes two main types: Photolithography and Soft lithography. Photolithography uses UV rays through direct exposure to the sun. The photoresist that is used to cover the substrate changes properties in UV and forms patterns on the substrate. In Soft Lithography, the master mold is made using photolithography, in which the PDMS is used to make the patterns, and the mold previously made by the photolithography is transferred to the PDMS stamp [93,94].

5.2.2 Molding-based Techniques

There are two types of molding-based techniques: injection molding and hot embossing. Both are used to produce many microfluidic devices. In injection molding, the pre-polymerized pellets are first heated in the barrels, then injected into the mold under pressure, where they solidify. On the other hand, in hot embossing, the thermoplastics are melted and put into the molds, where delicate structures are formed due to pressure and the high-temperature microfluidic device [95-97].

5.2.3 Direct Writing Techniques

Unlike the molding technique, mold is not required in this method, but microfluidic devices are produced directly. Laser ablation is a technique that uses a laser on a substrate, which melts and forms the structure by absorbing the laser energy. In 3D Printing, a laser or mold is not required in this 3d print to build the structure layer by layer. This very delicate structure of the microfluidic can be formed at a rapid speed [98,99].

Materials Research Foundations 191 (2026) 180-223
https://doi.org/10.21741/9781644904213-6

5.2.4 Etching Techniques

Etching is the process that is mainly used for silicon and glass because the chemicals used in this process are thermostable and inert, making them suitable for etching. Wet Etching is primarily used to make microfluidic devices from glass. It smooths out the rough surfaces in the microchannels, which helps with the smooth flow of the liquid within these channels. Dry Etching is used as anisotropic material. This method works better on the silicon due to the difference in temperature in the material, which is caused by the presence of the reactive ion exchange, which is done in this process [100-102].

5.3 Fluidic Control and Sample Handling

Microfluidic devices, like the name suggests, have a small amount of fluid ($10^{-9} - 10^{-18}$ $liters$) Present in them. The fluids' viscosity, flow, speed, and properties can affect the microfluidic devices' overall properties. The fluid flow can be laminar or turbulent, and this can be determined by the Reynolds number, which measures the ratio of inertia and viscous forces. Mostly, the microfluidic devices have laminar flow due to the very minute amounts of liquid present, which causes inertia to be overcome by viscosity [103].

Mostly in microfluidic devices, the fluids flow in an open/closed and two-phase system. The open/closed system is shown in Figure 9. An open microfluidic system is performed on the open surface using the wetting phenomena, according to which the liquid spreads on the surface due to pressure till equilibrium is reached. In the closed microfluidic **system**, the liquid is confined in the microfluidic channels, and the pressure from pumps and any other system is used to drive it through the microchannels. This confined system is mainly used in biomedical sensing. On the other hand, the two-phase microfluidic system is mainly used to separate biomolecules. It consists of two immiscible liquids, a liquid or a gas-liquid phase, flowing in the microchannels. This system is used for the precise control and manipulation of the liquid [104-106].

5.3.1 Types of Fluidic Control

There are two types of fluid control: active and passive.

5.3.1.1 Active Fluid Control System

In the active flow control of microfluidics, external pressure is applied to the fluid using pumps to move the fluid in a direction. Valves are mainly used to control the flow rate of the liquid and in which direction it flows. The electrophoresis in this technique produces polarization in the fluid, which then moves in the non-electric field, and the fluid moves in a controlled manner. This is a very safe method, and the fluid control is fast [107,108].

5.3.1.2 Passive Flow Control

Passive flow control of the devices does not need any external force; they work by other methods, such as tension and osmotic flow, which do not require energy and are much cheaper. In the osmotic pressure method, a semipermeable membrane is placed in the microfluidic device, and the difference in pressure on both sides of the membrane causes the fluid flow. In some microfluidic devices, gravity controls the flow of liquids. In other cases, the capillary system is used, like in plants, where the material moves in porous tubes from one place to another by capillary action. A pressure-driven system is the most commonly used passive system in which water applies pressure

Materials Research Foundations 191 (2026) 180-223 https://doi.org/10.21741/9781644904213-6

on each other and the walls of the microfluidic channels, and the flow pressure from passive pumps helps in the system's flow [109,110].

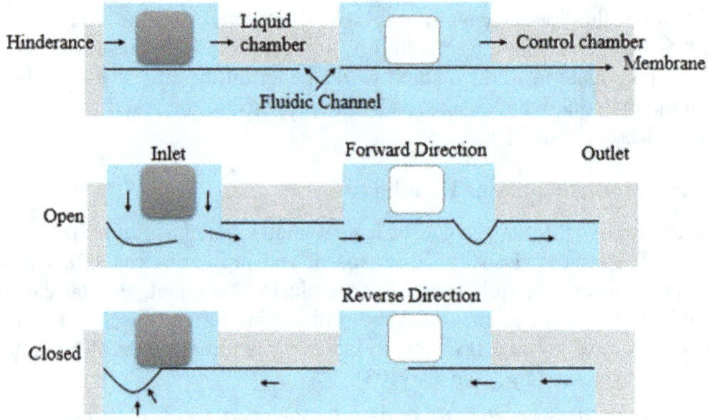

Figure 9 Schematic representation showing fluid flow directions, inlets, and outlets in open and closed microfluidic platforms.

5.3.1.3 Nanomaterial-Assisted Microfluidic Flow

Nanomaterials also help enhance the function of microfluidic devices by controlling their fluid flow. These nanomaterials can become attached to the microchannels of microfluidic devices, increasing the sensitivity of the microfluidic devices in the case of biosensing.

6. Nanomaterial-Based Detection Mechanisms in Microfluidic Biosensing

The use of nanomaterials in producing microfluidic biosensors is compelling because of their unique functionality. They enable powerful analytical platforms for diverse bioanalytical applications by integrating electrochemical, optical, magnetic, photothermal, distance, and paper-based readouts to microfluidic devices [111].

6.1 Electrochemical Detection Mechanism

Electrochemical detection in microfluidic biosensing is attractive in terms of simple devices, affordability, and suitability for on-site assessment, as shown in Figure 10. Modern electroanalytical techniques are highly responsive, specific, manageable, time-efficient and can be applied to various analytical purposes.

6.1.1 Amperometric Detection

Amperometry is a common electrochemical technique in which the amplitude of either oxidation or reduction current is calculated at a particular voltage over a period of time. There is an increased interest in developing nanomaterials to be used for amperometric techniques to control biomolecule binding specifically. Nanomaterials are used to amplify amperometric signals. **Amperometric biosensors** are used on hybrid and composite nanomaterials [112].

6.1.2 Impedimetric Detection

Electrical impedance (Z) is expressed as the ratio of $V(t)$ to $I(t)$ of an incremental change in the voltage and the corresponding change in the current, as shown in Eq. 1. Impedance Z could be defined from this definition as the quotient of the voltage-time function V(t) and the resulting current–time function $I(t)$.

$$Z = V(t) / I(t) = V_0 \sin(2\pi ft) / I_0 \sin(2\pi ft + \phi) = |Z| e^{j\phi b} \quad (1)$$

The role of nanomaterials in improving impedimetric biosensor technology has mainly been established. Their extensive surface area, ideal electronic and electrocatalytic properties, and biocompatibility make them crucial in reducing biosensor detection limits. Impedimetric biosensors have benefited from incorporating materials such as metal nanoparticles, nanowires, nanorods, carbon nanotubes, and graphene, which are used in applications like bacteria detection.

When impedimetric biosensors are combined with biological recognition agents such as bacteriophages and lectins, the sensitivity and capability of detecting bacteria are effectively improved. Combining nanomaterials with microfluidic systems enhances the detection of bacterial pathogens more effectively and accurately in impedimetric biosensors [113].

6.1.3 Potentiometric Detection

Under zero current conditions, such biosensors measure the potential difference between two electrodes and report the concentration of some specific ions or molecules. Tuning the lattice configuration is expected to improve the detection performance of metal compound-based biosensors [114].

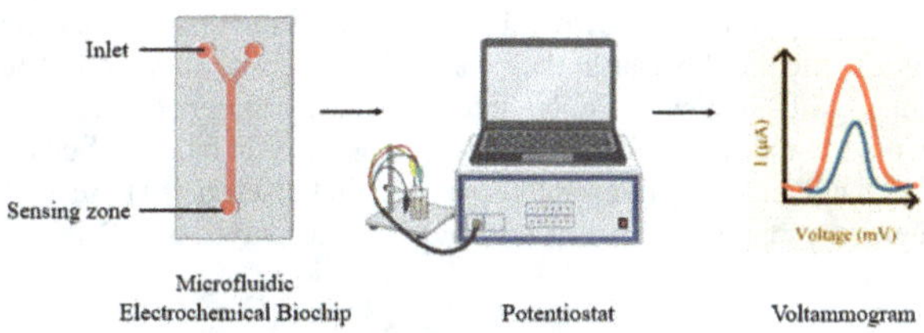

Figure 10 Electrochemical Microfluidic Biosensing: The setup illustrates a signal generated by a microfluidic chip, processed by a potentiostat, and visualized as a voltammogram output.

6.2 Optical Detection Mechanisms

Optical biosensors are based on light detection to detect any biological or chemical at high sensitivity and in real-time. There are various methods, such as fluorescence, colorimetry, or chemiluminescence, and each has its advantages [115].

6.2.1 Fluorescence Detection

Biosensors based on fluorescence detect and analyze chemical and biological materials based on fluorescent materials. Quantum dots (QDS) are nanomaterials with special fluorescence; they have the advantages of tunable excitation properties, higher quantum yield, and good stability in solution. Fluorescence Polarization-based sensing systems can use these nanomaterials as signal generation sources to generate fluorescent signals, and these nanomaterials can serve as signal amplifiers to increase the system's sensitivity. FP-based biosensing monitors molecular binding reactions by observing changes in polarization, which is determined by the size and rotation rate of the fluorescent elements [116,117].

6.2.2 Colorimetric Detection

Colorimetric biosensors use visible color change to detect target molecules. They are beautiful due to the simplicity and cost-effectiveness of colorimetric biosensors. The naked eye could induce different color changes, or simple instruments using nanomaterials, including gold, copper, and quantum dots. Utilizing nanomaterials integrated into colorimetric detection systems provides rapid, sensitive, and convenient detection of pathogens and other analytes based on unique catalytic activities, aggregation, destabilization, and similar properties of nanomaterials [118].

6.2.3 Chemiluminescence Detection

Chemiluminescence (CL) is commonly used for detection as it is sensitive and selective and does not produce excess signal background. Applying catalysts (use of nanomaterials), emitters, and carriers has improved CL systems' performance substantially by increasing emission intensity and sensitivity. Further developments in nanoparticles are constantly expanding sensitivity and accuracy of chemiluminescent probes, simplifying the detection of biomolecules [119].

6.3 Magnetic and Photothermal Detection

6.3.1 Magnetic Detection

Microfluidic biosensing systems, as shown in Figure 11, featuring magnetic nanoparticles (MNPs) have attracted much attention because they have the potential to manipulate MNPs within microfluidic channels with magnetic fields, detect MNPs by integrated magnetic microsensors, and conveniently modify the surfaces of MNPs as well as specific binding. Such systems can lead to the generation of handheld, sensitive, and cheap on-chip biosensors. These applications depend on the unique magnetic properties of superparamagnetism displayed by MNPs.

Materials Research Foundations 191 (2026) 180-223 https://doi.org/10.21741/9781644904213-6

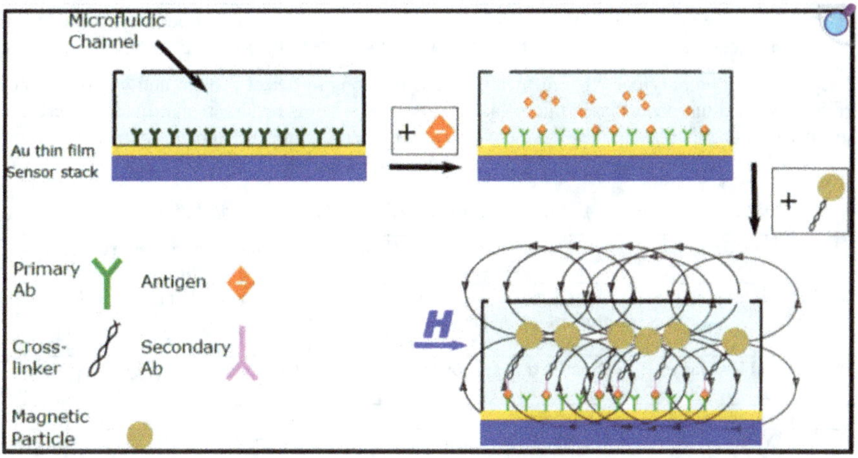

Figure 11 Schematic mechanism of a standard magnetic biosensor. Adapted from [1], licensed under CC BY 4.0 https://creativecommons.org/licenses/by/4.0/). No changes were made.

6.3.2 Photothermal Detection

Photothermal detection depends on the photothermal effect of nanomaterials, where light energy is converted into heat. This heat generation can be applied to sensing microfluidic systems in a label-free mode for biosensing. Gold nanoparticles (AuNPs) have been demonstrated to be used as a peroxidase-like enzyme and a photothermal substrate [120].

6.4 Distance-Based and Paper-Based Microfluidic Biosensors

6.4.1 Distance-based Microfluidic Biosensors

Distance-based microfluidic assays (d-μASs) possess instrument-free visual point-of-care testing, where the detection signal is read off at a readable distance. These reactions produce a visible front, and the distance the front travels is proportional to the analyte concentration [121].

6.4.2 Paper-based Microfluidic Biosensors

Microfluidic devices made of paper have the advantages of low cost, small size, and ease of portability for biosensing. The nanomaterials integrated with paper-based biosensors amplify their capability of converting pathogens' signals through unique mechanisms to provide an amplified colorimetric readout. This combination of nanomaterials and paper-based microfluidics makes a powerful analytical platform for various applications [122].

A new generation of paper-based microfluidic analytical devices combines features from two relatively well-developed microfluidics (multiplex ability) and the tractability and simplicity found in Lateral Flow Test Strip (LFTS) technology. In addition to LFTS biosensors, other paper biosensors are also contributing to in-field, low-cost, and highly responsive target molecule

Materials Research Foundations 191 (2026) 180-223 https://doi.org/10.21741/9781644904213-6

detection. As the paper-based microchannel sensor only consists of a singular part, it adds to its convenience compared to a strip-test biosensor without the need to assemble multiple pads [123].

They typically employ color-based measurement methods to detect protein and glucose. Using various NPs within paper-based microfluidic biosensor systems results in significantly increased sensitivity. For this reason, they have developed a modern approach to creating compact, highly responsive sensors, which are promising for various research fields.

Table 2 Comparison of Distance-Based and Paper-Based Microfluidic Biosensors

Feature	Distance-based Microfluidic Biosensors	Paper-based Microfluidic Biosensors
Detection Principle	Front distance is directly proportional to analyte concentration	Nanomaterial-enhanced color signal
Signal Readout	Distance travelled by the reaction front	Color change on paper substrate
Instrumentation	No instrument	Minimal
Material	Microfluidic channel	Paper
Advantages	Quantitative, portable	Simple, Low-cost, multiplex
Applications	Point-of-care diagnostics	Detection of protein, glucose, and pathogen

7. Applications of Nanomaterial-Based Microfluidic Biosensors

Nanomaterial microfluidic biosensors combine the advantages of microfluidics, i.e., small sample volume, fast processing time, and portability, with the extraordinary properties of nanomaterials to provide the rapid and efficient detection of target analytes. These biosensors are explored in terms of their applications for point-of-care diagnostics, environmental monitoring, food safety, biomedical research, and drug development [124].

7.1 Role in Point-of-Care Diagnostics

Early disease detection via microfluidic biosensors is crucial in improving patient outcomes. Since the early symptoms are inconspicuous, the chances of delayed diagnoses and treatments are high, hence the need for regular screening. Advantages offered by these devices include compact size, portability, small sample consumption, and increased manoeuvrability.

7.1.1 Cancer Diagnostics

Microfluidic systems based on nanomaterials have shown promise in their ability to sense tumor markers, for example, circulating tumor cells, microRNAs (miRNAs), proteins, and small extracellular vesicles (EVs). These biomarkers indicate how tumor cells behave, how cancer progresses, and whether or not a tumor will metastasis. Early detection of them significantly helps in improving patient survival rates. One example is an electrochemical immunosensor integrated into a microfluidic system, which can detect early melanoma at 10 cells/10 mL [125].

Materials Research Foundations 191 (2026) 180-223 https://doi.org/10.21741/9781644904213-6

7.1.2 Infectious-Disease Diagnostics

Microfluidics-based nano biosensors are used in the detection and measurement of biomolecules on a micro-scale, using fluid flow and nanotechnology principles. They offer fast and precise results, leading to early disease diagnosis. Additionally, they enable real-time monitoring of biomarkers and low levels of biomolecules accurately, thus decreasing the chances of wrong results. They can discern biomarkers in bodily fluids such as urine and saliva without requiring invasive procedures, which poses a lower incidence of infection and unfavorable complications. Not only this, they provide various benefits, such as simple manufacturing, low reagent usage, prompt response, improved detection performance, and real-time detection of target molecules [126].

7.2 Role in Environmental Monitoring

Biosensors based on nanomaterials in microfluidics perform continuous environmental assessment of pollutants and rapid and/or inexpensive detection of heavy metals, pesticides, and pathogens through real-time analysis [127].

7.2.1 Heavy Metal Detection

Lead (Pb), mercury (Hg), arsenic (As), cadmium (Cd), and chromium (Cr) are heavy metals that pose a significant threat to the environment and human health. The current sensors can be integrated into microfluidic devices that enable the detection of these ions. For instance, a portable microfluidic device with a DNAzyme fluorescent biosensor can detect lead concentrations in μM and nM. Multiple economical and portable LOC systems for Pb^{2+} detection have been developed, such as microfluidic LOCs with DNAzyme fluorescent biosensors. These devices are sensitive enough to measure Pb^{2+} at 5 ppb or 10 µM with the help of a range of methods, including oxygen plasma-treated PDMS microchannels and gold nanoprobes. Further, some systems are designed to be mobile and unpowered and use magnification for detection. Similarly, Hg2+ and Pb2+ ions have been detected using a PDMS microfluidic graphene oxide (GO)/aptamer sensor device. Hg2+ and Pb2+ have reached the detection level of 0.7 ppb and 0.53 ppb, respectively, and both are below the level set by the World Health Organization (WHO) [128].

Table 3 Application of nanomaterial-based microfluidic biosensors for heavy metal detection

Metal ion	Nanomaterial used	Microfluidic platform	Detection Limit (ppb)
Pb^{2+}	DNAzyme with gold nanoprobes	PDMS-based LOC device	5 / 10 µM
Hg^{2+}	Graphene oxide (GO) / aptamer	PDMS microfluidic sensor	0.7
Pb^{2+}	Graphene oxide (GO) / aptamer	PDMS microfluidic sensor	0.53

7.2.2 Pathogen Detection

For environmental monitoring of pathogens, they incorporate nanomaterial-based microfluidic biosensors that provide flexibility in automation, integration, miniaturization, and multiplexing. Large-scale pathogen screening employs nucleic acid-based methods, such as polymerase chain reaction (PCR) or loop isothermal amplification (LAMP) detection systems. These colorimetric

Materials Research Foundations 191 (2026) 180-223 https://doi.org/10.21741/9781644904213-6

sensors can be detected in real time without complex instrumentation and are easily integrated with Lab on a Chip using PDMS microfluidics. Gold nanoparticles, among other agents, are commonly used because they rely on an agent that changes color in the presence of a pathogen [128].

7.3 Role in Food Safety and Quality Control

Microfluidic biosensors have wide applications in food safety, particularly in detecting heavy metals, pesticides, pathogens, allergens, or genetically modified organisms (GMOS) to guarantee safe food production and supply [129].

7.3.1 Detection of Pathogens

Microfluidic biosensors composed of nanomaterials show enhanced sensitivity and specificity in detecting foodborne pathogens such as Escherichia coli, Salmonella, Listeria monocytogenes, and other harmful bacteria. These biosensors provide rapid, point-of-care analysis, allowing for detecting and tracing contamination through various food items, protecting consumer health.

7.3.2 Detection of Food Contaminants

Nanomaterial functionalized microfluidic biosensors can detect food contaminants, including heavy metals, pesticides, and hazardous food additives like lead, arsenic, and mercury. Their outstanding sensitivity means that testing becomes immediate and on-site, making this a valuable technique for controlling food safety and regulatory control in agriculture and food manufacturing [130].

7.3.3 Detection of Food Allergens

Nanomaterial sensors such as molecularly imprinted polymers (MIPs) can rapidly and reliably identify commonly encountered food allergens across various food types, such as soy, peanuts, or gluten. Thanks to their current accuracy, such sensors enable efficient management of foodstuff allergens and enhance food safety for people with dietary limitations [131].

7.3.4 Detection in Beverages and Packaging Materials

Using nanomaterials in microfluidic biosensors significantly improves the ability to test for contaminants, including microplastics and hazardous chemicals, in beverages such as soda, juice, and bottled water. Such biosensors are also used to detect when packaging materials expose beverages to fluid additives or plasticizers for the safety of consumers.

7.4 Role in Biomedical Research and Drug Development

Microfluidic Biosensors are also essential in biomedical research and drug development due to their advantages in reducing sample consumption, performing rapid analysis in single cells, and controlling precise preparation processes. They are used to precisely control and automate drug screening, delivery, and personalized medicine [132].

7.4.1 Drug Development, Screening, and Delivery

Microfluidic biosensor devices are used to develop, screen, and deliver drugs by controlling the environment and precisely controlling fluids. It allows the study of drugs on a cellular level and recreates in vivo physiological conditions on a chip. Mimicking entire organs for testing drug

Materials Research Foundations 191 (2026) 180-223 https://doi.org/10.21741/9781644904213-6

effects and speeding up presently established means of travel, such as shots, these devices are handy. With integrated fluid handling systems combined with microneedles, less painful and more focused drug delivery is achievable. Microfluidic technologies also present new possibilities for anti-cancer drug screening. Microchips for high-content screening of mixtures of complex neuroactive substances have been demonstrated to be sensitive, selective, and robust [133].

7.4.2 Cell Culture and Analysis

A controlled environment for cell culture and drug testing using microfluidic devices is an ideal situation with applications for several types of cells, such as cancer, hepatic, and bacterial. With a controlled environment, it can monitor cell cultures over a long period and isolate single cells for further culturing. These devices can reproduce in vivo conditions and have different environmental profiles; thus, they can be used as drug-delivering and cell-culturing platforms. Cell culturing and observation on microfluidic platforms can be accomplished by manipulating the residing cells individually. The drug level effect on chemotaxis can be investigated locally on cells in vitro [134].

7.4.3 Pharmacokinetic and Pharmacodynamic Studies

A significant advantage of microfluidics in PK and PD studies is that dynamic PK and PD profiles can be studied. Microfluidic devices can reproduce in vivo PK profiles in vitro, improving the prediction of in vivo efficacy in preclinical models. These systems enable the absorption, distribution, metabolism, and elimination, as well as the physiological and biological response of a drug candidate to the administered drug. PK's fundamental parameter is the drug concentration in blood, and PD parameters depend on the type of disease and can be well monitored using microfluidic devices [135].

8. Challenges and Future Perspectives of Advanced Nanomaterials in Microfluidic Biosensing

Although integrating nanomaterials in microfluidics has provided us with many benefits and enhanced the applications of microfluidics devices, critical hurdles impede their real-world adoption as commercial technologies. Among the most substantial limitations are problems of stability and reproducibility. Manufacturing the nanomaterial for integration is difficult on an industrial scale due to compatibility issues with microfluidic devices. It is also challenging to produce the nanomaterial that gives precise results during biosensing and produces uniform results. Sensor reproducibility is often poor due to the inconsistencies in nanomaterial synthesis, surface modification, and deposition techniques, which lead to different amounts of sensor output and poor batch-to-batch uniformity[136].

A second significant challenge is the scalability and low cost of nano-material-based biosensors. Most advanced nanomaterials are subjected to expensive reagents and sophisticated instrumentation required for high-precision synthesis. It is economically unsustainable for many applications, particularly in low-resource settings or for disposable, point-of-care, large-scale production. Prospective studies should aim to develop low-cost, green, and scalable synthesis strategies for these applications, such as solution-based, microwave-assisted growth or biogenic synthesis. Additionally, nanotechnology-integrated microfluidics can be manufactured on a large scale economically using additive manufacturing, inkjet printing, and roll-to-roll fabrication processes [137].

Materials Research Foundations 191 (2026) 180-223 https://doi.org/10.21741/9781644904213-6

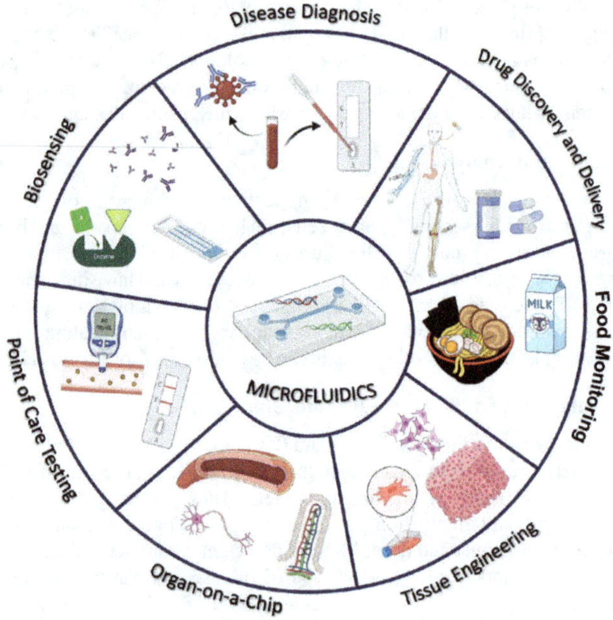

Figure 12 Applications of Nanomaterial-based Microfluidic Biosensors

In addition to technical constraints, ethics and regulatory constraints are emerging frontiers in deploying nanomaterial biosensors. It concerns the cytotoxicity, the environmental accumulation, and the long-term health effects of the interaction of engineered nanomaterials with biological systems. It also raises concerns regarding the privacy of patient data, patient consent, and proper ethical deployment in the context of a clinical or surveillance application. As part of the scientific community, regulatory agencies and policymakers need to establish ethical transparency and public trust by undertaking dialogue and setting out standardised testing protocols, toxicity assessments, and guidelines for using nanomaterials responsibly.

Some of the next-generation microfluidic biosensors include using nanomaterials, IoT, AI, and machine learning for the immediate processing of information, remote monitoring of conditions, and the predictive early diagnosis of diseases. The next stage of innovation might include autonomous sensor networks powered by energy-harvesting nanomaterials, intelligent nanomaterial functions, and materials that would be safe and eco-friendly. Integrating materials science, bioengineering, bioinformatics, and clinical medicine holds an enormous potential to shape a future in nanomaterial-based biosensing for practical and effective diagnosis.

9. Conclusion

The integration of advanced nanomaterials into microfluidic biosensor systems significantly enhances the efficiency and usability of diagnostic tools across healthcare, environmental monitoring, and food safety sectors. Leveraging key properties such as large surface area, tunable chemical reactivity, and high signal detectability, these systems offer exceptional sensitivity and selectivity, even for low-concentration and complex biological samples. While challenges remain such as issues of scalability, material stability, and system integration, the potential of nanomaterial-based microfluidic biosensors to meet the growing demands for early disease detection, real-time monitoring, and point-of-care testing is considerable. As we enter a new technological era, these biosensors are poised to become a transformative platform for accessible, cost-effective, and high-performance diagnostics.

Reference

[1] I. Giouroudi, G. Kokkinis, Recent advances in magnetic microfluidic biosensors, Nanomaterials (Basel). 7 (2017) 171. https://doi.org/10.3390/nano7070171

[2] V.R. Martínez, E. Lee, J.W. Oh, Exploring plasmonic standalone surface-enhanced raman scattering nanoprobes for multifaceted applications in biomedical, food, and environmental fields, Nanomaterials (Basel). 14 (2024) 1839. https://doi.org/10.3390/nano14221839

[3] P. Lisowski, P.K. Zarzycki, Microfluidic paper-based analytical devices (μPADs) and micro total analysis systems (μTAS): development, applications and future trends, Chromatographia. 76 (2013) 1201-1214. https://doi.org/10.1007/s10337-013-2413-y

[4] S.R.S. Pour, D. Calabria, A. Emamiamin, E. Lazzarini, A. Pace, M. Guardigli, M. Zangheri, M. Mirasoli, Microfluidic-based non-invasive wearable biosensors for real-time monitoring of sweat biomarkers, Biosensors (Basel). 14 (2024) 29. https://doi.org/10.3390/bios14010029

[5] S.A. Muhsin, Y.He, M. Al-Amidie, K. Sergovia, A. Abdullah, Y. Wang, O. Alkorji, R.A. Hulsey, G.L. Hunter, Z.K. Erdal, R.J. Pletka, H.S. George, X.F. Wan, M. Almasri, A microfluidic biosensor architecture for the rapid detection of COVID-19, Anal. Chim. Acta. 1275 (2023) 341378. https://doi.org/10.1016/j.aca.2023.341378

[6] S. Choi, M. Goryll, L.Y.M. Sin, P.K. Wong, J. Chae, Microfluidic-based biosensors toward point-of-care detection of nucleic acids and proteins, Microfluid. Nanofluidics. 10 (2011) 231-247. https://doi.org/10.1007/s10404-010-0638-8

[7] S.R.S. Pour, D. Calabria, A. Emamiamin, E. Lazzarini, A. Pace, M. Guardigli, M. Zangheri, M. Mirasoli, Electrochemical vs. optical biosensors for point-of-care applications: a critical review, Chemosensors. 11 (2023) 546. https://doi.org/10.3390/chemosensors11100546

[8] A. Chícharo, F. Cardoso, S. Cardoso, P.P. Freitas, Dynamical detection of magnetic nanoparticles in paper microfluidics with spin valve sensors for point-of-care applications, IEEE Trans. Magn. 50 (2014) 1-4. https://doi.org/10.1109/TMAG.2014.2325813

[9] C. Carrell, A. Kava, M. Nguyen, R. Menger, Z. Munshi, Z. Call, M. Nussbaum, C. Henry, Beyond the lateral flow assay: a review of paper-based microfluidics, Microelectron. Eng. 206 (2019) 45-54. https://doi.org/10.1016/j.mee.2018.12.002

[10] F. Özefe, Design and development of paper-based microfluidics for point-of-care applications (M.Sc. thesis), Izmir Institute of Technology, Izmir, Turkey, 2020.

Materials Research Foundations 191 (2026) 180-223 https://doi.org/10.21741/9781644904213-6

[11] S.M. Yang, S. Lv, W. Zhang, Y. Cui, Microfluidic point-of-care (POC) devices in early diagnosis: A review of opportunities and challenges, Sensors (Basel). 22 (2022) 1620. https://doi.org/10.3390/s22041620

[12] M. Johnston, H.C. Ates, R.T. Glatz, H. Mohsenin, R. Schmachtenberg, N. Göppert, D. Huzly, G.A. Urban, W. Weber, C. Dincer, Multiplexed biosensor for point-of-care COVID-19 monitoring: CRISPR-powered unamplified RNA diagnostics and protein-based therapeutic drug management, Mater. Today. 61 (2022) 129-138. https://doi.org/10.1016/j.mattod.2022.11.001

[13] A. Kumar, C. Hahn, S. Herchen, A. Soucy, E. Carpio, S. Harper, N. Rahmani, C. Anagnostopoulos, M. Faghri, A microfluidic paper-based lateral flow device for quantitative ELISA, Micro. 4 (2024) 22. https://doi.org/10.3390/micro4020022

[14] Y. Hou, Z. Liu, H. Huang, C. Lou, Z. Sun, X. Liu, J. Pang, S. Ge, Z. Wang, W. Zhou, H. Liu, Biosensor-based microfluidic platforms for rapid clinical detection of pathogenic bacteria, Adv. Funct. Mater. 35 (2025) 2411484. https://doi.org/10.1002/adfm.202411484

[15] L. Nie, F. Liu, P. Ma, X. Xiao, Applications of gold nanoparticles in optical biosensors, J. Biomed. Nanotechnol. 10 (2014) 2700-2721. https://doi.org/10.1166/jbn.2014.1987

[16] L. Catanzaro, V. Scardaci, M. Scuderi, M. Condorelli, L. D'Urso, G. Compagnini, Surface plasmon resonance of gold nanoparticle aggregates induced by halide ions, Mater. Chem. Phys. 308 (2023) 128245. https://doi.org/10.1016/j.matchemphys.2023.128245

[17] D. Brambilla, F. Panico, L. Zarini, A. Mussida, A.M. Ferretti, M. Aslan, M.S. Ünlü, M. Chiari, Copolymer-coated gold nanoparticles: enhanced stability and customizable functionalization for biological assays, Biosensors (Basel). 14 (2024) 319. https://doi.org/10.3390/bios14070319

[18] Y. Xu, H. Zhong, M. Shi, Z. Zheng, S. Liu, Q. Shou, H. Li, G. Yang, Z. Li, X. Xing, Microfiber-directed reversible assembly of Au nanoparticles for SERS detection of pollutants, Opt. Lett. 47 (2022) 2028-2031. https://doi.org/10.1364/OL.454581

[19] A. Halder, M. Zhang, Q. Chi, Electroactive and biocompatible functionalization of graphene for the development of biosensing platforms, Biosens. Bioelectron. 87 (2017) 764-771. https://doi.org/10.1016/j.bios.2016.09.030

[20] B. Zou, S. Lou, J. Duan, S. Zhou, Y. Wang, Design of raman reporter-embedded magnetic/plasmonic hybrid nanostirrers for reliable microfluidic SERS biosensors, Nanoscale. 15 (2023) 8424-8431. https://doi.org/10.1039/D3NR00303E

[21] J. Mondal, J.M. An, S.S. Surwase, K. Chakraborty, S.C. Sutradhar, J. Hwang, J. Lee, Y.-K. Lee, Carbon nanotube and its derived nanomaterials based high performance biosensing platform, Biosensors. 12 (2022) 731. https://doi.org/10.3390/bios12090731

[22] C. Aksu, P.D. Bradford, J.S. Jur, Microfluidic behavior of alumina nanotube-based pathways within hydrophobic CNT barriers, Langmuir. 36 (2020) 8792-8799. https://doi.org/10.1021/acs.langmuir.0c01096

[23] D. Rodrigues, A.I. Barbosa, R. Rebelo, I.K. Kwon, R.L. Reis, V.M. Correlo, Skin-integrated wearable systems and implantable biosensors: a comprehensive review, Biosensors (Basel). 10 (2020) 79. https://doi.org/10.3390/bios10070079

[24] H.S. Hwang, J.W. Jeong, Y.A. Kim, M. Chang, Carbon nanomaterials as versatile platforms for biosensing applications, Micromachines (Basel). 11 (2020) 814. https://doi.org/10.3390/mi11090814

[25] T. Li, Y. Liang, J. Li, Y. Yu, M.M. Xiao, W. Ni, Z. Zhang, G.J. Zhang, Carbon nanotube field-effect transistor biosensor for ultrasensitive and label-free detection of breast cancer exosomal miRNA21, Anal. Chem. 93 (2021) 15501-15507. https://doi.org/10.1021/acs.analchem.1c03573

[26] D. Soto, J. Orozco, Hybrid nanobioengineered nanomaterial-based electrochemical biosensors, Mol. 27 (2022) 3841. https://doi.org/10.3390/molecules27123841

[27] Y.W. Sun, D.G. Papageorgiou, C.J. Humphreys, D.J. Dunstan, P. Puech, J.E. Proctor, C. Bousige, D. Machon, A. San-Miguel, Mechanical properties of graphene, Appl. Phys. Rev. 8 (2021) 021310. https://doi.org/10.1063/5.0040578

[28] L.S. Hall, D. Hwang, B. Chen, B. Van Belle, Z.T. Johnson, J.A. Hondred, C.L. Gomes, M.D. Bartlett, J.C. Claussen, All-graphene-based open fluidics for pumpless, small-scale fluid transport via laser-controlled wettability patterning, Nanoscale. Horiz. 6 (2021) 76-85. https://doi.org/10.1039/D0NH00376J

[29] A. Ghanam, A.A. Lahcen, T. Beduk, H.N. Alshareef, A. Amine, K.N. Salama, Laser scribed graphene: a novel platform for highly sensitive detection of electroactive biomolecules, Biosens. Bioelectron. 168 (2020) 112509. https://doi.org/10.1016/j.bios.2020.112509

[30] S. Liu, J. Xiaohui, G.I.N. Waterhouse, Z.M. Zhang, L.M. Yu, A Cu_2O/PEDOT/graphene-modified electrode for the enzyme-free detection and quantification of glucose, J. Electroanal. Chem. 897 (2021) 115558. https://doi.org/10.1016/j.jelechem.2021.115558

[31] V. Dupuit, O. Terral, G. Bres, A. Claudel, B. Fernandez, A. Briançon-Marjollet, C. Delacour, A multifunctional hybrid graphene and microfluidic platform to interface topological neuron networks, Adv. Funct. Mater. 32 (2022) 2207001. https://doi.org/10.1002/adfm.202207001

[32] G. Seo, G. Lee, W. Kim, I. An, M. Choi, S. Jang, Y.J. Park, J.O. Lee, D. Cho, E.C. Park, Ultrasensitive biosensing platform for Mycobacterium tuberculosis detection based on functionalized graphene devices, Front. Bioeng. Biotechnol. 11 (2023) 1313494. https://doi.org/10.3389/fbioe.2023.1313494

[33] G. Ji, J. Tian, F. Xing, Y. Feng, Optical biosensor based on graphene and its derivatives for detecting biomolecules, Int. J. Mol. Sci. 23 (2022) 10838. https://doi.org/10.3390/ijms231810838

[34] T.T. Tung, M.J. Nine, M. Krebsz, T. Pasinszki, C.J. Coghlan, D.N.H. Tran, D. Losic, Recent advances in sensing applications of graphene assemblies and their composites, Adv. Funct. Mater. 27 (2017) 1702891. https://doi.org/10.1002/adfm.201702891

[35] F.P. García de Arquer, D.V. Talapin, E.H. Sargent, Semiconductor quantum dots: Technological progress and future challenges, Sci. 373 (2021) 6555. https://doi.org/10.1126/science.aaz8541

[36] Y. Cheng, S.D. Ling, Y. Geng, Y. Wang, J. Xu, Microfluidic synthesis of quantum dots and their applications in bio-sensing and bio-imaging, Nanoscale. Adv. 3 (2021) 2180-2195. https://doi.org/10.1039/D0NA00933D

[37] S. Jiao, L. Liu, J. Wang, K. Ma, J. Lv, A novel biosensor based on molybdenum disulfide (MoS_2) modified porous anodic aluminum oxide nanochannels for ultrasensitive microRNA-155 detection, Small. 16 (2020) e2001223. https://doi.org/10.1002/smll.202001223

[38] D. Mathur, M. Thakur, S.A. Díaz, K. Susumu, M.H. Stewart, E. Oh, S.A. Walper, I.L. Medintz, Hybrid nucleic acid-quantum dot assemblies as multiplexed reporter platforms for cell-free transcription translation-based biosensors, ACS Synth. Biol. 11 (2022) 4089-4102. https://doi.org/10.1021/acssynbio.2c00394

[39] N. Hildebrandt, C.M. Spillmann, W.R. Algar, T. Pons, M.H. Stewart, E. Oh, K. Susumu, S.A. Díaz, J.B. Delehanty, I.L. Medintz, Energy transfer with semiconductor quantum dot bioconjugates: A versatile platform for biosensing, energy harvesting, and other developing applications, Chem. Rev. 117 (2017) 536-711. https://doi.org/10.1021/acs.chemrev.6b00030

[40] G.K. Zath, R.A. Sperling, C.W. Hoffman, D.A. Bikos, R. Abbasi, A.R. Abate, D.A. Weitz, C.B. Chang, Rapid parallel generation of a fluorescently barcoded drop library from a microtiter plate using the plate-interfacing parallel encapsulation (PIPE) chip, Lab Chip. 22 (2022) 4512-4522. https://doi.org/10.1039/D2LC00909A

[41] Y. Yu, C. Zhang, X. Yang, L. Sun, F. Bian, Microfluidic synthesis of magnetic nanoparticles for biomedical applications, Small Methods. 9 (2025) e2401220. https://doi.org/10.1002/smtd.202401220

[42] S. Sharma, V. Bhatia, Magnetic nanoparticles in microfluidics-based diagnostics: an appraisal, Nanomedicine (Lond.). 16 (2021) 1329-1342. https://doi.org/10.2217/nnm-2021-0007

[43] B. Rezaei, P. Yari, S.M. Sanders, H. Wang, V.K. Chugh, S. Liang, S. Mostufa, K. Xu, J.P. Wang, J. Gómez-Pastora, K. Wu, Magnetic nanoparticles: a review on synthesis, characterization, functionalization, and biomedical applications, Small. 20 (2024) e2304848. https://doi.org/10.1002/smll.202304848

[44] D. Brambilla, A. Mussida, A.M. Ferretti, L. Sola, F. Damin, M. Chiari, Polymeric coating of silica microspheres for biological applications: suppression of non-specific binding and functionalization with biomolecules, Polymers. 14 (2022) 730. https://doi.org/10.3390/polym14040730

[45] J. Sobhanan, J.V. Rival, A. Anas, E.S. Shibu, Y. Takano, V. Biju, Luminescent quantum dots: synthesis, optical properties, bioimaging and toxicity, Adv. Drug Deliv. Rev. 197 (2023) 114830. https://doi.org/10.1016/j.addr.2023.114830

[46] Z. Liao, J. Wang, P. Zhang, Y. Zhang, Y. Miao, S. Gao, Y. Deng, L. Geng, Recent advances in microfluidic chip integrated electronic biosensors for multiplexed detection, Biosens. Bioelectron. 121 (2018) 272-280. https://doi.org/10.1016/j.bios.2018.08.061

[47] M. Mathew, C.S. Rout, Electrochemical biosensors based on $Ti_3C_2T_x$ MXene: future perspectives for on-site analysis, Curr. Opin. Electrochem. 30 (2021) 100782. https://doi.org/10.1016/j.coelec.2021.100782

[48] S.K. Krishnan, E. Singh, P. Singh, M. Meyyappan, H.S. Nalwa, A review on graphene-based nanocomposites for electrochemical and fluorescent biosensors, RSC Adv. 9 (2019) 9223-9248. https://doi.org/10.1039/C8RA09577A

Materials Research Foundations 191 (2026) 180-223 https://doi.org/10.21741/9781644904213-6

[49] B. Babamiri, R. Sadri, M. Farrokhnia, M. Hassani, M. Kaur, E.P.L. Roberts, M.M. Ashani, A.S. Nezhad, Molecularly imprinted polymer biosensor based on nitrogen-doped electrochemically exfoliated graphene/Ti_3CNT_x MXene nanocomposite for metabolites detection, ACS Appl. Mater. Interfaces. 16 (2024) 27714-27727. https://doi.org/10.1021/acsami.4c01973

[50] T.D. Schladt, K. Schneider, H. Schild, W. Tremel, Synthesis and bio-functionalization of magnetic nanoparticles for medical diagnosis and treatment, Dalton Trans. 40 (2011) 3355-3365. https://doi.org/10.1039/c0dt00689k

[51] M. Sharafeldin, K. McCaffrey, J.F. Rusling, Influence of antibody immobilization strategy on carbon electrode immunoarrays, Analyst. 144 (2019) 5108-5116. https://doi.org/10.1039/C9AN01093A

[52] K.M. Nell, S.A. Fontenot, T.G. Carter, M.G. Warner, C.L. Warner, R.S. Addleman, D.W. Johnson, Non-covalent functionalization of high-surface area nanomaterials: a new class of sorbent materials, Environ. Sci. Nano. 3 (2016) 160-171. https://doi.org/10.1039/C5EN00170F

[53] G. Karuppaiah, A. Vashist, M. Nair, M. Veerapandian, P. Manickam, Emerging trends in point-of-care biosensing strategies for molecular architectures and antibodies of SARS-CoV-2, Biosens. Bioelectron. X. 13 (2023) 100324. https://doi.org/10.1016/j.biosx.2023.100324

[54] B.D. Malhotra, M.A. Ali, Nanomaterials in biosensors: fundamentals and applications, in: Nanomaterials for Biosensors, Elsevier, 2018, pp. 1-74. https://doi.org/10.1016/B978-0-323-44923-6.00001-7

[55] Y. Wei, Q. Zeng, J. Huang, Q. Hu, X. Guo, L. Wang, An electro-responsive imprinted biosensor with switchable affinity toward proteins, Chem. Commun. (Camb.). 54 (2018) 9163-9166. https://doi.org/10.1039/C8CC05482G

[56] K. Thapa, W. Liu, R. Wang, Nucleic acid-based electrochemical biosensor: recent advances in probe immobilization and signal amplification strategies,

Wiley Interdiscip. Rev. Nanomed. Nanobiotechnol. 14 (2022) e1765.

[57] J. Kim, S. Noh, J.A. Park, S.C. Park, S.J. Park, J.H. Lee, J.H. Ahn, T. Lee, Recent advances in aptasensor for cytokine detection: a review, Sensors (Basel). 21 (2021) 8491. https://doi.org/10.3390/s21248491

[58] Z. Zhuo, Y. Yu, M. Wang, J. Li, Z. Zhang, J. Liu, X. Wu, A. Lu, G. Zhang, B. Zhang, Recent advances in SELEX technology and aptamer applications in biomedicine, Int. J. Mol. Sci. 18 (2017) 2142. https://doi.org/10.3390/ijms18102142

[59] B. Sequeira-Antunes, H.A. Ferreira, Nucleic acid aptamer-based biosensors: a review, Biomedicines. 11 (2023) 3201. https://doi.org/10.3390/biomedicines11123201

[60] A. Agha, W. Waheed, I. Stiharu, V. Nerguizian, G. Destgeer, E. Abu-Nada, A. Alazzam, A review on microfluidic-assisted nanoparticle synthesis, and their applications using multiscale simulation methods, Discov. Nano. 18 (2023) 18. https://doi.org/10.1186/s11671-023-03792-x

[61] J. Macko, N. Podrojková, R. Oriňáková, A. Orinak, New insights into hydrophobicity at nanostructured surfaces: experiments and computational models, Nanomater. Nanotechnol. 12 (2022) 184798042110623. https://doi.org/10.1177/18479804211062316

[62] A. Bezelya, B. Küçüktürkmen, A. Bozkır, Microfluidic devices for precision nanoparticle production, Micro. 3 (2023) 58. https://doi.org/10.3390/micro3040058

[63] H. Tavakoli, S. Mohammadi, X. Li, G. Fu, X. Li, Microfluidic platforms integrated with nano-sensors for point-of-care bioanalysis, Trends. Anal. Chem. 157 (2022) 116806. https://doi.org/10.1016/j.trac.2022.116806

[64] S. Prabha, D. Durgalakshmi, S. Rajendran, E. Lichtfouse, Plant-derived silica nanoparticles and composites for biosensors, bioimaging, drug delivery and supercapacitors: a review, Environ. Chem. Lett. 19 (2021) 1667-1691. https://doi.org/10.1007/s10311-020-01123-5

[65] B.A. Witika, P.A. Makoni, S.K. Matafwali, B. Chabalenge, C. Mwila, A.C. Kalungia, C.I. Nkanga, A.M. Bapolisi, R.B. Walker, Biocompatibility of biomaterials for nanoencapsulation: current approaches, Nanomaterials. 10 (2020) 1649. https://doi.org/10.3390/nano10091649

[66] O. Gideon, H.S. Samuel, I.A. Okino, Biocompatible materials for next-generation biosensors, Discov. Chem. 1 (2024) 34. https://doi.org/10.1007/s44371-024-00042-6

[67] G.M. Manoj, M. Shalini, K. Thenmozhi, V.K. Ponnusamy, H. Shankar, Recent advancements in the surface modification and functionalization of magnetic nanomaterials, Appl. Surf. Sci. Adv. 21 (2024) 100608. https://doi.org/10.1016/j.apsadv.2024.100608

[68] Q. Yang, Y. Wu, J. Chen, M. Lu, X. Wang, Z. Zhang, H. Xiong, J. Choo, L. Chen, Plasmonic nanomaterial-enhanced fluorescence and raman sensors: multifunctional platforms and applications, Coord. Chem. Rev. 507 (2024) 215768. https://doi.org/10.1016/j.ccr.2024.215768

[69] K. Kant, R. Beeram, Y. Cao, P.S.S. dos Santos, L. González-Cabaleiro, D. García-Lojo, H. Guo, Y. Joung, S. Kothadiya, M. Lafuente, Y. Leong, Y. Liu, Y. Liu, S.S.B. Moram, S. Mahasivam, S. Maniappan, D. Quesada-González, D. Raj, P. Weerathunge, X. Xia, Q. Yu, S. Abalde-Cela, R.A. Alvarez-Puebla, R. Bardhan, V. Bansal, J. Choo, L.C.C. Coelho, J.M.M.M. de Almeida, S. Gómez-Graña, M. Grzelczak, P. Herves, J. Kumar, T. Lohmueller, A. Merkoçi, J.L. Montaño-Priede, X.Y. Ling, R. Mallada, J. Pérez-Juste, M.P. Pina, S. Singamaneni, V.R. Soma, M. Sun, L. Tian, J. Wang, L. Polavarapu, I. Pastoriza Santos, Plasmonic nanoparticle sensors: current progress, challenges, and future prospects, Nanoscale. Horiz. 9 (2024) 2085-2166. https://doi.org/10.1039/D4NH00226A

[70] C. Lin, Y. Li, Y. Peng, S. Zhao, M. Xu, L. Zhang, Z. Huang, J. Shi, Y. Yang, Recent development of surface-enhanced raman scattering for biosensing, J. Nanobiotechnol. 21 (2023) 1-20. https://doi.org/10.1186/s12951-023-01890-7

[71] S. Mahanty, S. Majumder, R. Paul, R. Boroujerdi, E. Valsami-Jones, C. Laforsch, A review on nanomaterial-based SERS substrates for sustainable agriculture, Sci. Total. Environ. 950 (2024) 174252. https://doi.org/10.1016/j.scitotenv.2024.174252

[72] M.Q. Khan, J. Khan, M.A.H. Alvi, H. Nawaz, M. Fahad, M. Umar, Nanomaterial-based sensors for microbe detection: a review, Discov. Nano. 19 (2024) 120. https://doi.org/10.1186/s11671-024-04065-x

[73] S. Chen, S. Tian, Y. Wang, M. Li, D. Tang, Harnessing bifunctional nanozyme with potent catalytic and signal amplification for innovating electrochemical immunoassay, Biosens. Bioelectron. 278 (2025) 117340. https://doi.org/10.1016/j.bios.2025.117340

[74] Y. Gong, H. Yang, C. Ding, NIR-photoactivatable DNA nanomachines for spatiotemporally controllable monitoring of microRNA-21 in living cells based on signal amplification strategy, Biosens. Bioelectron. 267 (2025) 116755. https://doi.org/10.1016/j.bios.2024.116755

[75] A. Baranwal, R. Shukla, V. Bansal, Nanozyme-enhanced paper-based biosensor technologies, Trends. Anal. Chem. 172 (2024) 117573. https://doi.org/10.1016/j.trac.2024.117573

[76] L. Jiao, W. Xu, Y. Wu, H. Yan, W. Gu, D. Du, Y. Lin, C. Zhu, Single-atom catalysts boost signal amplification for biosensing, Chem. Soc. Rev. 50 (2021) 750-765. https://doi.org/10.1039/D0CS00367K

[77] M. Zhang, G. Wang, J. Chen, X. Lu, Single atom catalysts for sensors, Trends. Anal. Chem. 169 (2023) 117399. https://doi.org/10.1016/j.trac.2023.117399

[78] N. Joudeh, D. Linke, Nanoparticle classification, physicochemical properties, characterization, and applications: a comprehensive review for biologists, J. Nanobiotechnol. 20 (2022) 262. https://doi.org/10.1186/s12951-022-01477-8

[79] A. Bezelya, B. Küçüktürkmen, A. Bozkır, Microfluidic devices for precision nanoparticle production, Micro. 3 (2023) 822-866. https://doi.org/10.3390/micro3040058

[80] M. Erfan, M. Gnambodoe-Capochichi, Y.M. Sabry, D. Khalil, Y. Leprince-Wang, T. Bourouina, Spatiotemporal dynamics of nanowire growth in a microfluidic reactor, Microsyst. Nanoeng. 7 (2021) 77. https://doi.org/10.1038/s41378-021-00308-4

[81] K.R. Mitchell, J.E. Esene, A.T. Woolley, Advances in multiplex electrical and optical detection of biomarkers using microfluidic devices, Anal. Bioanal. Chem. 414 (2022) 167-180. https://doi.org/10.1007/s00216-021-03553-8

[82] M. Kumari, V. Gupta, N. Kumar, R.K. Arun, Microfluidics-based nanobiosensors for healthcare monitoring, Mol. Biotechnol. 66 (2023) 378-401. https://doi.org/10.1007/s12033-023-00760-9

[83] Z. Lian, Y. Chan, Y. Luo, X. Yang, K.S. Koh, J. Wang, G.Z. Chen, Y. Ren, J. He, Microfluidic formation of highly monodispersed multiple cored droplets using needle-based system in parallel mode, Electrophoresis. 41 (2020) 891-901. https://doi.org/10.1002/elps.201900403

[84] M. Vázquez, B. Paull, Review on recent and advanced applications of monoliths and related porous polymer gels in micro-fluidic devices, Anal. Chim. Acta. 668 (2010) 100-113. https://doi.org/10.1016/j.aca.2010.04.033

[85] P. Shrimal, G. Jadeja, S. Patel, A review on novel methodologies for drug nanoparticle preparation: Microfluidic approach, Chem. Eng. Res. Des. 153 (2020) 728-756. https://doi.org/10.1016/j.cherd.2019.11.031

[86] Y. Hu, B. Liu, Y. Wu, M. Li, X. Liu, J. Ding, X. Han, Y. Deng, W. Hu, C. Zhong, Facile high throughput wet-chemical synthesis approach using a microfluidic-based composition and

temperature controlling platform, Front. Chem. 8 (2020) 579828. https://doi.org/10.3389/fchem.2020.579828

[87] M. Erfan, M. Gnambodoe-Capochichi, Y.M. Sabry, D. Khalil, Y. Leprince-Wang, T. Bourouina, Spatiotemporal dynamics of nanowire growth in a microfluidic reactor, Microsyst. Nanoeng. 7 (2021) 77. https://doi.org/10.1038/s41378-021-00308-4

[88] J. Kim, Z. Li, I. Park, Direct synthesis and integration of functional nanostructures in microfluidic devices, Lab. Chip. 11 (2011) 1946-1951. https://doi.org/10.1039/c1lc20079h

[89] A.W. Moore, S.C. Jacobson, J.M. Ramsey, Microchip separations of neutral species via micellar electrokinetic capillary chromatography, Anal. Chem. 67 (1995) 4184-4189. https://doi.org/10.1021/ac00118a023

[90] A.G. Niculescu, C. Chircov, A.C. Bîrcă, A.M. Grumezescu, Fabrication and Applications of Microfluidic Devices: a review, Int. J. Mol. Sci. 22 (2021). https://doi.org/10.3390/ijms22042011

[91] J. Hwang, Y. Cho, M.S. Park, B.H. Kim, Microchannel fabrication on glass materials for microfluidic devices, Int. J. Precis. Eng. Manuf. 20 (2019) 1-2. https://doi.org/10.1007/s12541-019-00103-2

[92] X. Han, Y. Zhang, J. Tian, T. Wu, Z. Li, F. Xing, S. Fu, Polymer-based microfluidic devices: a comprehensive review on preparation and applications, Polym. Eng. Sci. 62 (2021) 3-24. https://doi.org/10.1002/pen.25831

[93] Darwin Microfluidics, Photolithography. https://blog.darwin-microfluidics.com/glossary/photolithography-microfluidics-explained/, 2025 (accessed 30 March 2025).

[94] S.R. Quake, A. Scherer, From micro- to nanofabrication with soft materials, Science. 290 (2000) 1536-1540. https://doi.org/10.1126/science.290.5496.1536

[95] Y. Kim, J. Song, Y. Lee, S. Cho, S. Kim, S.R. Lee, S. Park, Y. Shin, N. Jeon, High-throughput injection molded microfluidic device for single-cell analysis of spatiotemporal dynamics, Lab. Chip. 21 (2021) 3041-3051. https://doi.org/10.1039/D0LC01245A

[96] Y.J. Juang, Y.J. Chiu, Fabrication of polymer microfluidics: an overview, Polymers. 14 (2022) 2028. https://doi.org/10.3390/polym14102028

[97] S.S. Deshmukh, A. Goswami, Recent developments in hot embossing-a review, Mater. Manuf. Processes. 36 (2021) 501-543. https://doi.org/10.1080/10426914.2020.1832691

[98] S. Ravi-Kumar, B. Lies, X. Zhang, H. Lyu, H. Qin, Laser ablation of polymers: a review, Polym. Int. 68 (2019) 1391-1401. https://doi.org/10.1002/pi.5834

[99] A.V. Nielsen, M.J. Beauchamp, G.P. Nordin, A.T. Woolley, 3D printed microfluidics, Annu. Rev. Anal. Chem. 13 (2020) 45-65. https://doi.org/10.1146/annurev-anchem-091619-102649

[100] M. Kharlamov, O. Guseva, Features of the application of reactive ion etching of quartz in the production of pendulums of Q-flex accelerometers, Russ. Microelectron. 49 (2020) 184-194. https://doi.org/10.1134/S1063739720020055

[101] T.G. Konstantinova, M.M. Andronic, D.A. Baklykov, V.E. Stukalova, D.A. Ezenkova, E.V. Zikiy, M.V. Bashinova, A.A. Solovev, E.S. Lotkov, I.A. Ryzhikov, I.A. Rodionov, Deep multilevel wet etching of fused silica glass microstructures in BOE solution, Sci. Rep. 13 (2023) 5228. https://doi.org/10.1038/s41598-023-32503-w

[102] S. Aralekallu, R. Boddula, V. Singh, Development of glass-based microfluidic devices: a review on its fabrication and biologic applications, Mater. Des. 225 (2023) 111517. https://doi.org/10.1016/j.matdes.2022.111517

[103] A. Fabozzi, F.D. Sala, M. di Gennaro, M. Barretta, G. Longobardo, N. Solimando, M. Pagliuca, A. Borzacchiello, Design of functional nanoparticles by microfluidic platforms as advanced drug delivery systems for cancer therapy, Lab. Chip. 23 (2023) 1389-1409. https://doi.org/10.1039/D2LC00933A

[104] X. Han, J. Li, X. Tang, W. Li, H. Zhao, L. Yang, L. Wang, Droplet bouncing: fundamentals, regulations, and applications, Small. 18 (2022) e2200277. https://doi.org/10.1002/smll.202200277

[105] T. Ge, W. Hu, Z. Zhang, X. He, L. Wang, X. Han, Z. Dai, Open and closed microfluidics for biosensing, Mater. Today Bio. 26 (2024) 101048. https://doi.org/10.1016/j.mtbio.2024.101048

[106] J.K. Nunes, H.A. Stone, Introduction: microfluidics, Chem. Rev. 122 (2022) 6919-6920. https://doi.org/10.1021/acs.chemrev.2c00052

[107] J.Y. Qian, C.W. Hou, X.J. Li, Z.J. Jin, Actuation mechanism of microvalves: a review, Micromachines (Basel). 11 (2020) 172. https://doi.org/10.3390/mi11020172

[108] A. Ebrahimi, K. Icoz, R. Didarian, C.H. Shih, E.A. Tarim, B. Nasseri, A. Akpek, B. Cecen, A. Bal-Ozturk, K. Güleç, Y.C.E. Li, S. Shih, B.S. Tarim, H.C. Tekin, E. Alarçin, M. Tayybi-Azar, H. Ghorbanpoor, C. Özel, A.E. Sarıboyacı, F.D. Guzel, N. Bassous, S.R. Shin, H. Avci, Molecular separation by using active and passive microfluidic chip designs: a comprehensive review, Adv. Mater. Interfaces. 11 (2024) 2300492. https://doi.org/10.1002/admi.202300492

[109] E. Chappel, A review of passive constant flow regulators for microfluidic applications, Appl. Sci. 10 (2020) 8858. https://doi.org/10.3390/app10248858

[110] V. Narayanamurthy, Z.E. Jeroish, K.S. Bhuvaneshwari, P. Bayat, R. Premkumar, F. Samsuri, M.M. Yusoff, Advances in passively driven microfluidics and lab-on-chip devices: a comprehensive literature review and patent analysis, RSC Adv. 10 (2020) 11652-11680. https://doi.org/10.1039/D0RA00263A

[111] S. Malik, J. Singh, R. Goyat, Y. Saharan, V. Chaudhry, A. Umar, A.A. Ibrahim, S. Akbar, S. Ameen, S. Baskoutas, Nanomaterials-based biosensor and their applications: a review, Heliyon. 9 (2023) e19929. https://doi.org/10.1016/j.heliyon.2023.e19929

[112] A. Hayat, G. Catanante, J.L. Marty, Current trends in nanomaterial-based amperometric biosensors, Sensors (Basel). 14 (2014) 23439-23461. https://doi.org/10.3390/s141223439

[113] Y. Wang, Z. Ye, Y. Ying, New trends in impedimetric biosensors for the detection of foodborne pathogenic bacteria, Sensors (Basel). 12 (2012) 3449-3471. https://doi.org/10.3390/s120303449

[114] X. Liu, L. Huang, K. Qian, Nanomaterial-based electrochemical sensors: mechanism, preparation, and application in biomedicine, Adv. NanoBiomed. Res. 1 (2021) 2000104. https://doi.org/10.1002/anbr.202000104

[115] Y. Zhang, H. Tang, W. Chen, J. Zhang, Nanomaterials used in fluorescence polarization based biosensors, Int. J. Mol. Sci. 23 (2022) 8625. https://doi.org/10.3390/ijms23158625

[116] Z. Liu, Y. Zhou, J. Lu, T. Gong, E. Ibáñez, A. Cifuentes, W. Lu, Microfluidic biosensors for biomarker detection in body fluids: a key approach for early cancer diagnosis, Biomark. Res. 12 (2024) 1-29. https://doi.org/10.1186/s40364-024-00697-4

[117] D.M. Jameson, J.C. Croney, Fluorescence polarization: past, present and future, Comb. Chem. High Throughput Screen. 6 (2003) 167-176. https://doi.org/10.2174/138620703106298347

[118] Y. Choi, J.H. Hwang, S.Y. Lee, Recent trends in nanomaterials-based colorimetric detection of pathogenic bacteria and viruses, Small Methods. 2 (2018) 1700351. https://doi.org/10.1002/smtd.201700351

[119] C. Huang, W. Zhou, R. Wu, W. Guan, N. Ye, Recent advances in nanomaterial-based chemiluminescence probes for biosensing and imaging of reactive oxygen species, Nanomaterials. 13 (2023) 1726. https://doi.org/10.3390/nano13111726

[120] K. Khachornsakkul, T. Leelasattarathkul, Photothermal biosensing integrated with microfluidic paper-based analytical device for sensitive quantification of sarcosine, Talanta. 271 (2024) 125628. https://doi.org/10.1016/j.talanta.2024.125628

[121] M. Khan, B. Zhao, W. Wu, M. Zhao, Y. Bi, Q. Hu, Distance-based microfluidic assays for instrument-free visual point-of-care testing, Trends Anal. Chem. 162 (2023) 117029. https://doi.org/10.1016/j.trac.2023.117029

[122] X. Ge, A.M. Asiri, D. Du, W. Wen, S. Wang, Y. Lin, Nanomaterial-enhanced paper-based biosensors, Trends Anal. Chem. 58 (2014) 31-39. https://doi.org/10.1016/j.trac.2014.03.008

[123] W. Dungchai, O. Chailapakul, C.S. Henry, Electrochemical detection for paper-based microfluidics, Anal. Chem. 81 (2009) 5821-5826. https://doi.org/10.1021/ac9007573

[124] Y. Xiang, C. Hu, G. Wu, S. Xu, Y. Li, Nanomaterial-based microfluidic systems for cancer biomarker detection: recent applications and future perspectives, Trends Anal. Chem. 158 (2023) 116835. https://doi.org/10.1016/j.trac.2022.116835

[125] M. Kumari, V. Gupta, N. Kumar, R.K. Arun, Microfluidics-based nanobiosensors for healthcare monitoring, Mol. Biotechnol. 66 (2024) 378-401. https://doi.org/10.1007/s12033-023-00760-9

[126] J.J. Mim, M.A. Islam, S. Khan, F. Khan, M.A.I. Patwary, M.M.R. Arup, N. Hossain, Nanomaterial-based biosensors for early detection of Mpox virus: a comprehensive analysis, Sens. Actuators Rep. 9 (2024) 100263. https://doi.org/10.1016/j.snr.2024.100263

[127] M. Sharma, P. Mahajan, A.S. Alsubaie, V. Khanna, S. Chahal, A. Thakur, A. Yadav, A. Arya, A. Singh, G. Singh, Next-generation nanomaterials-based biosensors: real-time biosensing devices for detecting emerging environmental pollutants, Mater. Today Sustain. 29 (2025) 101068. https://doi.org/10.1016/j.mtsust.2024.101068

[128] M.K. Filippidou, S. Chatzandroulis, Microfluidic devices for heavy metal ions detection: a review, Micromachines. 14 (2023) 1520. https://doi.org/10.3390/mi14081520

[129] T. Kasputis, K.E. Hosmer, Y. He, J. Chen, Ensuring food safety: microfluidic-based approaches for the detection of food contaminants, Anal. Sci. Adv. 5 (2024) e2400003. https://doi.org/10.1002/ansa.202400003

[130] N. Tarannum, A. Gautam, T. Chauhan, D. Kumar, Nanomaterial based sensors for detection of food contaminants: a prospect, Sens. Technol. 2 (2024) 2373196. https://doi.org/10.1080/28361466.2024.2373196

[131] X. Zhao, M. Li, Y. Liu, Microfluidic-based approaches for foodborne pathogen detection, Microorganisms. 7 (2019) 381. https://doi.org/10.3390/microorganisms7100381

[132] Y. Liu, L. Sun, H. Zhang, L. Shang, Y. Zhao, Microfluidics for drug development: from synthesis to evaluation, Chem. Rev. 121 (2021) 7468-7529. https://doi.org/10.1021/acs.chemrev.0c01289

[133] S.E. Alavi, S. Alharthi, S.F. Alavi, S.Z. Alavi, G.E. Zahra, A. Raza, H.E. Shahmabadi, Microfluidics for personalized drug delivery, Drug Discov. Today. 29 (2024) 103936. https://doi.org/10.1016/j.drudis.2024.103936

[134] J. Sun, A.R. Warden, X. Ding, Recent advances in microfluidics for drug screening, Biomicrofluidics. 13 (2019) 064101. https://doi.org/10.1063/1.5121200

[135] D. Singh, S.P. Deosarkar, E. Cadogan, V. Flemington, A. Bray, J. Zhang, R.S. Reiserer, D.K. Schaffer, G.B. Gerken, C.M. Britt, E.M. Werner, F.D. Gibbons, T. Kostrzewski, C.E. Chambers, E.J. Davies, A.R. Montoya, J.H.L. Fok, D. Hughes, K. Fabre, M.P. Wagoner, J.P. Wikswo, C.W. Scott, A microfluidic system that replicates pharmacokinetic (PK) profiles in vitro improves prediction of in vivo efficacy in preclinical models, PLoS Biol. 20 (2022) e3001624. https://doi.org/10.1371/journal.pbio.3001624

[136] R. Mihyar, A.A. Shalmani, V. Wildt, M. Sheybanifard, A. Wang, J.N. May, S. Shahzad, E.M. Buhl, S. Rütten, D. Behrens, W. Walther, M. Tiboni, L. Casettari, J.F. Buyel, C.J.F. Rijcken, W.E. Hennink, S. von Stillfried, F. Kiessling, Y. Shi, J.M. Metselaar, Q. Peña, Microfluidic formulation, cryoprotection and long-term stability of paclitaxel-loaded π electron-stabilized polymeric micelles, J. Contr. Release. 375 (2024) 614-626. https://doi.org/10.1016/j.jconrel.2024.08.041

[137] X. Lai, B. Lu, P. Zhang, X. Zhang, Z. Pu, H. Yu, D. Li, Sticker microfluidics: a method for fabrication of customized monolithic microfluidics, ACS Biomater. Sci. Eng. 5 (2019) 6801-6810. https://doi.org/10.1021/acsbiomaterials.9b00953

Materials Research Foundations 191 (2026) 224-259
https://doi.org/10.21741/9781644904213-7

Chapter 7

Advanced Materials for Microfluidic Electrochemical Sensors

Izba Amjad[1], Shiza Noor[1], Tabina Amin[1], Hira Amjad[2], Maryam Ashiq[1], Tayyaba Asim[3], Adnan Mujahid[4] and Sadia Z. Bajwa [1*]

[1]National Institute for Biotechnology and Genetic Engineering College Pakistan Institute of Engineering and Applied Sciences (NIBGE-C, PIEAS), Faisalabad 38000, Pakistan

[2]Department of Biosciences, COMSATS University Islamabad

[3]School of Civil and Environmental Engineering, Institute of Environmental Science and Engineering, National University of Sciences and Technology, Islamabad 44000, Pakistan

[4]School of Chemistry, University of the Punjab, Quaid-i-Azam Campus, Lahore 54590, Pakistan

* sadya2002pk@yahoo.co.uk

Abstract

The integration of advanced nanomaterials into microfluidic electrochemical sensors enables sensitive, specific, and real-time point-of-care detection. These systems are pivotal for biomarker detection, environmental monitoring, food safety, and drug tracking. Nanomaterials such as metals, metal oxides, carbon-based, 2D materials, and conductive polymers offer superior electrical and catalytic properties, enhancing biomolecule immobilization and signal transduction. This chapter explores recent advances in nanomaterial-based microfluidic devices, emphasizing material design, signal amplification, and diverse applications. It also addresses challenges in sensitivity, reproducibility, and scalability, along with emerging trends like wearable devices and AI-integrated biosensors, advancing diagnostic capabilities in biomedical and environmental fields.

Keywords

Microfluidic Electrochemical Biosensors, Advanced Materials, Electrochemical Sensors, Nanomaterials

Contents

Materials Research Foundations 191 (2026) 224-259 https://doi.org/10.21741/9781644904213-7

Materials Research Foundations 191 (2026) 224-259 https://doi.org/10.21741/9781644904213-7

1. Introduction

Sensors are electrical devices that usually consists of recognition elements to a transducer to provide concurrent information on the composition of the system [1]. An ideal sensor should possess significant properties, as it should have the highest efficiency along with in-field operation capability and a cheaper cost. In addition, it must follow the standards set by WHO for portable sensors. With all these perspectives, sensors are promising devices for achieving Chips based point-of-care applications [2]. Microfluidic Sensors" are sensors that are utilized in a miniature microfluidic platform [3].

Microfluidics and nanofluidic technologies pave the way to manipulate and operate liquids and handle fluids in channels with very minor measurements, like a diameter of 1-10mm for sensing applications. In the era of advanced technologies, these sensors hold significant worth in healthcare and environmental monitoring with emerging properties like fast response, high sensitivity, and efficient output main benefits of these microfluidic systems include controlled liquid flow, low sample and reagent usage, fast analysis, less waste production and disposable content [4]. Microelectromechanical systems (MEMS) technology is the main foundation of microfabrication techniques. The major research of fluidic activities on a small scale and the technology of designing, calibrating, optimizing, executing, and fabricating such micro-devices for numerous point-of-care implementations are included in microfluidics. A basic test called POCT allows non-physicians to or unskilled individuals to perform medical tests in close proximity to patients with minimal resources, at a reasonable cost, with a short span of time [5].

There are unique benefits to using nanotechnology in bioanalytical devices when developing microfluidic devices. Selectivity and sensitivity can be boosted by improving significant properties of nanomaterials and combined with microfluidic sensors, which have a dedicated role in a wide range of applications. Nanomaterials have great promise in the sector due to their larger surface area, high electric conductivity quantum confinement effect and biodegradability features of microfluidic sensors [3]. In microfluidic sensors, gold and platinum nanoparticles are most commonly used as electrodes. Electrodes based on carbon nanomaterials have less likely cons and are more stable for a broad potential range than those based on metal nanoparticles [6]. Because of their remarkable mechanical, electrochemical, and electrical conductivity, carbon nanotubes (CNTs) have drawn the most attention of all of these materials. Their one-dimensional (1D) character presents a significant challenge during production, nevertheless. Graphene derivatives are being used most commonly among these carbon nanomaterials because of their increased availability of active sites and improved adhesion provided by their flat 2D structure. One of graphene's special and significant characteristics is its very thin (0.35 nm) and ultra-light honeycomb structure, with having planar density of about 0.77 mg. These graphene-based microfluidic sensors served as inspiration for the creation of the first microfluidic sensor employing MoS_2 for fluorescent DNA detection. A new addition to the 2D material family, MXene, is a unique material that may also assist in solving current problems with microfluidic sensors. This material is suitable for microfluidic sensors because of its exceptional electrical conductivity and significant biocompatibility [7].

In this chapter, we will discuss the advanced materials and fabrication that play a critical role in enhancing the performance and functionality of electrochemical microfluidic devices. Their superior electrical, mechanical, and chemical properties enable high sensitivity, rapid response and

Materials Research Foundations 191 (2026) 224-259 https://doi.org/10.21741/9781644904213-7

miniaturization. Supporting the development of these materials accelerates innovation in diagnostics, environmental monitoring, and lab-on-chip technologies.

2. Fundamentals of Microfluidic Electrochemical Sensors

Microfluidic electrochemical sensors allow for the detection of a different biomarkers in small sample volumes by fusing the analytical capability of electrochemistry with the precise fluid control of microfluidics. To perform electrochemical reactions, the oxidation and reduction will occur and transducer use to convert these reactions into detectable electrical signals. These types of sensors usually have microchannels and electrodes in which biological events are transformed into electrical impulses. In this case, an electrode is an important element that serves as a solid base for both electron transport and biomolecule immobilization. By connecting a chemically different layer to a transducer, electrochemical sensors are tiny devices that instantly offer information on a systems conformation. This process converts the chemical energy of the selected interface between the chemical species and the sensor into a signal that is useful for analysis. Since the methods and tools required are so simple, electrochemical sensors are the most common and traditional type of chemical sensors. These sensors have gained a lot of popularity recently since they are easy to automate and integrate into smaller locations without compromising analytical capabilities. Figure 1 is an explanation of the fundamentals of a microfluidic electrochemical biosensor. It combines principles of microfluidics (manipulating small fluid volumes in channels) and electrochemical sensing (detecting analytes via electrical signals) and their application in different fields of life. In addition, it is feasible for this trend to substitute the traditional lab-based instruments developed by the recognized in vitro diagnostic firms that claim that automation of analytes is suitable for assessment. Rapid analysis with a disposable, reusable, and economical platform for multiplexed point-of-care detections is made possible by the synergistic integration of electrochemical sensing systems with microelectrodes and miniaturized or microfluidic electrochemical devices in a single platform [8].

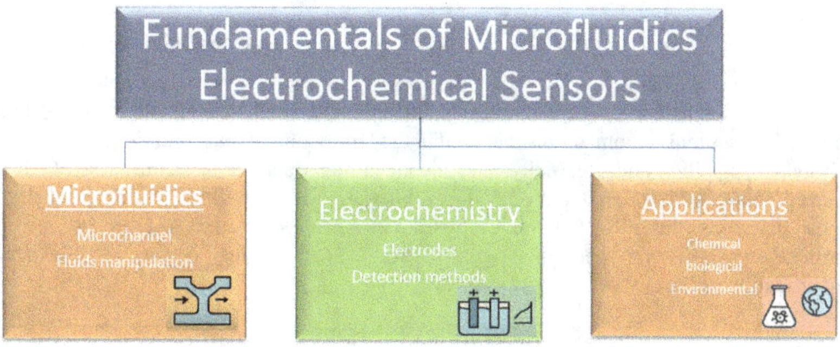

Figure 1 Fundamentals of Microfluidics Electrochemical Sensors

Figure 2: explaining the Structural variants of microchannel design, each with a distinct function in fluid handling, mixing, and separation. All variants are very beneficial for future challenges for lab on chip technology. Common techniques such as amperometry, voltammetry, and impedance spectroscopy are frequently used within microfluidic platforms to measure analytes in clinical, environmental, and industrial samples. Materials like PDMS, glass, silicon, and paper are typically used to fabricate these devices, enabling flexible, cost-effective designs. Applications range from medical diagnostics and food safety testing to environmental monitoring, with ongoing innovations focusing on improved integration with portable electronics, multiplexed analysis, and enhanced sensitivity through nanomaterial-based electrodes.

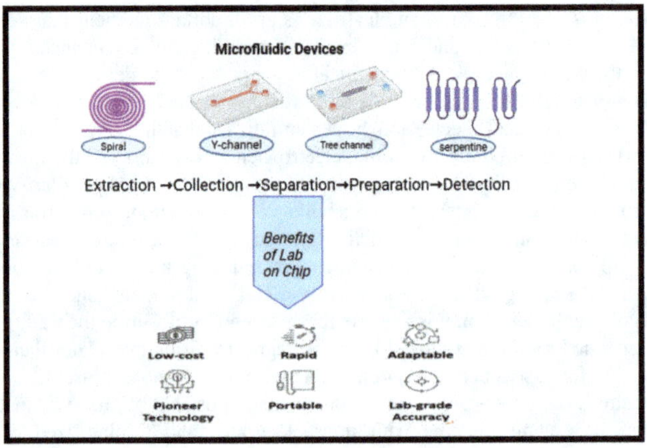

Figure 2 Structural variants of microchannel design, each with distinct applications in fluid handling, mixing, and separation

3. Types of Advanced Materials Used in Microfluidic Electrochemical Sensors

Microfluidic electrochemical sensors are the most advanced analytical technology due to their miniaturized architecture, high sensitivity, and ability to make detection on time. Figure 3 elaborated the various nanomaterials that used in the fabrication of microfluidic sensors affect their performance, stability, and specificity. The advanced materials that underpin these platforms are advanced nanomaterials, conductive polymers, biomaterials, as well as hybrid composites. This chapter discusses the most common classes of advanced materials used in microfluidic electrochemical sensors, and their contribution to improving detection efficiency and miniaturization of the devices.

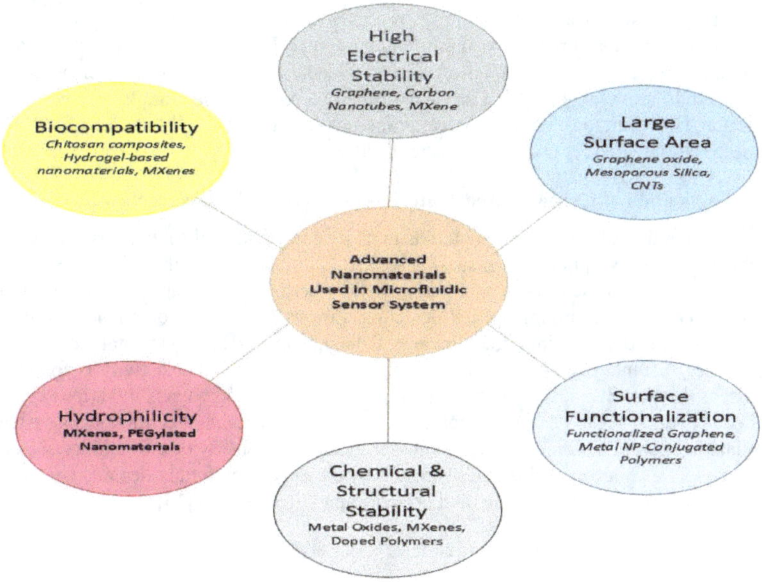

Figure 3 Advanced nanomaterials and their utilization to enhance the biosensor sensitivity and specificity.

3.1 Types of Advanced Nanomaterials

Advanced materials used in microfluidic electrochemical sensors help in improving their sensitivity, selectivity, and overall performance. The high reactivity and surface-to-volume ratios, along with many quantum effects, found in nanomaterials, aid the miniaturization of microfluidic sensors, along with increasing sensitivity and electron transfer. For such applications, these nanomaterials are vital for the development of efficient and reliable sensors [9].

3.2 Carbon-based Nanomaterials

Carbon-based nanomaterials are the main components in the development of microfluidic electrochemical sensor systems because of their unique properties like high surface area, excellent electrical conductivity and biocompatibility. Carbon nanotubes, graphene and carbon black among other materials are being extensively researched for their ability to improve the performance of electrochemical sensors [10,11]. Further different types of carbon-based nanomaterials are discussed below.

3.2.1 Types of Carbon-based Nanomaterials

The wide application of carbon nanotubes in electrochemical sensors is due to their great capability of electron transfer and large surface area for biomolecule immobilization. They are particularly

Materials Research Foundations 191 (2026) 224-259 https://doi.org/10.21741/9781644904213-7

suitable for promoting the electrochemical reactivity of biomolecules and proteins, which makes them ideal for biosensor applications [12]. In sensor applications, reduced graphene oxide (and even graphene itself) is currently becoming more popular. High electrical conductivity and large surface area are beneficial for the adsorption of molecules and electron transfer processes, and these materials are available [9]. Carbon black is a good material for its low cost and easy preparation. However, the electroanalytical properties and stability in dispersions of this molecule have been rediscovered for use in electrochemical sensors [13] .

3.2.2 Application of Carbon-based Materials in Microfluidic Electrochemical Sensors

The great versatility of carbon nanomaterials in many analytical applications has been shown and is enabled by their exceptional high surface area and compatibility with biological systems [11]. They are critical to biosensors in facilitating rapid electron transfer and maintaining biosensor stability in complex environments for detecting various biological compounds. And because precision and sensitivity are important in medical diagnostics and environmental monitoring, these are extremely useful. The application of their high sensitivity and selectivity is found in drug analysis, as they resolve adequate amounts of pharmaceutical compounds in various media [14]. Additionally, multi-walled carbon nanotubes are used in microchip electrophoresis systems that are an integral part of lab-on-a-chip technologies to improve sensitivity and stability. Detection of these analytes, such as hydrazine and isoniazid, by these systems further demonstrates the utility of carbon nanomaterials in miniature and portable analytical devices [13].

3.3 Metal and Metal oxide Nanoparticles

Due to their unique properties that improve sensor performance, metal and metal oxide nanoparticles have become integral components in the development of microfluidic electrochemical sensors. In this case, these nanoparticles are used for their capacity to improve conductivity, stability and sensitivity in many sensing applications [15]. These metal nanoparticles (such as zinc, copper, iron, nickel, and platinum) are used to increase the analytical performance of sensors. By enhancing the potential stability and electrical parameters of sensors, these nanoparticles improve the reliability and efficiency of sensors for the detection of different analytes [16,17].

Zinc oxide (ZnO), titanium dioxide (TiO_2), and iron oxide (Fe_3O_4) nanoparticles are among the most used nanostructured metal oxides in medical diagnostic biosensors due to their biocompatibility and catalytic properties [18]. In acetaminophen sensors, high sensitivity (91.0 μA cm^{-2} mM^{-1}) and low detection limit (0.23 μM) have been demonstrated for nickel oxide (NiO) nanoparticles [19]. In one study iron oxide (Fe_3O_4) nanodots attached to indium tin oxide electrodes, the use of iron oxide nanodots has produced low detection limits of 1.1 μM for hydrogen peroxide with rapid response time, strong adsorption capabilities [17].

These nanoparticles allow for miniaturization and integration of sensors in lab-on-a-chip systems, and also enhance the electron transfer rate in the sensors [20]. Different morphologies and surface functionalities of these materials enable tunable sensor responses, which are invaluable for the development of next-generation microfluidic electrochemical sensors for many diagnostics purposes (medical, environmental and food safety) [21].

Materials Research Foundations 191 (2026) 224-259
https://doi.org/10.21741/9781644904213-7

3.3.1 Application of metal and metal oxide in electrochemical sensing

Metal and metal oxide nanoparticles are critical in advancing the capabilities of microfluidic electrochemical sensors over a broad range of bioanalytical applications. One of the notable applications of metal oxide nanoparticles is their use in ion-selective electrodes (ISEs) as solid-state contacts. The incorporation of such an element enhances the potential stability and extends the operational lifespan of the sensors by several orders of magnitude [15] . Out of these, the one showing superior performance is the zinc oxide (ZnO) nanoparticle due to its high surface area, good electrical conductivity, and excellent chemical stability, synergistically enhancing signal transduction and minimizing the potential drift in prolonged measurement [22].

Noble metal nanocomposites, consisting of platinum (Pt) nanoparticles encapsulated within carbon nanotubes and nickel oxide (NiO), have also been fashioned as the electronic material in glucose sensing applications for achieving heightened sensitivity and selectivity [19]. The electrocatalytic activity of these hybrid systems is increased, allowing for precise detection of glucose at low concentrations required for real-time monitoring of glucose in compact, point-of-care, microfluidic platforms. For example, Pt-CNT-NiO-based sensors have been demonstrated capable of sensitivities in the range of several hundred $\mu A \cdot mM^{-1} \cdot cm^{-2}$, detection limits down to a few μM and fast and robust detection necessary for diabetic monitoring [20,23].

Metal and metal oxide nanoparticles have been extensively used for detection of biomolecules such as DNA strands, proteins, and enzymes beyond glucose analysis. They have the ability to mediate efficient electron transfer and amplifier electrochemical signals in order to reach extremely low detection limits in the nanomolar or even picomolar range [16]. For application in clinical diagnostics, genetic screening and proteomics, precise quantification of biological markers is essential, and these properties are indispensable. For this reason, the integration of these nanoparticles into the microfluidic electrochemical systems is gradually facilitating the growth of next generation biosensors specialized for fast and precise multiplexed bioanalytical assays [22-24] **(Figure 4)**

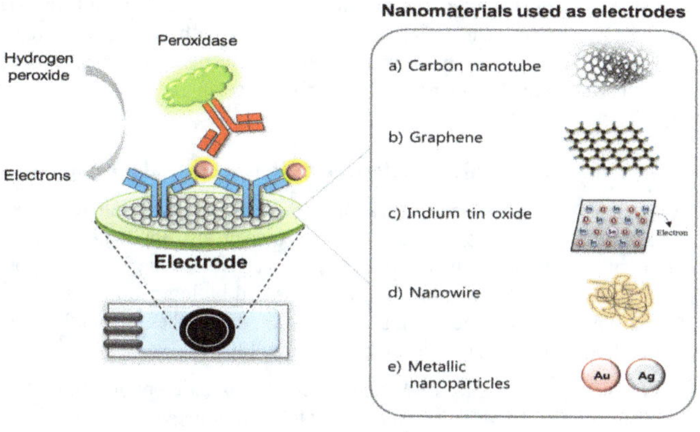

Figure 4 Nanomaterials used as electrodes to enhance the performance of Microfluidic Electrochemical Sensor[25]

Materials Research Foundations 191 (2026) 224-259
https://doi.org/10.21741/9781644904213-7

3.4 MXenes and 2D materials

Due to their highly promising applications as the next generation of electrochemical sensor typologies, MXenes, a class of 2D materials, composed of transition metal carbides, nitrides or carbonitrides, have been developing rapidly [26]. Their distinctive structural and physicochemical properties—such as excellent electrical conductivity, expansive surface area, hydrophilic nature, and remarkable mechanical strength—confer significant advantages in sensor design [27]. The abundance of surface functional groups (–OH, –O, –F) on MXenes are present which make it very effective for immobilization of biomolecules to improve specificity and sensitivity in target detection [28,29].

The suite of remarkable properties offered by MXenes makes them highly attractive material for the use in electrochemical sensing applications in microfluidic platforms. MXenes also have excellent electrical conductivity and a very large specific surface area, and such characteristics of MXenes allow for the efficient and rapid electron transfer while being high density of active surfaces for the analyte interaction [30,31], giving rise to high level of sensor sensitivity and detection performance. The high conductivity of Ti_3C_2Tx, one of the most robust MXenes studied, has been implicated in the improved signal-to-noise ratio and reduced detection limit of electrochemical IMMs in a variety of application involving biological sensing [28-32]

Inherent biocompatibility and hydrophilicity of MXenes are derived from their surface terminations (–OH, –O, –F) and strong interactions with aqueous environments and biological molecules [30] . These properties make them suitable for biosensors that utilize immobilized enzymes, antibodies or DNA, which ensures that the biological activity of the molecule is maintained during the detection process [26]. For instance, MXene based sensors have demonstrated good performance in glucose and DNA sensing where retaining bioactivity is crucial to allow accurate readings [29-33]

Moreover, MXenes are very versatile for sensor design. Being two dimensional and having reactive surface chemistry, they are easily combined with other functional materials, metal nanoparticles (e.g., Au; Pt) [33], conductive polymers (e.g., polyaniline; PEDOT:PSS) and carbon-based materials (e.g., graphene; carbon nanotubes) [28] . Usually, the hybrid composites have synergistic properties, like improved mechanical strength, enhanced electrocatalytic activity, and increased sensor stability which enhance their performance in the multiplexed and real time sensing systems [32].

3.4.1 Application of MXenes and 2D materials in electrochemical sensing

MXenes possess inherent electrochemical advantages, which make them inherently biocompatible and hydrophilic, traits that are of great value in biosensing applications. A noncovalent biomolecule immobilization, while maintaining the bioactivity of targets is supported by their ability to interact favorably with aqueous environments and biological molecules. The resulting hydrophilicity, combined with surface chemistry tunability, makes these sensors very sensitive and operable under conditions that include biological matrices [30].

Additionally, the structural versatility of MXenes allows synthetic opportunities to add them to previously described materials such as metal nanoparticles, conductive polymers and carbon based nanostructures to create multifunctional composites [27]. As a consequence, synergistic integrations among multiple sensors enable mechanical flexibility of sensors, improved charge

Materials Research Foundations 191 (2026) 224-259 https://doi.org/10.21741/9781644904213-7

transport characteristics, and well-defined surface functionalities in response to the analyte. The growing interest in MXenes as a foundational material for next generation microfluidic electrochemical devices is based in this sensor design adaptability [33].

3.5 Conductive Polymers

Due to their ability to act as Flexible and robust electrodes in the electrochemical sensors, conductive polymers (CPs) rise to the forefront as a key material in the development of flexible and robust microfluidic electrochemical sensors [34]. As these substrates possess high electrical conductivity, mechanical flexibility and capacity of forming composites, they are exceptionally suitable for applications in biomedical diagnostics and environmental monitoring [35].

The major advantage of CP lies in its intrinsic electrical conductivity, which can be greatly enhanced by doping processes. For example, the conductivity of CPs in their neutral state is generally between 10^{-6} and 10^{-10} $S{\cdot}cm^{-1}$, but can be as high as 10^5 $S{\cdot}cm^{-1}$ in the doped state [36]. With such high conductivity, fast electron transfer is advantageous to achieve high sensitivity and fast response time across electrochemical sensors [37]. In addition, the three-dimensional structures of CPs give ample active sites for analyte interaction and thus enhance their sensing capabilities.

CPs are equally noteworthy for their flexibility and stretchability. They can stay intact under various deformations like bending, twisting, or folding and sustain reasonable deformability. The development of wearable and portable sensor applications that require mechanical compliance requires this durability [35].

The goal of enhancing sensor performance has been achieved through integration of CPs with other materials. Improved conductivity and stability are achieved when composites are made with CPs and metal nanoparticles as well as two dimensional materials such as graphene, MXenes and metal organic frameworks (MOFs) [36]. The analytical performance of these sensors is superior due to the synergistic effects of their constituents, hence, the application of hybrid structures.

Additionally, CPs are being used to make stretchable and transparent sensors by combining them to carbon nanotubes, resulting in the improvement of CPs electric performance, and the mechanical compliance [38]. . Similarly, flexible and environmentally friendly sensors have been developed by amalgamating CPs with biodegradable materials such as silk. Moreover, these advancements in sensor functionality have been realized by CPs [39] and carbon nanotubes for the creation of stretchable and transparent sensors that perform well with regard to electrochemical performance and mechanical compliance. In addition to enhancing sensor functionality, these innovations meet the needs of the market to produce more sustainable and biocompatible materials in sensor technology.

Conductive polymers (CPs) like polyaniline (PANI), polypyrrole (Ppy) and poly(3,4-ethylenedioxythiophene) (PEDOT) have been essential in the field of electrochemical microfluidic biosensors because of their excellent electrical conductivity, chemical topography and efficient immobilization of biomolecules. These materials are useful in direct electron transfer, signal transduction, and miniaturization to integrate microfluidics and point-of-care diagnostics. CPs are highly sensitive, selective and operationally stable when engineered into nanostructures. Fabrication of composite with metal nanoparticles or carbon-based material making them a high-performance bio sensing. High compatibility with substrate also facilitates the creation of wearable

 Materials Research Forum LLC
Materials Research Foundations 191 (2026) 224-259 https://doi.org/10.21741/9781644904213-7

and implantable biosensors, and their capacity to create 3D electrode architectures enables to immobilize biomolecules as well as their real time and label-free detection in micro volumes of sample volumes [40]. New technologies have been developed in the recent past such as molecularly imprinted polymer systems and redox-polymer based sensors, which have enhanced rapid and highly selective metabolite sensing in microfluidic channels. Conductive polymers are used in glucose and catechol sensors, DNA, protein and whole-cell detection systems. CPs act as a matrix in enzyme immobilization, flexible electrodes in wearable biosensors to analyze analytes in sweat or blood. In summary, conductive polymers are the basis of current electrochemical microfluidic biosensors, which allow sensitive, selective, and miniaturized systems of analysis. The integration of these nanostructures, are anticipated to continue to improve the next generation of biosensing and bioelectronics devices [41].

3.6 Biomaterials and functionalized surfaces

In the development of microfluidic electrochemical sensors, biomaterials and functionalized surfaces have become important components in the development of microfluidic electrochemical sensors, due to the ability to increase sensitivity, selectivity, and overall device portability. Therefore, both these sensors are broadly employed in various application fields, namely medical diagnostics [42], environmental surveillance or food safety monitoring, where quick and precise detection of target analytes is mandatory. The synergy between engineered nanostructures and biomaterials used in sensing design allows for the development of real, high performance, microsystem sensors that can be used for on-site and in real time analysis [43].

This progress is built around the ability to incorporate and functionalize with functional nanomaterials (including carbon nanotubes, metal nanoparticles (gold, silver, platinum), and graphene derivatives) within the sensor matrix [43,44]. These nanomaterials have a high electrical conductivity, superior biocompatibility, large surface area. Such characteristics enhance electrochemical responses through acceleration of electron transfer kinetics and supply of lots of active sites for biomolecule immobilization [45]. Furthermore, they are nanoscale dimensions that seamlessly integrate within microfluidic platforms to produce compact, cost efficient and efficient sensing systems. Numerous studies have proven the ability of functionalized nanomaterials to increase the sensitivity [43] and specificity of biosensors due to their ability to provide high signal-to-noise ratio and thus detect trace amounts of analytes in bio and environmental matrices [44-46]

The sensors themselves are also highly dependent on microfluidic technology in terms of the performance, due to the ability to manipulate small fluid volumes with high precision [47]. Using this technology the amount of sample needed is reduced, while the throughput of screening and multiplex detection is increased [46,48]. The integration of living cells or biomolecules into microfluidic channels on substrates molecules also supports the use of microfluidic channels to avoid denaturation or degradation [44,49]. Furthermore, the union of microfluidics with the upcoming of advanced nanoelectrodes does not only boost the sensor sensitivity and stability but also simplifies the fabrication, scalability, and device reproducibility [50]. These are the features that are essential for transitioning laboratory-based sensors to point-of-care diagnostic tools [45]. Functionalized nanocomposites and microfluidic integration are found in the studies in Science of the Total Environment and Frontiers in Sensors to yield superior analytical platforms with fast response times and low power demands [47,48].

Materials Research Foundations 191 (2026) 224-259 https://doi.org/10.21741/9781644904213-7

3.7 Hybrid and Composite Materials

Their potential to greatly enhance sensor performance for many parameters including sensitivity, stability, selectivity, and structural integration, hybrid and composite materials have received interest in microelectrochemical sensor applications. Due to their adaptability and multifunctionality, they are good for such complex sensing environments as biomedical diagnostics, environmental monitoring and chemical analysis [51]

Carbon nanotube (CNT) based composites are the most explored hybrid systems in electrochemical sensing. Since CNTs are highly electrical conductive, have high aspect ratio and are chemical inert, they provide a highly efficient electron transfer platform [52,53]. Microelectrode fibers based on CNT hybrid nanocomposites have been used for the fabrication of microelectrodes for the electrochemical detection of small biomolecules such as dopamine and ferrocenemethanol with enhanced electrochemical sensitivity and response stability [54]. MWCNTs integrated with MXenes further enhance the charge transfer ability and surface activation synergistically that are particularly important in fabricating wearable biosensors for real-time physiological monitoring [53]. The combination of the two improves the energy storage capabilities, as well as the electrocatalytic properties of the sensors, leading to lowering the detection limit and enhancing the sensors' response time [55].

Hybrid sensor design also involves metallic nanostructures. These sensing platforms incorporate metal nanoparticles (made from for example gold, platinum, or cobalt) to target specific biomolecular interactions [51]. In addition to their role in increasing sensitivity, these metallic composites play a role in helping target analytes [56] to be selectively recognized by their contribution to facilitate electron transfer and catalytic reactions [57]. In particular, silver nanoparticles (AgNPs) have been widely used for their antimicrobial properties and their excellent conductivity. The combination of AgNPs synthesized through green chemistry approaches with matrices, such as chitosan, leads to significantly better electrochemical activity, signal stability and biocompatibility of the resulting sensors [58]. These hybrid systems show improved performance metrics that are clinically suitable for diagnostics and on-site environmental analysis. Table 1 represent huge potential of advanced nanomaterials towards in situ, efficient, portable, and precise sensing solutions of real world challenges utilizing hybrid materials [59].

Hybrid and composite materials have also become intrinsic in the development of electrochemical microfluidic biosensors, which are more sensitive, selective and multifunctional. They are very effective in analyte detection in miniaturized microfluidic systems due to their synergistic characteristics, i.e. high surface area, efficient electron transfer, increased biocompatibility, and controllable conductivity. Hybrid composites such as graphene-metal oxide hybrids, carbon nanoparticles with macrocycles and MOF-based nanocomposites amplify electrochemical signals and allow quick and sensitive detection of biomolecules such as glucose, cancer markers, and toxins [60]. Multiplexed and low-volume detection of multiple analyses with exquisite precision and rate may be possible through the integration of advanced nanomaterials such as MXenes, MOFs, and graphene-based composites. Also, composite materials like enzyme-hybrid conductive polymers and MOF-derived carbon nanocomposites improve the stability, antifouling of biosensors and their compatibility with biological conditions. In general, the combination of hybrid and composite materials is leading to the development of electrochemical microfluidic biosensors into compact, robust, and multifunctional diagnostic platforms, and the increasing use of

Materials Research Foundations 191 (2026) 224-259 https://doi.org/10.21741/9781644904213-7

biosensors in biomedical analysis and environmental monitoring, as well as in food safety detection [54].

Table 1: Different type of advanced materials used in microfluidic electrochemical sensors

Material Type	Examples	Key Properties	Sensor Role	Advantages	References
Carbon Nanomaterials	Graphene, CNTs, Carbon Dots	High surface area, conductivity	Electrode material	Enhanced signal, miniaturization	[10,12]
Metal NPs & Oxides	Au, Ag, TiO_2, Fe_3O_4	Catalytic activity, stability	Signal amplification	Low detection limits	[17,18,21]
MXenes & 2D	Ti_3C_2Tx, MoS_2, WS_2	High conductivity, tunable	Sensing layer	Flexibility, multi-target	[26,30]
Conductive Polymers	PANI, PEDOT:PSS	Electroactivity, flexibility	Electrode & transducer	Wearable sensor integration	[34,61]
Biomaterials	Enzymes, Antibodies	Specificity, biocompatibility	Biorecognition	High selectivity	[42,48]
Hybrids/Composites	Graphene-AuNPs, CNT-PEDOT	Multifunctionality	All sensor parts	Enhanced stability, multifunctional sensing	[51,54]

4. Electrochemical Detection Mechanisms in Microfluidics

Microfluidic technologies have revolutionized bio sensing by enabling the miniaturization, automation, and integration of multiple analytical functions into compact devices, often called lab-on-a-chip (LOC) systems. The ability to perform precise fluid handling within a platform at the microliter or nanoliter volume scale is desirable as such platforms can achieve rapid response times, small reagent consumption and have the possibility of being point of care diagnostics [62]. Electrochemical detection is one of the most widely used detection strategies integrated into microfluidic systems because of inherent advantages such as high sensitivity, low power requirements, ease of miniaturization, and compatibility with complex biological samples [63].

Electrochemical detection mechanisms are based on transforming biochemical or chemical interactions (e.g., their binding affinity, charge, etc.) into measurable electrical signals (e.g., current, voltage, impedance) using electrodes built within the microchannels. Because of the versatility of electrochemical methods, a very broad spectrum of targets (from small ions and metabolites to proteins, DNA and even whole cells) can be detected [64].

Depending on the information to be acquired, microfluidic biosensing can be carried out using voltammetry, amperometry, potentiometry, and electrochemical impedance spectroscopy (EIS). Different sensing applications are suited to different analytical capabilities of each technique. Voltammetric methods offer redox profiling of the samples with a high degree of detail, whereas amperometric sensors are useful in real-time quantifying enzyme-catalysed reactions. Impedimetric sensing is label−free and non−faradaic, and is especially appropriate for monitoring and detecting cellular activity and interface surfaces [65]. In contrast, potentiometric sensing is ideal for ion and pH monitoring.

Materials Research Foundations 191 (2026) 224-259 https://doi.org/10.21741/9781644904213-7

These techniques have been further improved recently by incorporating nanomaterials, redox mediators, signal amplification strategies, etc. The improvements have made it possible to produce microfluidic electrochemical sensors with increased sensitivity, selectivity and dynamic range, permitting ultra-sensitive diagnostics, real-time cellular metabolism monitoring and detection of several species in the same platform.

4.1 Voltammetric methods in microfluidics

The widespread detection technique called voltammetry measures electrode current responses to study redox events on surfaces by applying potential changes [66]. The detection method provides superior results when integrated with microfluidic systems because it restricts the measurement environment, enabling easier analyte transport and rapid mixing and requiring fewer samples [67]. When incorporated into microfluidic devices, voltammetric sensors provide simultaneous benefits in terms of high sensitivity levels and fast response, and broad analytical capabilities. The miniaturized electrode design of microchannels enables accurate detection through laminar microfluidic operations, which purposefully transfer analytes to the sensing interface [68]. The detection features of voltammetric methods will allow them to excel in 7657 automated research, immediate testing applications and point-of-care diagnostics procedures. Microfluidic systems employ three different voltammetric techniques: Cyclic Voltammetry, Differential Pulse Voltammetry, and Square Wave Voltammetry, distinguished by their unique operational properties.

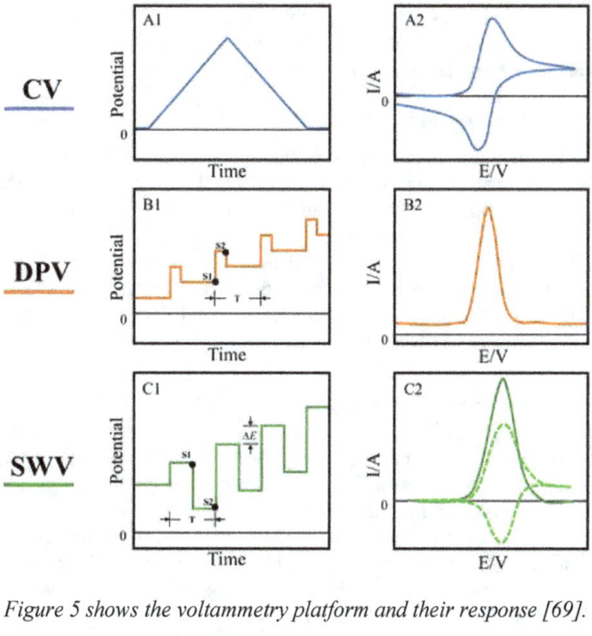

Figure 5 shows the voltammetry platform and their response [69].

Materials Research Foundations 191 (2026) 224-259 https://doi.org/10.21741/9781644904213-7

4.1.1 Cyclic Voltammetry (CV)

Cyclic Voltammetry (CV) is a very important and widely used voltammetric method in many fields of chemistry. It is commonly applied to investigate redox reactions, to evaluate the stability of reaction intermediates and products. Hence, this is often the first electroanalytical technique performed as it permits fast redox monitoring over a wide potential range [66-70].

Cyclic voltammetry operates on the principle that the applied potential is swept back and forth over a certain range, usually from negative to positive potentials and returning to its starting point. A voltammogram is generated by measuring the resulting current relative to the applied potential (Fig 5). The shape and features of this voltammogram reveal important electrochemical properties of the analyte, such as its redox potentials, reaction pathways, and reaction kinetics (Figure 5) [71,72].

CV is widely used in microfluidic systems as it is particularly useful for studying the behavior of redox active species in small volumes. Through this technique, scientists can observe electron transfer reaction dynamics with properties of many evaluation constituents, especially biomolecules such as neurotransmitters and DNA. Time-based measurement using CV becomes possible when applied to microfluidic channels during operation. This technique is a superior instrument for microfluidic systems and environmental detection procedures. Applying enzyme-modified electrodes using this technique allows the successful detection of glucose and lactate for portable biosensors [73]. Different fields within public health and food quality examination use smartphone-based detection systems built with graphene-altered screen-printed electrodes to evaluate glucose levels [74]. Fast scan cyclic voltammetry (FSCV) reaches high sensitivity and selectivity through its combination with enzyme-modified carbon fiber microelectrodes to monitor quick brain tissue glucose and lactate fluctuations [75].

4.1.2 Differential Pulse Voltammetry (DPV)

DPV is an electrochemical technique that gives better detection thresholds and improved signal-to-background ratio compared to basic voltammetry. Final current values are measured when a succession of small, fixed-amplitude voltage pulses is applied to a linear potential ramp (Figure 5) [66]. Current differences from successive pulse current measurements will give a voltammogram which suppresses background currents and enhances measurement sensitivity and resolution, particularly for low concentrations of analytes. Optimizing the DPV parameters, for example, pulse duration and amplitude, increases the detection capabilities [76]. Some trace analytes, like maleic hydrazide, have been successfully detected at nanomolar concentrations using the technique [77].

DPV detects low-concentration biomarkers, pollutants, and small molecules in analytical assays in a microfluidic device. The noise and interference reduction capabilities enable its performance to be more effective in environmental monitoring, clinical diagnostics, and in vivo physiological assessments. For instance, DPV was implemented as a low-cost biosensing system to detect 17β-estradiol in media samples with high detection accuracy [78]. Due to its precise measurement capabilities, DPV demonstrates excellent functionality for identifying minimal concentrations of medical testing substances such as cortisol and glucose. A small portable screening device which checks cortisol at the picomolar level, enabling medical screening within patient sites, has been developed by scientists [79]. Oxidative stress levels of biomarkers can be assessed through this technique, allowing researchers to understand sensor diagnostic abilities [80]. The experimental findings indicate that DPV represents a crucial method for enhancing microfluidic analytical systems because of its operational adaptability.

4.1.3 Square Wave Voltammetry (SWV)

The high sensitivity characteristics make square-wave voltammetry (SWV) an excellent electrochemical technique for analytical purposes in microfluidic systems. Forward and reverse currents are fully detected by applying squared voltage pulsed waves onto stationary potential conditions (Figure 5). The measured net current allows researchers to build graphs modelling current behavior concerning changing potentials for the speedy acquisition of data [81].

Real-time observation using SWV combined with high-throughput operation makes it a favorable technique for microfluidic platforms, since this method responds minimally to dissolve oxygen interference but functions properly at high frequencies. The miniaturized diagnostics employ SWV effectively, thanks to its detection performance reaching 2 pM recognition thresholds in the chip-based system [68]. Short biological processes, such as enzymatic reactions and neurotransmitter release analyzed better with SWV due to its capabilities to detect multiple analytes through microelectrode array technologies [82].

SWV is a fundamental technology component within portable diagnostic devices created for biomedical use. Researchers have used SWV to identify biomarkers such as glucose, lactate, urea, and uric acid in different biological solutions [73]. SWV shows high sensitivity, quick responses, and flexible operation as a portable system, making it ideal for diagnosing diseases when decentralized health care. Analysis through SWV functions as an important tool for proteins and nucleic acids research, which enables challenging bioanalytical evaluations and genetic screening procedures [84].

SWV-based sensors operate successfully as environmental and food safety watch systems to identify minute levels of contaminants, pollutants, and toxins through their high selective detection capability [83]. Combining SWV with microfluidics results in a strong system combining high performance with low price for diverse sensing applications.

4.2 Amperometric and Potentiometric Detection

Microfluidic platforms choose electrochemical biosensing techniques, including amperometry and potentiometry, because these methods offer a high sensitivity alongside simple operation and budget-friendly implementation. Such methods utilize biochemical reactions to generate electrical signals, enabling precise or basic analysis of substances, including ions and small molecules and metabolites and biomacromolecules [85]. Miniaturized system devices can work effectively with these detection methods because they make ideal candidates for continued observation and portable testing regardless of the device's location. Modern electrochemical sensor development includes wireless and portable options with low-volume production and affordable manufacturing processes suitable for health care and environmental testing applications [86].

4.2.1 Amperometric Detection

An electrode receives an applied steady potential during amperometry, while the interface current results from electroactive species redox reactions at the electrical interface. Quantitative analysis benefits from this technique because the measured current produces direct analytical information

Materials Research Foundations 191 (2026) 224-259 https://doi.org/10.21741/9781644904213-7

about the solution concentration. Technology finds widespread use in wastewater pharmaceutical residue surveillance and point-of-care medical diagnosis systems.

Amperometric detection is very versatile and highly sensitive, and thus, it can be integrated into many different microfluidic biosensing platforms. In microfluidic biosensors, amperometric detection often combines enzyme-functionalized electrodes (e.g., glucose or lactate oxidase) that catalyze a specific redox reaction, producing electrons as current (Figure 6) [73,87]. The microfluidic platforms provide a microscale environment within which electron transfer efficiency is improved, which decreases sample and reagent consumption, and also enhances signal-to-noise ratio. However, sensor sensitivity, stability, and longevity are greatly influenced by the enzyme immobilization techniques such as physisorption, covalent bonding, and incorporation in redox hydrogels [88]. These systems can run as single or multi-analyte detectors in compact, real-time monitoring formats for several applications.

Specifically, microfluidic amperometric biosensors have been developed to continuously monitor glucose and lactate, especially in interstitial fluid and saliva [89]. These devices use enzyme-based reactions using either glucose oxidase or lactate oxidase to catalyze hydrogen peroxide generation that is then electrochemically detected. Researchers have also successfully implemented amperometric sensors in hanging-drop microfluidic systems to measure real-time metabolism in 3D microtissue spheres [90]. This allows them to simultaneously monitor glucose uptake and lactate secretion during cancer model experiments under conditions that mimic human biology. Additionally, real-time monitoring of neurotransmitters such as dopamine and serotonin can be done by amperometric detection, especially in brain-on-chip models and microdialysis systems [91]. They are extremely sensitive to time and essential in neurophysiological studies.

4.2.2 Potentiometric Detection

Potentiometric sensors use an ion-selective working electrode and a reference electrode and measure the potential difference between them under zero current conditions. This produces a logarithmically dependent potential concerning the activity of a particular ion within the sample and is used for stable, selective, and quantitative analysis of ionic species.

Ion-selective electrodes (ISES) are usually fabricated using ion-sensitive membranes, conductive polymers, or nanostructured materials in microfluidic systems. Due to their miniaturized design and integration into lab-on-a-chip platforms, they can directly monitor ions in real time at their location of interest with little sample and reagent consumption.

The performance of ISEs has been greatly improved in recent years to enable trace-level detection in the micro- and nanomolar range. This includes innovations in all-solid-state sensors and their integration into paper-based microfluidic systems due to their low cost, disposability, and environmental friendliness. Potential stability has been improved for solid-state reference electrodes suitable for use on paper-based platforms for clinical and environmental monitoring [92].

Improvements continue to enhance sensitivity and analytical performance through backside calibration potentiometry, controlled current coulometry, and pulsed chronopotentiometry. This advancement has broadened the applicability of potentiometric sensors to include ultra-sensitive affinity bioanalysis by utilizing nanoparticle labels.

Materials Research Foundations 191 (2026) 224-259 https://doi.org/10.21741/9781644904213-7

4.3 Impedimetric Sensing

EIS is a powerful method that allows scientists to study the properties of interfaces and detect events involving biorecognition on electrodes. The measurements of electrochemical system impedance throughout multiple frequencies enable EIS to supply complete information about charge transfer resistance, mass transport, and interfacial electrochemical reactions [93,94]. The technique shows exceptional responsiveness to interface processes, providing non-destructive label-free measurement of biomolecular binding interactions, including DNA hybridization, cellular adhesion, and antigens meeting their antibodies.

EIS has become a fundamental tool for biosensor construction through the development of immunosensors and aptasensors, in addition to DNA hybridization sensors, because target binding causes changes in electrical impedance at electrode surfaces [93,95]. EIS provides effective recognition event monitoring by dispensing with labels and delivers information about reaction kinetics and interfacial processes [94].

The advancement in material development through scientific techniques has dramatically improved the functioning of biosensor instruments based on EIS methodology. Implementing carbon nanotubes with gold nanoparticles and layered double hydroxides (LDHs) nanomaterials improves sensor electron transfer and amplifies signals while delivering better reliability [65,96]. Surface modification using functional nanomaterials at electrodes enables quick charge transmission, which improves the precision of the detection of dilute analytes.

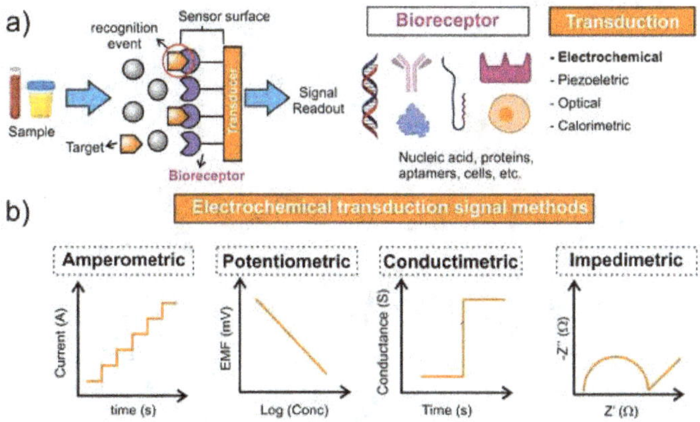

Figure 6 shows the examples of various electrochemical transduction signals and their pattern of response [99].

The addition of EIS to microfluidic systems has extended its practical application scope. Microfluidic systems deliver exact fluid operation with reduced sample volumes and faster test durations, enabling them to show outstanding characteristics for portable miniaturized biosensing technology [97]. Various studies report excellent performance from impedimetric microfluidic biosensors, including their application in protein detection without labels for cardiac troponin I

Materials Research Foundations 191 (2026) 224-259 https://doi.org/10.21741/9781644904213-7

diagnosis through DNA-based self-assembled monolayers positioned on gold electrodes and numerous other monitoring applications [98].

The combination of microfluidic systems with nanotechnology applied to electrode interfaces and EIS technology enables applications of exceptionally sensitive and inexpensive, and swift biosensing systems. Existing research leads to efforts to integrate EIS-based biosensors inside automated lab-on-a-chip platforms to improve their use in point-of-care diagnosis and real-time biological testing [96].

5. Recent Advances in Signal Amplification Strategies

Signal amplification is necessary to identify low-abundance biomarkers and analytes and increase the sensitivity of electrochemical biosensors. Innovations in nanotechnology in recent years, enzyme catalysis, nucleic acid engineering, and material science, have significantly expand the microfluidic capabilities. Table 2 shows the execution of specific target amplification and molecular amplification strategies to achieve ultrasensitive detection is enabled by broad reorganisation along target-based and signal-based approaches to rely on the synergistic effects of molecular amplification, nanomaterials, and catalytic reactions. In addition, the feasibility of incorporating amplification methods to emerging platforms like paper-based analytical devices (PADs) has opened up a wide range of applications in clinical, environmental, and industrial diagnostics by applying cost-effective, rapid, and highly sensitive diagnostics (Figure 5) [100].

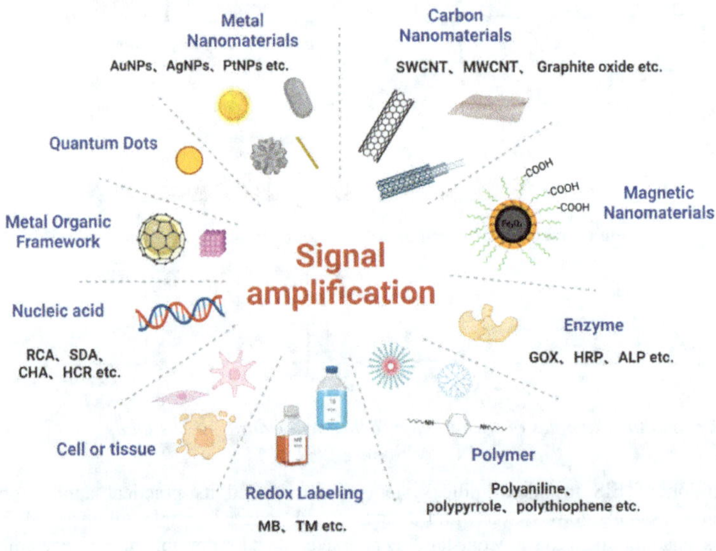

Figure 7 signal amplification strategy for microfluidic electrochemical sensors[108]

Materials Research Foundations 191 (2026) 224-259 | https://doi.org/10.21741/9781644904213-7

Table 2: summarizes different signal amplification strategies used in electrochemical microfluidic biosensors, mechanisms, and applications. This section discusses recent innovations in amplification techniques, as it is currently essential for microfluidic biosensors

Amplification Strategy	Mechanism	Typical Material	Effects on Sensor Performance
Nanomaterial-Based Amplification	Electron transfer is enhanced, and the biorecognition surface area is increased.	Gold nanoparticles, layered double hydroxides, carbon nanotubes	It gives better sensitivity, lower detection limits, and higher surface loading [101,102]
Enzyme and Nanozyme Amplification	Signal is amplified through substrate conversion in a catalytic reaction.	Horseradish peroxidase, glucose oxidase, Fe_3O_4 nanozymes	High specificity, catalytic signal boost, rapid and real-time detection [103]
Redox Cycling	Repetitive redox reactions between the analyte and mediator increase the measurable current	Ferrocene, quinones, methylene blue	Support signal amplification without target replication; Multiplexing capability [104]
DNA-Based and Hybrid Amplification	Hybridisation chain reactions or rolling circle replication to perform isothermal amplification	DNA hairpins, circular DNA template	Ultrasensitive detection of nucleic acids and mutations, enzyme-free signal enhancement [105]
Smart Material Integration and Signal Gating	Conformational or surface changes are triggered by environmental stimuli in the membrane to modulate a signal	pH-responsive hydrogels, shape-memory polymers	Minimised background noise, enhanced selectivity, self-regulated sensing [106,107]

5.1 Nanomaterial-Based Amplification

Amplification strategies based on nanomaterials have revolutionized electrochemical biosensing as they have enhanced the sensitivity in biomolecule detection to the ultrasensitive level. Among them, various nanomaterials such as metal nanoparticles, carbon-based nanostructures, and layered double hydroxides (LDHs) remarkably boost electrochemical signals through increased surface area, enhanced conductivity, and faster electron transfer rate [101]. These nanomaterials can be functionalized with nucleic acids or antibodies, rendering them as amplifying tags for DNA and protein detection platforms to yield enhanced and more reliable signal outputs [102].

The synergistic properties of catalytic activity, conductivity, and biocompatibility promote the catalytic reactions and signal transduction to enhance the biosensors' sensitivity [101]. As a result, powerful bioaffinity assays and label-free detection schemes have been developed by utilizing carbon nanotubes, nanowires, etc. For example, very recent innovations include using gold nanoparticles for cancer biomarker detection and copper-modified LDHs (Cu@LDH) for glucose sensing in microfluidic devices, demonstrating the increasing potential of nanomaterials-assisted amplification for point-of-care diagnostics [109].

Materials Research Foundations 191 (2026) 224-259 https://doi.org/10.21741/9781644904213-7

5.2 Enzyme and Nanozyme Amplification

The enzyme and nanozyme-based amplification strategies are key to achieving high sensitivity and specificity in electrochemical biosensors. Natural enzymes like oxidases are still essential to signal amplification in applications such as glucose or lactate monitoring because they are highly specific. However, due to their susceptibility to denaturation, short shelf life, and high production cost, nanozymes were developed as a robust alternative.

Nanozymes are nanomaterials engineered with enzymatic properties, which have the advantages of chemical stability, programmable catalytic efficiency, and resistance to harsh conditions [103,110]. Nanozymes are derived from metals, metal oxides, nanocarbons, and metal organic frameworks and have been used to detect different analytes, from small molecules to cells [111]. Moreover, they have low cost, are reusable, and bend with compatible biosensing formats, making them applicable, for example, in alkene oxide sensors for oxidative stress detection, immunoassay, and cancer diagnostics [112]. Despite the challenges of low catalytic efficiency and substrate specificity, nanozymes are becoming increasingly important in enhancing the application of biosensing technologies (Figure 6).

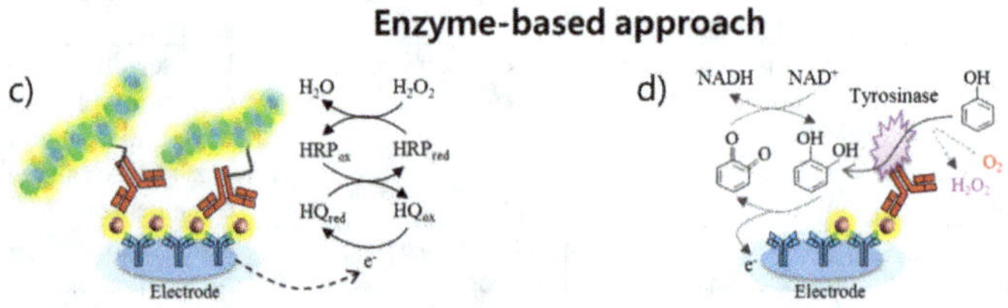

Figure 8 Enzyme based approach for signal amplification of a microfluidic device [108]

5.3 Redox Cycling and Electron Mediators

Redox cycling is a well-established signal amplification technique in electrochemical biosensing where the measurement signal is proportional to several electrons exchanged between analyte molecules and electrode surfaces. This process is frequently accomplished in the presence of electron mediators such as ferrocene and methylene blue, which increases the electron shuttling rate and detection sensitivity. Redox cycling enables several redox events per analyte, which can greatly enhance signal strength, even at low levels of the analyte. Utilizing this approach, DNA hybridization can be detected as low as 1 pM [104], and protein detection sensitivity is improved by 13,300-fold [113].

This method has been expanded to use with innovative redox-active platforms. One such example is the biobased redox capacitor built by grafting catechols on chitosan, which can signal amplify for the detection of bacterial metabolites, and reach a detection limit down to 50 nM for pyocyanin [114]. In addition, such capacitors can discriminate the redox-cycling species from the concentrated non-gassing redox-active compounds according to their unique electrochemical signatures [115]. Finally, redox cycling is useful in multiplexed nucleic acid detection and neurotransmitter monitoring in neurological studies because it can simultaneously act as a reductant and an oxidant.

Materials Research Foundations 191 (2026) 224-259 https://doi.org/10.21741/9781644904213-7

5.4 DNA-Based and Hybrid Amplification Strategies

Among the DNA-based amplification strategies, hybridization chain reaction (HCR) and rolling circle amplification (RCA) are specific and programmable, giving them potential for molecular diagnostics. HCR is an enzyme-free, isothermal method of self-assembling DNA hairpins triggered by target sequences, and used in several biosensors for food safety, environmental monitoring, and disease detection [105].

On the other hand, RCA uses a circular DNA template and a DNA polymerase to generate long single-stranded DNA products with repeated sequences. With high amplification efficiency and fidelity, it is specifically meant for point-of-care diagnostics and nucleic acid-based bioassays [116]. These DNA amplification methods are compatible with electrochemical, fluorescent, and colorimetric signal readouts for ultrasensitive detection of microRNAs, gene mutations, and infectious pathogens [117]. Moreover, the performance of this technology has continued to improve to the point where precise and low-abundance biomarkers are now being detected and used for genetic testing and disease monitoring for human subjects in the clinical research setting.

5.5 Smart Material Integration and Signal Gating

Recently, the design of advanced electrochemical sensors has incorporated smart materials, namely stimuli-responsive systems. These materials exhibit dynamic signal gating, enabling the sensors to respond to external stimuli, like pH, temperature, light, or biochemical cues, to control their real-time behaviors. The sensitivity and selectivity are enhanced, and background interference is minimized by their ability to modulate signal-on/off responses [106,107].

Such materials can be integrated to create flexible electrode surfaces and gating membranes with controllable permeability. In such microfluidic platforms, using these responsive components can result in a high degree of control for individual sample handling, adaptive flow control, and signal modulation in the application context [118]. Additionally, their bio-interactive design encourages active communication between the biomaterial and the biological environment, enabling applications in wearables, diagnostics, and tissue engineering [119].

6. Applications of Advanced Materials in Microfluidic Electrochemical Sensors

Nanotechnology advancements have led to a greater diversity in biosensor research. By adjusting size and shape, it is possible to improve biosensor performance and increase detection power by investigating metal- and metal oxide-based nanomaterials, nanorods, nanowires, nanoflakes, nano cones, nano spherical, carbon nanotubes, quantum dots, and nanocomposites [120].

6.1 Biomedical Applications

Early disease detection improves the likelihood of successful treatment and is essential for preventative measures. Biochemical cycles are impacted and negative health effects result when biomolecules are present in human blood at levels higher than normal. Therefore, for large-scale monitoring systems, reliable biomarker quantification and quick detection are very important. The ability of microfluidics technology to replicate conventional standard laboratory instruments on a little chip-based system has drawn a lot of interest from researchers and scientists particularly in electrochemistry and biochemistry studies. Almost every field uses microfluidics, including the

Materials Research Foundations 191 (2026) 224-259 — https://doi.org/10.21741/9781644904213-7

biochemical, pharmaceutical, biomedical, and electrochemical domains. Minimal sample volume, quick reaction, accuracy, multiplex operation, and quick evaluation are some of its benefits.

6.1.1 Cancer Diagnosis

Most carcinomas arise by excessive cell growth. A series of biological events occur, including the development of tumors, the migration of cancer cells, the spread of tumors, specific genetic alterations in the normal cell (initiation), and changes in the transcriptome and proteins. Typically, malignant cells differ from normal cells in terms of size, phenotype, and metabolism. In the meantime, the microenvironment surrounding the tumor also shifts. When immune cells encounter cancer cells, they release certain unique cytokines. Matrix metalloproteinases (MMPs) also cause major alterations in the extracellular matrix (ECM). Numerous cancer biomarkers develop during the development, growth, and spread of tumors, including circulating tumor DNA (ctDNA), microRNA, cancer antigens, cytokines, MMPs, exosomes, altered genomic material, and circulating tumor cells (CTCs) [121] . It has been shown that a dual- modality, low-cost, and highly effective sensor may detect cancer biomarkers by utilizing both SPR and electrochemical sensing techniques. Sufficient antibody binding to the sensor surface is made possible by the Au nanoposts increased surface area. The sensitivity of the electrochemical mode is several times higher than that of the bulk electrode without nanoposts because the nanoposts allow electrons to diffuse radially. Tracking the affiliations and dissociations of biomarker molecules at the sensor surface is made simple when the sensor is in the SPR mode. This sensor offers greater reproducibility than others that use synthesized nanomaterials because it uses homogenous and ordered nanostructures. In order to increase detection reliability and false alarm immunity for breast cancer detection, the sensor dual-modality performance offers a novel strategy [122].

6.1.2 Cardiovascular diseases Diagnosis

In the diagnosis and risk assessment of individuals with suspected or confirmed CVD, cardiac biomarkers are essential. The gold standard for identifying myocardial damage among the several biomarkers is cardiac troponins (cTn). Three different components of the regulating protein troponins—troponin C (cTnC), cTnI, and cTnT—are present in striated muscle fibers. cTnI and cTnT are unique to the heart muscle, but cTnC is expressed in both cardiac and skeletal muscle. cTnI and cTnT are extremely sensitive and specific indicators for identifying even slight myocardial damage because they are released into the bloodstream from the injured cardiomyocytes after myocardial injury, such as acute myocardial infarction. Within a few hours of the onset of symptoms, there is an increase in blood levels of cTnI and cTnT, which peaks between 12 and 24 hours later and stays high for a few days. Electrochemical biosensors have become a viable substitute for traditional immunoassays to identify cardiac biomarkers. These sensors combine the sensitivity and adaptability of electrochemical transduction techniques with the specificity of biological recognition elements (such as antibodies or aptamers). High sensitivity, quick response, miniaturization potential, and compatibility with point-of-care (POC) testing are only a few benefits of electrochemical biosensors. The performance of electrochemical biosensors has been further enhanced by the use of nanomaterials and microfluidic platforms, which has made it easier to create multiplexed, automated, and portable assays for the detection of cTnI and cTnT [123].

Materials Research Foundations 191 (2026) 224-259 https://doi.org/10.21741/9781644904213-7

6.1.3 Glucose Detection

A widely discussed topic is the use of electrochemical sensors with bulk electrodes to detect glucose in human blood serum. Few studies, meanwhile, have looked at the electro-analytical capabilities of an electrochemical glucose sensor embedded onto a microfluidic chip. For a quick microfluidic microfabricated chip that detects glucose in human blood serum samples using Pt nanoparticles as an effective biosensing platform.

The poly(dimethylsiloxane) (PDMS)-based microfluidic chip was made by soft lithography and consisted of a fused-silica capillary with a tapered tip, a Pt-black working electrode, a Pt wire counter electrode, and an Ag/AgCl reference electrode. With a repeatability of 2.65% RSD, the outcome of the microfluidic sensor demonstrated a linear response up to 43.5 mM glucose. The microfluidic sensor response may potentially be impacted by contaminants on the PDMS surface [124].

6.2 Food Safety

In the human body, pathogens are infectious mediators that include bacteria, viruses, fungus, prions, and protozoa that can cause illness. Foodborne infections can infect the body in a number of ways. In this case, foodborne illnesses spread by consuming poisoned food are the main causes of food safety problems. Food poisoning, diarrhea, dysentery, and sometimes mortality are the main signs of foodborne infections in humans. Thirteen million deaths globally are predicted by 2050 as a result of bacterial infections, which cause an astoundingly high number of deaths annually. Therefore, identifying and detecting foodborne pathogenic microorganisms is crucial. In addition, the main methods for identifying foodborne pathogenic bacteria are immunological analysis like PCR-ELISA, nucleic acid-based methods, and conventional culturing techniques [125].

These techniques are time-consuming, costly, require a very large device, and are unstable, therefore they don't offer the benefits needed for point-of-care applications. Foodborne pathogens can be detected and evaluated using sophisticated transduction systems called electrochemical sensors based on microfluidics. By directly converting the inbound electrical signal into an electric field, these allow the creation of tiny systems with simple instrumentation. Additionally, they have the following general advantages over analytical workstations: (i) similar influencing sensitivity, (ii) the ability to operate in murky medium, and (iii) the possibility of miniaturization, which allows for the analysis of even very small volumes. When it comes to identifying various foodborne pathogenic bacteria, including Salmonella, Lm, S. aureus, Escherichia coli (E. coli), and others, electrochemical biosensors are a great choice. Only certain electrochemical sensors based on microfluidics are known to be able to detect harmful bacteria at a limit of 1 CFU/mL. Because of their advantageous qualities, special nanomaterials are used in this case to produce and modify the sensing electrodes of the electrochemical biosensor for improved outcomes [126].

6.3 Environmental Toxicants

Numerous papers describe the effective combination of electrochemical biosensors and microfluidic systems, as well as their application in the detection of heavy metals and pesticides, among other environmental toxicants. For example, an electrochemical microfluidic paper-based analytical device (μPAD) was recently created to track the suppression of Escherichia coli respiratory activity after exposure to pesticides, heavy metals, and penicillin sodium [127].

Materials Research Foundations 191 (2026) 224-259 https://doi.org/10.21741/9781644904213-7

6.4 Wearable sensor

The material selection for the sensor device determines its wearability and it should be able to adjust to the biological, physical, and chemical characteristics of the skin. It is challenging to replicate the curved and uneven surfaces that make up the human body. Furthermore, skin is naturally malleable and viscoelastic. Wearable sensors should therefore effectively adapt to the characteristics of the skin and be paired with the capability of digitally processing the inputs. Researchers have spent a great deal of effort creating a material that conforms well to the surface of our skin. Advanced Materials that function as an interfacial substrate to human skin, such as polymers, textiles, and elastomers, are frequently chosen. These materials are perfect for creating wearable sensing devices because of their characteristics that adapt to human skin. Important characteristics of polymer-based materials include robustness, flexibility, and chemical resistance. Elastomers are also very conformal and stretchable. Wearable technology uses fabrics because of their softness and ability to absorb. The research team, led by Professor Ye, has successfully created a smartwatch sensor that uses a paper substrate as a microfluidic layer to analyze perspiration. The smartwatch's primary function is to track the electrolyte (Na+ and K+) levels in perspiration. For the analysis, Whatman filter paper is altered to function as a microfluidic channel. On PET substrates, corresponding ion-selective layers for K+ and Na+ are printed. A smartwatch sensor is used to conduct on-body real-time examination during stationary cycling. Sweat storage, transportation, and analysis can be effectively managed by integrating microfluidics into the worn sensor system [128].

7. Challenges and Future perspective

Although developments in improved materials are closely linked to the development of microfluidic electrochemical biosensors, there are numerous interrelated obstacles that cut across the technical, scientific, and translational domains. Reproducibility, long-term stability, and biocompatibility are major challenges when incorporating nanomaterials like graphene, MXenes, metal-organic frameworks, and conductive polymers into microfluidic configurations. In controlled environments, these materials frequently exhibit exceptional electrochemical qualities; however, fouling, signal drift, and non-specific adsorption cause performance loss in intricate biological matrices.

Fabrication introduces an additional level of complication because many high-performance materials are not suited for environmentally sustainable and scalable manufacturing as they require hazardous reagents, cleanroom-dependent processes, or harsh synthesis conditions. Clinical translation and industrialization are also constrained by unsolved regulatory and regulatory difficulties, especially with regard to new nanomaterials. Beyond these issues with materials and manufacturing, new requirements are brought about by the incorporation of biosensors into smart healthcare systems and Internet of Things (IoT) frameworks. Without sacrificing sensor sensitivity or miniaturization, real-time communication, low-power operation, wireless data transmission, and cybersecurity must be addressed.

In real-world healthcare settings, adoption is made more difficult by the absence of standardized communication protocols and the requirement for flawless compatibility with wearable technology, cellphones, and cloud infrastructure. The future has great promise in spite of these complex obstacles. The lifetime and reliability of biosensors could be significantly increased by newly developed smart materials with stimulus-responsive, antifouling, or self-healing properties.

Materials Research Foundations 191 (2026) 224-259 https://doi.org/10.21741/9781644904213-7

At the same time, developments in additive manufacturing, roll-to-roll processing, and 3D printing could make it possible to produce flexible, customisable biosensor platforms on a wide scale at a reasonable cost. Additionally, the integration of AI, machine learning, and materials science creates opportunities for predictive diagnoses and adaptive sensing based on ongoing, real-time biochemical data. Long-term, these developments may lead to independent, cloud-based biosensing systems that are essential to remote diagnostics, ecological tracking, personalized medication, and worldwide healthcare surveillance. However, achieving this goal would necessitate not only revolutionary advancements in fabrication and materials but also a comprehensive strategy that unifies standards, crosses disciplines, and synchronizes technical advancement with social and therapeutic demands.

Conclusion

Although incorporating innovative substances into microfluidic electrochemical biosensors has enormous promise to transform diagnostics, there are still many practical and technical obstacles to overcome. Important discoveries show that although nanomaterials like metal-organic frameworks, graphene, and MXenes offer improved sensitivity and specificity, problems with stability, repeatability, and biocompatibility in actual biological samples restrict their use. Such materials fabrication procedures sometimes entail exorbitant expenses, hazardous chemicals, and scalability limitations. Furthermore, non-specific adsorption and surface fouling seriously impair the precision and durability of biosensors. The shift from experimental to clinical applications is made more difficult by regulatory uncertainties and a lack of standardization. Additional difficulties arise when these biosensors are integrated with IoT and smart healthcare systems. These difficulties include the need for device interoperability, data security, low battery consumption, and real-time connectivity. However, there is a significant chance to overcome these obstacles and make microfluidic electrochemical biosensors a vital component of remote and customized healthcare thanks to the combination of smart materials, AI-driven analytics, and sophisticated production processes.

References

[1] M. Javaid, A. Haleem, R.P. Singh, S. Rab, R. Suman, Significance of sensors for industry 4.0: Roles, capabilities, and applications, Sens. Int 2 (2021) 100110. https://doi.org/10.1016/j.sintl.2021.100110

[2] J. Vetelino, A. Reghu, Introduction to sensors, CRC press2017. https://doi.org/10.1201/9781315218274

[3] S. Radhakrishnan, M. Mathew, C.S. Rout, Microfluidic sensors based on two-dimensional materials for chemical and biological assessments, Mater. Adv 3(4) (2022) 1874-1904. https://doi.org/10.1039/D1MA00929J

[4] M. Kumari, V. Gupta, N. Kumar, R.K. Arun, Microfluidics-based nanobiosensors for healthcare monitoring, Mol. Biotechnol 66(3) (2024) 378-401. https://doi.org/10.1007/s12033-023-00760-9

[5] A. St John, C.P. Price, Existing and emerging technologies for point-of-care testing, Clin. Biochem. Rev 35(3) (2014) 155.

Materials Research Foundations 191 (2026) 224-259 https://doi.org/10.21741/9781644904213-7

[6] L. Fritea, F. Banica, T.O. Costea, L. Moldovan, L. Dobjanschi, M. Muresan, S. Cavalu, Metal nanoparticles and carbon-based nanomaterials for improved performances of electrochemical (Bio) sensors with biomedical applications, Materials 14(21) (2021) 6319. https://doi.org/10.3390/ma14216319

[7] J. Sengupta, C.M. Hussain, MXene-based electrochemical biosensors: advancing detection strategies for biosensing (2020-2024), Biosens. J 15(3) (2025) 127. https://doi.org/10.3390/bios15030127

[8] J.M. Mohan, K. Amreen, A. Javed, S.K. Dubey, S. Goel, Emerging trends in miniaturized and microfluidic electrochemical sensing platforms, Curr. Opin. Electrochem 33 (2022) 100930. https://doi.org/10.1016/j.coelec.2021.100930

[9] E. Asadian, M. Ghalkhani, S. Shahrokhian, Electrochemical sensing based on carbon nanoparticles: A review, Sens. Actuators B: Chem 293 (2019) 183-209. https://doi.org/10.1016/j.snb.2019.04.075

[10] A.C. Power, B. Gorey, S. Chandra, J. Chapman, Carbon nanomaterials and their application to electrochemical sensors: a review, Nanotechnology Reviews 7(1) (2018) 19-41. https://doi.org/10.1515/ntrev-2017-0160

[11] J.F. Hernández-Rodríguez, F. D. Pelle, D. Rojas, D. Compagnone, A. Escarpa, Xurography-enabled thermally transferred carbon nanomaterial-based electrochemical sensors on polyethylene terephthalate-ethylene vinyl acetate films, Anal. Chem 92(19) (2020) 13565-13572. https://doi.org/10.1021/acs.analchem.0c03240

[12] M.B. Kulkarni, N.H. Ayachit, T.M. Aminabhavi, Recent advances in microfluidics-based electrochemical sensors for foodborne pathogen detection, Biosensors (Basel) 13(2) (2023). https://doi.org/10.3390/bios13020246. https://doi.org/10.3390/bios13020246

[13] F. Arduini, S. Cinti, V. Mazzaracchio, V. Scognamiglio, A. Amine, D. Moscone, Carbon black as an outstanding and affordable nanomaterial for electrochemical (bio) sensor design, Biosens. Bioelectron. 156 (2020) 112033. https://doi.org/10.1016/j.bios.2020.112033

[14] A. Cetinkaya, S.I. Kaya, G. Ozcelikay, F. Budak, S.A. Ozkan, Carbon nanomaterials-based novel hybrid platforms for electrochemical sensor applications in drug analysis, Critical Reviews in Analytical Chemistry 54(5) (2024) 1227-1242.

[15] Y. Yang, J. Luo, Electrochemical sensing of carbon coated metal oxide nanoparticles, academic journal of engineering and technology science 5(10) (2022) 16-19. https://doi.org/10.25236/AJETS.2022.051003

[16] M.A. Rashed, N. Nayem, M.H. Rahman, M. Faisal, J.S. Algethami, S. Alsareii, M.K. Islam, F.A. Harraz, A sensitive, selective non-enzymatic electrochemical detection and kinetic study of glucose over Pt nanoparticles/SWCNTs/NiO ternary nanocomposite, Journal of the Taiwan Institute of Chemical Engineers 151 (2023) 105113. https://doi.org/10.1016/j.jtice.2023.105113

[17] J.M. George, A. Antony, B. Mathew, Metal oxide nanoparticles in electrochemical sensing and biosensing: a review, Microchim. Acta 185 (2018) 1-26. https://doi.org/10.1007/s00604-018-2894-3

[18] K. Pietrzak, N. Krstulović, D. Blažeka, J. Car, S. Malinowski, C. Wardak, Metal oxide nanoparticles as solid contact in ion-selective electrodes sensitive to potassium ions, Talanta 243 (2022) 123335. https://doi.org/10.1016/j.talanta.2022.123335

[19] J. Yang, J.-H. Yu, J.R. Strickler, W.-J. Chang, S. Gunasekaran, Nickel nanoparticle-chitosan-reduced graphene oxide-modified screen-printed electrodes for enzyme-free glucose sensing in portable microfluidic devices, Biosens. Bioelectron. 47 (2013) 530-538. https://doi.org/10.1016/j.bios.2013.03.051

[20] W. Siangproh, W. Dungchai, P. Rattanarat, O. Chailapakul, Nanoparticle-based electrochemical detection in conventional and miniaturized systems and their bioanalytical applications: A review, Anal. Chim. Acta 690(1) (2011) 10-25. https://doi.org/10.1016/j.aca.2011.01.054

[21] M. Anu Prathap, B. Kaur, R. Srivastava, Electrochemical sensor platforms based on nanostructured metal oxides, and zeolite-based materials, Chem. Rec 19(5) (2019) 883-907. https://doi.org/10.1002/tcr.201800068

[22] F. Wang, S. Hu, Electrochemical sensors based on metal and semiconductor nanoparticles, Microchim. Acta 165 (2009) 1-22. https://doi.org/10.1007/s00604-009-0136-4

[23] A.S. Agnihotri, A. Varghese, Transition metal oxides in electrochemical and bio sensing: A state-of-art review, Applied Surface Science Advances 4 (2021) 100072. https://doi.org/10.1016/j.apsadv.2021.100072

[24] U. Shahzad, M. Saeed, H.M. Marwani, J.Y. Al-Humaidi, S.u. Rehman, R.H. Althomali, M.R. Awual, M.M. Rahman, Recent progress on potentiometric sensor applications based on nanoscale metal oxides: a comprehensive review, Critical Reviews in Analytical Chemistry (2024) 1-18. https://doi.org/10.1080/10408347.2024.2337876

[25] I.-H. Cho, J. Lee, J. Kim, M.-s. Kang, J.K. Paik, S. Ku, H.-M. Cho, J. Irudayaraj, D.-H. Kim, Current technologies of electrochemical immunosensors: Perspective on signal amplification, Sensors 18(1) (2018) 207. https://doi.org/10.3390/s18010207

[26] F.E. Ab Latif, A. Numan, N.M. Mubarak, M. Khalid, E.C. Abdullah, N.A. Manaf, R. Walvekar, Evolution of MXene and its 2D heterostructure in electrochemical sensor applications, Coordination Chemistry Reviews 471 (2022) 214755. https://doi.org/10.1016/j.ccr.2022.214755

[27] P.-S. Ganesh, S.-Y. Kim, Electrochemical sensing interfaces based on novel 2D-MXenes for monitoring environmental hazardous toxic compounds: A concise review, Journal of Industrial and Engineering Chemistry 109 (2022) 52-67. https://doi.org/10.1016/j.jiec.2022.02.006

[28] N.H. Solangi, N.M. Mubarak, R.R. Karri, S.A. Mazari, A.S. Jatoi, Advanced growth of 2D MXene for electrochemical sensors, Environmental Research 222 (2023) 115279. https://doi.org/10.1016/j.envres.2023.115279

[29] A. Rhouati, M. Berkani, Y. Vasseghian, N. Golzadeh, MXene-based electrochemical sensors for detection of environmental pollutants: A comprehensive review, Chemosphere 291 (2022) 132921. https://doi.org/10.1016/j.chemosphere.2021.132921

[30] M. Mathew, C.S. Rout, Electrochemical biosensors based on Ti3C2Tx MXene: future perspectives for on-site analysis, Curr. Opin. Electrochem 30 (2021) 100782. https://doi.org/10.1016/j.coelec.2021.100782

[31] A.Y.S. Tan, H.T.A. Awan, F. Cheng, M. Zhang, M.T. Tan, S. Manickam, M. Khalid, K. Muthoosamy, Recent advances in the use of MXenes for photoelectrochemical sensors, Chemical Engineering Journal 482 (2024) 148774. https://doi.org/10.1016/j.cej.2024.148774

[32] Z. Otgonbayar, W.-C. Oh, Comprehensive and multi-functional MXene based sensors: An updated review, FlatChem 40 (2023) 100524. https://doi.org/10.1016/j.flatc.2023.100524

[33] J. Zhu, E. Ha, G. Zhao, Y. Zhou, D. Huang, G. Yue, L. Hu, N. Sun, Y. Wang, L.Y.S. Lee, Recent advance in MXenes: a promising 2D material for catalysis, sensor and chemical adsorption, Coordination Chemistry Reviews 352 (2017) 306-327. https://doi.org/10.1016/j.ccr.2017.09.012

[34] Z.-H. Jin, Y.-L. Liu, J.-J. Chen, S.-L. Cai, J.-Q. Xu, W.-H. Huang, Conductive polymer-coated carbon nanotubes to construct stretchable and transparent electrochemical sensors, Analytical chemistry 89(3) (2017) 2032-2038. https://doi.org/10.1021/acs.analchem.6b04616

[35] H. Xu, J. Zhu, T. Zhao, S. Ni, Y. Yang, Q. Hu, X. Jin, Facile design of structurally robust, highly conductive and well-flexible hybrid film based on MXene, cellulose nanofiber and poly (3, 4-ethylenedioxythiphoenes): polystyrene sulfonate for supercapacitors, Journal of Colloid and Interface Science 670 (2024) 163-173. https://doi.org/10.1016/j.jcis.2024.05.079

[36] F. Gao, J. Song, Z. Xu, L. Xu, Y. Guo, L. Miao, X. Luo, All-polymer free-standing electrodes for flexible electrochemical sensors, Sens. Actuators B: Chem 334 (2021) 129675. https://doi.org/10.1016/j.snb.2021.129675

[37] Y. Pan, J. Zhang, X. Guo, Y. Li, L. Li, L. Pan, Recent advances in conductive polymers-based electrochemical sensors for biomedical and environmental applications, Polymers 16(11) (2024) 1597. https://doi.org/10.3390/polym16111597

[38] F. Cui, H. Jafarishad, Z. Zhou, J. Chen, J. Shao, Q. Wen, Y. Liu, H.S. Zhou, Batch fabrication of electrochemical sensors on a glycol-modified polyethylene terephthalate-based microfluidic device, Biosens. Bioelectron. 167 (2020) 112521. https://doi.org/10.1016/j.bios.2020.112521

[39] K. Gmucová, Fundamental aspects of organic conductive polymers as electrodes, Curr. Opin. Electrochem 36 (2022) 101117. https://doi.org/10.1016/j.coelec.2022.101117

[40] B. Lakard, Electrochemical biosensors based on conducting polymers: A review, Applied Sciences 10(18) (2020) 6614. https://doi.org/10.3390/app10186614

[41] N. Abhishek, A. Verma, A. Singh, T. Kumar, Metal-conducting polymer hybrid composites: A promising platform for electrochemical sensing, Inorganic Chemistry Communications 157 (2023) 111334. https://doi.org/10.1016/j.inoche.2023.111334

[42] N. Wongkaew, M. Simsek, C. Griesche, A.J. Baeumner, Functional nanomaterials and nanostructures enhancing electrochemical biosensors and lab-on-a-chip performances: recent progress, applications, and future perspective, Chemical reviews 119(1) (2018) 120-194. https://doi.org/10.1021/acs.chemrev.8b00172

[43] C. Zhu, G. Yang, H. Li, D. Du, Y. Lin, Electrochemical sensors and biosensors based on nanomaterials and nanostructures, Anal. Chem 87(1) (2015) 230-249. https://doi.org/10.1021/ac5039863

[44] X. Jia, S. Dong, E. Wang, Engineering the bioelectrochemical interface using functional nanomaterials and microchip technique toward sensitive and portable electrochemical biosensors, Biosens. Bioelectron. 76 (2016) 80-90. https://doi.org/10.1016/j.bios.2015.05.037

[45] G. Maduraiveeran, M. Sasidharan, V. Ganesan, Electrochemical sensor and biosensor platforms based on advanced nanomaterials for biological and biomedical applications, Biosens. Bioelectron. 103 (2018) 113-129. https://doi.org/10.1016/j.bios.2017.12.031

[46] T. Lin, Y. Xu, A. Zhao, W. He, F. Xiao, Flexible electrochemical sensors integrated with nanomaterials for in situ determination of small molecules in biological samples: a review, Anal. Chim. Acta 1207 (2022) 339461. https://doi.org/10.1016/j.aca.2022.339461

[47] W. Nawrot, K. Malecha, Biomaterial embedding process for ceramic-polymer microfluidic sensors, Sensors 20(6) (2020) 1745. https://doi.org/10.3390/s20061745

[48] S. Ahmed, S. Lim, Review of recent metamaterial microfluidic sensors, Sensors 18(1) (2018) 232. https://doi.org/10.3390/s18010232

[49] I.-H. Cho, D.H. Kim, S. Park, Electrochemical biosensors: Perspective on functional nanomaterials for on-site analysis, Biomaterials research 24(1) (2020) 6. https://doi.org/10.1186/s40824-019-0181-y

[50] A. Walcarius, S.D. Minteer, J. Wang, Y. Lin, A. Merkoçi, Nanomaterials for bio-functionalized electrodes: recent trends, J. Mater. Chem. B 1(38) (2013) 4878-4908. https://doi.org/10.1039/c3tb20881h

[51] K. Reddy, H. Bandal, M. Satyanarayana, K. Goud, K. Gobi, T. Jayaramudu, J. Amalraj, H. Kim, Recent trends in electrochemical sensors for vital biomedical markers using hybrid nanostructured materials, Advanced Science 7 (2020). https://doi.org/10.1002/advs.201902980

[52] R. Nishimoto, Y. Sato, J. Wu, T. Saizaki, M. Kubo, M. Wang, H. Abe, I. Richard, T. Yoshinobu, F. Sorin, Thermally drawn CNT-based hybrid nanocomposite fiber for electrochemical sensing, Biosens. J 12(8) (2022) 559. https://doi.org/10.3390/bios12080559

[53] J. Wang, M. Musameh, Carbon nanotube/teflon composite electrochemical sensors and biosensors, Anal. Chem 75(9) (2003) 2075-2079. https://doi.org/10.1021/ac030007+

[54] R. Batool, A. Rhouati, M.H. Nawaz, A. Hayat, J.L. Marty, A review of the construction of nano-hybrids for electrochemical biosensing of glucose, Biosensors 9(1) (2019) 46. https://doi.org/10.3390/bios9010046

[55] S. Zhang, M.A. Zahed, M. Sharifuzzaman, S. Yoon, X. Hui, S.C. Barman, S. Sharma, H.S. Yoon, C. Park, J.Y. Park, A wearable battery-free wireless and skin-interfaced microfluidics integrated electrochemical sensing patch for on-site biomarkers monitoring in human perspiration, Biosensors and Bioelectronics 175 (2021) 112844. https://doi.org/10.1016/j.bios.2020.112844

[56] Q. Zhang, S. Ma, K. Zhang, L. Zhang, C. Liu, H. Shi, C. Wang, N. Wang, A. Zhu, A facile integrated microfluidic chip based on Chitosan-Gold Nanoparticles-Anchored Three-Dimensional graphene fiber film for monitoring prostate specific antigen, Microchemical Journal 184 (2023) 108171. https://doi.org/10.1016/j.microc.2022.108171

[57] X. Li, Z. Chen, Y. Zhong, F. Yang, J. Pan, Y. Liang, Cobalt hexacyanoferrate modified multi-walled carbon nanotubes/graphite composite electrode as electrochemical sensor on microfluidic chip, Anal. Chim. Acta 710 (2012) 118-124. https://doi.org/10.1016/j.aca.2011.10.035

[58] M. Salve, A. Mandal, K. Amreen, P.K. Pattnaik, S. Goel, Greenly synthesized silver nanoparticles for supercapacitor and electrochemical sensing applications in a 3D printed microfluidic platform, Microchemical Journal 157 (2020) 104973. https://doi.org/10.1016/j.microc.2020.104973

[59] K.J. Klunder, K.M. Clark, C. McCord, K.E. Berg, S.D. Minteer, C.S. Henry, Polycaprolactone-enabled sealing and carbon composite electrode integration into electrochemical microfluidics, Lab on a Chip 19(15) (2019) 2589-2597. https://doi.org/10.1039/C9LC00417C

[60] D. Soto, J. Orozco, Hybrid Nanobioengineered nanomaterial-based electrochemical biosensors, Molecules 27(12) (2022) 3841. https://doi.org/10.3390/molecules27123841

[61] D. Zhao, Q. Zhang, W. Chen, X. Yi, S. Liu, Q. Wang, Y. Liu, J. Li, X. Li, H. Yu, Highly flexible and conductive cellulose-mediated PEDOT: PSS/MWCNT composite films for supercapacitor electrodes, ACS Appl. Mater. Interfaces 9(15) (2017) 13213-13222. https://doi.org/10.1021/acsami.7b01852

[62] G. Luka, A. Ahmadi, H. Najjaran, E. Alocilja, M. DeRosa, K. Wolthers, A. Malki, H. Aziz, A. Althani, M. Hoorfar, Microfluidics integrated biosensors: A leading technology towards lab-on-a-chip and sensing applications, Sensors 15(12) (2015) 30011-30031. https://doi.org/10.3390/s151229783

[63] C. Chaoran, Electrochemical biosensors based on microfluidic chips, Second International Conference on Biological Engineering and Medical Science (ICBioMed 2022), 2023, p. 126110Y. https://doi.org/10.1117/12.2669567

[64] M. Javanmard, H. Esfandyarpour, F. Pease, R. Davis, Electrical detection of proteins and DNA using bioactivated microfluidic channels: Theoretical and experimental considerations, J. Vac. Sci. Technol., B:Microelectron. Nanometer Struct. Process., Meas., Phenom 27(6) (2009) 3099-3103. https://doi.org/10.1116/1.3264675

[65] H.S. Magar, R.Y. Hassan, A. Mulchandani, Electrochemical impedance spectroscopy (EIS): Principles, construction, and biosensing applications, Sensors 21(19) (2021) 6578. https://doi.org/10.3390/s21196578

[66] S.P. Kounaves, Voltammetric techniques, Handbook of instrumental techniques for analytical chemistry (1997) 709-726.

[67] M.D. Dryden, D.D. Rackus, M.H. Shamsi, A.R. Wheeler, Integrated digital microfluidic platform for voltammetric analysis, Anal. Chem 85(18) (2013) 8809-8816. https://doi.org/10.1021/ac402003v

[68] J. Wang, R. Polsky, B. Tian, M.P. Chatrathi, Voltammetry on microfluidic chip platforms, Anal. Chem 72(21) (2000) 5285-5289. https://doi.org/10.1021/ac000484h

[69] J. Liu, Y. Xu, S. Liu, S. Yu, Z. Yu, S.S. Low, Application and progress of chemometrics in voltammetric biosensing, Biosensors 12(7) (2022) 494. https://doi.org/10.3390/bios12070494

[70] P.T. Kissinger, W.R. Heineman, Cyclic voltammetry, J. Chem. Educ 60(9) (1983) 702. https://doi.org/10.1021/ed060p702

[71] K.Y. Shin, W. Oum, D.J. Yu, S. Kang, E.B. Kim, H.W. Kim, Fundamentals of cyclic voltammetry, J. Sens. Sci. Technol 30(6) (2021) 384-387. https://doi.org/10.46670/JSST.2021.30.6.384

[72] S.A. Vijaya Durga CH, Ajitha A and Umamaheswara Rao V, An overview on cyclic voltammetry and its application in pharmaceutical analysis, IJCPS 5(2) (2014).

[73] W. Dungchai, O. Chailapakul, C.S. Henry, Electrochemical detection for paper-based microfluidics, Anal. Chem 81(14) (2009) 5821-5826. https://doi.org/10.1021/ac9007573

[74] D. Ji, L. Liu, S. Li, C. Chen, Y. Lu, J. Wu, Q. Liu, Smartphone-based cyclic voltammetry system with graphene modified screen printed electrodes for glucose detection, Biosens. Bioelectron. 98 (2017) 449-456. https://doi.org/10.1016/j.bios.2017.07.027

[75] L.Z. Lugo-Morales, P.L. Loziuk, A.K. Corder, J.V. Toups, J.G. Roberts, K.A. McCaffrey, L.A. Sombers, Enzyme-modified carbon-fiber microelectrode for the quantification of dynamic fluctuations of nonelectroactive analytes using fast-scan cyclic voltammetry, Anal. Chem 85(18) (2013) 8780-8786. https://doi.org/10.1021/ac4017852

[76] J. Pelzer, F. Scholz, G. Henrion, L. Nitschke, Optimization of parameters for differential pulse voltammetry at the hanging mercury drop electrode, Electroanalysis 1(5) (1989) 437-440. https://doi.org/10.1002/elan.1140010510

[77] M.S. Lin, B.I. Jan, A sensitive measurement of maleic hydrazide by a differential pulse voltammetry followed by a cleansing pulse train, J. Chin. Chem. Soc 46(6) (1999) 879-884. https://doi.org/10.1002/jccs.199900119

[78] Y. Dai, C.C. Liu, Detection of 17 β-estradiol in environmental samples and for health care using a single-use, cost-effective biosensor based on differential pulse voltammetry (DPV), Biosens. J 7(2) (2017) 15. https://doi.org/10.3390/bios7020015

[79] A.F.D. Cruz, N. Norena, A. Kaushik, S. Bhansali, A low-cost miniaturized potentiostat for point-of-care diagnosis, Biosens. Bioelectron. 62 (2014) 249-254. https://doi.org/10.1016/j.bios.2014.06.053

[80] M. Jamshidi, A. Walcarius, M. Thangamuthu, M. Mehrgardi, A. Ranjbar, Electrochemical approaches based on micro-and nanomaterials for diagnosing oxidative stress, Microchim. Acta 190(4) (2023) 117. https://doi.org/10.1007/s00604-023-05681-7

[81] E. Costa-Rama, M.T.F. Abedul, Electrochemical behavior of the redox probe hexaammineruthenium (III)([Ru (NH3) 6] 3+) using voltammetric techniques, Laboratory Methods in Dynamic Electroanalysis, Elsevier2020, pp. 25-33. https://doi.org/10.1016/B978-0-12-815932-3.00003-6

[82] A. Yakushenko, V. Schöps, D. Mayer, A. Offenhäusser, B. Wolfrum, On-chip fast scan cyclic voltammetry for selective detection of redox active neurotransmitters, Phys. Status Solidi (a) 211(6) (2014) 1364-1371. https://doi.org/10.1002/pssa.201330276

[83] A. Chen, B. Shah, Electrochemical sensing and biosensing based on square wave voltammetry, Anal.Methods 5(9) (2013) 2158-2173. https://doi.org/10.1039/c3ay40155c

[84] B. Dogan-Topal, S.A. Ozkan, B. Uslu, The analytical applications of square wave voltammetry on pharmaceutical analysis, Open Chem. Biomed. Meth. J 3(1) (2010) 56-73. https://doi.org/10.2174/1875038901003010056

[85] P.R. Tuladhar, W. Lowery, D.E. Cliffel, Parallel amperometric and potentiometric multianalyte detection in a microfluidic system, Electrochemical Society Meeting Abstracts 244, The Electrochemical Society, Inc., 2023, pp. 2996-2996. https://doi.org/10.1149/MA2023-02632996mtgabs

[86] R. Bocu, Extended review concerning the integration of electrochemical biosensors into modern IoT and wearable devices, Biosens. J 14(5) (2024) 214. https://doi.org/10.3390/bios14050214

[87] A.L. Campaña, S.L. Florez, M.J. Noguera, O.P. Fuentes, P. Ruiz Puentes, J.C. Cruz, J.F. Osma, Enzyme-based electrochemical biosensors for microfluidic platforms to detect pharmaceutical residues in wastewater, Biosens. J 9(1) (2019) 41. https://doi.org/10.3390/bios9010041

[88] S. Mross, S. Pierrat, T. Zimmermann, M. Kraft, Microfluidic enzymatic biosensing systems: A review, Biosens. Bioelectron. 70 (2015) 376-391. https://doi.org/10.1016/j.bios.2015.03.049

[89] R. Dudde, G. Piechotta, R. Hintsche, Micromachined amperometric cells for continuous monitoring of glucose and lactate, 1st Transdisciplinary Conference on Distributed Diagnosis and Home Healthcare, 2006. D2H2., IEEE, 2006, pp. 1-4. https://doi.org/10.1109/DDHH.2006.1624782

[90] P.M. Misun, J. Rothe, Y.R. Schmid, A. Hierlemann, O. Frey, Multi-analyte biosensor interface for real-time monitoring of 3D microtissue spheroids in hanging-drop networks, Microsyst. Nanoeng 2(1) (2016) 1-9. https://doi.org/10.1038/micronano.2016.22

[91] M.G. Boutelle, H. Allen, O. Hill, M. Berners, R. John, P. Dobson, P. Leigh, New technologies for amperometric biosensors, JMR 9(5-6) (1996) 664-671. https://doi.org/10.1002/(SICI)1099-1352(199634/12)9:5/6<664::AID-JMR318>3.0.CO;2-U

[92] R. Ding, M. Fiedoruk-Pogrebniak, M. Pokrzywnicka, R. Koncki, J. Bobacka, G. Lisak, Solid reference electrode integrated with paper-based microfluidics for potentiometric ion sensing, Sens. Actuators B: Chem 323 (2020) 128680. https://doi.org/10.1016/j.snb.2020.128680

[93] E.B. Bahadır, M.K. Sezgintürk, A review on impedimetric biosensors, Artif. Cells Nanomed. Biotechnol 44(1) (2016) 248-262. https://doi.org/10.3109/21691401.2014.942456

[94] E. Katz, I. Willner, Probing biomolecular interactions at conductive and semiconductive surfaces by impedance spectroscopy: routes to impedimetric immunosensors, DNA-sensors,

and enzyme biosensors, Electroanalysis 15(11) (2003) 913-947. https://doi.org/10.1002/elan.200390114

[95] J.-Y. Park, S.-M. Park, DNA hybridization sensors based on electrochemical impedance spectroscopy as a detection tool, Sensors 9(12) (2009) 9513-9532. https://doi.org/10.3390/s91209513

[96] Y.-S. Chen, C.-H. Huang, P.-C. Pai, J. Seo, K.F. Lei, A review on microfluidics-based impedance biosensors, Biosens. J 13(1) (2023) 83. https://doi.org/10.3390/bios13010083

[97] M.E. Strong, J.R. Richards, M. Torres, C.M. Beck, J.T. La Belle, Faradaic electrochemical impedance spectroscopy for enhanced analyte detection in diagnostics, Biosens. Bioelectron. 177 (2021) 112949. https://doi.org/10.1016/j.bios.2020.112949

[98] K. Alsabbagh, T. Hornung, A. Voigt, S. Sadir, T. Rajabi, K. Länge, Microfluidic impedance biosensor chips using sensing layers based on DNA-based self-assembled monolayers for label-free detection of proteins, Biosens. J 11(3) (2021) 80. https://doi.org/10.3390/bios11030080

[99] M. Frigoli, M.P. Krupa, G. Hooyberghs, J.W. Lowdon, T.J. Cleij, H. Diliën, K. Eersels, B. van Grinsven, Electrochemical sensors for antibiotic detection: A focused review with a brief overview of commercial technologies, Sensors 24(17) (2024) 5576. https://doi.org/10.3390/s24175576

[100] T.X. Hoang, L.M.T. Phan, T.A.T. Vo, S. Cho, Advanced signal-amplification strategies for paper-based analytical devices: A comprehensive review, Biomedicines 9(5) (2021) 540. https://doi.org/10.3390/biomedicines9050540

[101] J. Lei, H. Ju, Signal amplification using functional nanomaterials for biosensing, Chem. Soc. Rev 41(6) (2012) 2122-2134. https://doi.org/10.1039/c1cs15274b

[102] J. Wang, Nanomaterial-based amplified transduction of biomolecular interactions, Small 1(11) (2005) 1036-1043. https://doi.org/10.1002/smll.200500214

[103] A.M. Ashrafi, Z. Bytesnikova, J. Barek, L. Richtera, V. Adam, A critical comparison of natural enzymes and nanozymes in biosensing and bioassays, Biosens. Bioelectron. 192 (2021) 113494. https://doi.org/10.1016/j.bios.2021.113494

[104] A. Walter, J. Wu, G.-U. Flechsig, D.A. Haake, J. Wang, Redox cycling amplified electrochemical detection of DNA hybridization: application to pathogen E. coli bacterial RNA, Anal. Chim. Acta 689(1) (2011) 29-33. https://doi.org/10.1016/j.aca.2011.01.014

[105] A. Miti, G. Zuccheri, Hybridization chain reaction design and biosensor implementation, DNA Nanotechnology: Methods and Protocols (2018) 115-135. https://doi.org/10.1007/978-1-4939-8582-1_8

[106] Z. Liu, W. Wang, R. Xie, X.-J. Ju, L.-Y. Chu, Stimuli-responsive smart gating membranes, Chem. Soc. Rev 45 (2016) 460-475. https://doi.org/10.1039/C5CS00692A

[107] V. Bocharova, E. Katz, Switchable electrode interfaces controlled by physical, chemical and biological signals, Chem. Rec 12(1) (2012) 114-130. https://doi.org/10.1002/tcr.201100025

[108] X. Wang, D. Lu, Y. Liu, W. Wang, R. Ren, M. Li, D. Liu, Y. Liu, Y. Liu, G. Pang, Electrochemical signal amplification strategies and their use in olfactory and taste evaluation, Biosens. J 12(8) (2022) 566. https://doi.org/10.3390/bios12080566

[109] L. Bezinge, A. Suea-Ngam, A.J. deMello, C.-J. Shih, Nanomaterials for molecular signal amplification in electrochemical nucleic acid biosensing: Recent advances and future prospects for point-of-care diagnostics, Mol. syst. des. eng 5(1) (2020) 49-66. https://doi.org/10.1039/C9ME00135B

[110] Q. Wang, S. Liu, Z. Tang, Recent progress in the design of analytical methods based on nanozymes, J. Mater. Chem. B 9(39) (2021) 8174-8184. https://doi.org/10.1039/D1TB01521D

[111] W. Song, B. Zhao, C. Wang, Y. Ozaki, X. Lu, Functional nanomaterials with unique enzyme-like characteristics for sensing applications, J. Mater. Chem. B 7(6) (2019) 850-875. https://doi.org/10.1039/C8TB02878H

[112] H.Y. Shin, T.J. Park, M.I. Kim, Recent research trends and future prospects in nanozymes, J. Nanomater 1(2015) 756278. https://doi.org/10.1155/2015/756278

[113] M.R. Akanda, H. Ju, An integrated redox cycling for electrochemical enzymatic signal enhancement, Analytical Chemistry 89(24) (2017) 13480-13486. https://doi.org/10.1021/acs.analchem.7b03802

[114] E. Kim, T. Gordonov, W.E. Bentley, G.F. Payne, Amplified and in situ detection of redox-active metabolite using a biobased redox capacitor, Anal. Chem 85(4) (2013) 2102-2108. https://doi.org/10.1021/ac302703y

[115] Y. Liu, E. Kim, I.M. White, W.E. Bentley, G.F. Payne, Information processing through a bio-based redox capacitor: Signatures for redox-cycling, Bioelectrochemistry 98 (2014) 94-102. https://doi.org/10.1016/j.bioelechem.2014.03.012

[116] N.-I. Goo, D.-E. Kim, Rolling circle amplification as isothermal gene amplification in molecular diagnostics, Biochip J 10 (2016) 262-271. https://doi.org/10.1007/s13206-016-0402-6

[117] H. Chai, W. Cheng, D. Jin, P. Miao, Recent progress in DNA hybridization chain reaction strategies for amplified biosensing, ACS Appl. Mater. Interfaces 13(33) (2021) 38931-38946. https://doi.org/10.1021/acsami.1c09000

[118] B. Ziółkowski, M. Czugala, D. Diamond, Integrating stimulus responsive materials and microfluidics: The key to next-generation chemical sensors, J. Intell. Mater. Syst. Struct 24(18) (2013) 2221-2238. https://doi.org/10.1177/1045389X12459591

[119] S. Sakiyama-Elbert, J. Hubbell, Functional biomaterials: design of novel biomaterials, Annu. Rev. Mater. Res 31(1) (2001) 183-201. https://doi.org/10.1146/annurev.matsci.31.1.183

[120] N. Xiang, Z. Ni, Microfluidics for biomedical applications, MDPI, (2023) 161. https://doi.org/10.3390/books978-3-0365-7127-0

[121] F. Cui, Z. Zhou, H.S. Zhou, Measurement and analysis of cancer biomarkers based on electrochemical biosensors, J. Electrochem. Soc 167(3) (2019) 037525. https://doi.org/10.1149/2.0252003JES

[122] M.A. Ali, S. Tabassum, Q. Wang, Y. Wang, R. Kumar, L. Dong, Plasmonic-electrochemical dual modality microfluidic sensor for cancer biomarker detection, 2017 IEEE 30th International Conference on Micro Electro Mechanical Systems (MEMS), IEEE, 2017, pp. 390-393. https://doi.org/10.1109/MEMSYS.2017.7863423

[123] Z. Li, A. Zhong, Electrochemical detection strategies for cardiovascular disease biomarkers: Applications in troponin I and troponin T assays, Int. J. Electrochem. Sci (2024) 100675. https://doi.org/10.1016/j.ijoes.2024.100675

[124] A. Sharma, W.S. AlGhamdi, H. Faber, T.D. Anthopoulos, Based microfluidics devices with integrated nanostructured materials for glucose detection, Advanced Microfluidics Based Point-of-Care Diagnostics, CRC Press2022, pp. 191-228. https://doi.org/10.1201/9781003033479-8

[125] M.B. Kulkarni, N.H. Ayachit, T.M. Aminabhavi, Recent advances in microfluidics-based electrochemical sensors for foodborne pathogen detection, Biosens. J 13(2) (2023) 246. https://doi.org/10.3390/bios13020246

[126] O. Zolti, B. Suganthan, R.P. Ramasamy, Lab-on-a-chip electrochemical biosensors for foodborne pathogen detection: A review of common standards and recent progress, Biosens. J 13(2) (2023) 215. https://doi.org/10.3390/bios13020215

[127] T.O. Hara, B. Singh, Electrochemical biosensors for detection of pesticides and heavy metal toxicants in water: recent trends and progress, ACS. ES&T. Water 1(3) (2021) 462-478. https://doi.org/10.1021/acsestwater.0c00125

[128] B. Ramachandran, Y.-C. Liao, Microfluidic wearable electrochemical sweat sensors for health monitoring, Biomicrofluidics 16(5) (2022) 051501. https://doi.org/10.1063/5.0116648

Materials Research Foundations 191 (2026) 260-288 https://doi.org/10.21741/9781644904213-8

Chapter 8

Microfluidics-Based Immunosensors

M. Muntasa[1], Aimen Areej[2], Bushra Bibi[2], Sundus Anbreen[2], Aimen Fatima[1], Adeel Afzal[3], Asma Rehman[1], Waheed S. Khan[1] and Sadia Z. Bajwa[1*]

[1]Nanobiotech Group, National Institute for Biotechnology and Genetic Engineering (NIBGE) 38000, Faisalabad, Pakistan

[2]Department of Biotechnology, Quaid-e-Azam University, 15320, Islamabad

[3]School of Chemistry, University of the Punjab, Quaid-i-Azam Campus, Lahore 54590, Pakistan

*sadya2002pk@yahoo.co.uk

Abstract

Microfluidic immunosensors offer new possibilities for bioanalysis in a quick, sensitive, and compact way by fusing the precision of microfluidics with the selectivity of immunoassay. These systems work with a variety of substrates, including glass, paper, and polydimethylsiloxane (PDMS), using microscale fluid handling techniques like capillary, electrokinetic, and pressure-driven flow. ELISA-on-a-chip, electrochemical, and optical/SPR sensors are among the other formats that have been claimed to provide low-volume, real-time detection. Recent developments in fabrication, including photolithography, 3D printing, soft lithography, and paper-based methods, have significantly enhanced and extended the capabilities of capillary fluidic systems. Microfluidic immunosensors are transforming POC diagnostics, food safety, and biomedical research with their portability, cost-effectiveness, and integration advantages, despite sensitivity and scaling issues.

Keywords

Microfluidics, Immunosensor, Point of Care Diagnostic, Paper-Based Microfluidics, Capillary Flow, Lab-on-a-Chip

Contents

Materials Research Foundations 191 (2026) 260-288 https://doi.org/10.21741/9781644904213-8

1. Introduction

Microfluidic biosensing systems typically include a collection of fluid manipulation units that demonstrate and test a variety of biomolecules. To perform efficiently, microfluidic immunosensors require a low viscosity carrier fluid that flows easily through microscopic channels while remaining biocompatible to retain antibody-antigen interactions [1]. Surface tension is carefully controlled to prevent disruptive bubble formation, while exact pH/ionic balance maintains biomolecular stability throughout detection. These well-adjusted physical and chemical

Materials Research Foundations 191 (2026) 260-288 https://doi.org/10.21741/9781644904213-8

features work together to provide precise immunosensing in the microfluidic environment and allow for self-driving circuits. Both are essential for automated bioassay applications and are used in microfluidic biosensors [2]. Microfluidic biosensors combine microfluidic and biological techniques for the diagnosis of disease, such as liquid biopsy, pathogen identification, and local testing, as shown in Figure 1. It is known as an integrated lab-on-a-chip technology, which integrates chemical surface treatments with biological analysis in a tiny platform usually spanning a few square centimeters. It detects biological activity by converting protein interactions into quantitative outputs, which are then read out using sound, light, electrical impulses, or magnetic fields. These techniques combine molecular biology with engineering, allowing for a wide range of diagnostic data collection and interpretation methods [3].

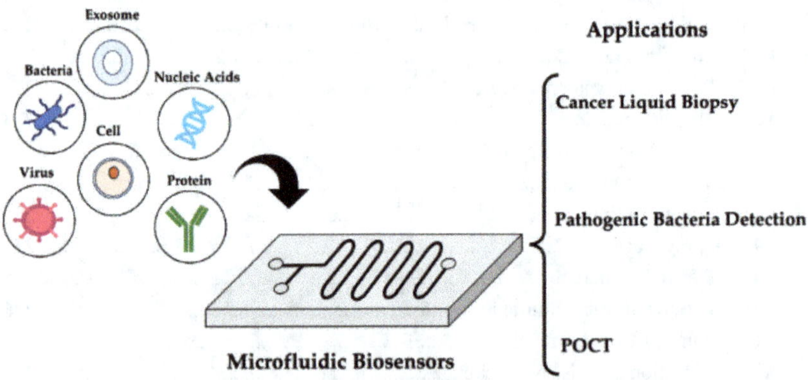

Fig.1 Microfluidic biosensors for identifying biological components, with emphasis on liquid biopsy applications in oncology as well as exosome characterization (covering microbial, genetic, cellular, viral, protein biomarkers), and point of care testing for rapid identification, demonstrating their significance in clinical analysis

The evolution of microfluidic devices is to mimic the functionality of the human organs on a chip. These complex systems are referred to as organ-on-a-chip (OoC) systems, which combine different known microfluidic elements into one in vitro system, enabling the simulation of both normal and diseased physiological functions Examples include gut-on-chip, lung-on-chip, blood vessel-on-chip, kidney-on-chip, and cancer-on-chip technologies. In the future, all these can be linked together to form a full human-on-a-chip model that recapitulates the coordinated actions of various organs. Such a system could significantly improve the ability to predict how new drugs will behave in the human body, potentially reducing or even replacing the need for animal testing. This innovation would mark the development of a sophisticated class of in vitro assays for preclinical research [4]. Other OoCs have used superior sensors to detect parameters like temperature, O_2/CO_2 concentrations, pH, and resistance to transepithelial electric current. Such in situ sensing systems give crucial data concerning the microenvironment, biological characteristics, and state of cells and tissues in the OoC [5].

Materials Research Foundations 191 (2026) 260-288 https://doi.org/10.21741/9781644904213-8

The microfluidics organ-on-a-chip is one of the personalized medicine paradigm shifts. OoCs offer unparalleled individualization by incorporating living cells from an induced pluripotent stem cell of a patient. OoCs loaded with liver cells of a patient can forecast drug metabolism and detect possible toxicity complications before clinical trials are conducted. The potential of OoC technology in conjunction with personalized cells has tremendous prospects in investigating the effect of drugs on the genetic makeup of human bodies. Modern drug testing provides accuracy for many biological processes, yet it is still unable to fully capture the intricacy of living systems. The trend may be altered through patient-specific methods that will bridge the difference between the isolated models and human physiology as shown in Fig.2 [6]. Following this strategy, the liver-on-a-chip platform integrates a collagen-based hydrogel that imitates the liver's native extracellular matrix, taking advantage of collagen's intrinsic abundance in the tissue. Through the use of biomimetic approaches, this design illustrates the true physiological intricacy of the organ and the complicated vascular architecture of the periportal zone, reconciling the limits of isolated research with whole-organism biology [5].

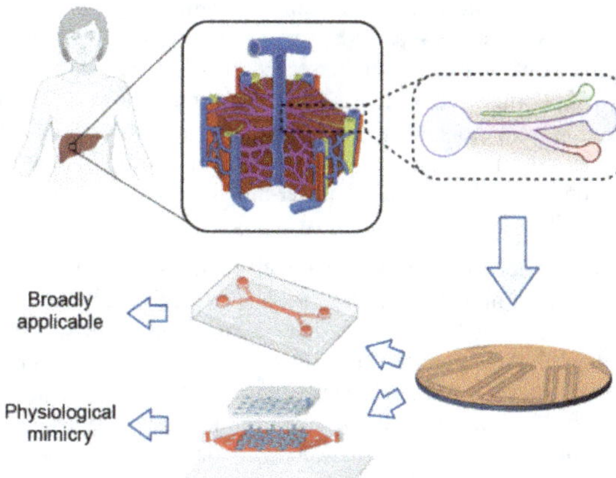

Fig. 2 This cutting-edge microfluidic liver platform uses precisely designed cellular configurations and biochemical conditions to replicate real liver physiology. The system's advanced architecture solves significant flaws in traditional laboratory models by enabling precise pathology investigations and pharmaceutical safety evaluation

The COVID-19 pandemic, caused by the airborne SARS-CoV-2 virus, brought into focus the threat of infectious disease spread. Throughout history, populations have been disproportionately hit by the likes of influenza, HIV, and Ebola disease, and their spread has been accelerated by urbanization and mobility [2]. The emergence of microfluidic technology has transformed traditional approaches to diagnosing infections, reducing the dependence on extensive specimen preparation, manual reagent handling, and complex detection processes. Microfluidic platforms have set a new benchmark for accuracy and precision in identifying pathogens. These systems have proven particularly effective in detecting infectious agents such as COVID-19, tuberculosis, and influenza, positioning them as a cutting-edge solution in modern diagnostics [7].

Materials Research Foundations 191 (2026) 260-288 https://doi.org/10.21741/9781644904213-8

In the case of SARS-CoV-2 virus detection, immunosensors use plasmonic amplification and absorption spectroscopy to increase sensitivity, and in the case of heterogeneous influenza detection, CRISPR-Cas12 is paired with quantum dot-labeled magnetic nanoparticles and electric field acceleration that facilitates greater amplification in signals [8]. Moreover, one of the leading causes of death worldwide is cancer, since it causes about 9.6 million deaths annually (WHO, 2018). Immunosensors use antibodies and antigens in the detection of cancer biomarkers such as proteins and nucleic acids. Even a single molecule in the picolitre dose can be detected by microfluidics forces using surface plasmon resonance and electrokinetic immersion. This mix, in comparison to traditional ELISA, reduces the time of the analyses of hours to minutes and is ultra-trace sensitive [9]. On lab-on-a-chip sensor, a nanostructured gold electrode and toxin-specific antibodies are conjoined to monitor the concentration of mycotoxin in corn and wheat. The toxin decreases signal strength with increase in its concentration due to competitive interaction with HRP-labeled tracer molecules. The portable device is used to test instant food safety using antibody specificity and sensitive nanomaterials [10].

Successful disease control and outbreak containment are highly reliant on early infectious disease diagnosis. In addition, early biosensor-based detection of infectious illness can minimize severity and mortality, particularly in remote and underserved populations, where access to centralized facilities is scarce [11].Point-of-care, rapid, and high-throughput COVID-19 detection is facilitated by microfluidic biosensors, which integrate nanomaterials with miniaturized fluidics for high-density, low-power, and low-cost devices. As compared to traditional lateral flow assays, gold nanoparticle or quantum dot-enhanced paper-based microchannels offer instrument-free laboratory-grade sensitivity [12].

Microfluidic technology offers a cost-effective and sample-efficient approach to molecular diagnostics. This technique minimizes infection risk and avoids invasive procedures by detecting biomarkers in non-invasive fluids like saliva and urine. Additionally, it supports high-throughput analysis, enabling the simultaneous processing of hundreds of samples with automated preparation and detection [13].To achieve increased sensitivity, microfluidics offers a closed biosensing platform which is stable. The impact of an open environment on the sensing performance highly degrades the performance of portable biosensors used on-site. The increased sensitivity and selectivity are provided by the incorporation of the microfluidic structures, where sample treatment and bio sensor functions can take place in a confined and somewhat stable environment [14].

2. Fundamentals of microfluidics

Microfluidics uses precise control of microscopic fluid volumes within hair-thin channels, allowing for complex biochemical studies on a miniature scale. This technology transforms diagnostics, drug research, and lab automation with unparalleled efficiency by leveraging unique microscale characteristics [1].

2.1 Principles of microfluidics

Microfluidic immunosensors use antibody-coated microchannels to detect biomarkers in small fluid quantities. They combine microfluidics, which allows for precise fluid flow control, with immunosensing, which involves binding antibodies to antigens, resulting in speedy and sensitive diagnoses. Detection methods include optical (such as fluorescence), electrochemical (changes in current), and mechanical (bending of cantilevers) signals [15].

Materials Research Foundations 191 (2026) 260-288 https://doi.org/10.21741/9781644904213-8

2.2 Types of microfluidic platforms

Selecting the right material is one of the most important step for device fabrication. At micro level, the properties, such as laminar flow rates, heat to mass transfer rates, and surface to volume ratio, are much more enhanced as compared to the macroscale and are therefore substantially affected by the type of material used. Various other characteristics, such as durability, biocompatibility, surface functionalization potential, and pressure conditions, are considered while choosing the right material as a microfluidic platform Commonly used microfluidic platforms are made up of glass, ceramic, silicon, metal, and polymer, and each one of these platforms has its merits and demerits depending on its use [16].

2.2.1 Glass

Glass based microfluidic system is widely recognized because of their high optical transparency chemical compatibility, and thermal resistance as shown Fig.3. It utilizes capillary-driven flow. Particularly, glass-based microfluidic platforms are valuable for diagnostic purposes and biological assays as compared to polydimethylsiloxane (PDMS) and polymethylmethacrylate (PMMA) materials. There are usually four types of glass materials used for the fabrication of microfluidic devices, such as soda-lime glass, borosilicate glass, quartz silica, and alkali-free glass. Glass-based system provides high optical transparency. Different fabrication techniques are used, including etching and silicon integration. In the etching method, microstructures are developed using glass, then it is bonded to another substrate to form a hybrid layer then active components and efficient valves are developed by fabricating the hybrid layer because of the high elasticity and ability of glass to resist deformation under pressure. Whereas in silicon integration, glass is embedded with silicon to enhance its application in diagnostics. Glass-based microfluidic devices have been used for monitoring the glutathione levels in the supernatants of cell culture. Thus, a glass-based microfluidic platform is an efficient system for diagnostic applications and biological assays [17]. A major drawback of glass is that it is difficult to fabricate in micro-scale chips, which significantly increases the cost due to the requirement of time-consuming labor and clean rooms for synthesis [18].

2.2.2 Paper

Paper-based microfluid platform consists of paper as a building block, a very inexpensive material of cellulose fibers with a high surface-to-volume ratio and 3D microstructures as shown in Fig. 3. The fluid flow mechanism of the paper-based microfluid platform is based on capillary action for fluid movement. Cellulose and nitrocellulose are usually used for microfluidic platforms. Cellulose paper is engineered into microstructures for the conductance of multiple bioreactions without cross-contamination. There are three well-used methods of manufacturing microstructures on paper. They include chemical modification, adsorption of reagents to the surface of the fiber, and plugging of the paper pores. Conductance microfluidic paper-based platforms have been used to conduct cancer research. The capture of targeted cancer cells is done using microfluidic paper-based electrochemiluminescence origami cyto-device (μ-PECLOC). Particularly, it is used with human acute promyelocytic leukemia cells, human breast adenocarcinoma cells, human acute lymphoblastic leukemia cells, and human chronic myelogenous leukemia cells. It enclosed cancer cells with good reproducibility, stability and accuracy [19]. One of the drawbacks of a paper-based system is its wet fragility and thickness requirements to become transparent. The invention of cheap optically transparent polymers such as polydimethylsiloxane (PDMS) largely replaced paper-based material, but they are still used for point-of-care diagnosis [20].

2.2.3 Polydimethylsiloxane (PDMS)

PDMS stands for polydimethylsiloxane, a polymer-based structural material used for in vivo and in vitro studies in biological applications as shown in Fig. 3. It uses capillary-driven and laminar fluid flow mechanisms. PDMS plays a vital role in microfluidic systems as it enables easy and quick integration into glass systems for conducting seamless biological investigations. Mechanical properties of PDMS assist in developing deformable channels. These channels mimic the physiological vessels, facilitating in vitro study of biological fluid-structure interactions (FSI). PDMS acts as a building block for studying vessel models and supporting cellular level microfluid studies, such as endothelial cells have been grown on channel walls for investigation purposes. Moreover, researchers reported the utilization of PDMS based microfluidics to study cellular angiogenesis and vessel sprouting. Different chemical modifications of PDMS tailor it for specific biological applications, like functionalization, enabling the change hydrophobic surface of PDMS into a hydrophilic surface for efficient sensing of cellular processes, such as studying the interaction between cytokine signals. Most commonly, UV, plasma, and salinization are used for modifying the surface of PDMS. UV modifications create a thin oligomer coating on the channel walls of PDMS-based microfluids, plasma modification adds SiOH polar groups to PDMS surface, whereas salinization produces silane monolayers that enhance cell adhesion of PDMS based organ chips. All of these modification improves the functionality of PDMS for applications such as studying surface plasmon resonance and chemical interaction of biomolecules, and protein crystallization. PDMS provides various advantages in microfluidic systems. It provides real-time monitoring, accurate quantification, and quantitative measurements of observed biological phenomena, including cellular coculture research [21]. Like any other material, PDMS has several disadvantages as well; the porosity of its walls causes water evaporation, changing the composition of fluids. Further, porous walls used for the adsorption of material on the surfaces are incompatible with organic solvents as their molecule swell the platform when adsorbed onto the surface [22]. Moreover comparison of different microfluidic substrates highlighting their advantages and shortcomings has been presented in Table 1.

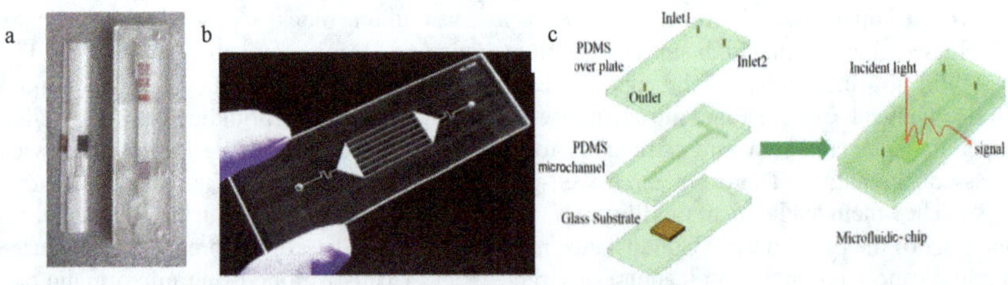

Fig. 3 Three designs for microfluidic platforms are presented a paper-based, which offers inexpensive disposability; b glass, which ensures optical clarity and chemical resistance;c PDMS, which offers flexibility and biocompatibility By utilizing their distinct material features, these platforms facilitate a wide range of applications, from sophisticated lab research to point-of-care diagnostics

Materials Research Foundations 191 (2026) 260-288 https://doi.org/10.21741/9781644904213-8

Table1 Comparison of Microfluidic Platforms

Material	Flow Mechanism	Key Properties	Applications	Fabrication Methods	Disadvantages	Reference
Glass	Capillary driven	High optical transparency Chemical compatibility Thermal resistance Durable	Diagnostic applications Biological assays Monitoring glutathione levels in cell culture supernatants	Etching Silicon integration	Difficult to fabricate at microscale Expensive Requires clean rooms and skilled labour	[17,18]
Paper	Capillary action	Inexpensive (cellulose-based) High surface-to-volume ratio 3D porous structure	Point-of-care diagnostics Cancer cell research	Chemical modification Reagent adsorption Pore plugging	Wet fragility Thickness limits transparency Largely replaced by polymers in advanced studies	[19,20]
PDMS	Capillary driven & laminar flow	Biocompatible polymer Transparent Deformable Tunable surface chemistry	Organ-on-chip Angiogenesis studies Cell culture Protein crystallization Cytokine signal interaction	Molding & soft lithography Surface functionalization (UV, plasma, salinization)	Porous walls cause water evaporation Incompatible with organic solvents Fluid composition may change due to adsorption	[21,22]

2.3 Fluid flow mechanisms

Fluid dynamics inside the working of a microfluidic device. A properly designed microfluidic device has enhanced functionality, adaptability, and performance due to these flows. Mimicking human biology, such as blood flow, using microfluidics helps in understanding cardiovascular diseases and drug development. To date, many techniques for the generation of periodic flow have been studied; understanding these flow mechanisms is essential for designing reliable microfluidic systems.

2.3.1 Capillary flow

The capillary fluid flow mechanism utilizes capillary action for the movement of fluid in the microchannels/capillaries, as shown in Fig. 4. It does not require any external pressure. Fluid flows because of the wetting properties and surface tension of the capillary to overcome the effects of liquid gravity and viscosity. Different factors can affect the rise of liquid in the microchannels, including the type of fluid, internal geometry, and surface properties of the capillary. It is an entirely autonomous fluid handling system. Microfluid based on capillary flow have been used for studying blood viscosity during embryonic development of zebrafish. Such microfluidic systems have potential for application in pharmaceutical and chemical industries as a portable system to analyze their physical and chemical properties [23].

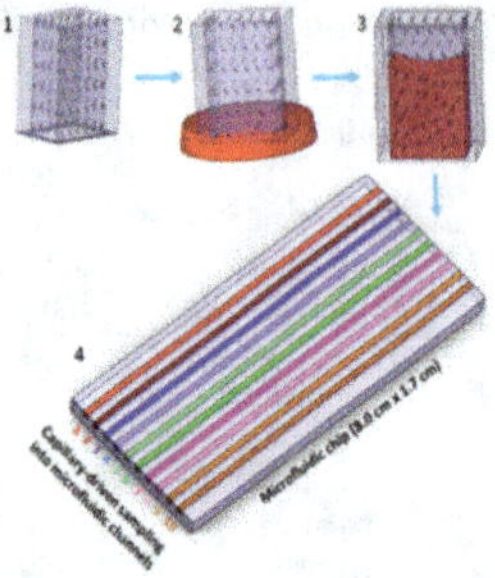

Fig. 4 The microfluidic device take advantage of fundamental capillary action, in which fluid travels through small channels due to surface tension and adhesion forces. Using miniature fluid dynamics, this novel microchannel device quickly identifies bacterial strains and evaluates the efficacy of antibiotics, providing lab-quality data in a small package

2.3.2 Electrokinetic fluid flow

In microfluidics system, an electric field is applied to charge the porous surface of the microfluid for initiating the movement of bulk ionic fluid in the channels. This electrokinetic flow is usually termed as electroosmotic or electroosmosis flow in nano and microfluids. Electrokinetic flow is applied for various technological processes, including the removal of heavy metals from contaminated soil and dewatering waste sludge for environmental remediation [24].

One of the vital parts of microfluid devices is the "micromixer" that contributes to the mixing of samples required for conducting biochemistry analysis, nucleic acid synthesis, and even facilitates drug delivery. A T-shaped microfluid system has been developed that depends on electrokinetic fluid flow mechanisms for mixing of species/samples to conduct biochemical analysis [25].

2.3.3 Pressure-driven fluid flow mechanism

In a microfluid platform, the pressure difference is applied to drive fluid across the microchannels of the microfluid device. It facilitates precise control of fluid flow rates at a microscopic level. Applied pressure is controlled by pressure-driven pumps such as peristaltic pumps or syringe pumps, etc. Traditionally, pressure-driven pumps have been available in a bulky size that limits their integration with microfluidic sensors. Moreover, external forces such as electricity are needed to produce the forces for driving fluid, hindering portable use of traditionally available commercial pumps. So, the structure of pressure pumps has been optimized for their easy integration into miniaturized microfluid biosensors to achieve convenient, accurate, and cost-effective biochemical analysis. For example, a portable plug-and-play syringe pump has been developed for accurate and controlled delivery of fluid. Microfluid sensors can use an air-pressure difference to control fluid flow. Researchers reported the evolution of a microfluidic sensor for conducting digital PCR, which is a self-priming digital microfluidic sensor. Gene templates and primers are injected using atmospheric pressure into the digital microfluid sensor. It is lyophilized and vacuum processed for preservation. Fluid automatically flows into the microchannels of the sensor due to air pressure. In this way, it is used for DNA detection and genetic analysis [26].

Materials Research Foundations 191 (2026) 260-288 https://doi.org/10.21741/9781644904213-8

3. Types of microfluidic immunosensors

Microfluidic immunosensors have microchannels for the movement of fluids and all parts of the immunoassay procedure and are used for testing at the microlevel. There are many types of such microfluidic immunosensors.

3.1 Enzyme-linked immunosorbent assay (ELISA) on a chip

Microfluidic technologies are a promising approach for solving diagnostic problems due to their integration and complete protocol on a chip [27]. A lab-on-chip provides services of a traditional laboratory setting with miniaturization by reducing the volume of reagents and incubation times. The enzyme-linked immunosorbent assay on a chip (ELISA on a chip) is an immunoassay on a chip method for the detection and measurement of particular biomolecules (antibodies or antigens) within complicated samples. It is based on the high-affinity antigen-antibody binding interaction between an antigen (Ag) and an antibody (Ab), where one is immobilized on a solid support (chip) [28]. Either the antigen or the antibody is coated onto the chip in a standard assay. The analyte of interest in the sample is added and then left to bind. It is washed, and then an enzyme-conjugated secondary reagent is added: it could be an antibody that recognizes the bound antigen, or an antigen that recognizes a bound antibody. For the reduction of the volume requirement of a sample following the ELISA technique, this immunoassay has also been performed on the microfluidic device by various researchers with different detection/quantitation methods. For this purpose, a component that is non-fluorescent but acts as an enzyme substrate is utilized, which is converted into a detectable fluorescent product subsequently during the assay [29]. This fluorescent reporter is detected off-chip or after the enzymatic reaction of the ELISA format in the analysis channel.

3.2 Electrochemical immunosensors

Electrochemical immunosensors are affinity-based biosensors in which a specific antibody or antigen (biorecognition element) is immobilized on an electrode surface to capture its complementary analyte [30]. Binding of the target antigen by the antibody (or vice versa) produces a change in the electrochemical properties at the electrode, generating a measurable electrical signal. This combines the high specificity of immunorecognition (antibody–antigen binding) with the sensitivity and simplicity of electrochemical detection. All immunosensors rely on the specific binding between an antibody (Ab) and its antigen (Ag) on a transducer surface. In an electrochemical immunosensor, this immunorecognition event is coupled to an electrical read-out [31]. Because most antibodies and antigens are not intrinsically electroactive, electrochemical immunosensing typically employs either a direct (label-free) or indirect (labeled) detection strategy.

Label-free immunosensor

In label-free (direct) immunosensors, the binding of the antigen to the immobilized antibody induces a physical or chemical change at the electrode interface (such as altered impedance, interfacial charge or surface mass) that is measured directly. Such sensors permit real-time, on-site monitoring without additional reagents, but they can suffer from non-specific adsorption and lower sensitivity [32].

Materials Research Foundations 191 (2026) 260-288
https://doi.org/10.21741/9781644904213-8

Label immunosensor

Labeled (indirect) immunosensors use a tag or enzyme attached to the antibody or antigen to generate a secondary electroactive signal.

3.2.1 Types of electrochemical immunosensors

Electrochemical immunosensors have types such as amperometric immunosensors, potentiometric immunosensors, impedimetric immunosensors, Conductometric immunosensors and Voltametric immunosensors as shown in Table 2.

Amperometric immunosensor

In amperometric immunosensors, a constant potential is applied, and the resulting current (from oxidation or reduction of an electroactive species) is measured. In practice, this usually means applying a fixed bias and detecting current due to a label (e.g., enzymatically produced hydrogen peroxide or a redox mediator [33].

Potentiometric immunosensor

In potentiometric immunosensors, the potential (voltage) change at an ion-selective or field-effect electrode is measured due to antigen binding. Typically, the electrode is held at near-zero current, and the antigen–antibody interaction alters the surface charge or ionic environment, causing a potential shift [34].

Impedimetric immunosensor

In impedimetric (Impedance) immunosensors, electrochemical impedance spectroscopy (EIS) is used to probe the electrode interface. An AC voltage of small amplitude (e.g., 5–10 mV) is applied over a range of frequencies, and the impedance (especially the charge-transfer resistance) is measured [35].

Conductometric immunosensor

In conductometric immunosensors, changes in the solution or film conductivity caused by the immunoreaction are detected. When an antigen–antibody complex forms (especially if enzymes are involved), the concentration of ions in solution can change, altering conductance [36].

Voltametric immunosensor

In voltametric immunosensors, it involves sweeping the electrode potential (e.g., cyclic voltammetry, linear-sweep, differential pulse, or square-wave voltammetry) and recording current voltage curves [37].

Table 2 Different types of electrochemical immunosensors

Immunosensor	Detection	Example	Reference
Ampometric immunosensor	Current is measured	Nanostructured gold electrode for the electrochemical detection of bound ferrocene-labeled antibody molecules.	[33]
Potentiometric immunosensor	Potential is measured	A polypyrrole-based potentiometric biosensor that detects enzyme labelled immuno-complexes	[34]
Impedimetric immunosensor	Impedance is measured	A label-free electrochemical impedimetric immunosensor for sensitive detection of ochratoxin A	[35]
Conductometric immunosensor	Conductivity is measured	Magnetite nanoparticles-based immunosensor for *Escherichia coli* detection	[36]
Voltammetric immunosensor	Voltage is measured	Graphene-based voltametric immunosensor for sensitive detection of the egg allergen ovalbumin	[37]

3.3 Optical and fluorescent-based immunosensors

All optical immunosensors share the same fundamental principle: a specific antibody–antigen binding event is transduced into a measurable optical signal. In a label-free optical sensor, the antigen binding changes the local refractive index or thickness of the biolayer, shifting a resonance condition or altering an interference pattern [42]. In labeled (signal-amplified) formats, an optical label is attached to one of the immunoassay components. A fluorescent immunosensor uses a fluorescent tag on the antibody or a sandwich detector; binding brings the fluorophore in proximity to the surface, and excitation light elicits an emission signal [43]. For instance, in SPR, the resonant angle of reflected light changes proportionally to the mass of bound antigen. Similarly, chemiluminescent immunosensors rely on enzyme-catalyzed light emission (e.g., luminol or alkaline phosphatase reactions) when antigen–antibody complexes form [44]. In all cases, the antibody's high specificity ensures that only the target antigen produces the optical response. Thus, immunosensing is often described in three modes: (1) label-free monitoring of the binding event [32], (2) direct labeled detection, and (3) indirect enzyme-linked detection producing color or luminescence [45].Optical immunosensors are surface plasmon resonance (SPR) immunosensors, fluorescence-based immunosensors, chemiluminescence-based immunosensors, and fiber-optic Immunosensors. In surface plasmon resonance immunosensors, a thin metal film (usually gold) is used on a prism or grating. Polarized light at a specific angle excites surface plasmons, and antigen binding shifts the resonance [46]. In fluorescence-based immunosensors, antibodies are conjugated to fluorescent dyes, quantum dots, or other fluorophores. When an antigen is captured, a light source (often a laser or LED) excites the fluorophore, and emitted light is measured [47]. In chemiluminescence-based immunosensors, chemiluminescent labels or enzyme systems are used that produce light upon immunoreaction [48]. In fiber-optic immunosensors, Optical fibers can serve as both the support and the transducer. In one design, a tapered or unclad fiber coated with antibodies is illuminated, creating an evanescent field that excites bound fluorescent labels [49].

3.4 Surface plasmon resonance (SPR) in microfluidics

Surface Plasmon Resonance (SPR) is an optical detection method that utilizes collective electronic oscillations (surface plasmons) at a metal–dielectric interface. Incident light (most commonly through a prism) in a standard SPR biosensor is coupled with a thin gold film to create an

Materials Research Foundations 191 (2026) 260-288 https://doi.org/10.21741/9781644904213-8

evanescent plasmon wave whose resonance condition is highly sensitive to the refractive index of the local surroundings. Molecular binding processes (e.g., protein–ligand interactions) at the metal surface alter the local refractive index and thereby shift the SPR resonance, allowing label-free, real-time monitoring of bimolecular interactions [50]. Commercial SPR platforms exploit this principle to provide high-throughput kinetic measurements of antibodies, proteins, nucleic acids, and other analytes. Integrating SPR sensors with microfluidic flow channels further enhances their utility: the microfluidic format automates fluid handling, greatly reduces sample and reagent volumes, and supports multiplexed assays for point-of-care diagnostics [51].

In a Kretschmann SPR sensor, p-polarized light is directed through a high-refractive-index prism onto a thin metal film (typically gold ~50 nm thick) on a glass slide. At a specific incident angle (the SPR angle), the evanescent field from total internal reflection couples to surface plasmons in the metal, producing a sharp dip in the reflected intensity. Surface plasmons have an electromagnetic field that oscillates along the metal–dielectric interface and decays exponentially into both media [52]. When analyte molecules bind to receptors on the metal surface, the local refractive index increases, causing a shift in the SPR resonance condition. By monitoring this shift in real time, one obtains a continuous sensor gram that quantifies the binding and dissociation of the target analyte [53].

4 Fabrication techniques for microfluidic immunosensors

Microfluidic immunosensor fabrication involves combining microfluidic systems with immunoassay platforms using methods designed to ensure accuracy, consistency, and cost-effectiveness. These fabrication approaches vary based on the intended application, the materials used, and the specific sensor design. Below is an overview of the primary techniques commonly utilized in the development of microfluidic immunosensors.

4.1 Soft lithography

Soft lithography is a frequently employed method for fabricating microfluidic immunosensors due to its affordability, flexibility, and compatibility with biological systems. This technique involves forming micro-scale structures using elastic materials such as polydimethylsiloxane (PDMS), making it particularly advantageous for developing devices that incorporate immunoassay functionalities.

The routine manufacturing procedure involves the following steps. First, a master mold is created by photolithography to create the intended microchannel patterns. PDMS is subsequently cast on top of this mold and left to cure, forming a soft elastomeric duplicate. The PDMS duplicate is then attached to a substrate, normally using oxygen plasma treatment, which assists in the creation of tightly sealed microfluidic channels. To facilitate specific binding in immunoassays, the interior walls of such microchannels are functionalized by immobilizing antibodies or other biomolecules [54].

An emerging innovation in this area involves microreactor-assisted soft lithography. This approach integrates the synthesis and patterning of nanostructured films directly within the microfluidic platform, effectively addressing one of the key challenges of conventional thin-film deposition—film instability during processing. Devices developed using this technique exhibit enhanced sensitivity and stability, making them ideal for real-time immunoassay applications [55]. In the future, research in soft lithography is expected to focus on the creation of materials with better

Materials Research Foundations 191 (2026) 260-288 https://doi.org/10.21741/9781644904213-8

elasticity, chemical resistance, and optical transparency, all of which are essential for conformity to biomedical application requirements. Moreover, remote diagnostics are promising directions that are related to the integration of soft lithography with digital technologies, e.g., smartphone-based readers or cloud-connected health monitoring.

4.2 Photolithography

Photolithography plays a crucial role in the development of precise miniature patterns essential for biosensor fabrication. The method is akin to a high-accuracy stencil system utilizing light. The process begins with a biosensor chip that is coated with a light-sensitive compound known as a photoresist, functioning similarly to a thin layer of UV-reactive paint. A patterned stencil, or mask, is then placed over this coated chip. Microfluidic immunosensors are often manufactured using photolithography, starting with substrate preparation. Common substrates include silicon, glass, or quartz, which are cleaned before photoresist application. A light-sensitive photoresist (SU-8) is applied to the substrate through spin coating, creating a thin, even layer. The thickness of the layer is determined by the spin speed and spin time, while soft baking dries off solvents and enhances sticking.

After that, a patterned shape photomask is overlaid onto the substrate and subjected to ultraviolet (UV) light, which either dissolves the photoresist, depending on whether the system is positive or negative. After exposure, the substrate is then developed using a chemical developer that selectively dissolves the exposed (or non-exposed) photoresist areas, leaving the patterned microstructures exposed. Post-processing hard bake is commonly applied to enhance both the chemical resistance and structural integrity of the end pattern. The method finds large-scale application due to its accuracy, reproducibility, and biocompatibility [56].

One of the recent breakthroughs in the area is that photolithography has been used in conjunction with nanotechnologies such as nanowires, 2D materials, and bio-functionalized nanoparticles. Another key development involves the automation and scaling of photolithographic techniques for large-scale manufacturing. As lab-on-a-chip technologies become more common in clinical and commercial settings, the scalability of photolithography supports the production of high-performance devices in large volumes. This is particularly important for point-of-care diagnostics, where affordability, consistency, and high throughput are crucial.

Additionally, there is growing attention to biocompatibility and electronic integration. Progress in biocompatible photoresists and flexible substrates will allow photolithographic techniques to be applied in developing wearable and implantable immunosensors, enabling real-time and continuous health monitoring [57].

Finally, photolithography is expected to continue advancing, particularly in its role in developing 3D microfluidic architectures. These structures are essential for recreating complex biological systems and enabling more accurate, in vivo-like immunoassays.

4.3 3D printing

3D printing represents a state-of-the-art technique for fabricating microfluidic devices, offering distinct benefits such as rapid prototyping, the ability to create intricate 3D geometries, and the use of diverse materials not compatible with conventional photolithographic methods. In microfluidics, 3D printing supports the direct construction of entire channel networks without the

Materials Research Foundations 191 (2026) 260-288 https://doi.org/10.21741/9781644904213-8

need for masks, cleanroom environments, or step-by-step layer alignment. This makes it particularly appealing for producing customizable, cost-efficient point-of-care diagnostic tools.

The workflow typically begins with computer-aided design (CAD) software, which is used to develop microfluidic device layouts and channels. Appropriate materials, like photopolymer resins or thermoplastics, are selected next [58]. Devices are then fabricated using techniques such as stereolithography (SLA) or digital light processing (DLP), both of which cure resins in a layer by layer fashion using UV light. After fabrication, steps like washing, curing, and surface treatment follow to assess structural performance and biocompatibility. Notably, sensors and electrodes can be integrated during printing, allowing real-time device assembly. This approach reduces both fabrication time and complexity significantly [59]. The use of 3D printing for lab-on-a-chip devices shows great potential in areas like cell culture systems and immunosensing. Its advantages including scalability, cost-effectiveness, and adaptability make it a promising alternative to traditional photolithographic techniques for biomedical and diagnostic applications.

One of the most impactful uses of 3D printing lies in point-of-care testing, especially in producing compact and disposable microfluidic immunoassay chips. For instance, stereolithographically fabricated microfluidic devices have been shown to provide multiplexed biomarker detection, making them well-suited for low-resource environments [60]. Another study successfully employed 3D printing to build multilayer chips with integrated valves and mixers, paving the way for dynamic control features to be embedded within the chip itself [61]. The biosensors have been utilized in the field of environmental and food safety; a 3D-printed electrochemical biosensor has been used to detect E. coli and other deadly pathogens in water sources [62]. Moreover, vascular-mimicking scaffolds have been developed to support endothelial cell growth under controlled flow, significantly enhancing technologies used in tissue modeling and drug testing.

4.4 Paper-based microfluidic fabrication

Paper-based microfluidic systems have emerged as promising platforms for immunosensing due to their affordability, ease of use, and effectiveness for point-of-care applications. They take advantage of paper's natural capillary action to transport fluids, eliminating the need for external pumps. This characteristic is particularly beneficial in low-resource environments [63].

4.4.1 Wax Printing

Wax printing is a common technique where patterns are applied to paper using solid wax, which is then melted to form hydrophobic boundaries. The untreated paper areas remain hydrophilic, allowing controlled fluid movement. This low-cost approach is appealing as it avoids the need for specialized equipment and is well-suited for a variety of diagnostic uses [64]. It can commonly be done by placing wax on chromatography paper and heating it to harden the patterns. This technique has been effectively employed to produce glucose and protein sensors, whereby it is easy to detect by using colorimetric assays [65].

4.4.2 Inkjet Printing

Inkjet printing involves accurately depositing biomolecules and reagent solutions on paper to form functional zones within microfluidic systems. This method can create detailed channel structures and support rapid prototyping in a single step [66]. The technology is ideal for applications requiring precise placement of biological materials. For example, it has been employed to fabricate

Materials Research Foundations 191 (2026) 260-288 https://doi.org/10.21741/9781644904213-8

ELISA devices capable of detecting malaria biomarkers, demonstrating its global diagnostic potential [67]. Another study used this approach to print antibodies on paper for sensitive Salmonella detection [68].

4.4.3 Spray Coating

In this technique, a paper is sprayed with a hydrophobic substance and toner is sprayed to outline the required design as a stencil. This quick and efficient process, taking approximately 10 minutes, yields devices suited for glucose detection and similar applications [69]. A recent study also reported the successful use of spray coating to fabricate electrochemical paper-based immunosensors [70].

4.4.4 Plastic Lamination

Plastic lamination techniques typically involve cutting paper substrates and bonding them with plastic films using heat. This process enhances the devices' resistance to chemicals and boosts their mechanical strength, making them reliable for diagnostic use [71]. Another common method uses a cost-effective and rapid approach to apply plastic films via a cut-and-laminate technique, enabling the production of durable paper-based microfluidic devices. These laminated devices have also been successfully adapted for ELISA-based assays, showing effective integration into bioassays [71]. Table 3 lists the materials, benefits, and most recent advancements of fabrication processes used in the creation of microfluidic immunosensors.

Table 3 Fabrication Techniques for Microfluidics Immunosensors

Technique	Key Materials	Fabrication Steps	Advantages	Recent Innovations	Reference
Soft Lithography	PDMS, master Mold (photoresist)	1. Create a master Mold via photolithography 2. Cast PDMS and cure 3. Bond to substrate with plasma 4. Functionalize channels	Biocompatible Flexible Low cost High resolution	Microreactor-assisted lithography for nanostructures	[54,55]
Photolithography	Silicon, glass, quartz, SU-8 photoresist	1. Apply photoresist via spin coating 2. Expose to UV with mask 3. Develop pattern 4. Post-bake for durability	High precision Scalable for mass production Compatible with nanotech	Smartphone integration for remote diagnostics	[56,57]
3D Printing	Photopoly-mer resins, thermoplastic	1. Design in CAD 2. Print via SLA/DLP 3. Wash, cure, and surface treat	Rapid prototyping No cleanroom needed	Integration with 2D nanomaterials and bio-nanoparticles 3D microfluidic architectures	[58-62]

Materials Research Foundations 191 (2026) 260-288 https://doi.org/10.21741/9781644904213-8

		4. Integrate electrodes/sensors	Complex 3D structures possible		
Paper-Based Microfluidics	Cellulose paper, wax, ink, plastic films	Varies by subtype	Ultra low-cost Easy fabrication	Multiplexed biomarker detection Multilayer chips with valves and mixers	[63]
Wax Printing	Wax chromatography paper	Print wax Heat to melt and form hydrophobic barriers	Simple No complex tools Suitable for colorimetric assays	Electrochemical paper sensors ELISA integration	[64,65]
Inkjet Printing	Biomólecules , inkjet printer	Print biomolecules reagents directly onto paper	High precision One-step fabrication Fast prototyping	Effective for enzyme-based color detection	[66-68]
Spray Coating	Hydrophobic spray, toner	Spray hydrophobic layer Apply toner as stencil	Fast (≈10 mins) Cost-effective Suitable for electrochemical sensors	Glucose sensors Electrochemical immunosensors	[69,70]
Plastic Lamination	Paper, plastic films	Cut paper Laminate with plastic using heat	ELISA-based paper devices Durable diagnostics	Efficient production of paper biosensors	[71]

5. Applications of microfluidic immunosensors

Rapid diagnosis is crucial for a cure at an early stage. Microfluid systems are grabbing the attention of researchers due to their advantages to the medical system, such as early diagnosis, reduced utilization of reagents and patient samples for performing medical analysis [72]. The quick diagnosis with microfluidics-based immunosensors is best shown by an example in which a novel microfluidic system coupled with immunological assay has been engineered to detect proteins of SARS-CoV-2, under 30 seconds, and antigens were detected. Also, when microfluid devices are paired with biosensors, they enable the quick processing of samples for the detection of pathogenic microorganisms. It greatly supports the timely decision making in diagnostic and therapeutic scenarios [73]. Analogously, microfluidic biosensors were used in the analysis of several cancer related biomarkers such as PSA, CEA, and CA-125 [74]. In one of the studies, a droplet based platform was utilized to conduct cytokine profiling at a single-cell level, showing immune heterogeneity through bead-based immunoassays [75]. Paper-based microfluidic biosensor based on DNAzymes for in situ monitoring of heavy metal ions (e.g., lead) in tap water has been

Materials Research Foundations 191 (2026) 260-288 https://doi.org/10.21741/9781644904213-8

fabricated, with field-sensitive high performance [76]. Along the same token, microfluidic biosensors have also been developed in food safety; for example, a chemiluminescent microfluidic immunoassay was developed to quickly identify Salmonella without the practice of conventional culture, substantially reducing the detection time [77]. Liver-on-chip has also been designed in drug testing research as a bio-mimetic system for the evaluation of drug toxicity on the basis of diagnostic biomarkers [78]. These varied experimental applications highlight the versatility and potential of integrated microfluidic-biosensor systems in addressing real-world challenges, while also paving the way for future advancements in personalized diagnostics and environmentally sustainable solutions.

5.1 Point-of-care diagnostics

There is increasing demand for clinically tested point-of-care diagnostics, particularly in resource-poor environments with compromised healthcare services. Early disease control based on knowledge about the disease will require more sophisticated technologies, e.g., microfluidic immunosensors, offering a convenient method for quick detection of various infections [2,79]. The COVID-19 pandemic has underscored the need to have fast and effective diagnostic tools. To detect SARS-COV-2 nucleocapsid in serum, a microfluidics immunosensor was developed by preparing immunomagnetic nano beads with increased detection sensitivity even from very limited samples (<50 μL). Since these immunosensors are supposed to detect the concentration of the specific protein so it is a quantitative diagnostic immunosensors. These results suggested that the performance of these microfluid immunosensors is very similar to that of PCR-based tests employed to detect corona virus positive patients [80]. Another is the detection of Zika virus using electrochemical microfluid immunosensors, as shown in Fig. 5. In this setup, a Gold electrode with Zika virus envelope protein coating serves as a capture probe. Surveillance of Zika and other flaviviruses, such as Dengue, is different as they share the same vector species. The following electrodes overcome this challenge by offering a highly promising detection device of the Zika virus in saliva and serum samples based on the detection of non-structural protein 1 (NS1) and envelope protein of domain III. Moreover, these microfluidic immunosensors are able to distinguish between Dengue and Zika virus-specific antibodies to provide more precise detection of Zika virus when compared with more conventional approaches like ELISA or PCR [81].

Moreover, a hand-held and inexpensive microfluidic immunosensor for quick detection of HIV has also been developed by the name of "POCKET immunoassay". The device is prepared by polydimethylsiloxane (PDMS) and is proposed to detect anti-gp41 of HIV. It reduces the silver ions, which are initiated by the presence of the marker anti-gp41. Antibodies employed in this system are labeled with gold colloids that catalyze the silver reduction process. The increased level of silver reduction, the greater the concentration of anti-gp-41, and successfully detects HIV positive samples [82]. It is an excellent platform for the detection of HIV in source-limiting settings, particularly where lab facilities and trained professionals may be limited.

Materials Research Foundations 191 (2026) 260-288 https://doi.org/10.21741/9781644904213-8

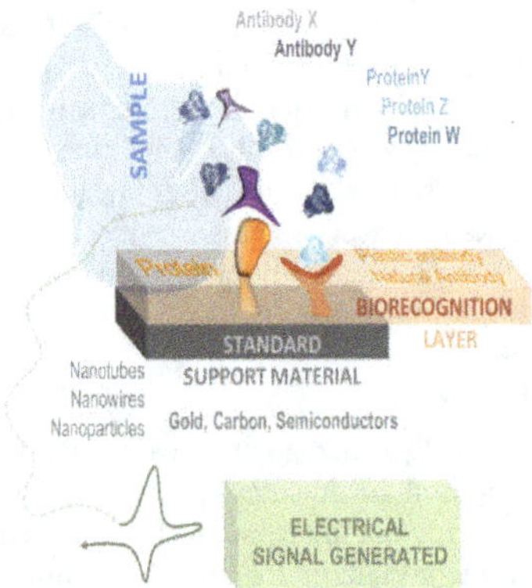

Fig.5 A miniature diagnostic tool that uses electrochemical sensing principles to detect Zika virus infections. By identifying unique viral components in physiological fluids, the device's gold sensor fragments coated with viral proteins distinguish Zika from other viruses like dengue

5.2 Detection and disease tracking

As explained above, microfluid immunosensors play invaluable roles in cancer biomarkers, cardiac markers, viruses, and bacterial detection. A lot of effort is being channeled towards the introduction of user-friendly and efficient detection platforms for diagnostics and therapeutics. One of the techniques is by applying multiplex microfluid immunosensors that can identify multiple viruses and bacteria at the same time through barcode multiplexing. Barcode multiplexing utilizes colored beads to distinguish between different pathogens in one sample [83]. Microfluidic immune sensors are also utilized for coronavirus detection and HIV detection. Moreover, Immunomagnetic nanoparticles have been utilized in a microfluid immunosensor, a high-throughput and cost-effective method of detecting carcinoembryonic antigen to screen for cancer [84].

5.3 Environmental monitoring and food safety

Food safety is one of the most significant health issues all over the world, as chemical and biological contaminants are significant hazards for human health. There is a need to detect food contaminants by employing instant and precise detection systems like microfluid biosensors. These devices offer improved food safety and can detect trace levels of contaminants based on their very sensitive detection abilities, thereby keeping the quality and safety of edible products intact. Food pathogens are the cause of some of the most lethal foodborne illnesses. One of the most notorious food-borne pathogens is *Salmonella,* which accounts for most gastrointestinal disorders in Europe. In 2018, more than half of the foodborne illnesses were caused by *Salmonella* outbreaks in Europe [85]. The microfluidic chip with sandwich shaped anti-Salmonella polyclonal antibodies has been developed for Salmonella detection and food safety [86]. Other diseases like *Vibrio cholera* can also be detected by fluorescence detection in microfluidic platforms [87]. A

Materials Research Foundations 191 (2026) 260-288 https://doi.org/10.21741/9781644904213-8

microfluid immunosensor integrated with a graphene layer and gold nanoporous structure is employed to quantify foodborne mycotoxin (T-2 toxin) [10].

5.4 Drug discovery and biomedical research

Microfluidic systems can also be applied for the manipulation of fluids, their microchannels can be employed for conducting experiments related to drug optimization and selection. Such devices speed up reactions to obtain results like target selection, lead identification, and drug optimization. It is also utilized in preclinical testing and dosage evaluation, as indicated in Fig. 6. In the selection of individual targets, protein-to-protein interactions should be considered. Microfluid biosensors are an efficient tool in the investigation of protein-protein interactions and can count proteins even in tiny amounts. Epitope-tagged human β2-adrenergic receptors (β2ARs), are an important drug target, as they are implicated in various cardiovascular and respiratory diseases and can be detected by a microfluid device [88].

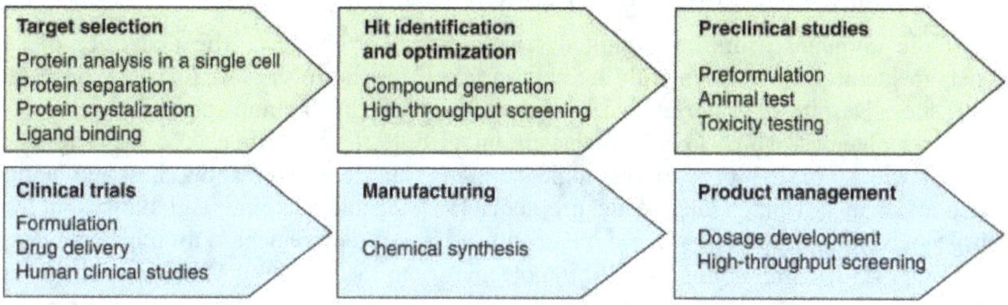

Fig. 6 The flowchart depicts the vast range of applications for microfluidic systems during the drug discovery process. These tiny platforms significantly shorten research timelines by allowing for precise protein manipulation, high-throughput chemical screening, and effective toxicity evaluation

Microfluidic devices have been used to prepare pirfenidone, which is an antioxidant as well as an anti-inflammatory drug [89]. Digital microfluidic systems dependent on electric forces have been used for biomedical and biological research, such as immunoassays, clinical diagnostics, single-cell analysis, and proteomic analysis [90]. Moreover, these digital microfluidic platforms assist in monitoring the drug screening and manipulation of hepatic organoids. In traditional drug screening methods heterogenic response of similar cells against a specific drug cannot be observed because the information recorded as response is the average of all cells. Whereas microfluidic chips provide a promising platform for analyzing the anti-drug response of a single cell. These chips are also used to analyze the drug response for a cancer cell line possessing different phenotypes. These microfluidic chips are also important for studying breast cancer cells in 2D and 3D culture conditions. The effect of human mammary fibroblasts on the metastasis and infiltration of breast cancer cells has been demonstrated [91].

Another problem of physiological irrelevancy in drug discovery has been resolved using microfluid immunosensors. Microfluidic albumin immunosensors were developed by immobilization of anti-HAS antibody and integrated with a liver-on-a-chip microphysiological system. It was used to provide continuous feedback for disease treatment and modeling. The

Materials Research Foundations 191 (2026) 260-288
https://doi.org/10.21741/9781644904213-8

effectiveness of the system was evaluated for 6 days. It was found to be efficient for live monitoring of drug treatment and disease modelling. Such systems provide real-time data for taking evidence-based decisions in the future for devising novel treatment strategies and conducting biomedical research.

6. Challenges and future prospects

Microfluidics-based immunosensors face challenges such as downsizing, ensuring consistent fluid management, and maintaining sensor stability under changing conditions. Future potential include breakthroughs in nanomaterials, AI-driven data analysis, and scalable production approaches for increased sensitivity and portability. However, establishing seamless integration into point-of-care diagnostics necessitates overcoming biocompatibility concerns, financial limitations, and regulatory obstacles to general implementation.

6.1 Challenges of sensitivity and specificity

Microfluidic immunosensors have limited sensitivity and specificity. The inabilities of small channels to identify biomarkers with the appropriate levels lie in the fact that they have fewer binding sites. Despite the interference of the assay system by the non-specific binding of the interfering molecules, which lowers diagnostic integrity [92]. Nevertheless, changes in sample matrices such as in PH or variable ionic strength, interfere with antibodies and antigens interactions, which worsen diagnostic problems [93]. Surface fouling and biofouling cause inconsistent results, so specific surface treatments are now a requirement to minimize inadvertent adsorption [94]. The immobilization of biomolecules onto the sensor surface is a critical step in biosensor development. However, due to their sensitivity and susceptibility to environmental conditions, biomolecules cannot be directly adsorbed onto electrode surfaces or other inorganic substrates, as this can lead to rapid denaturation. To preserve their biological activity, a biocompatible intermediate layer must be introduced between the sensing elements (or receptors) and the electrode surface [95]. While surface modification techniques can enhance signal detection, but can also introduce additional noise, complicating data interpretation. Therefore, careful optimization of flow dynamics and incubation times is essential for achieving accurate and reliable sensor performance [96].

6.2 Scalability challenges

Despite their potential, biosensors aren't yet widely used because of several limitations. Their adaptability is limited by the fact that many are overly specialized, identifying just particular targets. They require stronger designs and more sophisticated materials to function consistently in real-world situations. Although these sensors work well in laboratories, it is challenging to scale up production since they frequently need complex manufacturing and specialized settings. Researchers are constantly trying to make them more durable and adaptable. Overall, biosensor technology's full promise has yet to be fulfilled [95].

6.3 Future trends in microfluidic immunosensors

The most promising future research involves modifying nanomaterials, such as carbon nanotubes, gold nanoparticles, and graphene oxide, on paper substrates to enhance the sensitivity and specificity of immunosensors. Nanomaterials enable multiplexed signal transduction to allow the determination of low abundances of biomarkers [97]. There is also the development of multiplexing capabilities, whereby several biomarkers can be spotted at once on a single piece of paper. This is particularly useful in the diagnosis of diseases like cancer or in the treatment of co-infection [15]. Production wise, there are massive technologies of 3D printing based on paper, roll-to-roll, screen printing, and automated printing technologies to allow mass production of 3D printed and fabricated μPADs. The volatility and time-intensive characteristics of the original wax and inkjet processes will be overcome using these techniques, enabling large-scale production [98]. Intelligent microfluidics combines machine learning (ML) with microfluidics and is an emerging discipline. It reduces human effort, speeds up research, and improves accuracy in tasks like disease detection, drug development, and lab testing, making it a powerful tool for advancing science and healthcare. This combination has already demonstrated success in solving problems by using machine learning to automatically identify patterns, classify data, and optimize processes in microfluidic systems instead of relying on slow, manual analysis [99].

Conclusion

Through microfluidic immunosensors, biomedical research and diagnostics have undergone a revolution. These systems allow for the quick, sensitive, and high-throughput detection of infections, biomarkers, and therapeutic targets by utilizing precise fluid control at the microscale. This technology allows diverse analyses such as immunoassays, single-cell analysis, and real-time medication response monitoring, which overcome the constraints of traditional bulk analysis by offering insights at the cellular and molecular levels. The capacity to examine diverse cell populations with different morphologies is a significant benefit since it enables researchers to see how each cell reacts to medications rather than relying on average data. As nanomaterials, biosensors, and AI-driven data analysis evolve, microfluidic immunosensors have the potential to transform personalized medicine, point-of-care diagnostics, and targeted cancer therapy. Their adaptability and miniaturization make them critical tools for future biomedical advances.

References

[1] K.-K. Liu, R.-G. Wu, Y.-J. Chuang, H.S. Khoo, S.-H. Huang, F.-G. Tseng, Microfluidic systems for biosensing, Sensors 10(7) (2010) 6623-6661. https://doi.org/10.3390/s100706623

[2] T. Lehnert, M.A. Gijs, Microfluidic systems for infectious disease diagnostics, Lab Chip 24(5) (2024) 1441-1493. https://doi.org/10.1039/D4LC00117F

[3] S. Wang, X. Guan, S. Sun, Microfluidic biosensors: Enabling advanced disease detection, Sensors 25(6) (2025) 1936. https://doi.org/10.3390/s25061936

[4] E.K. Sackmann, A.L. Fulton, D.J. Beebe, The present and future role of microfluidics in biomedical research, Nature 507(7491) (2014) 181-189. https://doi.org/10.1038/nature13118

[5] J.W. Yang, D. Khorsandi, L. Trabucco, M. Ahmed, A. Khademhosseini, M.R. Dokmeci, J.Y. Ye, V. Jucaud, Liver-on-a-chip integrated with label-free optical biosensors for rapid and continuous monitoring of drug-induced toxicity, Small 20(48) (2024) 2403560. https://doi.org/10.1002/smll.202403560

Materials Research Foundations 191 (2026) 260-288 https://doi.org/10.21741/9781644904213-8

[6] F. Alghannam, M. Alayed, S. Alfihed, M.A. Sakr, D. Almutairi, N. Alshamrani, N. Al Fayez, Recent progress in PDMS-based microfluidics toward integrated organ-on-a-chip biosensors and personalized medicine, Biosensors 15(2) (2025) 76. https://doi.org/10.3390/bios15020076

[7] A.S. Kumar, S. Venkatesalu, S. Dilliyappan, A.P. Pasupulla, L. Prathap, T. Palaniyandi, G. Baskar, M. Ravi, A. Sugumaran, Microfluidics as diagnostic tools, Clin. Chim. Acta 556 (2024) 117841. https://doi.org/10.1016/j.cca.2024.117841

[8] E.A. Tarim, B. Karakuzu, C. Oksuz, O. Sarigil, M. Kizilkaya, M.K.A. Al-Ruweidi, H.C. Yalcin, E. Ozcivici, H.C. Tekin, Microfluidic-based virus detection methods for respiratory diseases, Emergent Mater. 4 (2021) 143-168. https://doi.org/10.1007/s42247-021-00169-7

[9] S. Kothawade, V. Padwal, Cutting-edge 3D printing in immunosensor design for early cancer detection, Microchim. Acta 192(1) (2025) 42. https://doi.org/10.1007/s00604-024-06880-6

[10] L.N.F. Solis, G.J.S. Junior, M. Bertotti, L. Angnes, S.V. Pereira, M.A. Fernández-Baldo, M. Regiart, Electrochemical microfluidic immunosensor with graphene-decorated gold nanoporous for T-2 mycotoxin detection, Talanta 273 (2024) 125971. https://doi.org/10.1016/j.talanta.2024.125971

[11] A.C. Sajjan, H. Jha, T. Borulkar, S. Regalla, R. Yadav, Systematic review: Biosensors for early detection of infectious diseases, J. Contemp. Clin. Pract. 11 (2025) 195-202.

[12] W.R. de Araujo, H. Lukas, M.D. Torres, W. Gao, C. de la Fuente-Nunez, Low-cost biosensor technologies for rapid detection of COVID-19 and future pandemics, ACS Nano 18(3) (2024) 1757-1777. https://doi.org/10.1021/acsnano.3c01629

[13] K. Wu, X. He, J. Wang, T. Pan, R. He, F. Kong, Z. Cao, F. Ju, Z. Huang, L. Nie, Recent progress of microfluidic chips in immunoassay, Front. Bioeng. Biotechnol. 10 (2022) 1112327. https://doi.org/10.3389/fbioe.2022.1112327

[14] F. Bahavarnia, M. Hasanzadeh, D. Sadighbayan, F. Seidi, Recent progress and challenges on the microfluidic assay of pathogenic bacteria using biosensor technology, Biomimetics 7(4) (2022) 175. https://doi.org/10.3390/biomimetics7040175

[15] B.G. Rosa, O.E. Akingbade, X. Guo, L. Gonzalez-Macia, M.A. Crone, L.P. Cameron, P. Freemont, K.-L. Choy, F. Güder, E. Yeatman, Multiplexed immunosensors for point-of-care diagnostic applications, Biosens. Bioelectron. 203 (2022) 114050. https://doi.org/10.1016/j.bios.2022.114050

[16] A.G. Niculescu, C. Chircov, A.C. Bîrcă, A.M. Grumezescu, Fabrication and applications of microfluidic devices: A review, Int. J. Mol. Sci. 22(4) (2021) 2011. https://doi.org/10.3390/ijms22042011

[17] S. Aralekallu, R. Boddula, V. Singh, Development of glass-based microfluidic devices: A review on its fabrication and biologic applications, Mater. Des. 225 (2023) 111517. https://doi.org/10.1016/j.matdes.2022.111517

[18] S.M. Scott, Z. Ali, Fabrication methods for microfluidic devices: An overview, Micromachines 12(3) (2021) 319. https://doi.org/10.3390/mi12030319

[19] M. Dou, S.T. Sanjay, M. Benhabib, F. Xu, X. Li, Low-cost bioanalysis on paper-based and hybrid microfluidic platforms, Talanta 145 (2015) 43-54. https://doi.org/10.1016/j.talanta.2015.04.068

[20] M.M. Gong, D. Sinton, Turning the page: Advancing paper-based microfluidics for broad diagnostic application, Chem. Rev. 117(12) (2017) 8447-8480. https://doi.org/10.1021/acs.chemrev.7b00024

[21] K. Raj M, S. Chakraborty, PDMS microfluidics: A mini review, J. Appl. Polym. Sci. 137(27) (2020) 48958. https://doi.org/10.1002/app.48958

[22] J. Friend, L. Yeo, Fabrication of microfluidic devices using polydimethylsiloxane, Biomicrofluidics 4(2) (2010) 026502. https://doi.org/10.1063/1.3259624

[23] S.-u. Hassan, A. Tariq, Z. Noreen, A. Donia, S.Z. Zaidi, H. Bokhari, X. Zhang, Capillary-driven flow microfluidics combined with smartphone detection: An emerging tool for point-of-care diagnostics, Diagnostics 10(8) (2020) 509. https://doi.org/10.3390/diagnostics10080509

[24] C. Yang, Y. Kang, X. Huang, Electrokinetic flow in porous media, Encycl. Microfluid. Nanofluid. (2015) 795-806. https://doi.org/10.1007/978-1-4614-5491-5_418

[25] A. Alizadeh, W.L. Hsu, M. Wang, H. Daiguji, Electroosmotic flow: From microfluidics to nanofluidics, Electrophoresis 42(7-8) (2021) 834-868. https://doi.org/10.1002/elps.202000313

[26] Y. Liu, H. Shen, X. Yang, S. Kang, L. Cai, T. Tian, R. Su, C. Yang, Z. Zhu, Recent progress in microfluidic biosensors with different driving forces, TrAC Trends Anal. Chem. 158 (2023) 116894. https://doi.org/10.1016/j.trac.2022.116894

[27] S. Battat, D.A. Weitz, G.M. Whitesides, An outlook on microfluidics: The promise and the challenge, Lab Chip 22(3) (2022) 530-536. https://doi.org/10.1039/D1LC00731A

[28] F. Hoshino, O. Watanabe, X. Wu, Y. Takimoto, T. Osawa, Low-cost and easy-to-use "on-chip ELISA" for developing health-promoting foods, Lipid Hydroperoxide-Derived Modification Biomol. (2014) 151-161. https://doi.org/10.1007/978-94-007-7920-4_13

[29] E. Eteshola, M. Balberg, Microfluidic ELISA: On-chip fluorescence imaging, Biomed. Microdevices 6(1) (2004) 7-9. https://doi.org/10.1023/B:BMMD.0000013360.65653.c2

[30] F. Mollarasouli, S. Kurbanoglu, S.A. Ozkan, The role of electrochemical immunosensors in clinical analysis, Biosensors 9(3) (2019) 86. https://doi.org/10.3390/bios9030086

[31] P. Skládal, Advances in electrochemical immunosensors, Electroanalysis 9(10) (1997) 737-745. https://doi.org/10.1002/elan.1140091002

[32] R. Cancelliere, E. Paialunga, A. Grattagliano, L. Micheli, Label-free electrochemical immunosensors: A practical guide, TrAC Trends Anal. Chem. (2024) 117949. https://doi.org/10.1016/j.trac.2024.117949

[33] W. Shi, Z. Ma, A novel label-free amperometric immunosensor for carcinoembryonic antigen based on redox membrane, Biosens. Bioelectron. 26(6) (2011) 3068-3071. https://doi.org/10.1016/j.bios.2010.11.048

[34] D. Purvis, O. Leonardova, D. Farmakovsky, V. Cherkasov, An ultrasensitive and stable potentiometric immunosensor, Biosens. Bioelectron. 18(11) (2003) 1385-1390. https://doi.org/10.1016/S0956-5663(03)00066-6

[35] M. Prodromidis, Impedimetric immunosensors-A review, Electrochim. Acta 55(14) (2010) 4227-4233. https://doi.org/10.1016/j.electacta.2009.01.081

[36] M. Hnaiein, W. Hassen, A. Abdelghani, C. Fournier-Wirth, J. Coste, F. Bessueille, D. Leonard, N. Jaffrezic-Renault, A conductometric immunosensor based on functionalized magnetite nanoparticles for E. coli detection, Electrochem. Commun. 10(8) (2008) 1152-1154. https://doi.org/10.1016/j.elecom.2008.04.009

[37] M. Giannetto, L. Elviri, M. Careri, A. Mangia, G. Mori, A voltammetric immunosensor based on nanobiocomposite materials for the determination of alpha-fetoprotein in serum, Biosens. Bioelectron. 26(5) (2011) 2232-2236. https://doi.org/10.1016/j.bios.2010.09.040

[38] L.X. Tiefenauer, S. Kossek, C. Padeste, P. Thiébaud, Towards amperometric immunosensor devices, Biosens. Bioelectron. 12(3) (1997) 213-223. https://doi.org/10.1016/S0956-5663(97)85339-0

[39] A.-E. Radi, X. Muñoz-Berbel, V. Lates, J.-L. Marty, Label-free impedimetric immunosensor for sensitive detection of ochratoxin A, Biosens. Bioelectron. 24(7) (2009) 1888-1892. https://doi.org/10.1016/j.bios.2008.09.021

[40] M. Hnaiein, W.M. Hassen, A. Abdelghani, C. Fournier-Wirth, J. Coste, F. Bessueille, D. Leonard, N. Jaffrezic-Renault, A conductometric immunosensor based on functionalized magnetite nanoparticles for E. coli detection, Electrochem. Commun. 10(8) (2008) 1152-1154. https://doi.org/10.1016/j.elecom.2008.04.009

[41] S. Eissa, L. L'Hocine, M. Siaj, M. Zourob, A graphene-based label-free voltammetric immunosensor for sensitive detection of the egg allergen ovalbumin, Analyst 138(15) (2013) 4378-4384. https://doi.org/10.1039/c3an36883a

[42] Z. An, C.-H. Jang, Label-free optical detection of aflatoxin by using a liquid crystal-based immunosensor, Microchem. J. 142 (2018) 335-342. https://doi.org/10.1016/j.microc.2018.07.013

[43] P. Damborský, J. Švitel, J. Katrlík, Optical biosensors, Essays Biochem. 60(1) (2016) 91-100. https://doi.org/10.1042/EBC20150010

[44] J. Lin, H. Ju, Electrochemical and chemiluminescent immunosensors for tumor markers, Biosens. Bioelectron. 20(8) (2005) 1461-1470. https://doi.org/10.1016/j.bios.2004.05.008

[45] A. Hayat, L. Barthelmebs, J.L. Marty, Enzyme-linked immunosensor based on superparamagnetic nanobeads for easy and rapid detection of okadaic acid, Anal. Chim. Acta 690(2) (2011) 248-252. https://doi.org/10.1016/j.aca.2011.02.031

[46] J. Mitchell, Small molecule immunosensing using surface plasmon resonance, Sensors 10(8) (2010) 7323-7346. https://doi.org/10.3390/s100807323

[47] A.Q. Rani, B. Zhu, H. Ueda, T. Kitaguchi, Recent progress in homogeneous immunosensors based on fluorescence or bioluminescence using antibody engineering, Analyst 148(7) (2023) 1422-1429. https://doi.org/10.1039/D2AN01913B

[48] M.M. Calabretta, M. Zangheri, D. Calabria, A. Lopreside, L. Montali, E. Marchegiani, I. Trozzi, M. Guardigli, M. Mirasoli, E. Michelini, Paper-based immunosensors with bio-chemiluminescence detection, Sensors 21(13) (2021) 4309. https://doi.org/10.3390/s21134309

[49] C.M. Maragos, V.S. Thompson, Fiber-optic immunosensor for mycotoxins, Nat. Toxins 7(6) (1999) 371-376. https://doi.org/10.1002/1522-7189(199911/12)7:6<371::AID-NT86>3.0.CO;2-8

[50] S.H. Cho, S. Choi, J.M. Suh, H.W. Jang, Advancements in surface plasmon resonance sensors for real-time detection of chemical analytes: Sensing materials and applications, J. Mater. Chem. C 13(13) (2025) 6484-6507. https://doi.org/10.1039/D4TC04890C

[51] Y. Luo, F. Yu, R.N. Zare, Microfluidic device for immunoassays based on surface plasmon resonance imaging, Lab Chip 8(5) (2008) 694-700. https://doi.org/10.1039/b800606g

[52] L. Malic, T. Veres, M. Tabrizian, Nanostructured digital microfluidics for enhanced surface plasmon resonance imaging, Biosens. Bioelectron. 26(5) (2011) 2053-2059. https://doi.org/10.1016/j.bios.2010.09.001

[53] R. Kurita, Y. Yokota, Y. Sato, F. Mizutani, O. Niwa, On-chip enzyme immunoassay of a cardiac marker using a microfluidic device combined with a portable surface plasmon resonance system, Anal. Chem. 78(15) (2006) 5525-5531. https://doi.org/10.1021/ac060480y

[54] J.C. Love, J.R. Anderson, G.M. Whitesides, Fabrication of three-dimensional microfluidic systems by soft lithography, MRS Bull. 26(7) (2001) 523-528. https://doi.org/10.1557/mrs2001.124

[55] B. Chun, V.V.K. Doddapaneni, M. Lucero, C. Pan, Z. Gao, Z. Feng, R. Malhotra, C.-H. Chang, Microreactor-assisted soft lithography of nanostructured antimony sulfide thin film patterns: Nucleation, growth and application in solid-state batteries, Energy Adv. 3(9) (2024) 2200-2211. https://doi.org/10.1039/D4YA00436A

[56] A.J. Dy, A. Cosmanescu, J. Sluka, J.A. Glazier, D. Stupack, D. Amarie, Fabricating microfluidic valve master molds in SU-8 photoresist, J. Micromech. Microeng. 24(5) (2014) 057001. https://doi.org/10.1088/0960-1317/24/5/057001

[57] G.M. Whitesides, The origins and the future of microfluidics, Nature 442(7101) (2006) 368-373. https://doi.org/10.1038/nature05058

[58] N. Bhattacharjee, A. Urrios, S. Kang, A. Folch, The upcoming 3D-printing revolution in microfluidics, Lab Chip 16(10) (2016) 1720-1742. https://doi.org/10.1039/C6LC00163G

[59] H. Gong, B.P. Bickham, A.T. Woolley, G.P. Nordin, Custom 3D printer and resin for 18 μm × 20 μm microfluidic flow channels, Lab Chip 17(17) (2017) 2899-2909. https://doi.org/10.1039/C7LC00644F

[60] K. Kadimisetty, S. Malla, K.S. Bhalerao, I.M. Mosa, S. Bhakta, N.H. Lee, J.F. Rusling, Automated 3D-printed microfluidic array for rapid nanomaterial-enhanced detection of multiple proteins, Anal. Chem. 90(12) (2018) 7569-7577. https://doi.org/10.1021/acs.analchem.8b01198

[61] R. Su, J. Wen, Q. Su, M.S. Wiederoder, S.J. Koester, J.R. Uzarski, M.C. McAlpine, 3D printed self-supporting elastomeric structures for multifunctional microfluidics, Sci. Adv. 6(41) (2020) eabc9846. https://doi.org/10.1126/sciadv.abc9846

[62] M. Xu, R. Wang, Y. Li, Electrochemical biosensors for rapid detection of Escherichia coli O157:H7, Talanta 162 (2017) 511-522. https://doi.org/10.1016/j.talanta.2016.10.050

[63] A.W. Martinez, S.T. Phillips, G.M. Whitesides, Three-dimensional microfluidic devices fabricated in layered paper and tape, Proc. Natl. Acad. Sci. U.S.A. 105(50) (2008) 19606-19611. https://doi.org/10.1073/pnas.0810903105

[64] W. Dungchai, O. Chailapakul, C.S. Henry, A low-cost, simple, and rapid fabrication method for paper-based microfluidics using wax screen-printing, Analyst 136(1) (2011) 77-82. https://doi.org/10.1039/C0AN00406E

[65] X. Li, X. Liu, Fabrication of three-dimensional microfluidic channels in a single layer of cellulose paper, Microfluid. Nanofluidics 16 (2014) 819-827. https://doi.org/10.1007/s10404-014-1340-z

[66] X. Li, J. Tian, G. Garnier, W. Shen, Fabrication of paper-based microfluidic sensors by printing, Colloids Surf. B Biointerfaces 76(2) (2010) 564-570. https://doi.org/10.1016/j.colsurfb.2009.12.023

[67] S. Kumar, P. Bhushan, S. Bhattacharya, Diagnosis of communicable diseases using paper microfluidic platforms, Point-of-Care Diagnostics-New Progresses and Perspectives, IAPC Open Book and Monograph Platform (OBP), Zagreb, Croatia (2016). https://doi.org/10.5599/obp.11.2

[68] Y.-C. Yu, Z. Wang, X. Ji, E.J. Williamson, H.M. Cordoba, A.M. Ulloa-Gomez, A.J. Deering, G.T.-C. Chiu, J.P. Allebach, L.A. Stanciu, Application of a dual-modality colorimetric analysis method to inkjet printing lateral flow detection of Salmonella typhimurium, Microchim. Acta 191(9) (2024) 559. https://doi.org/10.1007/s00604-024-06633-5

[69] Y.-J. Juang, S.-K. Hsu, Fabrication of paper-based microfluidics by spray on printed paper, Polymers 14(3) (2022) 639. https://doi.org/10.3390/polym14030639

[70] A.M.N. Gomez, Development of Electrochemical and Colorimetric Sensing Platforms for Agriculture and Healthcare Applications, Purdue University Graduate School, 2022.

[71] N. Kumawat, S.S. Soman, S. Vijayavenkataraman, S. Kumar, Rapid and inexpensive process to fabricate paper-based microfluidic devices using a cut and heat plastic lamination process, Lab Chip 22(18) (2022) 3377-3389. https://doi.org/10.1039/D2LC00452F

[72] Y.Q. Almajidi, S.M. Algahtani, O.S. Alsawad, H.S. Budi, S. Mansouri, I. Ali, M.M. Al-Hamdani, R.M. Romero-Parra, Recent applications of microfluidic immunosensors, Microchem. J. 190 (2023) 108733. https://doi.org/10.1016/j.microc.2023.108733

[73] Y. Wang, J. Chen, Y. Zhang, Z. Yang, K. Zhang, D. Zhang, L. Zheng, Advancing microfluidic immunity testing systems: New trends for microbial pathogen detection, Molecules 29(14) (2024) 3322. https://doi.org/10.3390/molecules29143322

[74] B. Asci Erkocyigit, O. Ozufuklar, A. Yardim, E. Guler Celik, S. Timur, Biomarker detection in early diagnosis of cancer: Recent achievements in point-of-care devices based on paper microfluidics, Biosensors 13(3) (2023) 387. https://doi.org/10.3390/bios13030387

[75] T. Khajvand, P. Huang, L. Li, M. Zhang, F. Zhu, X. Xu, M. Huang, C. Yang, Y. Lu, Z. Zhu, Interfacing droplet microfluidics with antibody barcodes for multiplexed single-cell protein secretion profiling, Lab Chip 21(24) (2021) 4823-4830. https://doi.org/10.1039/D1LC00567G

[76] J. Liu, Y. Bi, W. Tai, Y. Wei, Q. Zhang, A. Liu, Q. Hu, L. Yu, The development of a paper-based distance sensor for the detection of Pb^{2+} assisted with the target-responsive DNA hydrogel, Talanta 257 (2023) 124344. https://doi.org/10.1016/j.talanta.2023.124344

[77] J. Liu, I. Jasim, Z. Shen, L. Zhao, M. Dweik, S. Zhang, M. Almasri, A microfluidic-based biosensor for rapid detection of Salmonella in food products, PLoS One 14(5) (2019) e0216873. https://doi.org/10.1371/journal.pone.0216873

[78] H.M.U. Farooqi, M.A.U. Khalid, K.H. Kim, S.R. Lee, K.H. Choi, Real-time physiological sensor-based liver-on-chip device for monitoring drug toxicity, J. Micromech. Microeng. 30(11) (2020) 115013. https://doi.org/10.1088/1361-6439/ababf4

[79] J. Li, P.B. Lillehoj, Microfluidic magneto immunosensor for rapid, high-sensitivity measurements of SARS-CoV-2 nucleocapsid protein in serum, ACS Sens. 6(3) (2021) 1270-1278. https://doi.org/10.1021/acssensors.0c02561

[80] P. Li, Z. Zhang, Q. Zhang, N. Zhang, W. Zhang, X. Ding, R. Li, Current development of microfluidic immunosensing approaches for mycotoxin detection via capillary electromigration and lateral flow technology, Electrophoresis 33(15) (2012) 2253-2265. https://doi.org/10.1002/elps.201200050

[81] G. Cabral-Miranda, A.R. Cardoso, L.C. Ferreira, M.G.F. Sales, M.F. Bachmann, Biosensor-based selective detection of Zika virus specific antibodies in infected individuals, Biosens. Bioelectron. 113 (2018) 101-107. https://doi.org/10.1016/j.bios.2018.04.058

[82] S. Tharakan, O. Faqah, W. Asghar, A. Ilyas, Microfluidic devices for HIV diagnosis and monitoring at point-of-care (POC) settings, Biosensors 12(11) (2022) 949. https://doi.org/10.3390/bios12110949

[83] F. Chen, Q. Hu, H. Li, Y. Xie, L. Xiu, Y. Zhang, X. Guo, K. Yin, Multiplex detection of infectious diseases on microfluidic platforms, Biosensors 13(3) (2023) 410. https://doi.org/10.3390/bios13030410

[84] H. Hu, G. Cai, Z. Gao, C. Liang, F. Yang, X. Dou, C. Jia, J. Zhao, S. Feng, B. Li, A microfluidic immunosensor for automatic detection of carcinoembryonic antigen based on immunomagnetic separation and droplet arrays, Analyst 148(9) (2023) 1939-1947. https://doi.org/10.1039/D2AN01922A

[85] O. Ehuwa, A.K. Jaiswal, S. Jaiswal, Salmonella, food safety and food handling practices, Foods 10(5) (2021) 907. https://doi.org/10.3390/foods10050907

[86] G. Cai, L. Zheng, M. Liao, Y. Li, M. Wang, N. Liu, J. Lin, A microfluidic immunosensor for visual detection of foodborne bacteria using immunomagnetic separation, enzymatic catalysis and distance indication, Microchim. Acta 186 (2019) 1-9. https://doi.org/10.1007/s00604-019-3883-x

[87] B. Nasseri, N. Soleimani, N. Rabiee, A. Kalbasi, M. Karimi, M.R. Hamblin, Point-of-care microfluidic devices for pathogen detection, Biosens. Bioelectron. 117 (2018) 112-128. https://doi.org/10.1016/j.bios.2018.05.050

[88] L. Kang, B.G. Chung, R. Langer, A. Khademhosseini, Microfluidics for drug discovery and development: From target selection to product lifecycle management, Drug Discov. Today 13(1-2) (2008) 1-13. https://doi.org/10.1016/j.drudis.2007.10.003

[89] F. Bastan, M. Kazemeini, Enhanced production of Pirfenidone through a microfluidic system: A novel and thorough chemical kinetics investigation, Chem. Eng. Process. Process Intensif. 204 (2024) 109928. https://doi.org/10.1016/j.cep.2024.109928

[90] J.-L. He, D.-S. Wang, S.-K. Fan, Opto-microfluidic immunosensors: From colorimetric to plasmonic, Micromachines 7(2) (2016) 29. https://doi.org/10.3390/mi7020029

[91] J. Sun, A.R. Warden, X. Ding, Recent advances in microfluidics for drug screening, Biomicrofluidics 13(6) (2019) 061503. https://doi.org/10.1063/1.5121200

[92] C.D. Flynn, D. Chang, A. Mahmud, H. Yousefi, J. Das, K.T. Riordan, E.H. Sargent, S.O. Kelley, Biomolecular sensors for advanced physiological monitoring, Nat. Rev. Bioeng. 1(8) (2023) 560-575. https://doi.org/10.1038/s44222-023-00067-z

[93] J. Van Houten, A. Dosajh, A.F. Ogata, Multiplexed electrochemical detection of biomarkers in biological samples, Anal. Bioanal. Chem. (2023) 73-120. https://doi.org/10.1039/BK9781839169366-00073

[94] E. Mauriz, Low-fouling substrates for plasmonic sensing of circulating biomarkers in biological fluids, Biosensors 10(6) (2020) 63. https://doi.org/10.3390/bios10060063

[95] M. Song, X. Lin, Z. Peng, S. Xu, L. Jin, X. Zheng, H. Luo, Materials and methods of biosensor interfaces with stability, Front. Mater. 7 (2021) 583739. https://doi.org/10.3389/fmats.2020.583739

[96] A.I. Barbosa, N.M. Reis, Continuous flow of reagents interferes with the kinetics and equilibrium of antibody-antigen binding in microfluidic heterogeneous immunoassays, Sens. Actuators B Chem. 395 (2023) 134454. https://doi.org/10.1016/j.snb.2023.134454

[97] S. Kumar, W. Ahlawat, R. Kumar, N. Dilbaghi, Graphene, carbon nanotubes, zinc oxide and gold as elite nanomaterials for fabrication of biosensors for healthcare, Biosens. Bioelectron. 70 (2015) 498-503. https://doi.org/10.1016/j.bios.2015.03.062

[98] Y. He, Y. Wu, J.-Z. Fu, W.-B. Wu, Fabrication of paper-based microfluidic analysis devices: A review, RSC Adv. 5(95) (2015) 78109-78127. https://doi.org/10.1039/C5RA09188H

[99] X. Ma, G. Guo, X. Wu, Q. Wu, F. Liu, H. Zhang, N. Shi, Y. Guan, Advances in integration, wearable applications, and artificial intelligence of biomedical microfluidics systems, Micromachines 14(5) (2023) 972. https://doi.org/10.3390/mi14050972

Materials Research Foundations 191 (2026) 289-316 https://doi.org/10.21741/9781644904213-9

Chapter 9

Microfluidic Technology for Bioanalysis

Vipul D. Prajapati[1*], Hiral Vadgama[2]

[1]Department of Pharmaceutics, S.S.R. College of Pharmacy (Permanently Affiliated to Savitribai Phule Pune University), Sayli, Silvassa, U. T. of D.N.H. and D.D., India

[2]Department of Pharmaceutical Quality Assurance, S.S.R. College of Pharmacy (Permanently Affiliated to Savitribai Phule Pune University), Sayli, Silvassa, U. T. of D.N.H. and D.D., India

*vippra2011@gmail.com

Abstract

Microfluidics, commonly referred to as "lab-on-a-chip" technology, enables crucial manipulation of fluids at the micro level of scale, allowing high-throughput biological and chemical analyses. This chapter elaborates the application of microfuidic techniques in estimation of bioanalytics like nucleic acid (DNA/RNA) analysis, protein assays, and point-of-care diagnostics. Integrating microfluidics into drug delivery systems has paved the path for advanced treatments, including target specific delivery of drugs and gene therapy. Additionally, organ-on-a-chip models offer enhanced methods for analyte testing and tissue-engineering. Although fabrication complexity poses challenges, trends like artificial intelligence and nanotechnology offer promising advancements. Microfluidics continues to drive bioanalysis and therapeutic innovation, advancing personalized and regenerative medicine.

Keywords

Bioanalysis, Digital Microfluidic, Lab-on-a-chip, Microfluidics, Paper-based Microfluidic Platforms

List of Abbreviations

PDMS	: Polydimethylsiloxane
LOC	: Lab-on-a-chip
mTAS	: Micro total analysis systems
MEMS	: Microelectromechanical systems
DMF	: Digital microfluidics
OoC	: Organs-on-a-chip
2D	: Two-dimensional
3D	: Three-dimensional
PMR	: Polymerase chain reaction
DNA	: Deoxyribonucleic acid
RNA	: Ribonucleic acid
HTS	: High-throughput screening
ELISAs	: Enzyme-linked immunosorbent assays
POC	: Point-of-care

Materials Research Foundations 191 (2026) 289-316 https://doi.org/10.21741/9781644904213-9

DDSs	: Drug delivery systems
COVID-19	: Coronavirus disease of 2019
CRISPR	: Clustered regularly interspaced short palindromic repeats
LDH	: Lactate Dehydrogenase
AI	: Artificial Intelligence
ML	: Machine Learning
FPDAs	: Field programmable droplet arrays
IoT	: Internet of thinking
ANNs	: Artificial neural networks
NNs	: Neural networks

Contents

Materials Research Foundations 191 (2026) 289-316
https://doi.org/10.21741/9781644904213-9

1. Introduction

Microfluidics or micrototal analysis systems (referred to as "lab-on-a-chip (LOC)") is the growing technology deal with manipulating fluids at a microscale, usually within channels of dimensions of tens to hundreds of micrometers. By miniaturizing experimental processes into an integrated, compact device, this technology allows for efficient, high-throughput analysis of biological and chemical samples [1]. Microfluidics has its roots in the domains of analytical chemistry and microelectronics, and it emerged in the late eighteenth and early ninetieth. Instruments for capillary electrophoresis and chromatography were the main focus of early systems. But as biological sciences developed, especially during the Human Genome Project, the need for small, high-throughput instruments increased, which sped up the use of microfluidics in biological research and diagnostics. A key advance was made possible by the advent of soft lithography techniques, particularly those utilizing polydimethylsiloxane (PDMS), which made microfluidic devices much more accessible to researchers and allowed for their quick and affordable manufacturing [2,3].

The foundation of microfluidics is the design and application of instruments that can manage and controlling extremely minute fluidic volumes, often in the nanolitre or picolitre range. These systems use networks of valves, micropumps, microchannels, and micromixers to precisely regulate fluid flow and reactions [4]. The ability to control and flexibilate fluid dynamics with such accuracy is the primary need for bioanalytical applications, in case of deoxyribonucleic acid (DNA) sequencing, protein assays, and point-of-care diagnostics [5]. Compared to conventional techniques, microfluidic devices offer a clear benefit in that they enable quick analysis, need less reagents, and are less expensive [6]. Since its founding in the 1990s, the entire division of microfluidics has expanded quickly and had a big influence on other fields like biotechnology, chemistry, and biomedicine because of its ability to process microvolumes of materials with fewer reagents and faster reaction times [7].

In the context of bioanalysis, microfluidics has introduced a paradigm shift. It enables rapid and sensitive diagnostic testing, precise manipulation of single cells, miniaturized assays for detecting nucleic acids and proteins, and high-throughput screening in drug discovery [8,9]. All such properties make the microfluidic system related devices especially useful in point-of-care settings and resource-constrained contexts since they not only decrease reagent use and analysis time but also enhance accuracy and portability [10]. This chapter highlight the key platforms and applications that are shaping modern bioanalysis, and delve into the cutting-edge developments in this rapidly evolving field. By examining the challenges and future prospects of microfluidic technology, it uncovers the full potential in revolutionizing the landscape of bioanalysis.

2. Types of Microfluidic Bioanalytical Platforms

The various types of Microfluidic bioanalytical systems are illustrated in Fig. 1. Each type of platform has distinct advantages for various bioanalytical tasks, ranging from diagnostics to molecular biology. These systems enable precise manipulation and analysis of biological samples in a controlled microenvironment, resulting in increased efficiency, speed, and accuracy.

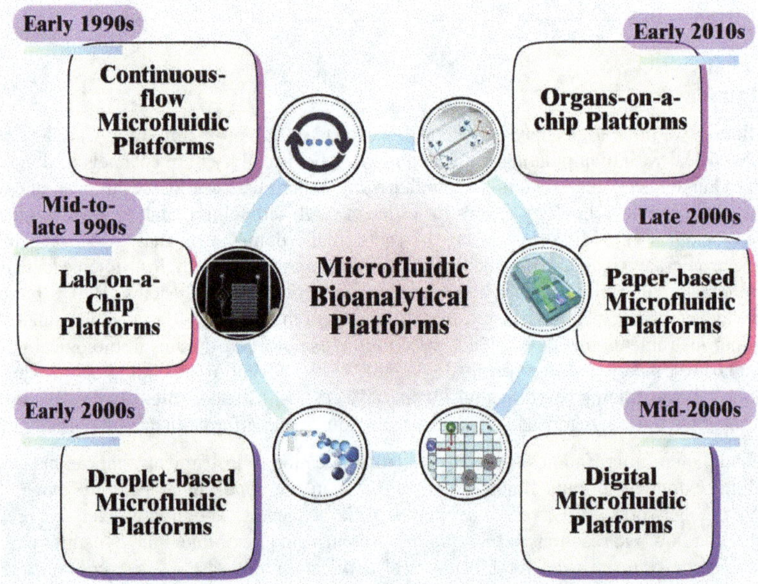

Figure 1. *Diagrammatic illustration showing the different microfluidic bioanalytical platforms with the year of invention.*

2.1 Continuous-flow Microfluidic Platforms

Continuous-flow microfluidic platforms are designed to transfer fluids through microchannels without interruption using pressure, electrokinetic forces, or capillary action. These platforms use constant fluid flow driven by external pumps or pressure systems through microchannels. Reactions occur as fluids move continuously through zones designed for mixing, separation, or detection.These systems are especially suitable for continuous operations such sample mixing, separation, and reagent supply. Continuous-flow systems may perform high-throughput bioanalyses, such as enzyme assays or cell-based research, with great precision since they control the flow in real time [11].

These platforms are frequently utilized in applications like drug screening, environmental monitoring, and disease identification. They can also use a variety of detecting modalities, including as optical or electrochemical sensors, to monitor chemical or biological interactions in real time. Continuous-flow microfluidics are extremely useful for executing multi-step bioanalytical procedures such as DNA amplification or protein detection, where chemicals must be introduced sequentially [12].

2.2 Lab-on-a-Chip Platforms

LOCs, which range in size from millimeters to square centimeters, are tiny microfluidic devices that combine several laboratory operations onto a single chip. Integrates sequential experimental operations including the sample preparation, reaction, separation, and estimation onto a singular microchip. These platforms automate complex analytical procedures at a micro-scale. Currently, physics, engineering, chemistry (including biochemistry), and biology are all involved in the development and implementation of microfluidics technology [13]. The exact control of fluids at the microscale is the foundation of LOC platforms. LOCs are sometimes reported as the micro total analysis systems (mTAS) due to their capacity of changing the minute level of fluid quantities, frequently in picoliters or too smaller. Microelectromechanical systems (MEMS) are what LOCs are. Because microfabrication and microfluidics enable the creation of intricate microfluidic lines/channel and chambers in a single chip, they have made it feasible to miniaturize analytical procedures, from sample preparation to detection [14].

These platforms are especially helpful in point-of-care diagnostics because they provide quick, affordable, and portable ways to analyze blood samples, diagnose illnesses, and run genetic tests. LOC systems are being utilized more and more in forensic research, clinical diagnostics, and environmental monitoring because they allow for the on-site examination of intricate biological material [15].

2.3 Droplet-based Microfluidic Platforms

Droplet-based microfluidic platforms are extremely effective at encapsulating small volumes of fluids, such as biological samples, into discrete droplets. Droplet flow has several benefits against the standard single-phase microfluidics, including the removal of Taylor dispersion, elevated blending, separation of the droplet contents from the textures, and the ability to contain and address individual cells or biomolecules [16]. These droplets execute thousands of distinct reactions at once because they operate as isolated microreactors [17,18]. Applications needing high-throughput analysis, like drug discovery, enzyme kinetics, and single-cell analysis, are best suited for droplet-

Materials Research Foundations 191 (2026) 289-316 https://doi.org/10.21741/9781644904213-9

based systems. One subclass of microfluidics technology that has the potential to completely transform medication delivery methods is droplet-based microfluidics [19].

The ability of droplet-based systems to carry out multiplexed assays in a compact manner, reducing reagent use and enabling a variety of biological investigations to be carried out simultaneously, is one of their main advantages. Droplet microfluidics is frequently used for high level throughput screening of bio-analytics, proteomics, and cellular genomics. It offers sensitive and rapid examination at the level of individual cells [20].

2.4 Digital Microfluidic Platforms

Digital microfluidics (DMF) utilises the electrical field for changing the individual fluid droplets on an open surface. DMF provides precise, programmable control over droplets, allowing for actions like mixing, splitting, and transport, in contrast to continuous-flow or droplet-based systems [21]. According to, DMF is very adaptable and may be excecuted to determine the varieties of bioanalytical tasks with little sample and reagent use.

DMF systems are used in diagnostic tests, proteomics, and high-throughput screening. Since the same droplet array can be employed for a variety of tasks, DMF is very adaptable for both clinical and scientific purposes. DMF devices increase our understanding of cellular heterogeneity by offering an automated phenomena for highly sensitive proteome analysis of the smallest cell populations [22].

2.5 Paper-based Microfluidic Platforms

They are easy, affordable devices made from cellulose and other porous materials. Because capillary action promotes fluid mobility, these platforms do not dependent on the external energy sources. This makes them especially helpful in scenarios like field diagnostics and point-of-care estimating where resources are limited [23]. When liquid is introduced, the paper's inherent wicking properties draw it through the channels and onto regions of dried reagents, typically producing a color shift that indicates the presence of an analyte [24].

The potential of paper-based microfluidics to carry out inexpensive diagnostic procedures, including pregnancy tests, pathogen detection, and glucose monitoring, has drawn interest. They are perfect for use in environmental and global health monitoring since they are frequently combined with biosensors to enable quick real-time detection of biomarkers or infectious pathogens [25].

2.6 Organs-on-a-chip Platforms

This is a biomimetic system that mimics the physiological processes of organs on a microfluidic chip and is ranked among the top 10 emerging technologies [26]. Living cells are cultured in interconnected chambers to simulate tissue interfaces, flow dynamics, and physiological responses. By carefully adjusting and managing crucial factors like concentration differences, shear stress, cellular patterns, tissue borders, and tissue-organ interactions, this cutting-edge platform replicates the dynamic microenvironment of organs or tissues [27]. Advanced microfluidic systems called organs-on-a-chip (OoC) platforms are made to resemble the composition and functionality of human organs. These platforms mimic organ-level functions, including those of the heart, liver, kidneys, and lungs, by combining biological cells with microengineering. Organs-on-a-chip systems are widely utilized in toxicity research, disease

Materials Research Foundations 191 (2026) 289-316 https://doi.org/10.21741/9781644904213-9

modeling, and drug testing because conventional two-dimensional (2D) cell cultures cannot faithfully mimic the complexity of living tissues.

These platforms have transformed in vitro testing by offering more accurate human biological models, which are essential for researching medication reactions, disease progression, and customized medicine. These gadgets aid in the creation of safer medications by simulating organ microenvironments and predicting how people will react to medications or environmental pollutants [28]. The brief summary of the key microfluidic techniques is depicted in a Table 1.

Table 1. *Description of the key microfluidic techniques used in bioanalysis with its salient reported features.*

Microfluidic platform	Quantification limit	Runtime	Sample	Resolution/ Throughput	References
Continuous-Flow	~nM to μM (e.g., 1 nM - 10 μM for small molecules)	Seconds to minutes	Blood, serum, environmental samples	Continuous processing; flow rates ~1–100 μL/min; single-cell to bulk analysis	[29]
Lab-on-a-Chip	~pM to nM (e.g., 10 pM for nucleic acids)	10 minutes to 2 hours	Nucleic acids, proteins, cells	Multiplexed assays; throughput up to 96–384 assays per run	[30]
Droplet-Based	~pM to nM (e.g., 10 pM for proteins)	Minutes to hours	Single cells, reagents in droplets	Throughput: 1000–10,000 droplets/sec; volume ~pL–nL per droplet	[31,32]
Digital Microfluidics	~fM to pM (e.g., 100 fM for proteins)	10 minutes to 1 hour	Serum, plasma, cell lysates	Multiplexed; sample volumes ~nL to μL; high sensitivity	[33]
Paper-Based	~μM to mM (e.g., 10 μM for glucose)	5–30 minutes	Urine, blood, saliva	Moderate throughput; low-cost; qualitative to semi-quantitative	[34,35]
Organs-on-a-Chip	Cell-level detection (single-cell resolution)	Hours to days	Cultured human tissue cells	High resolution; mimics organ function; multiplexed sensors	[36]

3. Microfluidics for Biomolecule Analysis

Microfluidic technology has significantly expanded the capabilities of biological and chemical analysis by providing platforms for rapid, efficient, and precise assays as depicted in a Fig. 2. Several key applications of microfluidics in biological and chemical analyses, ranging from nucleic acid studies to diagnostic testing are described.

Materials Research Foundations 191 (2026) 289-316 https://doi.org/10.21741/9781644904213-9

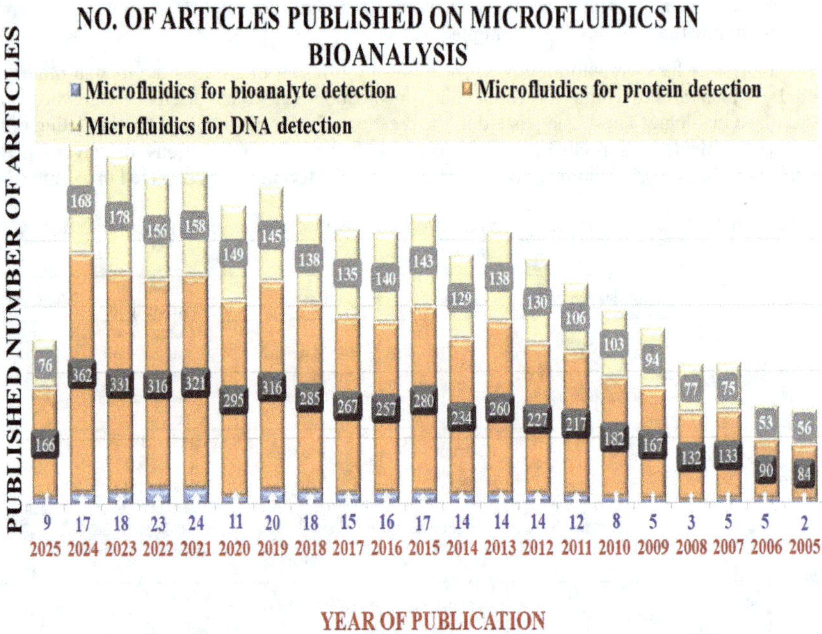

Figure 2. An overview of the tentative number of publications in the PubMed database from the last two decades, particularly addressing the terms "Microfluidic technology for bioanalyte detection", "Microfluidics for protein detection", and "Microfluidics for DNA detection".

3.1 Microfluidics in Nucleic acid Analysis

Microfluidic platforms offer faster and more precise processing of genetic material, making them ideal for DNA and ribonucleic acid (RNA) analysis. Critical applications including DNA sequencing, gene expression profiling, and polymerase chain reaction (PCR) are made easier by these gadgets. Microfluidics' tiny volumes and quick heat cycling capabilities improve reaction kinetics while consuming less sample and reagent [37]. For extremely quick and on-site genetic diagnostics, integration with methods like PCR and isothermal amplification (like LAMP) is possible. For extremely quick and on-site genetic diagnostics, integration with methods like PCR and isothermal amplification (like LAMP) is possible. The workflow of microfluidics in DNA/RNA analysis is illustrated in Fig. 3. Applications in genotyping, gene expression analysis, and infectious disease diagnostics are advanced by droplet microfluidics, which in particular provides high-throughput single-molecule detection and digital PCR capabilities [38]. Additionally, integration with next-generation sequencing has enabled the development of compact, high-throughput sequencing platforms that provide insights into genomics and transcriptomics [39].

Materials Research Foundations 191 (2026) 289-316 | https://doi.org/10.21741/9781644904213-9

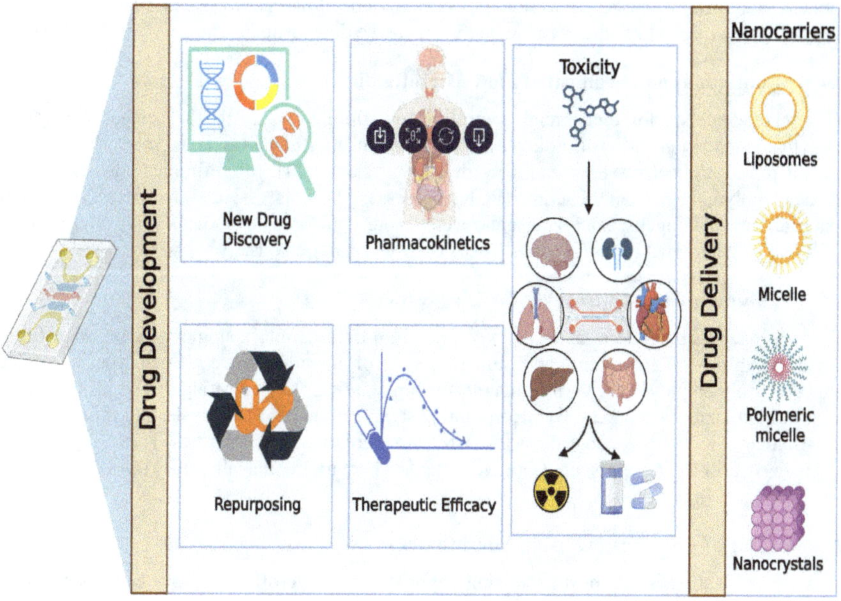

Figure 3. *Diagrammatic illustration of workflow of microfluidics in DNA/RNA analysis.*

3.2 Microfluidics for Protein and Enzyme Assays

Microfluidics has enhanced protein and enzyme assays, which are essential to biochemical research, by providing fine-grained control over reaction conditions. Proteins, enzymes, and metabolites in biological materials can be found and measured using highly sensitive assays carried out by microfluidic devices. For example, the miniaturization of enzyme-linked immunosorbent assays (ELISAs) onto microfluidic systems has reduced assay time and reagent costs while preserving excellent sensitivity and specificity [40]. Studies on protein crystallization and folding also employ microfluidics, which offers a deeper understanding of the structures and functions of proteins. Microfluidic environments provide precise temporal control and mixing efficiency that are advantageous for enzyme kinetics experiments [41]. These systems are instrumental in exploring protein interactions and activity under controlled conditions .

3.3 Immunoassays on a Chip

Immunoassays, which are frequently employed in clinical diagnostics to identify antigens and antibodies, have been successfully downsized into "lab-on-a-chip" devices. High sensitivity and short assay times are made possible by these microfluidic immunoassays, which enable multiplexed investigation of cytokines, hormones, and biomarkers. Because of this, they are perfect for uses like identifying hormones, infections, and biomarkers in clinical samples [42]. Microfluidics and immunoassays have been combined to provide point-of-care applications and

Materials Research Foundations 191 (2026) 289-316 https://doi.org/10.21741/9781644904213-9

increase diagnostic accuracy. Technologies like microbead-based ELISAs and lateral flow tests have been downsized to make portable and inexpensive diagnostic instruments possible [43].

3.4 Continuous and Segmented Flow Microfluidics

This method allows for the crucial controlling the fluid flows as the basic need for various analytical applications. Microfluidic platforms can form and study artificial membranes, which are crucial in biophysical research and drug delivery systems [44]. Microfluidics enables fast and efficient polymerase chain reaction (PCR) processes. PCR is essential for detecting genetic material from pathogens, analyzing gene expression, and forensic studies. With microfluidics, PCR can be performed in few seconds by the conventional methods.

3.5 Lab-on-a-Chip Approaches for Diagnostics

This approaches integrate number of experimental processes into a single specific microfluidic device, enabling the automation and miniaturization of diagnostic assays. Such types of devices are commonly employed for various diagnostic applications, from detecting infectious diseases to monitoring metabolic markers. By incorporating sample preparation, analysis, and estimation into the singular chip, these systems allow for faster and more reliable diagnostics. These platforms are particularly utilized in the resource-limited settings where traditional laboratory infrastructure may be unavailable [45].

3.6 Point-of-Care Testing using Microfluidics

This type of testing is the most promising applications of microfluidics, as it enables rapid and decentralized testing for a range of conditions, from infectious diseases to chronic illnesses. Microfluidic point-of-care (POC) devices are portable, simple in use, and capable of depicting the results in few seconds, making them ideal for use in clinics, emergency rooms, and even in-home settings. These devices can perform complex diagnostic tests with minimal operator input, ensuring faster treatment decisions. Examples include glucose monitors, pregnancy tests, and Coronavirus disease of 2019 (COVID-19) diagnostic assays [46].

3.7 Drug Delivery Systems and Advanced Therapeutics

Microfluidics has transformed the development of drug delivery systems (DDSs) by providing for exact control over the time, dose, and location of drug release. Applications in this field consist of varieties of DDSs.

3.7.1 Targeted Drug Delivery

Microfluidic devices can be fabricated in sutable ways to release the specific molecules at specific target sites, such as cancerous tumors, while reducing the responses on the rest of healthy tissues [47]. By enabling the creation of complex delivery vehicles like nanoparticles, liposomes, and micelles that may be functionalized with targeting ligands like antibodies, peptides, or small molecules, microfluidic devices have greatly improved the accuracy of targeted drug administration as depicted in Fig. 4 [48]. To guarantee that the medication is only released at the intended location, these vehicles have the ability to interact to particular receptors on the surface of sick cells. The creation of homogeneous, monodisperse drug carriers with regulated size, shape, and drug-loading capacity is made easier by microfluidic platforms. Microfluidics' exact control

Materials Research Foundations 191 (2026) 289-316
https://doi.org/10.21741/9781644904213-9

over fluid dynamics guarantees carrier production repeatability, which is essential for clinical applications [49].

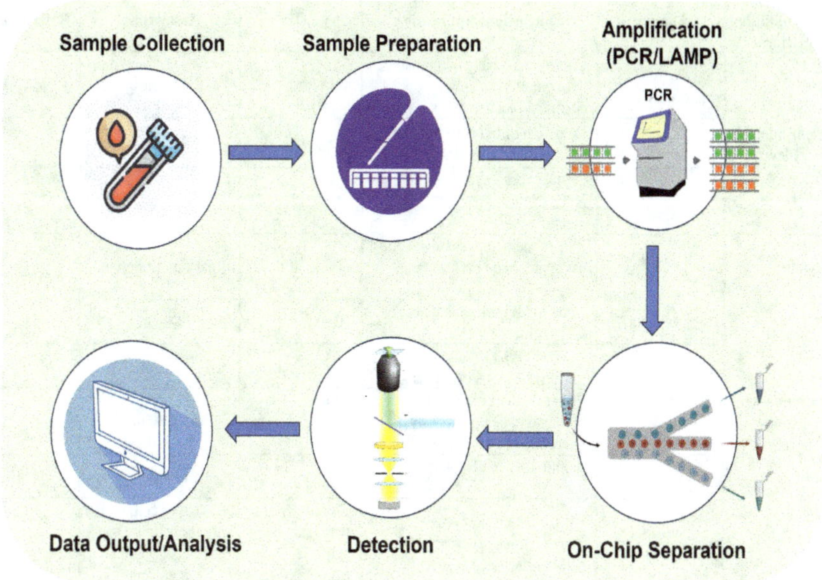

Figure 4. *Microfluidics in drug development and various drug delivery vehicles for targeted drug release.*

3.7.2 Advanced Therapies

These systems are also being investigated for the creation of next-generation treatments, such as the delivery of biologics, which include proteins and peptides, and necessitate exact dosage control [50]. Genetic material (DNA, RNA, or clustered regularly interspaced short palindromic repeats (CRISPR) components) [51] can be efficiently delivered to particular cells or tissues thanks to microfluidic devices [52,53]. Researchers can develop homogeneous gene delivery vectors including viral particles or lipid-based nanoparticles, that encapsulate the genetic material and shield it from deterioration by employing microfluidics. By engineering these carriers to target particular cells, gene therapy treatments can be made more precise and safer. Table 2 depicts types of microfluidic platforms along with analytes used and application.

Table 2. *Comparison of various types of microfluidic platforms againts the type of analyte and conclusion/application.*

Type of Microfluidic platform	Analyte	Conclusion/ Application	LOD	Assay time	References
Continuous-Flow	Small molecules	Controlled mixing combined with parallel chemical/biochemical sensors	~nM–μM	Minutes	[54]
	DNA	Detection of hybridization and electrophoretic separation	Not specified	Minutes	[55]
	Proteins	Droplet shrinkage for sample preconcentration and quantification	~1 μg/mL	<10 min	[56]
Lab-on-a-Chip	DNA	Electrochemical gene detection using graphene FET and programmable flow	~0.642 aM	Real-time (minutes)	[57]
		Hepatitis B multiplex detection with a glucose meter	10 pM	~Minutes	[58]
	Proteins	hCG and PSA immunoassays for the diagnosis of cancer	~pg/mL range	15 min	[59]
		Quantification of cancer biomarkers using antibody microarrays	Sub-ng/mL	~10–30 min	[60]
	Glucose	Serum enzymatic colorimetric tests	200 μM	~3.5 min (210 sec)	[61]
	Cells	Utilized in clinical settings for integrated lab processes and cell-based assays.	~1 cell per well (single-cell level resolution)	Assay per cell: ~<10 minutes; Entire run: High-throughput real-time	[62]
Droplet-Based	Glucose	Real-time metabolic monitoring of organs on a chip	μM range	2.5 min	[63]
	LDH (Lactate Dehydrogenase)	Monitors liver damage and cytotoxicity in organ-on-a-chip devices	Low μM	2.5 min	[63]
	Total Bile Acids	Evaluates the metabolism and liver function of spheroid cultures.	Low μM	2.5 min	[63]
	Dopamine & ascorbic acid	Drug tests using droplet electrochemistry	20 μM (DA) / 41 μM (AA)	Seconds	[64]
Digital Microfluidics	Proteins (Thrombin)	Gold electrode-based aptamer-based impedimetric protein detection	0.01 nM	<10 min	[65]
	Proteins (Multiplexed panel)	Digital immunoassay using machine learning for bedside biomarker analysis	<10 pg/mL	30 min total, 5 min incubation	[66]
	Serum	High-plex single-molecule detection in a sample of less than 15 μL	<5 pg/mL	40 min	[67]
	Cells	Monitors the fluid's interaction with cells	~3 μm particle size	Real-time	[68]
	DNA (hybridization)	Label-free detection with a digital microfluidic device that	500 pM	<1 min	[69]

Materials Research Foundations 191 (2026) 289-316 | https://doi.org/10.21741/9781644904213-9

		incorporates a nanostructured SPRi biosensor			
Paper-Based	Glucose	Diabetes screening using the colorimetric test of starch-iodine	0.05 mM	~5 min	[70]
	Uric Acid	Platform for multi-analyte POC testing	45 ng/mL	~Minutes (room temp)	[71]
	Urea	Human plasma detection using stimulus-sensitive valves	2.5 μg/mL	~Minutes	[71]
	Triglycerides	Human plasma detection using stimulus-sensitive valves	1.5 μg/mL	~Minutes	[71]
Organs-on-a-Chip	Lactate	Evaluates the metabolic activity of liver microtissues and cancer in real time	Not explicitly stated; validated against standard lab kits	~Continuous real-time monitoring	[72]
	Glucose	Monitors the intake of nutrients in living tissue cultures	Low μM	2.5 min (droplet-based)	[63]
	Secreted Cytokines	Cytokine secretion at single-cell resolution for immune profiling	pg/mL range	Minutes to hours	[73]

4. Integration with Omics Technologies

The integration of microfluidics with omics technologies has grown as a transformative development in the modern bioanalysis. By enabling the manipulation and analysis of minute biological samples in precisely controlled microscale environments, microfluidic systems allow researchers to achieve levels of efficiency, sensitivity, and throughput that are difficult to match with the general analytical techniques.

4.1 Microfluidics in Genomics and Transcriptomics

Microfluidic devices have made major advancements in the processing and analysis of nucleic acids possible in the area of transcriptomics and genomics. The creation of droplet-based microfluidics, which encapsulate individual cells or molecules in tiny aqueous compartments to enable high-throughput cellular RNA sequencing, has been one of the most noteworthy developments. These platforms improve reproducibility and reduce sample loss by simplifying and automating difficult workflows including cell lysis, nucleic acid extraction, amplification, and barcoding into on-chip procedures [74].

The development of spatial transcriptomics, which uses microfluidics to map gene expression patterns inside tissue sections at subcellular resolution, has also benefited from microfluidics. Furthermore, accurate genomic modification and single-cell screening are made possible by the incorporation of gene-editing technologies like CRISPR into microfluidic devices [75,76]. Together, these developments offer a better comprehension of cellular diversity and regulatory networks, which are essential for both basic biology and medical diagnosis [77,78].

4.2 Applications in Proteomics and Metabolomics

By tackling some of the most enduring issues in the discipline, such as sample preparation difficulty, low sensitivity, and excessive reagent consumption, microfluidic technologies have also significantly advanced proteomics and metabolomics. Microfluidic devices facilitate workflows

Materials Research Foundations 191 (2026) 289-316 https://doi.org/10.21741/9781644904213-9

like enzymatic digestion, separation, and sample cleanup within integrated channels by enabling the exact and controlled handling of proteins and metabolites. These procedures are frequently directly connected to mass spectrometry, which improves sensitivity and makes it possible to analyse complicated biological samples in high resolution [79].

Microfluidic platforms are very useful in metabolomics because they can quickly process and analyse volatile and labile metabolites under strictly controlled circumstances. These technologies are very helpful for tracking cellular conditions in almost real time and enable dynamic, time-resolved characterization of metabolic reactions [80,81]. The compatibility of microfluidics with low-volume, low-abundance clinical samples further strengthens its role in translational and personalized medicine research.

4.3 High-Throughput Screening for Biomedical Research

Microfluidic platforms, which provide unmatched scalability and efficiency, have become essential in high-throughput screening (HTS). With these devices, thousands of parallel tests can be carried out with excellent temporal precision and little reagent use. Pneumatic valve arrays, digital microfluidics, or droplet-based formats—microfluidic HTS platforms allow for the quick and automated implementation of intricate experimental procedures [82,83]. Applications including drug discovery, functional genomics, and biomarker validation have found this to be especially helpful. To find possible therapeutic targets or clarify gene functions, microfluidic chips, for example, can be used to screen libraries of small compounds or genetic perturbations, such as CRISPR-based knockouts. Researchers can reduce costs and human labor while increasing throughput and data richness by combining microfluidics with robotic automation and machine learning [84,85].

5. Emerging Innovations in Microfluidics

The advancements in this field are revolutionizing the ways in which microfluidic platforms are designed, operated, and applied in bioanalysis, diagnostics, and personalized medicine.

5.1 Artificial Intelligence and Machine Learning Integration

Artificial intelligence (AI) is a very useful technique for efficiently handling large datasets. Researchers have used AI approaches in a variety of domains and applications during the past few decades, such as the creation of virtual assistants, driverless cars, disease detection, and picture processing and recognition. By identifying patterns in a variety of datasets, algorithms namely artificial neural networks (ANNs) and deep NNs mimic human intelligence. In order to create logical connections between data points and facilitate activities like recognition, reasoning, and knowledge acquisition, these algorithms use hypotheses during the data collection process [86]. For instance, AI can be used to optimize experimental parameters in real-time, improve pattern recognition for disease diagnostics, and automate complex bioassays. AI, for instance, can be used to automate intricate bioassays, enhance pattern recognition for illness diagnosis, and optimize experimental parameters in real-time. In customized medicine, where patient-specific data can be processed and used to more effectively modify treatments, AI-enhanced microfluidic devices are probably going to be essential [87]. Moreover, Machine learning (ML) techniques facilitate the analysis of complex datasets generated by microfluidic experiments, aiding in the rapid identification of biomarkers and disease signatures [88,89].

Materials Research Foundations 191 (2026) 289-316 https://doi.org/10.21741/9781644904213-9

5.2 Digital Microfluidics and Programmable Systems

A paradigm shift in fluid handling is represented by digital microfluidics (DMF), which uses electrowetting to manage discrete droplets on an array of electrodes. Because droplets can be precisely moved, merged, divided, or combined under programmable control, this technique provides unmatched versatility in carrying out intricate biochemical processes. Field Programmable Droplet Arrays (FPDAs), which allow for reconfigurable and high-throughput experiments on a single chip, are the result of recent developments [90]. These programmable systems are instrumental in automating laboratory workflows, reducing reagent consumption, and enhancing the scalability of diagnostic tests [91].

5.3 Wearable and Implantable Bioanalytical Microdevices

The development of wearable and implantable microfluidic devices has created new opportunities for focused therapy and ongoing health monitoring. In order to provide real-time relevant data on physiological variables like electrolyte balance, glucose levels, and stress signs, wearable microfluidic sensors can non-invasively collect and analyse sweat, interstitial fluid, or other biofluids. Conversely, implantable microfluidic systems can monitor tissue reactions, dispense medications at regulated rates, or support regenerative medicine applications. In order to guarantee patient comfort and compliance, these devices are made to be biocompatible and minimally intrusive. A big step toward proactive and individualized healthcare treatments is the incorporation of microfluidics into wearable and implantable forms [92].

6. Challenges and troubleshooting

Microfluidic technology has received broad application in bioanalysis due to its precision and efficiency. However, there are significant obstacles and limits involved with its development and execution. Here is a thorough synopsis:

6.1 Technical Complexity

Advanced fabrication techniques like microlithography and micromachining are used in the creation of microfluidic devices, and they call for specific equipment and expertise [93]. The creation of specialized devices for particular uses can be costly and time-consuming, even if microfluidics offers extremely controlled settings for bioanalysis. It's still difficult to put together modular systems that satisfy certain experimental conditions [94]. Advances in three-dimensional (3D) printing and fast prototyping are addressing these difficulties by delivering cost-effective and scalable alternatives. Device prototyping and testing are made simpler by 3D printing's ability to swiftly and easily fabricate complex microfluidic structures without the requirement for specialized conditions [95].

6.2 Integration and Interfacing Issues

In order to accomplish multiple tasks, microfluidic devices frequently need to integrate different parts, such as sensors, pumps, and valves. The integration of previously described components into a working microfluidic device needs careful leveling of the material compatibility, bonding procedures, and desired packaging. Soft lithography, direct as well as thermal bonding, and microfluidic interconnections are the examples of common integration techniques. Integrating

Materials Research Foundations 191 (2026) 289-316 https://doi.org/10.21741/9781644904213-9

microfluidic devices with external analytical equipment, such as optical detectors or mass spectrometers, can be challenging and may require substantial modification.

Innovations in the growth of all-in-one lab-on-chip systems and microfluidic-electronic integration are tackling these issues. Compact detection modules and wireless connectivity are two examples of new interface technologies that facilitate smooth integration with external systems, increasing the usability and compatibility of microfluidic platforms with current laboratory infrastructure [96].

6.3 Reproducibility and Scalability

Ensuring that microfluidic devices generate consistent findings across numerous trials is a problem, especially in complicated bioanalytical applications where tiny alterations in device shape or flow conditions can create variations [97]. It is challenging to scale and mass-produce microfluidic devices for the industrial or clinical usage due to a lack of uniformity. Because each gadget is frequently customized for a particular use, there is less interoperability between them and production costs go up [98].

Standardized microfluidic systems that can be modified for various uses are being developed. Manufacturers can scale up production while preserving device design flexibility by creating modular, plug-and-play solutions. Additionally, industry standardization will make it easier for regulators to approve clinical use [99].

6.4 Ethical Considerations in Bioanalytical Miniaturization

The ethical implications of microfluidics' downsizing of bioanalytical systems are significant, especially in relation to data privacy, consent, and the possible decrease in animal testing. Microfluidic technologies hold the potential to alleviate ethical problems related to animal suffering by substituting OoC platforms for conventional animal models. For instance, the three-R (Replacement, Reduction, and Refinement) philosophy has been adhered to by developing microfluidic 3D models to lessen the sacrifice of experimental animals in the scientific invention [100]. Furthermore, the incorporation of AI and internet of thinking (IoT) into microfluidic devices raises new ethical issues regarding privacy and data security, calling for strong frameworks to guarantee responsible use [101].

6.5 Sample and Reagent Limitations

When working with extremely diluted samples or when higher volumes are required to provide meaningful findings, microfluidics' minimum reagent volumes can be a drawback. Certain bioanalytical applications that require larger sample concentrations might not always be appropriate for the microfluidic environment [4,102]. Protein extraction or separation are two examples of the complicated pre-treatment procedures that are frequently needed to prepare biological material for microfluidic analysis. The entire operation is made more complex and time-consuming by these pre-treatments [103]. The development of advanced biosensors that offer higher sensitivity is improving the detection limits in microfluidic assays.

6.6 Regulatory and Commercialization Hurdles

Microfluidic devices, particularly those used in medical diagnostics, may have a drawn-out and complicated regulatory approval process. Additionally, overcoming commercialization obstacles

Materials Research Foundations 191 (2026) 289-316 https://doi.org/10.21741/9781644904213-9

like cost-effectiveness and mass production is necessary to move from research prototypes to products that are ready for the market.

Early cooperation with regulatory agencies is essential to addressing regulatory difficulties. Using reliable, repeatable models for preclinical validation will also hasten approvals. Collaborations between academic institutions, business, and government organizations can expedite the commercialization process, facilitating the more effective transition of microfluidic technologies from the experimenal area to the commercial area [104].

6.7 Complexity of Data Analysis

Mixing, incubating, and separating are examples of multistep bioanalytical procedures that can produce complicated data sets that call for sophisticated analysis methods. As a result, interpreting the findings of microfluidic assays takes more time and effort. The design, layout planning, and sample preparation procedures in digital microfluidic devices pose algorithmic difficulties, especially in high-throughput bioanalysis [105].

6.8 Cost and Commercialization

Long-term efficiency is provided by microfluidic systems, but the initial expense of creating and manufacturing these devices can be unaffordable, particularly for small-scale research centers or academic labs [106]. Significant regulatory obstacles and expenses related to increasing production and guaranteeing adherence to industry standards must be overcome in order to bring microfluidic devices from the research stage to the market.

7. Future Trends and Developments

With new developments and trends, microfluidic technology is still developing and has the potential to completely transform bioanalysis, diagnostics, and biomedical research. As the field develops, a number of emerging themes are influencing microfluidics' course and enhancing its applicability to a wide range of fields and sectors.

7.1 Autonomous, Self-Learning Bioanalysis Systems

Using microfluidic platforms in conjunction with AI and ML is the entry of gate towards the self-optimizing, autonomous systems. In chemical synthesis and biological tests, these systems can improve efficiency and reproducibility by adaptively controlling experimental settings in real-time. To illustrate the promise of AI-driven microfluidics in speeding up research and development procedures, autonomous microreactor devices that use Bayesian optimization have been created to investigate ideal circumstances for ultrafast synthesis [107].

Additionally, the development of intelligent downstream processing in microfluidics emphasizes the significance of continuous, integrated methodologies for biologics production [108]. Innovations like the Drop Micro Fluid Agents (DMFAs) framework leverage large language models to automate the fabrication and optimization of droplet microfluidic apparatus, showcasing the potential of AI in streamlining microfluidic research and development [109].

Materials Research Foundations 191 (2026) 289-316 https://doi.org/10.21741/9781644904213-9

7.2 Multiplexed and High-throughput Microfluidic Platforms

A noteworthy breakthrough in microfluidics is the creation of highly multiplexed and high-throughput platforms. These technologies allow multiple samples or targets to be examined simultaneously on a single chip, which significantly elevates the efficiency and data throughput. Multiplexed microfluidic devices, which enable the faster, parallel testing of several chemicals or genetic markers, are becoming more and more essential in applications like drug development and genomic screening. These technologies reduce processing times and reagent costs while enabling comprehensive analysis in a few seconds required for traditional processes [110].

7.3 Three-dimensional Printing and Advanced Fabrication Techniques

The art and fabrication of devices is changing as a result of the introduction of 3D printing in microfluidics. While soft lithography and other traditional methods are frequently used to create microfluidic devices, 3D printing enables quick prototyping, customisation, and the creation of intricate structures that were previously challenging to accomplish. More intricate fluidic channels, chambers, and sensors can be merged into 3D-printed microfluidic toys, which is essential for cutting-edge uses like tissue engineering and organ-on-a-chip. Researchers may quickly iterate and improve gadget designs thanks to 3D printing's versatility [111].

7.4 Convergence with Nanotechnology and Synthetic Biology

A major contribution from nanotechnology is anticipated in the upcoming generation of microfluidic devices. In order to improve sensitivity, detection limits, and functionality, microfluidic systems are progressively incorporating nanoscale materials and architectures, such as nanowires, nanoparticles, and nanofibers. By enhancing biosensor performance, nanomaterials can detect biomolecules at incredibly low concentrations [112]. Because of its high level of sensitivity and specificity, low sample volume requirements, and real-time biomarker monitoring capabilities, microfluidic-based nanobiosensors lower the likelihood of false positives. Rapid identification and early treatment application are made possible by these sensors. Additionally, more accurate control over fluid dynamics and surface interactions can be made possible by combining nanotechnology and microfluidics, which is the heart for usage in drug molecules desired delivery, regular environment system monitoring, and cancerous cell‘s diagnosis [113].

Additionally, the combination of synthetic biology with microfluidics has been hailed as a biotechnology secret weapon that enables the highly precise design and operation of intricate biological circuits. Recent developments also examine how synthetic biology and nanotechnology might be combined to create novel medical applications, highlighting the contribution of microfluidics to these multidisciplinary strategies [114].

7.5 Sustainable Microfluidics

Microfluidic development is increasingly focusing on sustainability, with initiatives to cut down on waste, energy use, and dependency on hazardous ingredients. In an effort to lessen the environmental impact of single-use disposable devices, researchers are paying more attention to the use of biodegradable components for the fabrication of actual level of microfluidic apparatus. The creation of reusable or recyclable devices and the application of less energy-intensive production techniques are also key components of sustainable microfluidics. Sustainable

Materials Research Foundations 191 (2026) 289-316 https://doi.org/10.21741/9781644904213-9

microfluidics is anticipated to emerge as a crucial field of innovation as worries about the environmental impact of lab-generated waste increase [115].

Conclusion

Bioanalysis has changed as a result of microfluidic technology, which offers unmatched benefits in terms of accuracy, speed, and compactness. Microfluidic systems have made it possible to automate and scale down complicated scientific procedures, increasing their efficiency and accessibility, from the analysis of DNA and RNA to the detection of proteins and cells. Microfluidics' uses in the biological and chemical sciences are growing as a result of its incorporation into domains including drug discovery, customized medicine, and diagnostics. Because microfluidics can automate, optimize, and reduce the size of biological and chemical processes, it is a major advancement in bioanalysis.

By increasing data analysis, detection sensitivity, and device complexity, emerging technologies like artificial intelligence, nanotechnology, and 3D printing are expected to improve microfluidic systems. Microfluidics is a crucial instrument for upcoming scientific and medical breakthroughs because of its expanding potential for use in fields including sustainable diagnostics, high-throughput screening, and personalized therapies.

References

[1] G.M. Whitesides, The origins and the future of microfluidics, Nature. 442 (2006) 368-373. https://doi.org/10.1038/nature05058

[2] W. Qiu, C. Wu, Z. Wu, Surface modification of PDMS in microfluidic devices, in: A. Tiwari, M.D. Soucek (Eds.), Concise Encyclopedia of High-Performance Silicones, John Wiley & Sons, Inc. Hoboken, New Jersey, and Scrivener Publishing LLC, Salem, Massachusetts, 2014, pp. 141-150. https://doi.org/10.1002/9781118938478.ch10

[3] S. Banik, A. Uchil, T. Kalsang, S. Chakrabarty, M.A. Ali, P. Srisungsitthisunti, K.K. Mahato, S. Surdo, N. Mazumder, The revolution of PDMS microfluidics in cellular biology, Crit. Rev. Biotechnol. 43 (2023) 465-483. https://doi.org/10.1080/07388551.2022.2034733

[4] S.F. Berlanda, M. Breitfeld, C.L. Dietsche, P.S. Dittrich, Recent advances in microfluidic technology for bioanalysis and diagnostics, Anal. Chem. 93 (2021) 311-331. https://doi.org/10.1021/acs.analchem.0c04366

[5] Y. Song, Y. Zhou, K. Zhang, Z. Fan, F. Zhang, M. Wei, Microfluidic programmable strategies for channels and flow, Lab Chip. 24 (2024) 4483-4513. https://doi.org/10.1039/D4LC00423J

[6] M. Mohammadi, S.A. Qadir, A.M. Faraj, O.H. Shareef, H. Mahmoodi, F. Mahmoudi, S. Moradi, Navigating the future: Microfluidics charting new routes in drug delivery, In press, Journal Pre-Proof, Int. J. Pharm. 124142 (2024) 1-71. https://doi.org/10.1016/j.ijpharm.2024.124142

[7] O. Geschke, Microfluidics for bioapplications, Anal. Bioanal. Chem. 395 (2009) 619-619. https://doi.org/10.1007/s00216-009-3002-7

[8] B.L. Gray, New opportunities for polymer nanocomposites in microfluidics and biomedical MEMS: An introduction to cutting-edge composite polymer materials for use in microfluidics and biomedical MEMS, IEEE Nanotechnol. Mag. 8 (2014) 6-16. https://doi.org/10.1109/MNANO.2014.2309495

[9] S. Hengoju, O. Shvydkiv, M. Tovar, M. Roth, M.A. Rosenbaum, Advantages of optical fibers for facile and enhanced detection in droplet microfluidics, Biosens. Bioelectron. 200 (2022) 1-18. https://doi.org/10.1016/j.bios.2021.113910

[10] Y. Gao, Y. Wang, Y. Wang, P. Magaud, Y. Liu, F. Zeng, J. Yang, L. Baldas, Y. Song, Nanocatalysis meets microfluidics: A powerful platform for sensitive bioanalysis, TrAC-Trends Anal. Chem. 158 (2023) 1-21. https://doi.org/10.1016/j.trac.2022.116887

[11] E.X. Ng, M.N. Hsu, G. Sun, C.H. Chen, Single-cell assays using integrated continuous-flow microfluidics, Methods Enzymol. 628 (2019) 59-94. https://doi.org/10.1016/bs.mie.2019.06.014

[12] M. Karle, J. Miwa, G. Roth, R. Zengerle, F.V. Stetten, A novel microfluidic platform for continuous DNA extraction and purification using laminar flow magnetophoresis, IEEE Int. Conference MEMS, (2009) 276-279. https://doi.org/10.1109/MEMSYS.2009.4805372

[13] L.C. Duarte, F. Figueredo, C.L.S. Chagas, E. Corton, W.K.T. Coltro, A review of the recent achievements and future trends on 3D printed microfluidic devices for bioanalytical applications, Anal. Chim. Acta. 1299 (2024) 1-16. https://doi.org/10.1016/j.aca.2024.342429

[14] A.Z. Qamar, M.H. Shamsi, Desktop fabrication of lab-on-chip devices on flexible substrates: A brief review, Micromachines. 11 (2020) 1-17. https://doi.org/10.3390/mi11020126

[15] G. Kaur, M. Tomar, V. Gupta, Development of a microfluidic electrochemical biosensor: Prospect for point-of-care cholesterol monitoring, Sens. Actuators B: Chem. 261 (2018) 460-466. https://doi.org/10.1016/j.snb.2018.01.144

[16] M.G. Pollack, A.D. Shenderov, R.B. Fair, Electrowetting-based actuation of droplets for integrated microfluidics, Lab Chip. 2 (2002) 96-101. https://doi.org/10.1039/b110474h

[17] K.S. Elvira, F. Gielen, S.S.H. Tsai, A.M. Nightingale, Materials and methods for droplet microfluidic device fabrication, Lab Chip. 22 (2022) 859-875. https://doi.org/10.1039/D1LC00836F

[18] S.K. Cho, H. Moon, C.J. Kim, Creating, transporting, cutting, and merging liquid droplets by electrowetting-based actuation for digital microfluidic circuits, J. Microelectromech. Syst. 12 (2003) 70-80. https://doi.org/10.1109/JMEMS.2002.807467

[19] R. Fan, J. Wu, S. Duan, L. Jin, H. Zhang, C. Zhang, A. Zheng, Droplet-based microfluidics for drug delivery applications, Int. J. Pharm. 663 (2024) 1-14. https://doi.org/10.1016/j.ijpharm.2024.124551

[20] H. Li, Y. Ma, R. Fu, J. Peng, Y. Zhai, J. Li, W. Xu, S. Hu, H. Ma, A.R. Wheeler, S. Zhang, Droplet-based microfluidics with mass spectrometry for microproteomics, Eng. 43 (2024) 37-53. https://doi.org/10.1016/j.eng.2024.08.018

[21] J.K. Nunes, H.A. Stone, Introduction: Microfluidics, Chem. Rev. 122 (2022) 6919-6920. https://doi.org/10.1021/acs.chemrev.2c00052

[22] Z. Yang, K. Jin, Y. Chen, Q. Liu, H. Chen, S. Hu, Y. Wang, Z. Pan, F. Feng, M. Shi, H. Xie, H. Ma, H. Zhou, AM-DMF-SCP: Integrated single-cell proteomics analysis on an active-matrix digital microfluidic chip, JACS Au. 4 (2024) 1811-1823. https://doi.org/10.1021/jacsau.4c00027

[23] B. Sharma, A. Sharma, Microfluidics: Recent advances toward lab-on-chip applications in bioanalysis, Adv. Eng. Mater. 24 (2022) 1-21. https://doi.org/10.1002/adem.202100738

[24] L. Bezinge, C. Shih, D.A. Richards, A.J. deMello, Electrochemical paper-based microfluidics: Harnessing capillary flow for advanced diagnostics, Small. 20 (2024) 1-20. https://doi.org/10.1002/smll.202401148

[25] D.D. Liana, B. Raguse, J.J. Gooding, E. Chow, Recent advances in paper-based sensors, Sensors. 12 (2012) 11505-11526. https://doi.org/10.3390/s120911505

[26] S.N. Bhatia, D.E. Ingber, Microfluidic organs-on-chips, Nat. Biotechnol. 32 (2014) 760-772. https://doi.org/10.1038/nbt.2989

[27] R. Booth, H. Kim, Characterization of a microfluidic in vitro model of the blood-brain barrier (μBBB), Lab Chip. 12 (2012) 1784-1792. https://doi.org/10.1039/c2lc40094d

[28] H. Kandarova, Organ-on-a-chip as novel tox testing tools, in: P. Wexler (Eds.), Encyclopedia of Toxicology (4th Ed.), Academic press, Elsevier Inc., Amsterdam, Netherlands, 2024, pp.173-176. https://doi.org/10.1016/B978-0-12-824315-2.01037-X

[29] G.M. Whitesides, The origins and the future of microfluidics, Nature. 442 (2006) 368-373. https://doi.org/10.1038/nature05058

[30] S.K. Sia, L.J. Kricka, Microfluidics and point-of-care testing, Lab Chip. 8 (2008) 1982-1983. https://doi.org/10.1039/b817915h

[31] S.Y. Teh, R. Lin, L.H. Hung, A.P. Lee, Droplet microfluidics, Lab Chip. 8 (2008) 198-220. https://doi.org/10.1039/b715524g

[32] M. Tang, G. Wang, S.K. Kong, H.P. Ho, A review of biomedical centrifugal microfluidic platforms, Micromachines. 7 (2016) 1-29. https://doi.org/10.3390/mi7020026

[33] F. Piraino, F. Volpetti, C. Watson, S.J. Maerkl, A digital-analog microfluidic platform for patient-centric multiplexed biomarker diagnostics of ultralow volume samples, ACS Nano. 10 (2016) 1699-1710. https://doi.org/10.1021/acsnano.5b07939

[34] C. Cheng, A.W. Martinez, J. Gong, C.R. Mace, S.T. Phillips, E. Carrilho, K.A. Mirica, G.M. Whitesides, Paper-based ELISA, Angew. Chem. Int. Ed. 49 (2010) 4771-4774. https://doi.org/10.1002/anie.201001005

[35] S. Siavashy, M. Soltani, M. Ahmadi, B. Landi, H. Mehmanparast, F. Ghorbani-Bidkorbeh, A comprehensive review of one decade of microfluidic platforms applications in synthesis of enhanced carriers utilized in controlled drug delivery, Adv. Mater. Technol. 7 (2022) 1-23. https://doi.org/10.1002/admt.202101615

[36] D. Huh, B.D. Matthews, A. Mammoto, M. Montoya-Zavala, H.Y. Hsin, D.E. Ingber, Reconstituting organ-level lung functions on a chip, Sci. 328 (2010) 1662-1668. https://doi.org/10.1126/science.1188302

[37] A.K. White, K.A. Heyries, C. Doolin, M. VanInsberghe, C.L. Hansen, High-throughput microfluidic single-cell digital polymerase chain reaction, Anal. Chem. 85 (2013) 7182-7190. https://doi.org/10.1021/ac400896j

[38] A. Sun, P. Voparilova, X. Liu, B. Kou, T. Reznicek, T. Lednicky, S. Ni, J. Kudr, O. Zitka, Z. Fohlerova, P. Pajer, H. Zhang, P. Neuzil, An integrated microfluidic platform for nucleic acid testing, Microsyst. Nanoeng. 10 (2024) 1-14. https://doi.org/10.1038/s41378-024-00677-6

[39] S.L. Spurgeon, R.C. Jones, R. Ramakrishnan, High throughput gene expression measurement with real time PCR in a microfluidic dynamic array, PLoS One. 3 (2008) 1-7. https://doi.org/10.1371/journal.pone.0001662

[40] M. Herrmann, T. Veres, M. Tabrizian, Enzymatically-generated fluorescent detection in micro-channels with internal magnetic mixing for the development of parallel microfluidic ELISA, Lab Chip. 6 (2006) 555-560. https://doi.org/10.1039/b516031f

[41] A.R.D Boer, B. Bruyneel, J.G. Krabbe, H. Lingeman, W.M.A. Niessen, H. Irth, A microfluidic-based enzymatic assay for bioactivity screening combined with capillary liquid chromatography and mass spectrometry, Lab Chip. 5 (2005) 1286-1292. https://doi.org/10.1039/b506559c

[42] S.K. Vashist, J.H.T. Luong, Lab-on-a-chip (LOC) immunoassays, in: S.K. Vashist, J.H.T. Luong (Eds.), Handbook of immunoassay technologies, Academic press, Elsevier Inc., United Kingdom, 2018, pp. 415-431. https://doi.org/10.1016/B978-0-12-811762-0.00015-3

[43] D.M. Rissin, C.W. Kan, T.G. Campbell, S.C. Howes, D.R. Fournier, L. Song, T. Piech, P.P. Patel, L. Chang, A.J. Rivnak, E.P. Ferrell, J.D. Randall, G.K. Provuncher, D.R. Walt, D.C. Duffy, Single-molecule enzyme-linked immunosorbent assay detects serum proteins at subfemtomolar concentrations, Nat. Biotechnol. 28 (2010) 595-599. https://doi.org/10.1038/nbt.1641

[44] C.E. Stanley, R.C.R. Wootton, A.J. DeMello, Continuous and segmented flow microfluidics: Applications in high-throughput chemistry and biology, Chimia. 66 (2012) 88-98. https://doi.org/10.2533/chimia.2012.88

[45] R. Daw, J. Finkelstein, Lab on a chip, Nature. 442 (2006) 367-367. https://doi.org/10.1038/442367a

[46] M. Yuce, E. Filiztekin, K.G. Ozkaya, COVID-19 diagnosis - A review of current methods, Biosens. Bioelectron. 172 (2021) 1-15. https://doi.org/10.1016/j.bios.2020.112752

[47] Y. Liu, L. Sun, H. Zhang, L. Shang, Y. Zhao, Microfluidics for drug development: From synthesis to evaluation, Chem. Rev. 121 (2021) 7468-7529. https://doi.org/10.1021/acs.chemrev.0c01289

[48] J.P. Martins, D. Liu, F. Fontana, M.P.A. Ferreira, A. Correia, S. Valentino, M. Kemell, K. Moslova, E. Mäkilä, J. Salonen, J. Hirvonen, B. Sarmento, H.A. Santos, Microfluidic nanoassembly of bioengineered chitosan-modified FcRn-targeted porous silicon nanoparticles

@ hypromellose acetate succinate for oral delivery of antidiabetic peptides, ACS Appl. Mater. Interfaces 10 (2018) 44354-44367. https://doi.org/10.1021/acsami.8b20821

[49] Y. Liu, K.M. Castro Bravo, J. Liu, Targeted liposomal drug delivery: A nanoscience and biophysical perspective, Nanoscale Horiz. 6 (2021) 78-94. https://doi.org/10.1039/D0NH00605J

[50] R. Riahi, A. Tamayol, S.A.M. Shaegh, A.M. Ghaemmaghami, M.R. Dokmeci, A. Khademhosseini, Microfluidics for advanced drug delivery systems, Curr. Opin. Chem. Eng. 7 (2015) 101-112. https://doi.org/10.1016/j.coche.2014.12.001

[51] Z. Razavi, M. Soltani, H.P. Toroudi, P. Chen, CRISPR-microfluidics nexus: Advancing biomedical applications for understanding and detection, Sens. Actuators A: Phys. 376 (2024) 1-19. https://doi.org/10.1016/j.sna.2024.115625

[52] N. Kimura, M. Maeki, A. Ishida, H. Tani, M. Tokeshi, One-step production using a microfluidic device of highly biocompatible size-controlled noncationic exosome-like nanoparticles for RNA delivery, ACS Appl. Bio. Mater. 4 (2021) 1783-1793. https://doi.org/10.1021/acsabm.0c01519

[53] Y. Suzuki, H. Onuma, R. Sato, Y. Sato, A. Hashiba, M. Maeki, M. Tokeshi, M.E.H. Kayesh, M. Kohara, K.T. Kohara, H. Harashima, Lipid nanoparticles loaded with ribonucleoprotein-oligonucleotide complexes synthesized using a microfluidic device exhibit robust genome editing and hepatitis B virus inhibition, J. Control. Release 330 (2021) 61-71. https://doi.org/10.1016/j.jconrel.2020.12.013

[54] K. Hiramoto, K. Ino, Y. Nashimoto, K. Ito, H. Shiku, Electric and electrochemical microfluidic devices for cell analysis, Fron. Chem. 7 (2019) 1-9. https://doi.org/10.3389/fchem.2019.00396

[55] M.C. Draper, X. Niu, S. Cho, D.I. James, J.B. Edel, Compartmentalization of electrophoretically separated analytes in a multiphase microfluidic platform, Anal. Chem. 84 (2012) 5801-5808. https://doi.org/10.1021/ac301141x

[56] M.R.G. Kopp, M. Linsenmeier, B. Hettich, S. Prantl, S. Stavrakis, J.C. Leroux, P. Arosio, Microfluidic shrinking droplet concentrator for analyte detection and phase separation of protein solutions, Anal. Chem. 92 (2020) 5803-5812. https://doi.org/10.1021/acs.analchem.9b05329

[57] A. Purwidyantri, A. Ipatov, T. Domingues, J. Borme, M. Martins, P. Alpuim, M. Prado, Programmable graphene-based microfluidic sensor for DNA detection, Sens. Actuators B: Chem. 367 (2022) 1-10. https://doi.org/10.1016/j.snb.2022.132044

[58] Q. Wang, H. Wang, X. Yang, K. Wang, F. Liu, Q. Zhao, P. Liu, R. Liu, Multiplex detection of nucleic acids using a low-cost microfluidic chip and a personal glucose meter at the point-of-care, Chem. Commun. 50 (2014) 3824-3826. https://doi.org/10.1039/C4CC00133H

[59] F. Haghayegh, R. Salahandish, A. Zare, M. Khalghollah, A. Sanati-Nezhad, Immuno-biosensor on a chip: A self-powered microfluidic-based electrochemical biosensing platform for point-of-care quantification of proteins, Lab Chip. 22 (2022) 108-120. https://doi.org/10.1039/D1LC00879J

[60] C.K. Tang, A. Vaze, M. Shen, J.F. Rusling, High-throughput electrochemical microfluidic immunoarray for multiplexed detection of cancer biomarker proteins, ACS Sens. 1 (2016) 1036-1043. https://doi.org/10.1021/acssensors.6b00256

[61] M. Grumann, J. Steigert, L. Riegger, I. Moser, B. Enderle, K. Riebeseel, G. Urban, R. Zengerle, J. Ducree, Sensitivity enhancement for colorimetric glucose assays on whole blood by on-chip beam-guidance, Biomed. Microdevices. 8 (2006) 209-214. https://doi.org/10.1007/s10544-006-8172-x

[62] J.C. Briones, W.V. Espulgar, S. Koyama, H. Takamatsu, M. Saito, E. Tamiya, A high-throughput single-cell assay on a valve-based microfluidic platform applied to protein quantification, immune response monitoring, and drug discovery, Methods Mol. Biol. 2689 (2023) 119-142. https://doi.org/10.1007/978-1-0716-3323-6_10

[63] D.F.C. Alcantar, Y.D. Han, J. Choi, J.L.G. Cordero, A. Revzin, Automated droplet-based microfluidic platform for multiplexed analysis of biochemical markers in small volumes, Anal. Chem. 91 (2019) 5133-5141. https://doi.org/10.1021/acs.analchem.8b05689

[64] A.S. Ngam, P. Rattanarat, O. Chailapakul, M.S. Art, Electrochemical droplet-based microfluidics using chip-based carbon paste electrodes for high-throughput analysis in pharmaceutical applications, Anal. Chim. Acta. 883 (2015) 45-54. https://doi.org/10.1016/j.aca.2015.03.008

[65] Y. Wang, Z. Ye, J. Ping, S. Jing, Y. Ying, Development of an aptamer-based impedimetric bioassay using microfluidic system and magnetic separation for protein detection, Biosens. Bioelectron. 59 (2014) 106-111. https://doi.org/10.1016/j.bios.2014.03.015

[66] Y. Song, J. Zhao, T. Cai, S.-H. Su, E. Sandford, C. Flora, B.H. Singer, M. Ghosh, S.W. Choi, M. Tewari, K. Kurabayashi, Machine learning-based protein microarray digital assay analysis, bioRxiv. 226448 (2020) 1-25. https://doi.org/10.1101/2020.08.04.236448

[67] Y. Song, J. Zhao, T. Cai, A. Stephens, S.H. Su, E. Sandford, C. Flora, B.H. Singer, M. Ghosh, S.W. Choi, M. Tewari, K. Kurabayashi, Machine learning-based cytokine microarray digital immunoassay analysis, Biosens. Bioelectron. 180 (2021) 1-11. https://doi.org/10.1016/j.bios.2021.113088

[68] N. Talukder, A. Furniturewalla, T. Le, M. Chan, S. Hirday, X. Cao, P. Xie, Z. Lin, A. Gholizadeh, S. Orbine, M. Javanmard, A portable battery powered microfluidic impedance cytometer with smartphone readout: Towards personal health monitoring, Biomed. Microdevices. 19 (2017) 36-37. https://doi.org/10.1007/s10544-017-0161-8

[69] L. Malic, T. Veres, M. Tabrizian, Nanostructured digital microfluidics for enhanced surface plasmon resonance imaging, Biosens. Bioelectron. 26 (2011) 2053-2059. https://doi.org/10.1016/j.bios.2010.09.001

[70] M.M. Liu, X. Lian, H. Liu, Z.Z. Guo, H.H. Huang, Y. Lei, H.P. Peng, W. Chen, X.H. Lin, A.L. Liu, X.H. Xia, A colorimetric assay for sensitive detection of hydrogen peroxide and glucose in microfluidic paper-based analytical devices integrated with starch-iodide-gelatin system, Talanta. 200 (2019) 511-517. https://doi.org/10.1016/j.talanta.2019.03.089

[71] M.D.F. Ramos, M.B. Banuelos, L.F.C. Vallvey, Point-of-care assay platform for uric acid, urea, and triglycerides with a microfluidic paper device (μPAD) controlled by stimulus-sensitive valves, Talanta 254 (2023) 1-6. https://doi.org/10.1016/j.talanta.2022.124189

[72] P.M. Misun, J. Rothe, Y.R.F. Schmid, A. Hierlemann, O. Frey, Multi-analyte biosensor interface for real-time monitoring of 3D microtissue spheroids in hanging-drop networks, Microsyst. Nanoeng. 2 (2016) 1-9. https://doi.org/10.1038/micronano.2016.22

[73] L. Cong, J. Wang, X. Li, Y. Tian, S. Xu, C. Liang, W. Xu, W. Wang, S. Xu, Microfluidic Droplet-SERS platform for single-cell cytokine analysis via a cell surface bioconjugation strategy, Anal. Chem. 94 (2022) 10375-10383. https://doi.org/10.1021/acs.analchem.2c01249

[74] Y. Shang, Z. Wang, L. Xi, Y. Wang, M. Liu, Y. Feng, J. Wang, Q. Wu, X. Xiang, M. Chen, Y. Ding, Droplet-based single-cell sequencing: Strategies and applications, Biotechnol. Adv. 77 (2024) 1-17. https://doi.org/10.1016/j.biotechadv.2024.108454

[75] Y. Li, Z. Huang, L. Xu, Y. Fan, J. Ping, G. Li, Y. Chen, C. Yu, Q. Wang, T. Song, T. Lin, M. Liu, Y. Xu, N. Ai, X. Meng, Q. Qiao, H. Ji, Z. Qin, S. Jin, N. Jiang, M. Wang, S. Shu, F. Zhang, W. Zhang, G.-H. Liu, L. Chen, L. Jiang, UDA-seq: Universal droplet microfluidics-based combinatorial indexing for massive-scale multimodal single-cell sequencing, Nat. Methods. 22 (2025) 1192-1212. https://doi.org/10.1038/s41592-024-02586-y

[76] P. Datlinger, A.F. Rendeiro, C. Schmidl, T. Krausgruber, P. Traxler, J. Klughammer, L.C. Schuster, A. Kuchler, D. Alpar, C. Bock, Pooled CRISPR screening with single-cell transcriptome readout, Nat. Methods. 14 (2017) 297-301. https://doi.org/10.1038/nmeth.4177

[77] J. Zhu, K. Pang, B. Hu, R. He, N. Wang, Z. Jiang, P. Ji, F. Zhao, Custom microfluidic chip design enables cost-effective three-dimensional spatiotemporal transcriptomics with a wide field of view, Nat. Genet. 56 (2024) 2259-2270. https://doi.org/10.1038/s41588-024-01906-4

[78] E. Papalexi, R. Satija, Single-cell RNA sequencing to explore immune cell heterogeneity, Nat. Rev. Immunol. 18 (2018) 35-45. https://doi.org/10.1038/nri.2017.76

[79] H. Li, Y. Ma, R. Fu, J. Peng, Y. Zhai, J. Li, W. Xu, S. Hu, H. Ma, A.R. Wheeler, S. Zhang, Droplet-based microfluidics with mass spectrometry for microproteomics, Eng. 43 (2024) 37-53. https://doi.org/10.1016/j.eng.2024.08.018

[80] Y. Liu, X. Chen, Y. Zhang, J. Liu, Advancing single-cell proteomics and metabolomics with microfluidic technologies, Analyst. 144 (2019) 846-858. https://doi.org/10.1039/C8AN01503A

[81] Y. Zhu, R. Zhao, P.D. Piehowski, R.J. Moore, S. Lim, V.J. Orphan, L.P. Tolic, W.-J. Qian, R.D. Smith, R.T. Kelly, Subnanogram proteomics: Impact of LC column selection, MS instrumentation and data analysis strategy on proteome coverage for trace samples, Int. J. Mass Spectrom. 427 (2018) 4-10. https://doi.org/10.1016/j.ijms.2017.08.016

[82] Y. Shang, Z. Wang, L. Xi, Y. Wang, M. Liu, Y. Feng, J. Wang, Q. Wu, X. Xiang, M. Chen, Y. Ding, Droplet-based single-cell sequencing: Strategies and applications, Biotechnol. Adv. 77 (2024) 1-17. https://doi.org/10.1016/j.biotechadv.2024.108454

[83] W.M. Zhou, Y.Y. Yan, Q.R. Guo, H. Ji, H. Wang, T.T. Xu, B. Makabel, C. Pilarsky, G. He, X.Y. Yu, J.Y. Zhang, Microfluidics applications for high-throughput single cell sequencing, J. Nanobiotechnol. 19 (2021) 1-21. https://doi.org/10.1186/s12951-021-01045-6

Materials Research Foundations 191 (2026) 289-316 https://doi.org/10.21741/9781644904213-9

[84] E. Hedlund, Q. Deng, Single-cell RNA sequencing: Technical advancements and biological applications, Mol. Aspects Med. 59 (2018) 36-46. https://doi.org/10.1016/j.mam.2017.07.003

[85] D.G. Alonso, M. Yu, H. Qu, L. Ma, F. Shen, Advances in microfluidics-based technologies for single cell culture, Adv. Biosyst. 3 (2019) e1900003. https://doi.org/10.1002/adbi.201900003

[86] W. Ali, W. Tian, S.U. Din, D. Iradukunda, A.A. Khan, Classical and modern face recognition approaches: A complete review, Multimed. Tools Appl. 80 (2021) 4825-4880. https://doi.org/10.1007/s11042-020-09850-1

[87] S.R. Dabbagh, F. Rabbi, Z. Dogan, A.K. Yetisen, S. Tasoglu, Machine learning-enabled multiplexed microfluidic sensors, Biomicrofluidics. 14 (2020) 1-12. https://doi.org/10.1063/5.0025462

[88] N. Pouyanfar, S.Z. Harofte, M. Soltani, S. Siavashy, E. Asadian, F.G. Bidkorbeh, R. Keçili, C.M. Hussain, Artificial intelligence-based microfluidic platforms for the sensitive detection of environmental pollutants: Recent advances and prospects, Trends Environ. Anal. Chem. 34 (2022) 1-15. https://doi.org/10.1016/j.teac.2022.e00160

[89] J. Liu, H. Du, L. Huang, W. Xie, K. Liu, X. Zhang, S. Chen, Y. Zhang, D. Li, H. Pan, AI-powered microfluidics: Shaping the future of phenotypic drug discovery, ACS Appl. Mater. Interfaces 16 (2024) 38832-38851. https://doi.org/10.1021/acsami.4c07665

[90] D. Wang, K. Jin, J. Ji, C. Hu, M. Du, Y. Belgaid, S. Shi, J. Li, S. Hu, A. Nathan, J. Yu, H. Ma, Active-matrix digital microfluidics design for field programmable high-throughput digitalized liquid handling, iScience. 27 (2024) 1-13. https://doi.org/10.1016/j.isci.2024.109324

[91] K. Guo, Z. Song, J. Zhou, B. Shen, B. Yan, Z. Gu, H. Wang, An artificial intelligence-assisted digital microfluidic system for multistate droplet control, Microsyst. Nanoeng. 10 (2024) 1-12. https://doi.org/10.1038/s41378-024-00775-5

[92] B. Wang, Y. Li, M. Zhou, Y. Han, M. Zhang, Z. Gao, Z. Liu, P. Chen, W. Du, X. Zhang, X. Feng, B.F. Liu, Smartphone-based platforms implementing microfluidic detection with image-based artificial intelligence, Nat. Commun. 14 (2023) 1-18. https://doi.org/10.1038/s41467-023-36017-x

[93] J. Khandurina, A. Guttman, Bioanalysis in microfluidic devices, J. Chromatogr. A. 943 (2002) 159-183. https://doi.org/10.1016/S0021-9673(01)01451-0

[94] A.M. Tentori, A.E. Herr, Photopatterned materials in bioanalytical microfluidic technology, J. Micromech. Microeng. 21 (2011) 1-28. https://doi.org/10.1088/0960-1317/21/5/054001

[95] K.M. Leong, A.Y. Sun, M.L. Quach, C.H. Lin, C.A. Craig, F. Guo, T.R. Robinson, M. Chang, A.O. Olanrewaju, Democratizing access to microfluidics: Rapid prototyping of open microchannels with low-cost LCD 3D printers, ACS Omega. 9 (2024) 45537-45544. https://doi.org/10.1021/acsomega.4c07776

[96] M. Sekhwama, K. Mpofu, S. Sudesh, P.M. Kufa, Integration of microfluidic chips with biosensors, Discov. Appl. Sci. 6 (2024) 1-32. https://doi.org/10.1007/s42452-024-06103-w

[97] E.M. Miller, A.R. Wheeler, Digital bioanalysis, Anal. Bioanal. Chem. 393 (2009) 419-426. https://doi.org/10.1007/s00216-008-2397-x

[98] A. Ros, W. Hellmich, J. Regtmeier, T.T. Duong, D. Anselmetti, Bioanalysis in structured microfluidic systems, Electrophoresis. 27 (2006) 2651-2658. https://doi.org/10.1002/elps.200500923

[99] S. Farley, K. Ramsay, K.S. Elvira, A plug-and-play modular microcapillary platform for the generation of multicompartmental double emulsions using glass or fluorocarbon capillaries, Lab Chip. 21 (2021) 2781-2790. https://doi.org/10.1039/D1LC00126D

[100] T. Limongi, F. Guzzi, E. Parrotta, P. Candeloro, S. Scalise, V. Lucchino, F. Gentile, L. Tirinato, M.L. Coluccio, B. Torre, M. Allione, M. Marini, F. Susa, E.D. Fabrizio, G. Cuda, G. Perozziello, Microfluidics for 3D cell and tissue cultures: Microfabricative and ethical aspects updates, Cells. 11 (2022) 1-16. https://doi.org/10.3390/cells11101699

[101] S. Apoorva, N.-T. Nguyen, K.R. Sreejith, Recent developments and future perspectives of microfluidics and smart technologies in wearable devices, Lab Chip. 24 (2024) 1833-1866. https://doi.org/10.1039/D4LC00089G

[102] M.A. Farias, S. Dziomba, N.T. Tran, Microfluidic strategies for extraction and preconcentration of proteins and peptides, in: M.H. Chaudhery (Eds.), Handbook on miniaturization in analytical chemistry, Elsevier Inc., United Kingdom, 2020, pp.35-75. https://doi.org/10.1016/B978-0-12-819763-9.00003-9

[103] A.B. Kanu, Recent developments in sample preparation techniques combined with high-performance liquid chromatography: A critical review, J. Chromatogr. A. 1654 (2021) 1-21. https://doi.org/10.1016/j.chroma.2021.462444

[104] V. Iyer, Z. Yang, J. Ko, R. Weissleder, D. Issadore, Advancing microfluidic diagnostic chips into clinical use: A review of current challenges and opportunities, Lab Chip. 22 (2022) 3110-3121. https://doi.org/10.1039/D2LC00024E

[105] B.B. Bhattacharya, S. Roy, S. Bhattacharjee, Algorithmic challenges in digital microfluidic biochips: Protocols, design, and test, in: P. Gupta, C. Zaroliagis (Eds.), Applied algorithms, ICAA 2014, Lecture notes in computer science, Springer, Cham., Switzerland, 2014, pp. 1-16. https://doi.org/10.1007/978-3-319-04126-1_1

[106] N. Xiang, Z. Ni, Innovations in microfluidics to enable novel biomedical applications, Biosensors. 14 (2024) 1-3. https://doi.org/10.3390/bios14100507

[107] G.N. Ahn, J.H. Kang, H.J. Lee, B.E. Park, M. Kwon, G.S. Na, H. Kim, D.H. Seo, D.P. Kim, Exploring ultrafast flow chemistry by autonomous self-optimizing platform, Chem. Eng. J. 453 (2023) 1-8. https://doi.org/10.1016/j.cej.2022.139707

[108] V. Sharma, A. Mottafegh, J.U. Joo, J.H. Kang, L. Wang, D.P. Kim, Toward microfluidic continuous-flow and intelligent downstream processing of biopharmaceuticals, Lab Chip. 24 (2024) 2861-2882. https://doi.org/10.1039/D3LC01097J

[109] H. Zhang, J. Yang, R. Sun, S. Han, Z. Yang, L. Teng, Microfluidics for nano-drug delivery systems: From fundamentals to industrialization, Acta Pharm. Sin. B. 13 (2023) 3277-3299. https://doi.org/10.1016/j.apsb.2023.01.018

Materials Research Foundations 191 (2026) 289-316 https://doi.org/10.21741/9781644904213-9

[110] M.O. Ozen, K. Sridhar, M.G. Ogut, A. Shanmugam, A.S. Avadhani, Y. Kobayashi, J.C. Wu, F. Haddad, U. Demirci, Total microfluidic chip for multiplexed diagnostics (ToMMx), Biosens. Bioelectron. 150 (2020) 1-7. https://doi.org/10.1016/j.bios.2019.111930

[111] R. Su, F. Wang, M.C. McAlpine, 3D printed microfluidics: Advances in strategies, integration, and applications, Lab Chip. 23 (2023) 1279-1299. https://doi.org/10.1039/D2LC01177H

[112] M.L. Cunha, S.S. da Silva, M.C. Stracke, D.L. Zanette, M.N. Aoki, L. Blanes, Sample preparation for lab-on-a-chip systems in molecular diagnosis: A review, Anal. Chem. 94 (2022) 41-58. https://doi.org/10.1021/acs.analchem.1c04460

[113] M. Kumari, V. Gupta, N. Kumar, R.K. Arun, Microfluidics-based nanobiosensors for healthcare monitoring, Mol. Biotechnol. 66 (2024) 378-401. https://doi.org/10.1007/s12033-023-00760-9

[114] M.J. Doktycz, M.L. Simpson, Nano-enabled synthetic biology, Mol. Syst. Biol. 3 (2007) 1-10. https://doi.org/10.1038/msb4100165

[115] R. Lausecker, V. Badilita, U. Gleibner, U. Wallrabe, Introducing natural thermoplastic shellac to microfluidics: A green fabrication method for point-of-care devices. Biomicrofluidics, 10 (2015) 1-12. https://doi.org/10.1063/1.4955062

Microfluidics: Synthesis, Device Fabrication and Applications Materials Research Forum LLC
Materials Research Foundations 191 (2026) 317-342 https://doi.org/10.21741/9781644904213-10

Chapter 10

Microfluidics for Separation of Biomolecules

Vipul D. Prajapati[1,*], Shivang Kahar[2]

[1]Department of Pharmaceutics, S.S.R. College of Pharmacy (Permanently Affiliated to Savitribai Phule Pune University), Sayli, Silvassa, U. T. of D.N.H. and D.D., India.

[2]Department of Pharmaceutical Quality Assurance, S.S.R. College of Pharmacy (Permanently Affiliated to Savitribai Phule Pune University), Sayli, Silvassa, U. T. of D.N.H. and D.D., India.

*vippra2000@yahoo.com

Abstract

Microfluidic technologies have transformed biomolecular separation by offering precise, rapid, and miniaturized solutions. Leveraging microscale fluid dynamics—such as laminar flow and enhanced diffusion—these systems enable efficient separation through mechanisms like electrophoresis, di electrophoresis, affinity capture, and hydrodynamic filtration. They use small sample volumes to separate a variety of biomolecules, such as proteins, nucleic acids, cells, and exosomes. Advances in chip design, surface functionalization, and channel geometry have enhanced sensitivity and specificity. Integration with real-time detection methods like fluorescence, electrochemical sensors, and surface plasmon resonance has streamlined applications in diagnostics, proteomics, and environmental monitoring. Portable, ultra-sensitive, single-cell analysis tools are becoming possible thanks to emerging trends like droplet microfluidics, paper-based devices, and lab-on-a-chip (LOC) platforms. As microfluidics moves toward AI-driven automation and personalized medicine, addressing challenges in fabrication, cost, and regulation will be key to broader commercial and clinical adoption.

Keywords

Microfluidics, Biomolecules, Lab-on-a-Chip, Electrophoresis, Hydrodynamic Filtration, Affinity Capture, Droplet Microfluidics, Paper-Based Diagnostics, Single-Cell Analysis, Proteomics, Genomics, Personalized Medicine

List of Abbreviations

μPADs	: Microfluidic Paper-Based Analytical Devices
LOC	: Lab-on-a-Chip
DMF	: Digital Microfluidics
qPCR	: Quantitative Polymerase Chain Reaction
ctDNA	: Circulating Tumour DNA
IVD	: In Vitro Diagnostics
AI	: Artificial Intelligence
PCR	: Polymerase Chain Reaction
SPRi	: Surface Plasmon Resonance Imaging
SERS	: Surface-Enhanced Raman Spectroscopy

Materials Research Foundations 191 (2026) 317-342 https://doi.org/10.21741/9781644904213-10

CVD	: Cardiovascular Disease
FDA	: Food and Drug Administration
GMP	: Good Manufacturing Practice
PDMS	: Polydimethylsiloxane
IoT	: Internet of Things
2D	: Two-Dimensional
3D	: Three-Dimensional
ATPS	: Aqueous two-phase systems
LLPS	: Liquid-liquid phase separation
CTCs	: Circulating tumour cells
SPR	: Surface plasmon resonance
DLD	: Deterministic lateral displacement
SSAWs	: Standing surface acoustic waves
DPCR	: Digital Polymeric Chain Reaction
DNA	: Deoxyribonucleic acid
RNA	: Ribonucleic acid

Contents

Materials Research Foundations 191 (2026) 317-342 https://doi.org/10.21741/9781644904213-10

1. Introduction

Biomolecule separation is a cornerstone of modern biotechnology, diagnostics, and pharmaceutical research, providing the means to isolate and analyse specific components within complex biological mixtures. Biological samples typically comprise diverse molecular species—such as proteins, nucleic acids, metabolites, and cells—which necessitates high-resolution separation strategies to study individual constituents effectively [1]. Traditional techniques including chromatography, electrophoresis, and centrifugation have been instrumental in this process. However, these techniques frequently require costly equipment, lengthy processing times, and high sample volumes, which restricts their use in point-of-care and high-throughput settings [2]. In response, microfluidic technologies have emerged as powerful alternatives by miniaturizing and integrating analytical functions into compact lab-on-a-chip devices, enhancing speed, sensitivity, and automation [3].

1.1 History of biomolecule separation

The history of biomolecule separation spans over two centuries and has evolved significantly alongside developments in chemistry, molecular biology, and biotechnology. The earliest attempts reported date back to 1789, when Antoine François, comte de Fourcroy used simple precipitation techniques to isolate proteins from the biological materials. Although rudimentary and non-specific, these efforts laid the groundwork for biochemical purification methods and reflected the early understanding of biomolecular composition [4].

Materials Research Foundations 191 (2026) 317-342 https://doi.org/10.21741/9781644904213-10

A major turning point came in 1903 with the invention of chromatography by Russian botanist Mikhail Tswett, who used adsorption chromatography to separate plant pigments. High-performance liquid chromatography (HPLC), ion exchange, gel filtration, affinity, and other techniques were developed as a result of this technique, which ultimately became fundamental in the separation of proteins, nucleic acids, and other biomolecules. Using these methods, scientists were able to separate molecules according to characteristics like size, charge, and biological binding specificity [5]. In the mid-20th century, electrophoresis emerged as another transformative technology. Initial methods, like moving boundary electrophoresis, were soon replaced by the zone electrophoresis using gels. Between the 1940s and 1960s, researchers such as Oliver Smithies developed gel-based methods (e.g., starch and polyacrylamide gels) that significantly improved resolution and allowed for the separation of charged molecules by size. This era was crucial to the rise of molecular biology, as it enabled discrete visualization and analysis of biomolecules like DNA and proteins [6]. The 1970s to 1990s saw extensive innovation in chromatographic techniques. HPLC enabled high-resolution, rapid separation using pressurized solvent systems, while new affinity-based methods improved the specificity of purification. During the 1990s, perfusion chromatography was introduced, using porous bead matrices to enhance the flow rate and efficiency of protein separations. These advances were essential for both analytical and large-scale industrial applications [7].

In the late 1990s and 2000s, separation techniques began integrating with detection technologies, notably biosensors. Coupling methods such as HPLC with biosensors (e.g., immunosensors, enzyme-based systems) enhanced sensitivity and selectivity, allowing researchers to detect target molecules in highly complex mixtures. These innovations bridged analytical chemistry and bioassays, offering powerful tools for diagnostics and pharmacological studies [8]. The simplicity, affordability, and effectiveness of aqueous two-phase systems (ATPS) in extracting proteins, polysaccharides, and secondary metabolites from biological materials led to their rise in popularity. These systems have been enhanced with ionic liquids and deep eutectic solvents to improve selectivity and yield [9]. Additionally, researchers have developed innovative nanoporous materials and biosilica-based syringe filters that allow rapid and instrument-free separation of DNA and proteins, especially useful for point-of-care diagnostics in cancer research [10,11]. Parallel to these materials-based advancements, microfluidics has transformed the separation of chiral and small biomolecules. These LOC devices utilize capillary electrophoresis and other miniaturized techniques to offer rapid, high-efficiency separations in compact formats suitable for clinical and research settings [12]. In tandem, computational tools and artificial intelligence are increasingly being used to model and optimize separation processes, reducing the trial-and-error of experimental design and improving efficiency in large-scale bioprocessing [12]. Figure 1 depicts the timeline history of biomolecule separation.

Materials Research Foundations 191 (2026) 317-342 https://doi.org/10.21741/9781644904213-10

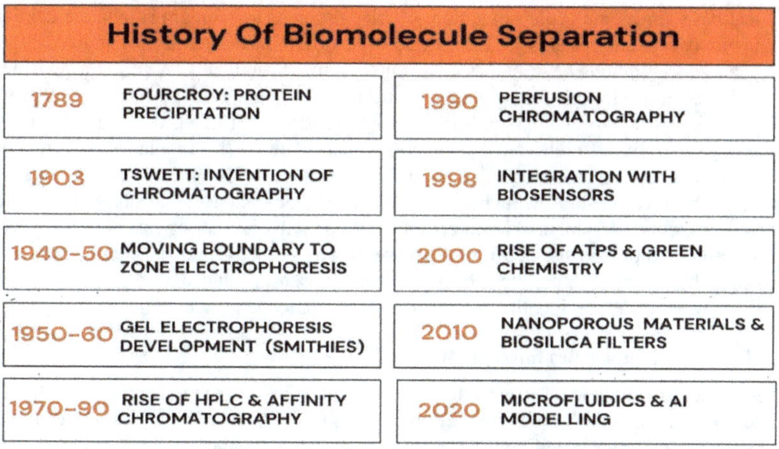

Figure 1. *An outlook image showing the timeline history of biomolecule separation using different techniques in analysis.*

1.2 Principle of Biomolecule Separation

The principle of biomolecule separation involves exploiting the inherent physical and chemical differences between molecules to isolate and purify them from complex biological mixtures. These principles include variations in size, charge, polarity, solubility, specific binding affinities, and phase behaviour, which are foundational to the design of separation techniques in biochemistry and biotechnology. One of the most fundamental approaches is the size-based separation, where molecules are separated according to their molecular weight or spatial dimensions. Techniques like gel filtration chromatography and gel electrophoresis rely on this principle. Due to their exclusion from the stationary phase's pores, larger molecules flow through the column more quickly in gel filtration, whereas smaller molecules travel longer routes. Similar to this, in electrophoresis, an electric field causes smaller molecules to move through a gel matrix more quickly [5]. Another important idea is the charge-based separation, which makes use of the net charge of biomolecules at a specific pH. Charged resins are used in the ion-exchange chromatography to draw in and hold on to oppositely charged molecules, which can be eluted by adjusting the mobile phase's pH or ionic strength. Another technique for sorting proteins according to their isoelectric point—the pH at which they have no net charge—is isoelectric focusing affinity-based separation [13], which takes advantage of particular biological interactions like receptor-ligand binding, enzyme-substrate, or antigen-antibody interactions. Affinity chromatography employs immobilized ligands on a stationary phase to selectively capture target biomolecules from a mixture, resulting in highly specific purification. This is particularly valuable for isolating tagged proteins, enzymes, and antibodies [14]. Techniques namely hydrophobic interaction chromatography (HIC) and reverse-phase high-performance liquid chromatography (RP-HPLC) are used in polarity and hydrophobicity-based separation. These methods distinguish biomolecules by their interaction with hydrophobic stationary phases, where nonpolar molecules are retained longer and eluted under gradient conditions [10]. Solubility differences also serve as

Materials Research Foundations 191 (2026) 317-342 https://doi.org/10.21741/9781644904213-10

a practical basis for biomolecule separation. Methods such as ammonium sulphate precipitation and aqueous two-phase systems (ATPS) exploit the differential solubility of molecules under varying salt concentrations or in biphasic solvent systems. Specifically, ATPS is a scalable and eco-friendly technique for one-step protein and nucleic acid separation [15]. In recent years, phase separation principles, especially liquid-liquid phase separation (LLPS), have gained attention. LLPS is driven by weak multivalent interactions that result in the spontaneous formation of biomolecular condensates, which compartmentalize proteins and RNA in cells without membrane boundaries. This mechanism is being harnessed for selective enrichment and study of intrinsically disordered proteins and other biomolecules [16]. Finally, advances in biosensor-assisted separations have improved the sensitivity and selectivity of detection. By coupling traditional chromatographic or electrophoretic techniques with biosensors that detect specific molecular interactions, researchers can identify and quantify trace components in complex samples [8].

1.3 Types of biomolecules for separation

Different types of biomolecules used for the separation purpose are illustrated in **figure 2**. Biomolecule separation techniques are used to isolate specific types of biological molecules from the complex mixtures for specific purposes such as research, diagnostics, and therapeutics. The main types of biomolecules commonly separated include proteins, nucleic acids (DNA/RNA), amino acids and peptides, polysaccharides, lipids, and small metabolites. Each class of biomolecule has unique structural and chemical properties that dictate the choice of separation method. Proteins are among the most frequently separated biomolecules due to their central role in enzymatic activity, structural functions, and disease markers. Usually, variations in size, charge, hydrophobicity, and binding affinity are affected in their separation efficiency. Commonly employed methods include affinity chromatography, gel filtration, and ion-exchange chromatography. Affinity-based methods are particularly effective because they leverage specific interactions, such as enzyme-substrate or antibody-antigen binding, to selectively isolate target proteins [5]. Nucleic acids, including DNA and RNA, are typically separated using gel electrophoresis, ultracentrifugation, or chromatographic methods. Their separation depends largely on size and charge. Emerging methods such as nanofluidic single-molecule sorting offer extremely high resolution for DNA, enabling manipulation and analysis at the single-molecule level [17]. Amino acids and peptides are smaller than the proteins and often require techniques with higher resolution, such as reverse-phase HPLC or capillary electrophoresis. In chiral studies, microfluidic techniques using chiral selectors are especially useful for resolving enantiomers of amino acids and other biomolecules. These systems can achieve high sensitivity and are ideal for small sample volumes [18]. Polysaccharides, due to their structural diversity and large size, are often separated using aqueous two-phase systems (ATPS) or membrane-based techniques. Their solubility and hydrophilicity make these systems efficient, environmentally friendly, and scalable. Such methods are particularly popular in industrial biotechnology and food processing [19]. Lipids, being hydrophobic, are separated using solvent-based extractions or thin-layer chromatography (TLC) and normal-phase HPLC. These methods rely on the differential solubility and polarity of lipid species and are critical in both clinical lipidomic and metabolic research. Finally, small metabolites and other bioactive molecules are separated primarily by gas chromatography (GC) or liquid chromatography-mass spectrometry (LC-MS), or liquid chromatography-mass spectrometry. These methods provide the high sensitivity and specificity needed to analyze compounds with low abundance in intricate biological matrices [12].

Materials Research Foundations 191 (2026) 317-342
https://doi.org/10.21741/9781644904213-10

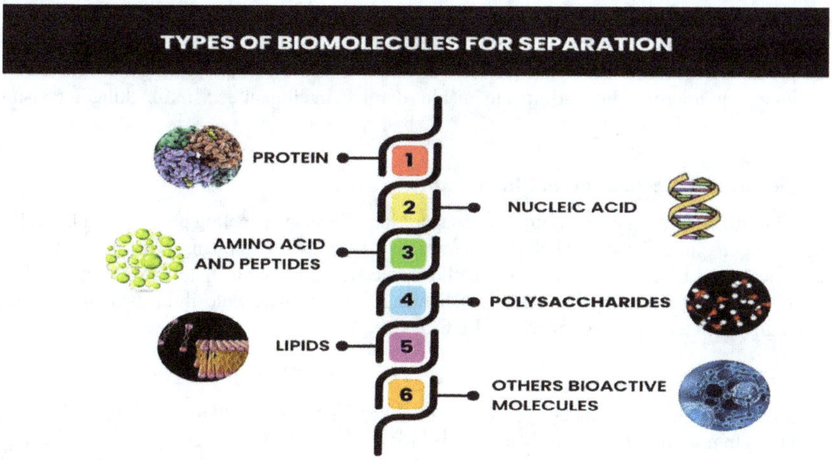

Figure 2. Schematic diagram exhibiting about various types of biomolecules used for separation.

1.4 Overview of Biomolecule Separation

Biomolecule separation refers to the isolation of targeted biological molecules from the heterogeneous mixtures, enabling functional studies and analytical applications across genomics, proteomics, and metabolomics. Popular methods that are suited to particular molecular properties like size, charge, or binding affinity include gel electrophoresis, size-exclusion chromatography, ion-exchange chromatography, and affinity chromatography [20]. These methods are crucial for identifying, quantifying, and characterizing molecules, underpinning major advances in molecular biology and drug development [1]. One major challenge, however, is the complexity and similarity of biomolecules—particularly proteins, which often share close molecular weights and may undergo post-translational modifications that further complicate their separation. As a result, hybrid strategies similar to two-dimensional electrophoresis and LC-MS have become indispensable for achieving the required analytical resolution [1].

1.5 Role of Microfluidics in Modern Bioanalysis

Microfluidics have transformed bioanalysis by allowing complex laboratory processes to be conducted on miniature chip-based platforms [2]. These systems, typically operating in the nanolitre to microlitre range, consist of intricately patterned microchannels that manipulate fluid movement with high precision [21]. They are excellent for sensitive and point-of-care applications due to their quick processing times and low reagent and sample consumption [20]. Microfluidics offers remarkable control over the physical forces—such as pressure, magnetic gradients, and electric fields—that can be utilized to separate molecules according to their size, charge, or affinity [22]. Technologies like free-flow electrophoresis and droplet-based sorting have expanded the potential for continuous, high-throughput analysis [23]. When coupled with biosensors, these devices enable real-time detection and characterization of separated species within the same

Materials Research Foundations 191 (2026) 317-342 https://doi.org/10.21741/9781644904213-10

platform, streamlining workflows in diagnostics and biomedical research [24]. Additionally, microfluidic systems are highly scalable, supporting parallelization and multiplexing to process multiple samples simultaneously [25]. Recent advancements are now integrating machine learning algorithms into microfluidic platforms to enable adaptive, intelligent decision-making in real-time assays.

2. Separation Mechanisms in Microfluidics

Microfluidic platforms have revolutionized separation science by enabling precise, rapid, and low-volume processing of complex biological samples. Various mechanisms—based on electrical, mechanical, and biochemical principles—allow selective separation of biomolecules, particles, and cells within microchannels. These mechanisms exploit microscale fluid dynamics, making them highly efficient and integrable into LOC device.

2.1 Electrophoresis

A classical technique called electrophoresis uses an applied electric field to cause charged particles to migrate. In microfluidics, it has been adapted for capillary electrophoresis (CE), allowing high-resolution separation in microchannels with minimal sample volumes. At microscale, electrophoretic mobility becomes more predictable due to uniform field distribution and stable laminar flows [26]. It is particularly effective for DNA, RNA, and protein analysis due to differences in charge-to-mass ratios [26]. Reduced Joule heating in microfluidic systems allows for stable temperature conditions, preserving biological sample integrity [26]. This technique is frequently integrated with laser-induced fluorescence detection to provide high sensitivity for biomolecular quantification.

2.2 Di-electrophoresis

Di-electrophoresis (DEP) manipulates the particles using non-uniform electric fields, causing movement based on induced polarization rather than net charge. It is exceptionally suited for label-free cell sorting. Microfluidic DEP can distinguish between healthy and diseased cells by exploiting their dielectric properties [27]. Positive DEP (attraction to high field regions) and negative DEP (repulsion) enable bidirectional control in separation workflows [28]. The technique has been utilized for bacterial detection, cancer cell enrichment, and stem cell isolation, due to its high selectivity [29]. DEP-based platforms can be integrated with microelectrodes fabricated by photolithography, allowing on-chip real-time manipulation and analysis [30].

2.3 Affinity-Based Separation

Affinity separation leverages the specific interaction between a target molecule and a recognition element, such as antibodies, aptamers, or lectins. In microfluidics, affinity elements are immobilized onto microchannel walls or micropillars to capture specific analytes with high selectivity [31]. For example, antibody-coated microchannels can trap circulating tumour cells (CTCs) with >95% specificity under continuous flow conditions [32]. Aptamer-functionalized surfaces are also effective in selectively capturing proteins and small molecules, offering reversibility and regeneration advantages over antibodies [33]. This technique's unmatched specificity makes it popular in biosensing, point-of-care (POC) diagnostics, and targeted drug delivery [34].

Materials Research Foundations 191 (2026) 317-342 | https://doi.org/10.21741/9781644904213-10

2.4 Size-Based Filtration

Size-based separation utilizes physical constraints like micropillar arrays or nanoporous membranes to differentiate particles by hydrodynamic radius. One popular technique for directing particles into size-specific flow paths is the deterministic lateral displacement (DLD), which makes use of staggered pillar arrays [35]. This approach allows high-throughput sorting of cells, viruses, and exosomes without requiring surface markers or label. Size-based filtration is also used for plasma separation, where blood cells are excluded by microfilters allowing only plasma to pass [36]. Nanoporous membranes can be embedded within microchannels for virus filtration and protein fractionation with ultrahigh precision [37].

2.5 Hydrodynamic Separation

Hydrodynamic separation uses the principles of laminar flow and particle-fluid interactions to achieve sorting in microchannels without any external fields. Particles of different sizes naturally migrate to different streamlines due to shear-induced lift forces in inertial microfluidic systems [38]. Pinched-flow fractionation (PFF) is another technique where particles are directed to narrow zones and sorted as the flow path widens downstream [39]. This passive method enables continuous separation of blood components, bacteria, and microparticles without clogging [40]. Hydrodynamic separation is advantageous for label-free, energy-efficient sorting and is easily scalable for clinical applications [41]. Table 1 depicts about the main microfluidic separation techniques and their targets.

Table 1. *Microfluidic Separation Techniques and Their Targets.*

Separation Techniques	Working Principle	Target Biomolecules	Advantages	References
Electrophoresis	Electric field-driven migration	DNA, RNA, Proteins	High resolution, established technique	[26]
Di electrophoresis	Motion in non-uniform electric field	Cells, Exosomes, Organelles	Label-free, precise separation	[27–30]
Affinity-Based Capture	Biomolecular binding interactions	Specific proteins, Exosomes	High specificity and sensitivity	[31–34]
Hydrodynamic Filtration	Size-based flow partitioning	Cells, Particles, Organelles	Passive, continuous-flow compatible	[35–41]
Droplet Microfluidics	Compartmentalized fluid manipulation	Enzymes, Single cells, DNA	High throughput, minimal cross-contamination	[76,78]

3. Biomolecules Targeted for Separation

The various types of biomolecules targeted using typical microfluidic chip for separation are illustrated in figure 3. A wide variety of biomolecules can be isolated and purified using microfluidic technologies because of their high specificity, low sample requirements, and compatibility with POC and real-time analysis systems [3]. These systems offer label-free or affinity-based mechanisms to isolate molecules ranging from macromolecular proteins to extracellular vesicles like exosomes, with high throughput and reproducibility.

 Materials Research Forum LLC
https://doi.org/10.21741/9781644904213-10

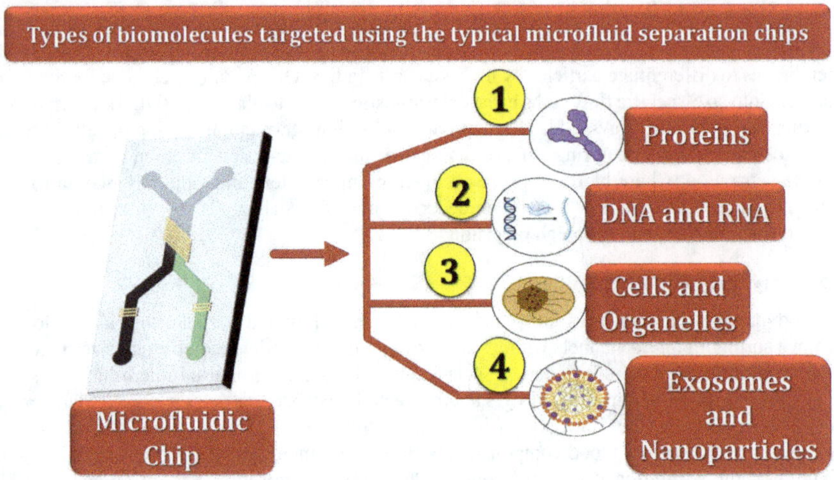

Figure 3. *Schematic diagram of various types of biomolecules targeted using typical microfluidic chip.*

3.1 Proteins

Protein separation in microfluidics is critical for diagnostics, biomarker validation, and functional proteomics. Due to their heterogeneity in charge, size, and hydrophobicity, the proteins are ideal candidates for microfluidic electrophoresis, isoelectric focusing, and immunoaffinity-based separation [42]. A notable application involves on-chip isoelectric focusing (IEF) to separate proteins based on their isoelectric points with high resolution and reproducibility [42]. Microfluidic platforms using immunoaffinity columns can selectively bind proteins based on antigen-antibody interactions, enabling high selectivity even in complex mixtures [43]. Microfluidic capillary electrophoresis has been used to separate denatured proteins in under a minute with sub-nanolitre volumes, improving resolution and efficiency over traditional slab gels [44]. A breakthrough in label-free protein detection was the integration of on-chip surface plasmon resonance (SPR), enabling separation and real-time quantification of bound proteins [44]. Microfluidic western blotting offers a compact and rapid alternative to conventional blotting, reducing analysis time to under 30 minutes and allowing quantification from nanolitre-scale samples [45]. Aptamer-based microchips have also been used for specific capture of target proteins from serum, offering regeneration capability and enhanced [46].

3.2 Nucleic acids

In order to isolate and process deoxyribonucleic acid (DNA) and ribonucleic acid (RNA) for forensics, gene expression profiling, and molecular diagnostics, microfluidic is serving a crucial role. Microfluidic gel electrophoresis chips are commonly used for sizing DNA fragments, useful in genotyping, forensic analysis, and cancer diagnostics [47].

Materials Research Foundations 191 (2026) 317-342 https://doi.org/10.21741/9781644904213-10

Dielectrophoretic traps have been shown to separate DNA strands by length in free solution without requiring gels or sieving matrices [48]. Microfluidic PCR (μPCR) platforms integrate amplification and separation steps on a single chip, allowing for ultra-fast thermal cycling and post-PCR analysis [49]. For RNA, silica-coated microbeads have been embedded into microchannels for on-chip RNA purification, enabling mRNA capture and elution with yields comparable to conventional spin columns [50].

3.3 Cells and Organelles

Microfluidic platforms offer a label-free, high-throughput alternatives to conventional cytometry for isolating single cells, rare cell types, and intracellular organelles. Inertial microfluidics effectively separates white blood cells, red blood cells, and platelets from whole blood by sorting cells by size and deformability using flow-induced forces [38]. Using standing surface acoustic waves (SSAWs), acoustic separation has been used to gently isolate cells without labels while reducing mechanical stress [51]. For mitochondria, dielectrophoretic in nano slit devices has enabled separation from cytoplasmic debris with high yield, facilitating downstream proteomics [28].

3.4 Exosomes and Nanoparticles

Extracellular vesicles, also known as exosomes, are membrane-bound vesicles that range in size from 30 to 150 nm and transport lipids, proteins, and nucleic acids. Microfluidics have emerged as the key tools for their label-free isolation, characterization, and clinical analysis. Exosomes can be separated from blood and urine based on the size using the deterministic lateral displacement (DLD) in nanoscale pillar arrays with little pre-processing [52]. On-chip immunomagnetic enrichment has been demonstrated using antibody-functionalized magnetic beads to isolate CD63+ exosomes from plasma, significantly reducing processing time and improving purity [53]. Micro vortex-based separation enables continuous-flow nanoparticle enrichment using dean flow dynamics, suitable for sorting synthetic nanoparticles and liposomes [54]. **Table 2** indicates the types of biomolecules targeted and suitable microfluidic techniques.

Table 2. *Types of Biomolecules targeted and Suitable Microfluidic Techniques.*

Biomolecule Type	Common Techniques	Typical Sample Source	Detection Methods	Reference
Proteins	CE, Affinity, SPR	Serum, plasma, cell lysate	Fluorescence, SPR, MS	[42–46]
DNA/RNA	CE, DEP, μPCR	Blood, saliva, swabs	Fluorescence, gel imaging	[47–50]
Cells	DEP, Hydrodynamic, SSAW	Whole blood, tissue	Microscopy, impedance cytometry	[28,51]
Exosomes/Nanoparticles	DLD, Immunomagnetic capture	Plasma, urine	Fluorescence, NTA, ELISA	[52–54]

4. Design and Fabrication of Microfluidic Separation Devices

Microfluidic separation devices' performance, sensitivity, and suitability for a range of chemical and biological tests are greatly influenced by their design. This section explores the foundational design principles including chip layout, channel geometry, detection system integration, and

Materials Research Foundations 191 (2026) 317-342 https://doi.org/10.21741/9781644904213-10

surface chemistry—all of which contribute to the specificity and efficiency of biomolecule separation.

4.1 Chip Layout and Channel Design

Effective microfluidic chip layout begins with channel geometry, which influences fluid resistance, flow behaviour, and particle separation profiles. Laminar flow, which dominates in microfluidics, allows controlled diffusion and minimal cross-contamination [2]. Serpentine channels are often used to enhance mixing through induced chaotic advection without external energy [55]. Deterministic lateral displacement (DLD) arrays, where micropillars are offset in a periodic arrangement, enable size-based separation of particles and cells with high precision [38]. Pinched-flow fractionation channels use controlled narrowing and expansion to separate particles by size in a continuous flow, allowing high-throughput operation without clogging [39].

4.2 Integration of Detection Methods

Detection methods are crucial for identifying, quantifying, and verifying separated analytes. To maintain system compactness and allow for real-time monitoring, these techniques must be reduced in size and smoothly incorporated into the chip layout. Because of its sensitivity and suitability for DNA, protein, and cell assays, fluorescence detection is the most popular method in microfluidics [20]. Surface plasmon resonance (SPR) integrated within microchannels allows label-free, real-time biosensing with applications in protein-protein and antigen-antibody interaction monitoring [56]. Electrochemical sensors, including impedance spectroscopy and amperometric detectors, are increasingly integrated for POC applications, particularly for glucose, lactate, and pathogen detection [57]. Optofluidic platforms combining optics and microfluidics can enable high-speed imaging and multi-channel detection for cell sorting and cytometry [57].

4.3 Surface Modification and Functionalization

Surface properties are essential in microfluidics because they affect fluid dynamics, adsorption, and biomolecule interactions. Surface modification tailors' microchannel walls for non-fouling, selective binding, or enhanced wettability. PEGylation is widely used to reduce nonspecific protein adsorption by grafting polyethylene glycol chains on PDMS or glass surfaces [58]. Plasma oxidation is commonly employed on PDMS to introduce silanol groups (Si–OH) for subsequent functionalization with silane or polymer coatings [59]. Antibody and aptamer functionalization of surfaces allows specific affinity-based capture of target analytes or cells. Microfluidic chips functionalized with anti-EpCAM antibodies have been used to selectively isolate circulating tumour cells [60]. Nanostructured coatings (e.g., gold nanoparticles, graphene oxide) enhance electrochemical sensitivity and surface area, improving detection performance in biosensors [61].

5. Applications of Microfluidic Separation Devices

Due to their distinct benefits, which include portability, low reagent consumption, low fabrication costs, and high integration, microfluidic separation technologies have experienced rapid growth in a variety of application domains. From healthcare to environmental monitoring, these devices are increasingly replacing bulky, lab-based systems with compact, automated platforms. Microfluidic devices fabrication and their application in the common disease diagnosis and drug deliveries are illustrated in **figure 4**.

Materials Research Foundations 191 (2026) 317-342 https://doi.org/10.21741/9781644904213-10

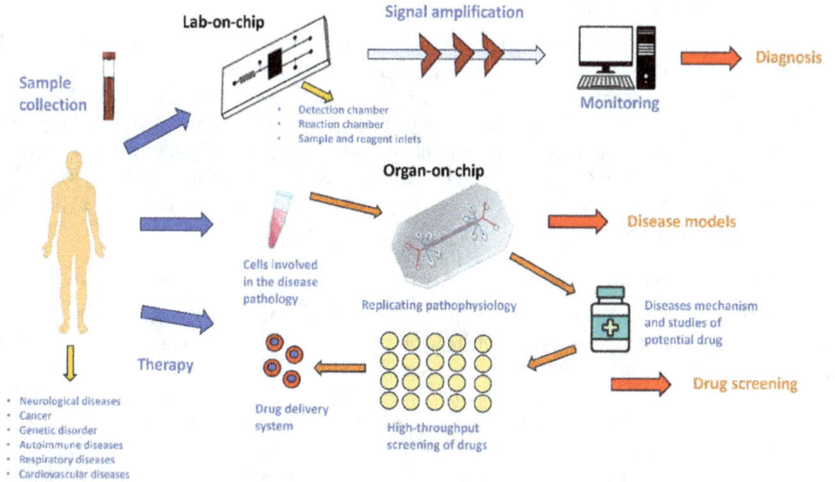

Figure 4. *Flow diagram of Microfluidic devices fabrication and application in disease diagnosis and drug delivery.*

5.1 Clinical Diagnostics

Microfluidic separation systems have had a profound impact on clinical diagnostics, enabling rapid and sensitive detection of biomarkers from the small sample volumes. These platforms integrate separation and detection modules into a single chip for automated, multiplexed analysis. Immunoaffinity-based microfluidics is used to isolate disease-specific markers, such as cancer biomarkers, from blood samples with high specificity [62]. Devices have been developed to detect cardiac troponin for early myocardial infarction diagnosis, delivering results in under 15 minutes from whole blood [63]. Microfluidic PCR systems have been implemented for detecting infectious agents such as human immunodeficiency virus (HIV) and *Mycobacterium tuberculosis*, reducing the overall diagnostic time significantly [64].

5.2 Proteomics and Genomics

Proteomics and genomics have been transformed by microfluidics, which offers the platforms for high-throughput DNA, RNA, and protein analysis, separation, and sample processing. To determine protein expression profiles in proteomics, microfluidic devices provide on-chip two-dimensional (2D) gel electrophoresis and western blotting [45]. Digital PCR (dPCR) in microfluidic droplets enables single-molecule sensitivity and precise quantification of any types of genetic materials—particularly useful in cancer mutation detection [65]. Microfluidic platforms allow rapid and automated RNA-sequencing library preparation, reducing the manual steps and possible contamination [66].

Materials Research Foundations 191 (2026) 317-342
https://doi.org/10.21741/9781644904213-10

5.3 Point-of-Care Testing

Point-of-Care (POC) diagnostic relies heavily on the microfluidic devices, which provides a quick, portable, and easy-to-use tools for use in clinics, remote locations, and even households. Paper-based microfluidics (μPADs) are widely used in the low-resource settings for rapid glucose, malaria, and dengue testing without the need for complex instrumentation [67]. Smartphone-integrated microfluidic platforms now enable image-based detection of pathogens and allergens using optical sensors and machine learning algorithms [68]. Colorimetric assays on lateral flow microdevices have been commercialized for COVID-19 antigen detection, combining speed and ease of interpretation [69].

5.4 Environmental and Food Testing

Microfluidic separation technologies, which provide real-time analysis of contaminants, toxins, and microbial populations, are finding increasing use in food safety testing and environmental monitoring. Electrochemical and fluorescent sensors embedded on a chip have been used in microfluidics to detect heavy metal ions (such as Pb^{+2} and Hg^{+2}) in water [70]. In food testing, microfluidic devices enable on-site detection of pesticides and pathogens on fruits, vegetables, and meat surfaces via immunoassays [71]. Label-free optical microfluidic sensors have also been applied for detecting bacterial contamination in milk and toxin residues in seafood, ensuring compliance with safety regulations [72].

6. Advantages and Challenges of Microfluidic Separation Devices

In the fields of analytical science and biomedical engineering, microfluidics has revolutionary potential. These gadgets, which are frequently credit card-sized, can perform intricate tasks such as biomolecule separation, detection, and analysis with exceptional precision. However, this powerful technology also presents the number of technical and practical challenges, especially when moving from laboratory prototypes to real-world applications.

6.1 Miniaturization and Efficiency

One of the key advantages of microfluidics lies in its miniaturized scale, which enables fast reaction kinetics, enhanced control of fluid dynamics, and significant reductions in reagent consumption. The reduced channel dimensions enhance mass and heat transfer efficiency, leading to faster biomolecular interactions and decreased diffusion distances [73]. The integration of several laboratory processes (such as mixing, reaction, and separation) into a single platform is made possible by microfluidic systems, which also provide spatial and temporal control over fluids, enhancing overall operational efficiency [2]. Their compactness also enables portable diagnostics, useful in resource-limited settings and in field-based environmental monitoring [74]. The precision of analytical results may be impacted by issues like non-specific adsorption of biomolecules brought on by the higher surface-area-to-volume ratio [75].

6.2 Sample Volume and Throughput

Microfluidic systems are made to work with incredibly tiny sample volumes, typically between nanolitres and picolitres, which is particularly beneficial for analysing scarce or precious biological samples such as cerebrospinal fluid or single-cell lysates [76]. By processing multiple samples or reactions at once, these devices facilitate parallel operations, increasing overall analytical throughput and decreasing turnaround time [77]. Droplet-based microfluidics further enhances throughput by enabling compartmentalized, high-speed reactions in discrete volumes, making it ideal for digital PCR and enzyme assays [78]. However, achieving high throughput requires complex design strategies such as parallelized channel networks and precise flow control mechanisms, which can complicate fabrication and system calibration [79].

6.3 Device Complexity and Cost

Despite the advantages, the complexity and cost of device fabrication remain major obstacles to the widespread adoption of microfluidic separation systems. Many devices still rely on cleanroom-based fabrication techniques such as photolithography and plasma bonding, which are expensive and time-consuming [80]. Functional modules like heaters, valves, and sensors that are integrated into a single chip increase design complexity and frequently call for interdisciplinary knowledge [21]. Moreover, while soft lithography using PDMS has simplified prototyping, PDMS is not always suitable for long-term or commercial applications due to its poor chemical resistance and moisture absorption [81]. Although they frequently lack the resolution required for sub-micrometre channel features, emerging methods like 3D printing and laser micromachining provide scalable and affordable alternatives [82]. Table 3 summarizes about the most common advantages of microfluidics in biomolecular separation.

Table 3. *Advantages of Microfluidics in Biomolecular Separation.*

Features	Impact on Separation	References
Miniaturization	Reduces reagent/sample volume, increases portability	[73,74]
Laminar Flow Control	Enhances separation precision and repeatability	[81,82]
Integrated Detection	Enables real-time, on-chip analysis	[20,80]
Surface Functionalization	Allows selective capture of target molecules	[78,79]
Multiplexing Capabilities	Processes multiple samples in parallel	[76,77]
Capillary Flow (μPADs)	Enables power-free operation in remote settings	[84–86]

7. Recent Advances and Emerging Trends in Microfluidics

Recent progress in microfluidics has expanded its application domain beyond conventional lab settings, enabling rapid diagnostics, environmental monitoring, and drug discovery. These innovations emphasize simplification, automation, and real-time analysis. This section explores the most transformative and current trends shaping microfluidics research and development. Table 4 depicts about the recent advances and emerging trends in microfluidics along with their advantages.

Materials Research Foundations 191 (2026) 317-342
https://doi.org/10.21741/9781644904213-10

Table 4. *Recent Advances and Emerging Trends in Microfluidics and Their Advantages.*

Emerging Trend	Feature/Innovation	Applications	Advantage	References
Paper-based microfluidics	Capillary-driven flow	POC diagnostics, environmental testing	Low-cost, portable	[84,85]
Droplet microfluidics	Discrete droplet manipulation	Single-cell analysis, drug screening	High throughput, minimal reagent	[87,88]
Lab-on-a-chip (LOC)	Fully integrated systems	Molecular diagnostics	Automation	[90,91]

7.1 Paper-Based Microfluidics

Paper-based microfluidics (μPADs) offer an ultra-low-cost, biodegradable, and electricity-free platform suitable for diagnostic use in under-resourced settings. Unlike conventional chips, paper devices leverage capillary action for fluid transport, eliminating the need for external pumps [83]. These devices have been applied for heavy metal detection in water using electrochemical readouts from wax-printed electrodes [84]. Recent work demonstrated about the plasma separation from whole blood on paper-based chips using simple mechanical filtering zones without centrifugation [85]. These platforms are also increasingly paired with smartphone-assisted detection, allowing image-based quantification for remote diagnostics [86].

7.2 Droplet and Digital Microfluidics

The ability to manipulate distinct, homogeneous volumes of liquids within an immiscible carrier fluid (such as oil) makes the droplet microfluidics perfect for single-cell sequencing, enzyme assays, and large-scale chemical reactions. Recent innovations include on-chip droplet generation and fusion techniques using active pneumatic control, allowing highly tuneable droplet operations for biochemical screening [87]. Electrowetting-based digital microfluidics (DMF) has emerged for programmable droplet actuation without fixed channels. It is particularly effective for clinical diagnostics and point-of-care genetic assays [88]. DMF has also been used for real-time drug screening in cancer therapeutics using integrated cell culture and imaging modules [77]. To address droplet uniformity challenges, advanced flow-focusing geometries were introduced for stable generation of monodisperse emulsions with high precision [89].

7.3 Lab-on-a-Chip Integration

LOC platforms are the convergence point of microfluidics, biosensing, and microelectronics, creating the compact systems capable of full analytical workflows on a chip. One significant breakthrough involved the integration of sample preparation, nucleic acid extraction, and qPCR onto a centrifugal LOC device for viral detection, offering complete automation [90]. Emerging LOC systems are also being developed for neurotoxicity testing using 3D neural tissue grown inside perfused microchambers, simulating organ-level responses [91]. These days, some LOCs use microfluidic impedance cytometry, which allows for high-throughput single-cell classification without labelling [92]. Integration with IoT and cloud-based platforms allows data streaming and remote access to diagnostic results in real time, greatly improving telehealth possibilities [93].

Materials Research Foundations 191 (2026) 317-342 https://doi.org/10.21741/9781644904213-10

8. Future Prospects of Microfluidic Separation Technologies

As microfluidics transitions from laboratory prototypes to widespread deployment, the field is being shaped by rapid advances in personalized medicine, AI-driven automation, and efforts to meet regulatory and commercial challenges. These factors can determine how microfluidic systems are integrated into future biomedical, industrial, and consumer technologies.

8.1 Personalized Medicine

Microfluidics is poised to be a cornerstone in personalized healthcare, enabling rapid, point-of-care diagnostics tailored to individual genomic or proteomic profiles. By facilitating single-cell analysis, microfluidics enabling insights into tumour heterogeneity and immune responses, critical for tailoring cancer therapies [94]. Devices have been developed to rapidly profile circulating tumour DNA (ctDNA) and exosomes from blood, helping guide therapy choices in real time [95]. Microfluidic drug screening platforms using patient-derived cells now enable clinicians to predict drug efficacy and toxicity in-vitro before patient administration [96].

8.2 Automation and AI Integration

Real-time data collection, processing, and interpretation are being revolutionized by the combination of microfluidic devices with artificial intelligence (AI) and machine learning. Automated image recognition algorithms in microfluidic cytometers allow the identification of rare cells or abnormal morphologies without manual labelling [97]. AI has also been used for design optimization, predicting microchannel geometries that maximize separation efficiency under various conditions [98]. Real-time feedback-controlled systems in droplet microfluidics are now using AI to adjust flow rates and droplet size autonomously, minimizing human intervention [99].

8.3 Commercialization and Regulatory Considerations

Despite technological maturity, commercialization of microfluidic devices faces challenges related to cost, reproducibility, and regulatory pathways. Standardization and quality control are critical barriers, as many lab-scale devices lack scalable and consistent manufacturing protocols [100]. From a regulatory standpoint, the FDA's classification of lab-on-a-chip systems often places them under in-vitro diagnostics (IVDs), requiring thorough validation and clinical trials [101]. Efforts are underway to develop open-source microfluidic platforms and mass-manufacturable polymer devices to support commercial expansion while ensuring compliance [102].

Conclusion

The advancement of microfluidic technologies has revolutionized the landscape of biomolecular separation, offering unprecedented precision, speed, and miniaturization. From their foundational principles rooted in fluid dynamics to their intricate fabrication processes and diverse material platforms, microfluidic systems have evolved into powerful analytical tools that can combine intricate lab operations onto a single chip. Using techniques like electrophoresis, di electrophoresis, affinity-based capture, and hydrodynamic filtration, these systems achieve highly effective separations by taking advantage of the special behaviors of fluids at the microscale, such as laminar flow, improved diffusion control, and surface-driven phenomena. Microfluidics now enables the selective isolation of a broad spectrum of biomolecules—including proteins, nucleic acids, cells, organelles, and exosomes—using minimal sample volumes and reagents. The design innovations in chip architecture, channel geometry, and surface functionalization have further amplified device sensitivity and specificity. Meanwhile, Workflows in environmental monitoring, proteomics, genomics, and diagnostics have been streamlined by the integration of detection techniques like surface plasmon resonance, fluorescence, and electrochemical sensors, which have enabled real-time, on-chip analysis of separated components.

Recent developments in lab-on-a-chip, droplet microfluidics, and paper-based diagnostics have broadened the range of applications to include ultra-sensitive platforms for single-cell and molecular-level analyses as well as field-deployable devices. Microfluidics is also essential to the creation of POC testing systems, which allow for quick and dispersed healthcare delivery. Looking ahead, the field is moving toward AI-driven automation, real-time feedback control, and integration with personalized medicine. The promise of tailoring diagnostics and therapeutics to individual patients using microfluidic devices representing a paradigm shifts in clinical care. However, widespread adoption requires overcoming significant challenges, including fabrication complexity, cost constraints, and the need for robust regulatory frameworks. Efforts in mass production techniques, standardization protocols, and open-source platforms are critical to ensuring the transition of microfluidics from research labs to commercial and clinical environments.

In summary, microfluidic separation technology stands at the intersection of biology, engineering, and data science, offering a transformative approach to molecular analysis. Its continued evolution is poised to drive major breakthroughs in healthcare, biotechnology, and environmental science—ushering in a new era of miniaturized, intelligent, and accessible analytical systems.

References

[1] R. Aebersold, M. Mann, Mass spectrometry-based proteomics, Nature. 422 (2003) 198-207. https://doi.org/10.1038/nature01511

[2] G.M. Whitesides, The origins and the future of microfluidics, Nature. 442 (2006) 368-373. https://doi.org/10.1038/nature05058

[3] E.K. Sackmann, A.L. Fulton, D.J. Beebe, The present and future role of microfluidics in biomedical research, Nature. 507 (2014) 181-189. https://doi.org/10.1038/nature13118

[4] Y. Saylan, A. Denizli, Supermacroporous composite cryogels in biomedical applications, Gels. 5 (2019) 1-20. https://doi.org/10.3390/gels5020020

[5] M. Glad, P.O. Larsson, New methods for separation and recovery of biomolecules, Curr. Opin. Biotechnol. 2 (1991) 1-6. https://doi.org/10.1016/S0958-1669(05)80148-X

[6] H.H.H. Chiang, The laboratory technology of discrete molecular separation: The historical development of gel electrophoresis and the material epistemology of biomolecular science, 1945-1970, J. Hist. Biol. 42 (2009) 495-527. https://doi.org/10.1007/s10739-008-9169-5

[7] W. Worthy, New separation method offers speed, resolution, Chem. Eng. News. 69 (1991) 25-26. https://doi.org/10.1021/cen-v069n046.p025

[8] H.A. Fishman, D.R. Greenwald, R.N. Zare, Biosensors in chemical separations, Annu. Rev. Biophys. Biomol. Struct. 27 (1998) 165-198. https://doi.org/10.1146/annurev.biophys.27.1.165

[9] A. Segaran, L.S. Chua, Review of recent applications and modifications of aqueous two-phase system for the separation of biomolecules, Int. J. Biol. Macromol. 276 (2024) 1-11. https://doi.org/10.1016/j.ijbiomac.2024.133856

[10] Y. Song, H. Bao, X. Shen, X. Li, X. Liang, S. Wang, Emerging nanoporous materials for biomolecule separation, Adv. Funct. Mater. 32 (2022) 1-16. https://doi.org/10.1002/adfm.202113153

[11] Y.O. Jang, G.S. Noh, H. Liu, B. Koo, Z. Qiao, Y. Shin, Dimethyl 3,3′-dithiobispropionimidate-functionalized diatomaceous earth particles for efficient biomolecule separation, Sci. Rep. 10 (2020) 1-9. https://doi.org/10.1038/s41598-020-72913-8

[12] S. Umasekar, N. Virivinti, Advances in modeling techniques for the production and purification of biomolecules: a comprehensive review, J. Chromatogr. B, 1232 (2024) 1-13. https://doi.org/10.1016/j.jchromb.2023.123945

[13] C. Acquah, E.M. Obeng, D. Agyei, C.M. Ongkudon, C.K.S. Moy, M.K. Danquah, Nano-doped monolithic materials for molecular separation, Separations. 4 (2017) 1-22. https://doi.org/10.3390/separations4010002

[14] K. Shimura, K. Kasai, Affinophoresis in two-dimensional agarose gel electrophoresis: Specific separation of biomolecules by a moving affinity ligand, Anal. Biochem. 161 (1987) 200-206. https://doi.org/10.1016/0003-2697(87)90672-5

[15] K.S. Khoo, H.Y. Leong, K.W. Chew, J.W. Lim, T.C. Ling, P.L. Show, H.W. Yen, Liquid biphasic system: A recent bioseparation technology, Processes. 8 (2020) 1-22. https://doi.org/10.3390/pr8020149

[16] Y. Shin, Rich phase separation behavior of biomolecules, Mol. Cells. 45 (2022) 6-15. https://doi.org/10.14348/molcells.2021.0204

[17] T. Yamamoto, T. Fujii, Nanofluidic single-molecule sorting of DNA: A new concept in separation and analysis of biomolecules towards ultimate level performance, Nanotechnol. 21 (2010) 1-20. https://doi.org/10.1088/0957-4484/21/39/395502

[18] L. Zhang, Q.G. Tan, J.Q. Fan, C. Sun, Y.T. Luo, R.P. Liang, J.D. Qiu, Microfluidics for chiral separation of biomolecules, TrAC Trends Anal. Chem. 158 (2023) 1-15. https://doi.org/10.1016/j.trac.2022.116842

[19] A. Ebrahimi, K. Icoz, R. Didarian, C. Shih, E.A. Tarim, B. Nasseri, A. Akpek, B. Cecen, A. Bal-Ozturk, K. Güleç, Y.E. Li, S. Shih, B.S. Tarim, H.C. Tekin, E. Alarçin, M. Tayybi-Azar, H. Ghorbanpoor, C. Özel, A. Eker Sarıboyacı, F.D. Guzel, N. Bassous, S.R. Shin, H. Avci, Molecular separation by using active and passive microfluidic chip designs: A comprehensive review, Adv. Mater. Interfaces. 11 (2024) 1-45. https://doi.org/10.1002/admi.202300492

[20] P.S. Dittrich, A. Manz, Lab-on-a-chip: Microfluidics in drug discovery, Nat. Rev. Drug Discov. 5 (2006) 210-218. https://doi.org/10.1038/nrd1985

[21] S.K. Sia, G.M. Whitesides, Microfluidic devices fabricated in poly(dimethylsiloxane) for biological studies, Electrophoresis. 24 (2003) 3563-3576. https://doi.org/10.1002/elps.200305584

[22] H.A. Stone, A.D. Stroock, A. Ajdari, Engineering flows in small devices: Microfluidics toward a lab-on-a-chip, Annu. Rev. Fluid Mech. 36 (2004) 381-411. https://doi.org/10.1146/annurev.fluid.36.050802.122124

[23] N.L. Jeon, S.K.W. Dertinger, D.T. Chiu, I.S. Choi, A.D. Stroock, G.M. Whitesides, Generation of solution and surface gradients using microfluidic systems, Langmuir. 16 (2000) 8311-8316. https://doi.org/10.1021/la000600b

[24] M.S. Chiriacò, M. Bianco, A. Nigro, E. Primiceri, F. Ferrara, A. Romano, A. Quattrini, R. Furlan, V. Arima, G. Maruccio, Lab-on-chip for exosomes and microvesicles detection and characterization, Sensors, 18 (2018) 1-41. https://doi.org/10.3390/s18103175

[25] T. Thorsen, S.J. Maerkl, S.R. Quake, Microfluidic large-scale integration, Science. 298 (2002) 580-584. https://doi.org/10.1126/science.1076996

[26] B.M. Chien, D.M. Lubman, Analysis of the fragments from collision-induced dissociation of electrospray-produced peptide ions using a quadrupole ion trap storage/reflectron time-of-flight mass spectrometer, Anal. Chem. 66 (1994) 1630-1636. https://doi.org/10.1021/ac00082a006

[27] N. Abd Rahman, F. Ibrahim, B. Yafouz, Dielectrophoresis for biomedical sciences applications: A review, Sensors. 17 (2017) 1-27. https://doi.org/10.3390/s17030449

[28] P. Gascoyne, S. Shim, Isolation of circulating tumor cells by dielectrophoresis, Cancers. 6 (2014) 545-579. https://doi.org/10.3390/cancers6010545

[29] Z.D.E. Shee, E.E.M. Noor, M.F. Baig, Protein Manipulation via Dielectrophoresis: Theoretical principles and emerging microfluidic platforms, Micromachines. 16 (2025) 1-25. https://doi.org/10.3390/mi16050531

[30] R. Pethig, Dielectrophoresis: Status of the theory, technology, and applications, Biomicrofluidics. 4 (2010) 1-35. https://doi.org/10.1063/1.3456626

[31] Y. Zhang, V. Lyons, D. Pappas, Fundamentals of affinity cell separations, Electrophoresis. 39 (2018) 732-741. https://doi.org/10.1002/elps.201700311

[32] E.L. Rodriguez, S. Poddar, S. Iftekhar, K. Suh, A.G. Woolfork, S. Ovbude, A. Pekarek, M. Walters, S. Lott, D.S. Hage, Affinity chromatography: A review of trends and developments over the past 50 years, J. Chromatogr. B Analyt. Technol. Biomed. Life Sci. 1157 (2020) 1-41. https://doi.org/10.1016/j.jchromb.2020.122332

[33] B. Mattiasson, T.G.I. Ling, Efforts to integrate affinity interactions with conventional separation technologies, J. Chromatogr. B Biomed. Sci. Appl. 376 (1986) 235-243. https://doi.org/10.1016/S0378-4347(00)80840-3

[34] D.S. Hage, J.A. Anguizola, C. Bi, R. Li, R. Matsuda, E. Papastavros, E. Pfaunmiller, J. Vargas, X. Zheng, Pharmaceutical and biomedical applications of affinity chromatography: Recent trends and developments, J. Pharm. Biomed. Anal. 69 (2012) 93-105. https://doi.org/10.1016/j.jpba.2012.01.004

[35] L.R. Huang, E.C. Cox, R.H. Austin, J.C. Sturm, Continuous particle separation through deterministic lateral displacement, Science. 304 (2004) 987-990. https://doi.org/10.1126/science.1094567

[36] R.B. Aim, S. Vigneswaran, H. Prasanthi, V. Jegatheesan, Influence of particle size and size distribution in granular bed filtration and dynamic microfiltration, Water Sci. Technol. 36 (1997) 207-215. https://doi.org/10.2166/wst.1997.0121

[37] H.C. Kim, J. Park, Y. Cho, H. Park, A. Han, X. Cheng, Lateral-flow particle filtration and separation with multilayer microfluidic channels, J. Vac. Sci. Technol. B. 27 (2009) 3115-3119. https://doi.org/10.1116/1.3258155

[38] D.D. Carlo, Inertial microfluidics, Lab Chip. 9 (2009) 3038-3046. https://doi.org/10.1039/b912547g

[39] M. Yamada, M. Nakashima, M. Seki, Pinched flow fractionation: Continuous size separation of particles utilizing a laminar flow profile in a pinched microchannel, Anal. Chem. 76 (2004) 5465-5471. https://doi.org/10.1021/ac049863r

[40] J.L. Iguain, J. Kurchan, Putting hydrodynamic interactions to work: Tagged-particle separation, Europhys. Lett. 63 (2003) 715-721. https://doi.org/10.1209/epl/i2003-00572-5

[41] A. Karimi, S. Yazdi, A.M. Ardekani, Hydrodynamic mechanisms of cell and particle trapping in microfluidics, Biomicrofluidics. 7 (2013) 1-23. https://doi.org/10.1063/1.4799787

[42] R.J. Meagher, Y.K. Light, A.K. Singh, Rapid, continuous purification of proteins in a microfluidic device using genetically-engineered partition tags, Lab Chip. 8 (2008) 527-532. https://doi.org/10.1039/b716462a

[43] Y.C. Wang, M.H. Choi, J. Han, Two-dimensional protein separation with advanced sample and buffer isolation using microfluidic valves, Anal. Chem. 76 (2004) 4426-4431. https://doi.org/10.1021/ac0497499

[44] S.G. Anema, The use of "lab-on-a-chip" microfluidic SDS electrophoresis technology for the separation and quantification of milk proteins, Int. Dairy J. 19 (2009) 198-204. https://doi.org/10.1016/j.idairyj.2008.11.001

[45] W. Pan, W. Chen, X. Jiang, Microfluidic western blot, Anal. Chem. 82 (2010) 3974-3976. https://doi.org/10.1021/ac1000493

[46] M. Gong, I. Nikcevic, K.R. Wehmeyer, P.A. Limbach, W.R. Heineman, Protein-aptamer binding studies using microchip affinity capillary electrophoresis, Electrophoresis. 29 (2008) 1415-1422. https://doi.org/10.1002/elps.200700777

[47] R. Wu, Y. Seah, Z. Wang, Microfluidic chip for stacking, separation and extraction of multiple DNA fragments, J. Chromatogr. A. 1437 (2016) 219-225. https://doi.org/10.1016/j.chroma.2016.01.076

[48] Y. Sun, Z. Lu, M. Miller, T. Perroud, Y. Tong, Application of microfluidic chip electrophoresis for high-throughput nucleic acid fluorescence fragment analysis assays, NAR Genom. Bioinform. 5 (2023) 1-7. https://doi.org/10.1093/nargab/lqad011

[49] T. Demeke, G.R. Jenkins, Influence of DNA extraction methods, PCR inhibitors and quantification methods on real-time PCR assay of biotechnology-derived traits, Anal. Bioanal. Chem. 396 (2010) 1977-1990. https://doi.org/10.1007/s00216-009-3150-9

[50] T. Nguyen, S. Kwak, S.J. Karpowicz, Re-use of commercial microfluidics chips for DNA, RNA, and protein electrophoresis, Biotechniques. 57 (2014) 267-271. https://doi.org/10.2144/000114232

[51] P. Augustsson, J.T. Karlsen, H.-W. Su, H. Bruus, J. Voldman, Iso-acoustic focusing of cells for size-insensitive acousto-mechanical phenotyping, Nat. Commun. 7 (2016) 1-9. https://doi.org/10.1038/ncomms11556

[52] D. Wang, S. Yang, N. Wang, H. Guo, S. Feng, Y. Luo, J. Zhao, A novel microfluidic strategy for efficient exosome separation via thermally oxidized non-uniform deterministic lateral displacement (DLD) arrays and dielectrophoresis (DEP) synergy, Biosensors. 14 (2024) 1-14. https://doi.org/10.3390/bios14040174

[53] H. Xu, C. Liao, P. Zuo, Z. Liu, B.C. Ye, Magnetic-based microfluidic device for on-chip isolation and detection of tumor-derived exosomes, Anal. Chem. 90 (2018) 13451-13458. https://doi.org/10.1021/acs.analchem.8b03272

[54] Y. Kim, B. Lee Chung, M. Ma, W.J.M. Mulder, Z.A. Fayad, O.C. Farokhzad, R. Langer, Mass production and size control of lipid-polymer hybrid nanoparticles through controlled microvortices, Nano Lett. 12 (2012) 3587-3591. https://doi.org/10.1021/nl301253v

[55] A.D. Stroock, S.K.W. Dertinger, A. Ajdari, I. Mezić, H.A. Stone, G.M. Whitesides, Chaotic mixer for microchannels, Science. 295 (2002) 647-651. https://doi.org/10.1126/science.1066238

[56] A. Hemmi, T. Usui, A. Moto, T. Tobita, N. Soh, K. Nakano, H. Zeng, K. Uchiyama, T. Imato, H. Nakajima, A surface plasmon resonance sensor on a compact disk-type microfluidic device, J. Sep. Sci. 34 (2011) 2913-2919. https://doi.org/10.1002/jssc.201100446

[57] B. Alhalaili, I.N. Popescu, C.O. Rusanescu, R. Vidu, Microfluidic devices and microfluidics-integrated electrochemical and optical (bio)sensors for pollution analysis: A review, Sustainability. 14 (2022) 1-38. https://doi.org/10.3390/su141912844

[58] M. Eichler, C.P. Klages, K. Lachmann, Surface functionalization of microfluidic devices, in: A. Dietzel (Eds.), Microsystems for pharmatechnology, Springer Cham., Switzerland, 2016, pp. 59-97. https://doi.org/10.1007/978-3-319-26920-7_3

[59] Y.H. Tennico, M.T. Koesdjojo, S. Kondo, D.T. Mandrell, V.T. Remcho, Surface modification-assisted bonding of polymer-based microfluidic devices, Sens. Actuators B: Chem. 143 (2010) 799-804. https://doi.org/10.1016/j.snb.2009.10.001

[60] J. Zhang, W. Sheng, Z.H. Fan, An ensemble of aptamers and antibodies for multivalent capture of cancer cells, Chem. Commun. 50 (2014) 6722-6725. https://doi.org/10.1039/c4cc02002b

[61] D.G. Rackus, M.D.M. Dryden, J. Lamanna, A. Zaragoza, B. Lam, S.O. Kelley, A.R. Wheeler, A digital microfluidic device with integrated nanostructured microelectrodes for electrochemical immunoassays, Lab Chip. 15 (2015) 3776-3784. https://doi.org/10.1039/C5LC00660K

[62] S.T. Sanjay, G. Fu, M. Dou, F. Xu, R. Liu, H. Qi, X.J. Li, Biomarker detection for disease diagnosis using cost-effective microfluidic platforms, Analyst. 140 (2015) 7062-7081. https://doi.org/10.1039/C5AN00780A

[63] Z. Beiram Zadeh, S.M. Hosseini, J. Mohammadnejad, M. Tanhaei, S. Ramakrishna, Label-free detection of cardiac biomarkers: A review on microfluidic electrochemical biosensors, ACS Appl. Bio Mater. 6 (2023) 2622-2635. https://doi.org/10.1021/acsabm.3c00257

[64] A. Vafadar, S. Takallu, S.K. Alashti, S. Rashidi, S. Bahrani, A. Tajbakhsh, E. Mirzaei, A. Savardashtaki, Advancements in microfluidic platforms for rapid biomarker diagnostics of infectious diseases, Microchem. J. 208 (2025) 1-20. https://doi.org/10.1016/j.microc.2024.112296

[65] M.K. Steinbach, J. Leipert, T. Matzanke, A. Tholey, Digital microfluidics for sample preparation in low-input proteomics, Small Methods. 9 (2025) 1-16. https://doi.org/10.1002/smtd.202400495

[66] X. Xu, Q. Zhang, J. Song, Q. Ruan, W. Ruan, Y. Chen, J. Yang, X. Zhang, Y. Song, Z. Zhu, C. Yang, A highly sensitive, accurate, and automated single-cell RNA sequencing platform with digital microfluidics, Anal. Chem. 92 (2020) 8599-8606. https://doi.org/10.1021/acs.analchem.0c01613

[67] E. Wang, Z. Guo, R. Tang, Y.H. Lo, Using airflow-driven, evaporative gradients to improve sensitivity and fluid control in colorimetric paper-based assays, Lab Chip. 21 (2021) 4249-4261. https://doi.org/10.1039/D1LC00542A

[68] S. Duan, T. Cai, F. Liu, Y. Li, H. Yuan, W. Yuan, K. Huang, K. Hoettges, M. Chen, E.G. Lim, C. Zhao, P. Song, Automatic offline-capable smartphone paper-based microfluidic device for efficient biomarker detection of Alzheimer's disease, Anal. Chim. Acta. 1308 (2024) 1-9. https://doi.org/10.1016/j.aca.2024.342575

[69] S. Lee, L. Bi, H. Chen, D. Lin, R. Mei, Y. Wu, L. Chen, S.W. Joo, J. Choo, Recent advances in point-of-care testing of COVID-19, Chem. Soc. Rev. 52 (2023) 8500-8530. https://doi.org/10.1039/D3CS00709J

[70] P. Aryal, E. Brack, T. Alexander, C.S. Henry, Capillary flow-driven microfluidics combined with a paper device for fast user-friendly detection of heavy metals in water, Anal. Chem. 95 (2023) 5820-5827. https://doi.org/10.1021/acs.analchem.3c00378

[71] R. Sivakumar, N.Y. Lee, Recent progress in smartphone-based techniques for food safety and the detection of heavy metal ions in environmental water, Chemosphere. 275 (2021) 1-16. https://doi.org/10.1016/j.chemosphere.2021.130096

[72] R. Pavadai, N. Pavadai, R. Palanisamy, A. Amalraj, M. Arivazhagan, G. Honnu, S. Kityakarn, S. Thongmee, J. Khumphon, D. Khamboonrueang, C. Issro, Recent trends and future challenges of metal-organic framework-based optical and electrochemical sensing platforms for ultra-sensitive detection of biomarkers and environmental contaminants: A review (Year-2023 & 2024), Microchem. J. 210 (2025) 1-26. https://doi.org/10.1016/j.microc.2025.112937

[73] T.M. Squires, S.R. Quake, Microfluidics: Fluid physics at the nanoliter scale, Rev. Mod. Phys. 77 (2005) 977-1026. https://doi.org/10.1103/RevModPhys.77.977

[74] J.K. Luo, Y.Q. Fu, Y. Li, X.Y. Du, A.J. Flewitt, A.J. Walton, W.I. Milne, Moving-part-free microfluidic systems for lab-on-a-chip, J. Micromech. Microeng. 19 (2009) 1-14. https://doi.org/10.1088/0960-1317/19/5/054001

[75] A. Beck, F. Obst, M. Busek, S. Grunzner, P. Mehner, G. Paschew, D. Appelhans, B. Voit, A. Richter, Hydrogel patterns in microfluidic devices by do-it-yourself UV-photolithography suitable for very large-scale integration, Micromachines. 11 (2020) 1-20. https://doi.org/10.3390/mi11050479

[76] J.E. Urrutia Gomez, M. Zhou, N.K. Mandsberg, J.A. Serna, J. von Padberg, S. Liu, M. Reischl, P.A. Levkin, A.A. Popova, Highly parallel and high-throughput nanoliter-scale liquid, cell, and spheroid manipulation on droplet microarray, Adv. Funct. Mater. 35 (2025) 1-15. https://doi.org/10.1002/adfm.202410355

[77] A. Ali-Cherif, S. Begolo, S. Descroix, J. Viovy, L. Malaquin, Programmable magnetic tweezers and droplet microfluidic device for high-throughput nanoliter multi-step assays, Angew. Chem. Int. Ed. Engl. 51 (2012) 10765-10769. https://doi.org/10.1002/anie.201203862

[78] R. Zilionis, J. Nainys, A. Veres, V. Savova, D. Zemmour, A.M. Klein, L. Mazutis, Single-cell barcoding and sequencing using droplet microfluidics, Nat. Protoc. 12 (2017) 44-73. https://doi.org/10.1038/nprot.2016.154

[79] Y. Zhu, Q. Fang, Analytical detection techniques for droplet microfluidics-A review, Anal. Chim. Acta. 787 (2013) 24-35. https://doi.org/10.1016/j.aca.2013.04.064

[80] P. Mukherjee, F. Nebuloni, H. Gao, J. Zhou, I. Papautsky, Rapid prototyping of soft lithography masters for microfluidic devices using dry film photoresist in a non-cleanroom setting, Micromachines. 10 (2019) 1-11. https://doi.org/10.3390/mi10030192

[81] N. Bhattacharjee, A. Urrios, S. Kang, A. Folch, The upcoming 3D-printing revolution in microfluidics, Lab Chip. 16 (2016) 1720-1742. https://doi.org/10.1039/C6LC00163G

[82] T. Nguyen, T. Sarkar, T. Tran, S.M. Moinuddin, D. Saha, F. Ahsan, Multilayer soft photolithography fabrication of microfluidic devices using a custom-built wafer-scale PDMS slab aligner and cost-efficient equipment, Micromachines. 13 (2022) 1-10. https://doi.org/10.3390/mi13081357

[83] N. Nuchtavorn, M. Macka, A novel highly flexible, simple, rapid and low-cost fabrication tool for paper-based microfluidic devices (μPADs) using technical drawing pens and in-house formulated aqueous inks, Anal. Chim. Acta. 919 (2016) 70-77. https://doi.org/10.1016/j.aca.2016.03.018

[84] M. Yuan, C. Li, Y. Zheng, H. Cao, T. Ye, X. Wu, L. Hao, F. Yin, J. Yu, F. Xu, A portable multi-channel fluorescent paper-based microfluidic chip based on smartphone imaging for simultaneous detection of four heavy metals, Talanta. 266 (2024) 1-9. https://doi.org/10.1016/j.talanta.2023.125112

[85] X. Yang, O. Forouzan, T.P. Brown, S.S. Shevkoplyas, Integrated separation of blood plasma from whole blood for microfluidic paper-based analytical devices, Lab Chip. 12 (2012) 274-280. https://doi.org/10.1039/C1LC20803A

[86] Y. Manmana, M. Macka, N. Nuchtavorn, Distance-based paper microfluidic devices for rapid visual quantification of heavy metals in herbal supplements and cosmetics, RSC Adv. 14 (2024) 36142-36151. https://doi.org/10.1039/D4RA05358C

[87] S.C.C. Shih, P.C. Gach, J. Sustarich, B.A. Simmons, P.D. Adams, S. Singh, A.K. Singh, A droplet-to-digital (D2D) microfluidic device for single cell assays, Lab Chip. 15 (2015) 225-236. https://doi.org/10.1039/C4LC00794H

[88] N. Shembekar, C. Chaipan, R. Utharala, C.A. Merten, Droplet-based microfluidics in drug discovery, transcriptomics and high-throughput molecular genetics, Lab Chip. 16 (2016) 1314-1331. https://doi.org/10.1039/C6LC00249H

[89] D. Pekin, V. Taly, Multiplex detection of KRAS mutations using passive droplet fusion, Methods Mol. Biol. 1547 (2017) 133-142. https://doi.org/10.1007/978-1-4939-6734-6_11

[90] J. Zhu, S. Pan, H. Chai, P. Zhao, Y. Feng, Z. Cheng, S. Zhang, W. Wang, Microfluidic impedance cytometry enabled one-step sample preparation for efficient single-cell mass spectrometry, Small. 20 (2024) 1-42. https://doi.org/10.1002/smll.202310700

[91] N.M. Rodriguez, W.S. Wong, L. Liu, R. Dewar, C.M. Klapperich, A fully integrated paperfluidic molecular diagnostic chip for the extraction, amplification, and detection of nucleic acids from clinical samples, Lab Chip. 16 (2016) 753-763. https://doi.org/10.1039/C5LC01392E

[92] M.A. Tariq, Advancing diagnostic capabilities through organ-on-a-chip technology, Future Biotechnol. 4 (2024) 126. https://doi.org/10.54393/fbt.v4i02.126

[93] V. Iyer, D.A. Issadore, F. Aflatouni, The next generation of hybrid microfluidic/integrated circuit chips: Recent and upcoming advances in high-speed, high-throughput, and multifunctional lab-on-IC systems, Lab Chip. 23 (2023) 2553-2576. https://doi.org/10.1039/D2LC01163H

[94] H. Pei, M. Yu, D. Dong, Y. Wang, Q. Li, L. Li, B. Tang, Phenotype-related drug sensitivity analysis of single CTCs for medicine evaluation, Chem. Sci. 11 (2020) 8895-8900. https://doi.org/10.1039/C9SC05566E

[95] T. Mulholland, M. McAllister, S. Patek, D. Flint, M. Underwood, A. Sim, J. Edwards, M. Zagnoni, Drug screening of biopsy-derived spheroids using a self-generated microfluidic concentration gradient, Sci. Rep. 8 (2018) 1-12. https://doi.org/10.1038/s41598-018-33055-0

[96] N. Hao, J.X.J. Zhang, Microfluidic screening of circulating tumor biomarkers toward liquid biopsy, Sep. Purif. Rev. 47 (2018) 19-48. https://doi.org/10.1080/15422119.2017.1320763

[97] C. Barnes, A.R. Sonwane, E.C. Sonnenschein, F. Del Giudice, Machine learning enhanced droplet microfluidics, Phys. Fluids. 35 (2023) 1-13. https://doi.org/10.1063/5.0163806

[98] A. Lashkaripour, C. Rodriguez, N. Mehdipour, R. Mardian, D. McIntyre, L. Ortiz, J. Campbell, D. Densmore, Machine learning enables design automation of microfluidic flow-focusing droplet generation, Nat. Commun. 12 (2021) 1-14. https://doi.org/10.1038/s41467-020-20284-z

[99] S. Dubey, P. Vishwakarma, T. Ramarao, S.K. Dubey, S. Goel, A. Javed, Artificial intelligence-based droplet size prediction for microfluidic system, Int. J. Numer. Methods Heat Fluid Flow. 34 (2024) 3045-3078. https://doi.org/10.1108/HFF-07-2023-0361

[100] D.R. Reyes, H. van Heeren, S. Guha, L. Herbertson, A.P. Tzannis, J. Ducree, H. Bissig, H. Becker, Accelerating innovation and commercialization through standardization of microfluidic-based medical devices, Lab Chip. 21 (2021) 9-21. https://doi.org/10.1039/D0LC00963F

[101] H. Cong, N. Zhang, Perspectives in translating microfluidic devices from laboratory prototyping into scale-up production, Biomicrofluidics. 16 (2022) 1-15. https://doi.org/10.1063/5.0079045

[102] S. Dekker, W. Buesink, M. Blom, M. Alessio, N. Verplanck, M. Hihoud, C. Dehan, W. Cesar, A.L. Nel, A.V.D. Berg, M. Odijk, Standardized and modular microfluidic platform for fast Lab on Chip system development, Sens. Actuators B: Chem. 272 (2018) 468-478. https://doi.org/10.1016/j.snb.2018.04.005

Materials Research Foundations 191 (2026) 343-363 https://doi.org/10.21741/9781644904213-11

Chapter 11

Microfluidic in Organic Synthesis

Tufan Singha Mahapatra[1], Sourav Ghosh[1], Saheli Roy[1], Susanta Ghanta[2,*]

[1]Department of Chemistry, Faculty of Science and Technology, ICFAI University Tripura, Agartala-799210, Tripura (W), India

[2]Department of Chemistry, National Institute of Technology Agartala, Jirania, Agartala, Tripura 799046, India

*susanta.ghanta82@gmail.com

Abstract

Microfluidic technology has emerged as a powerful platform for advancing organic synthesis by offering precise control over reaction parameters within miniaturized flow systems. Unlike conventional batch processes, microfluidic systems enable rapid mixing, efficient heat and mass transfer, and continuous reagent delivery, which collectively enhance reaction rates, selectivity, and safety. These advantages make microfluidics highly attractive for both academic research, where rapid optimization, mechanistic studies, and high-throughput screening are essential, and industrial applications requiring scalable, reproducible, and sustainable processes. Moreover, microfluidic platforms inherently support green chemistry principles by minimizing waste, reducing reagent use, and enabling safer reaction conditions. This chapter reviews the fundamental principles of microfluidic flow chemistry, its advantages, and explores its application across various types of organic reactions, discussing its role in shaping the future of sustainable and scalable chemical synthesis. This chapter also discusses the industrial uses and perspectives of microfluidic devices.

Keywords

Microfluid, Microreactor Devices, Organic Synthesis, Advantages, Scale-up of Microflow Reactions

Contents

Materials Research Foundations 191 (2026) 343-363 https://doi.org/10.21741/9781644904213-11

1. Introduction

Organic synthesis is a cornerstone of modern chemistry, fundamental to the development of new pharmaceuticals, agrochemicals, materials, and fine chemicals [1-4]. Traditionally carried out in batch reactors, organic synthesis has long faced challenges such as poor heat and mass transfer, limited scalability, labor-intensive optimization, and safety concerns, particularly for hazardous or highly exothermic reactions [5-7]. In recent years, microfluidic technology has emerged as a transformative tool that addresses many of these limitations, enabling both research laboratories and industrial sectors to rethink the conduct of organic synthesis [8,9]. Microfluidics, the science and technology of manipulating fluids within microscale channels (typically 10–500 μm in width), has emerged as a transformative approach in modern organic synthesis. It provides high control over reaction conditions due to its inherent physical advantages. These include enhanced surface-to-volume ratios, precise temperature control, rapid mixing, and laminar flow behavior. Such conditions enable chemists to fine-tune parameters like residence time, concentration, and temperature with exceptional accuracy far surpassing what is typically achievable in traditional batch systems [10-14].

In the context of laboratory organic synthesis, microfluidic platforms have become invaluable tools for method development, reaction discovery, and high-throughput screening [15-19]. The miniaturization and automation of synthetic processes accelerate the optimization of reaction

Materials Research Foundations 191 (2026) 343-363 https://doi.org/10.21741/9781644904213-11

conditions, reduce reagent consumption, and allow for parallel experimentation. Moreover, the integration of in-line analytical techniques such as FTIR, UV-Vis, and NMR spectroscopy enables real-time monitoring and control of chemical transformations, offering deeper mechanistic insights and reducing experimental turnaround time [20-24]. In the industrial perspective, microfluidic reactors have paved the way for continuous-flow synthesis, a process paradigm that enhances reproducibility, safety, and scalability [25-28]. The ability to perform reactions in a controlled, continuous manner allows for greater consistency in product quality and facilitates the safe handling of reactive intermediates or toxic reagents. Modular microreactor systems can be scaled by numbering up, rather than scaling up, thus allowing seamless transition from lab-scale experiments to pilot-scale or full-scale production without the usual challenges associated with batch-to-flow translation. Companies are increasingly adopting microfluidic technologies to achieve faster time-to-market, greener processes, and greater compliance with safety and regulatory standards. Moreover, microfluidics aligns closely with the principles of green chemistry [29,30]. By enabling more selective reactions, reducing solvent and reagent excess, and operating under milder conditions, microfluidic systems help minimize the environmental footprint of synthetic processes [31-33]. Their efficiency in heat and material usage not only contributes to safer and cleaner operations but also offers cost advantages and regulatory compliance in environmentally conscious manufacturing environments.

As a result of these combined advantages, microfluidics is increasingly regarded as a pivotal technology in both exploratory research and industrial-scale organic synthesis. This chapter examines the fundamental principles, recent advancements, and diverse applications of microfluidic systems in organic chemistry, while highlighting their increasing impact on the future of sustainable, scalable, and data-driven chemical synthesis.

2. Microreactors and Microfluidic Devices

2.1 Conceptualization

Microfluidic reactors are miniaturized systems that precisely control fluid flow within micrometer-scale channels. Nowadays, their ability to tightly regulate reaction conditions, such as temperature and concentration gradients, makes them ideal for exploring reaction mechanisms, optimising yields, and scaling up synthesis in a controlled manner. This correlation has led to significant advancements in continuous-flow organic synthesis, including applications in pharmaceuticals and research-oriented analytical chemistry [34]. There are various types of microreactors used for different applications, including capillary reactors, plate reactors, and laminar reactors [35]. When utilised as single channels, microreactors can provide exceptional rates of mass and heat transfer in multiphase systems, enabling faster and more efficient organic reactions compared to conventional batch processes. To attain the required throughput for large-scale industrial production, microreactors need to be enhanced by boosting the number of parallel channels. Initial research highlighted the benefits of microreactors in terms of upscaling, which involves deploying multiple parallel microchannel processing units. This idea later became a significant subject of detailed industrial evaluation concerning process enhancement. The analytical system, like a lab-on-a-chip (LOC) that comprises pumps, valves, mixers, reactors, and separators, or a micro-total-analysis system (μ-TAS), tends to be intricate. To reach the desired results, the process must employ a network system.

Materials Research Foundations 191 (2026) 343-363 https://doi.org/10.21741/9781644904213-11

Microfluidic devices are tiny systems that operate small volumes of chemical fluids (typically in the range of 10-100 μm). These devices offer precise control of fluid mixing, reactions, and chemical analysis. Recently, a variety of microfluidic devices have been developed using materials such as polymers [36], ceramics [37], and silicon [38] for organic reactions. Their configurations are influenced by the characteristics of the materials used and the types of responses, yet they are still in the initial stages of development [11]. These reactors are significant in chemical synthesis by enhancing conditions, making them promising for rapid organic synthesis.

2.2 Main Advantages

The utilization of microreactors has several benefits that make it a potentially useful tool for organic chemists. Microfluidic devices offer greater control over reaction conditions, enhancing productivity and security, and making chemistry faster, simpler, and more repeatable [39]. Accurate temperature control has been one of the most important properties of these devices due to their high surface-to-volume ratio, that enables rapid heat transfer at lower temperature gradients and enables accurate temperature control. Compared to conventional lab setups, these devices consume a significantly smaller amount of reagent, allowing for better results to be obtained with fewer resources. The micro-channels' tiny fluid passage distances result in quicker reaction kinetics and shorter reaction times. Reactions are more effective due to the enhanced mixing made possible by the micro-channels' architecture [40]. Multiple reactions, complex chemical processes, and high-throughput screening are made possible by the smooth automation of microfluidic devices. Another very important aspect of using these devices is the controlled flow of a limited reactant, which reduces the possibility of mishaps and makes it safer to react with high-energy or delicate compounds.

2.3 Scale-up of Microflow Reactions

Unlike normal batch reactors, in microreactors, the steady-state reaction conditions obtained are the same regardless of the amount of reactants flowing. Moreover, as microreactors operate continuously, it is possible to scale up reactions from small to large quantities by simply introducing more reagents and allowing the reaction to proceed for a longer period. Another way to scale up is called "parallel pipe lining", which involves multiple reactors working in parallel to give more product output [35]. However, because this process needs several reactors to operate at similar conditions (temperature and pressure), the latter approach is quite difficult. Even distribution of reactants among all reactors also requires several pumps or accurate flow dividers, which is also challenging.

Another term, microwave heating, has widespread use in organic chemistry synthesis; however, this process has a limited capacity for scaling up production. However, with the introduction of these new microfluidic devices, which can sustain very elevated temperatures and pressures comparable to those achieved in microwave reactors, along with continuous flow systems, these devices become attractive alternatives to conventional microwave technology [41].

The following sections of this chapter provide specific examples of microreactor technology-oriented organic synthesis, including a variety of organic transformations analysed with continuous flow and the benefits of using these microdevices.

Materials Research Foundations 191 (2026) 343-363 https://doi.org/10.21741/9781644904213-11

3. Microreactors and Microfluidic Devices Used in Organic Chemistry

3.1 Liquid–Liquid Reactions

3.1.1 Photochemical Reactions

These processes are essential for organic reactions initiated by light energy, commonly utilizing UV or visible light. To scale up photochemical reactions, long irradiation periods are often required [42]. However, achieving this can be challenging as it becomes difficult to maintain irradiation efficiency, which may lead to the breakdown of starting materials and the formation of unwanted byproducts. Incessant microflow appliances in photochemistry help ensure the reaction receives uniform and consistent light throughout. Shallow reactors maximize irradiation penetration, enabling shorter residence times and avoiding the issues commonly associated with traditional laboratory reaction vessels [43].

Fukuyama et al. explored [2+2] cycloaddition reactions between vinyl acetates and many cyclohexenone derivatives utilizing a microflow system equipped with a glass microreactor (1000 μm wide and 500 μm deep) and a 300 W high-pressure mercury lamp as the light source [44] (Scheme 1). Yields were enhanced, and the duration of the reactions was significantly shortened compared to the same reactions conducted in bulk processing utilizing a Pyrex flask. Two micro-photoreactors connected in series, when operated at double the flow rate of a single reactor, produced similar performance outcomes.

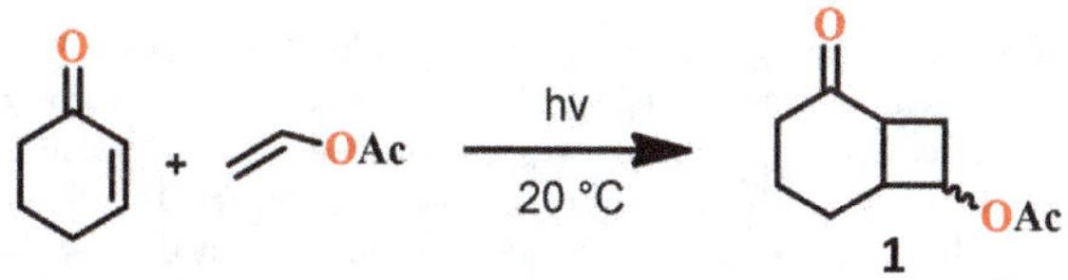

Scheme 1. [2+2] Cycloaddition reaction between vinyl acetate and cyclohexanone substrate.

Another photo-cycloaddition described by Booker-Milburn et al. [45] utilizing a Pyrex-filtered 400 W medium-pressure mercury lamp with an FEP (fluorinated ethylene propylene) flow reactor under flow circumstances, the intramolecular [5+2] cycloaddition of 2 to produce 3 in 63% yield, a vital step in the preparation of Stemona alkaloid (±)-neostenine (Scheme 2) [9].

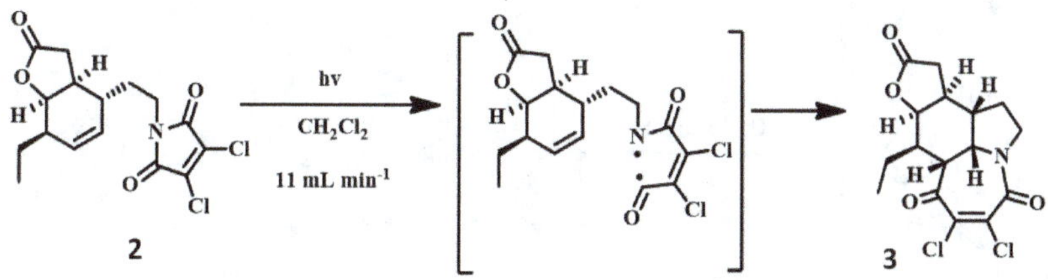

Scheme 2. Key step in the synthesis of (±)-neostenine.

Materials Research Foundations 191 (2026) 343-363
https://doi.org/10.21741/9781644904213-11

3.1.2 Heterocycle Synthesis

Microflow devices have proven to be highly effective for synthesizing heterocyclic compounds due to their cost-effectiveness and compact design. Acke and Stevens [46] described the multicomponent approach to prepare isochromen-1-one derivatives 5 (Scheme 3) utilizing a microflow device [9]. However, during the synthesis process, the final product tends to crystallize and accumulate within the microreactor. To address this issue, they introduced an immiscible perfluorinated solvent between the micromixer and the residence time unit (RTU), forming an oily barrier between the wall of the microreactor channel and the reaction plug. Such a layer helps prevent the degradation of the reactor walls and allows the reaction to proceed without blockages [47].

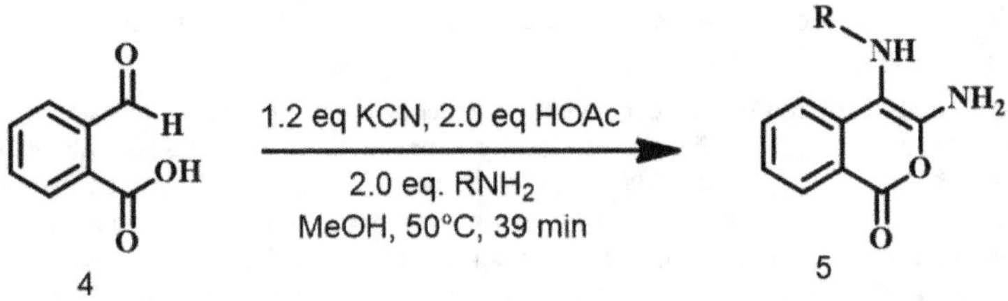

Scheme 3. Synthesis of isochromen-1-one derivatives.

Haswell et al. demonstrate the synthesis of 1,2-pyrazoles (5) through electroosmotic flow, utilizing a glass microreactor (Scheme 4a) [48]. Compared to traditional batch synthesis methods, the yields were significantly improved. This enhancement is likely due to the improved temperature control provided by the microflow device. As illustrated in Scheme 4b, this flow approach was also employed to synthesize isoxazole (7) and substituted pyrazole (6). The product yields in both reactions were markedly higher than those obtained through batch processes. Specifically, the amount of benzylated pyrazole enhanced to 100% from 76%, while the yield of isoxazole improved from 52% to 98% [9, 48].

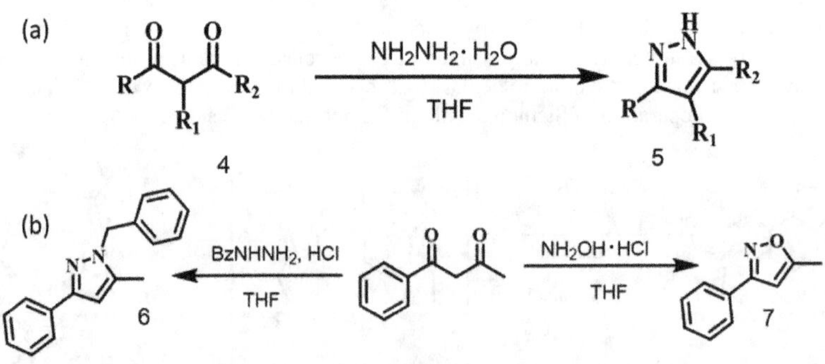

Scheme 4. (a) Synthesis of 1,2-pyrazole derivatives 5; (b) Synthesis of isoxazole (7) and pyrazole (6).

Materials Research Foundations 191 (2026) 343-363 https://doi.org/10.21741/9781644904213-11

3.1.3 Synthesis of Bio-oligomers

Microreactor synthesis can also be used for the preparation of bio-oligomers (β-peptides and oligosaccharides). Seeberger et al. prepare β-amino acid oligomers using a microreactor strategy, demonstrating that yields are superior to those obtained through solid-phase peptide synthesis (SPPS) due to their shortened reaction time, higher reaction temperature, and prevention of product precipitation during the reaction [49]. This precipitation typically produced a gel-like reaction mixture in batch mode, which resulted in inadequate blending and extended reaction times. Applying a fluorous benzylic ester to protect the first amino acid made it much easier to purify β-peptides in the later steps. Oligosaccharide synthesis has also been carried out using microreactor devices.

3.1.4 Multistep Reactions

Multiple reaction steps can be merged into a single continuous process by employing continuous flow methods for multi-step synthesis, facilitating faster access to target compounds. Performing multiple reaction steps simultaneously in flow allows for rapid optimization, significantly reduces reaction times, and frequently removes the need for lengthy procedures after the reaction, including the purification and segregation of intermediates. For instance, Cosford et al. recently reported a multistep preparatory method of bis-substituted 1,2,4-oxadiazoles (10) that utilized three consecutive microdevices [50] (scheme 5). The ability to superheat the solvent in the used microreactors greatly shortened reaction times. This approach reduced the synthesis time for 1,2,4-oxadiazoles from 3 days to ~35 minutes.

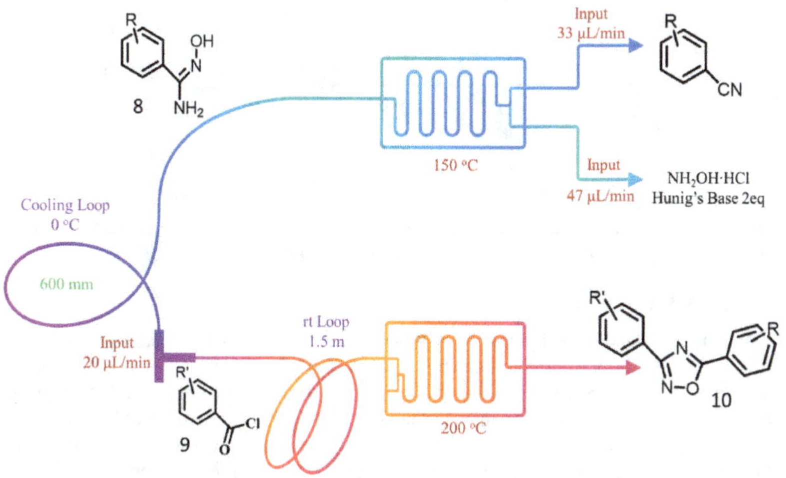

Scheme 5. Synthesis of 1,2,4-oxadiazoles (10)

Materials Research Foundations 191 (2026) 343-363
https://doi.org/10.21741/9781644904213-11

3.1.5 Free Radical Reactions

The popularity of radical reactions in organic synthesis is increasing due to their high adaptability and excellent tolerance for various functional groups. However, controlling these radical reactions can be challenging. Tris(trimethylsilyl)silane (TTMSS) has been utilized as a reducing agent in a microreactor to convert dithiocarbonate derivatives and halogen compounds into the corresponding alkanes [51]. At a temperature of 130 °C, reactions in toluene were completed in just 5 minutes (Scheme 6a). Similar conditions were employed to hydrosilylate alkynes and olefins, yielding excellent amounts of the corresponding silylated compounds. When compared to batch operations, upgraded ratios of Z/E were observed for the hydrosilylation of alkynes (Scheme 6b).

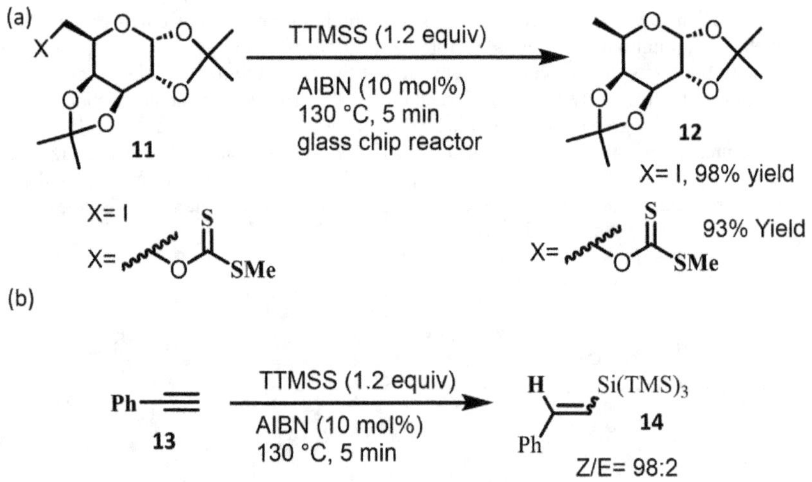

Scheme 6. Radical-based reactions conducted within the microreactor, where AIBN = 2,2'-azobis(2-methyl propionitrile).

3.2 Liquid–Gas Reactions

Gas-liquid micro reaction technology has recently gained significant attention due to its vast potential in various important chemical reactions and gas-liquid mass transfer processes. Gas-liquid microreactors can be categorized into four primary types based on how the two phases come into contact [52]. These types are: (a) Microchannel contactors (which include both single and parallel microchannel contactors), (b) Falling film microreactors, (c) Microreactors with phase interfaces stabilized by physical structures (also known as mesh microreactors) (d) Micro packed bed reactors, (e) Micro packed bed reactor.

Materials Research Foundations 191 (2026) 343-363 https://doi.org/10.21741/9781644904213-11

3.2.1 Direct fluorination

In conventional reactors, performing direct fluorination with elemental fluorine is often challenging due to the large amounts of byproducts generated and the risk of explosions. Gas-liquid microreaction technology can significantly enhance these reactions by providing better control over gas-liquid mixing and more effective temperature management. For instance, the direct fluorination of toluene in microreactors has demonstrated several technological advantages. Jähnisch et al. reported achieving yields of up to 28% for monofluorinated toluene in a falling film microreactor [53]. When comparing this microreactor to a laboratory bubble column, the space-time yields were significantly higher. Several other chemical compounds, such as 4-nitrotoluene and 1,3-dicarbonyl, have also been effectively fluorinated directly in a contactors, in addition to toluene. The process demonstrated effective annular flow. For instance, ethyl-2-chloro-3-oxobutanoate was fluorinated in a contactor having triple microchannel, the conversion of the substrate reached 59%, and 74% of crude product consisted of monofluorinated species after solvent removal. To compare, a laboratory stirred vessel yielded only 15% conversion, with 85% of the product containing monofluorinated species [54].

3.2.2 Liquid phase hydrogenation

A significant type of gas-liquid reaction commonly found in industry is liquid phase hydrogenation. Conventional reactors typically operate at high pressures to enhance the rapid transfer of hydrogen into the liquid phase. However, microreactors present a promising alternative due to their enhanced mass transfer rates, which can result in higher conversion rates or yields.

For instance, Kobayashi et al. explored the hydrogenation of benzylacetone using a palladium catalyst within a solitary microchannel that was 45 cm long, 668 μm wide, and 100 μm deep [55]. When functioning in annulated flow with a liquid space velocity of ~11 h^{-1} (which fluctuates according to the volume of the microchannel), the production of 4-phenyl-2-butanone could achieve a maximum of 97%. The microreactor operated successfully over a continuous five-hour period. However, the durability of the immobilized palladium catalyst over an extended period remains uncertain.

3.2.3 Direct synthesis of H_2O_2

In another work, researchers utilized a ten channel micro-packed bed reactor to directly combine H_2 and O_2 over Pd catalysts for hydrogen peroxide production [56]. The reaction occurred safely within the reactor, at H_2 : O_2 ratios 1:1, (an explosive ratio) and pressures of 2-3 MPa. It is important to note that hydrogen's explosion limits in O_2 range from 4%-94% by volume. Compared to conventional reactors, such bed reactor significantly enhanced the transfer rates of nitrogen and hydrogen to the liquid, resulting in a faster reaction rate.

3.2.4 Chemical absorption

Chemical factories often use chemical absorption processes, such as purifying exhaust gases. However, traditional absorption devices tend to occupy a significant amount of space owing to their comparatively low mass transfer rates. In contrast, gas-liquid microreactors demonstrate substantially improved mass transfer, indicating that they can achieve much greater absorption efficiency while being smaller in size. For example, research conducted by TeGrotenhuis et al. showed promising results in gas absorption when using a microreactor with a micro-porous plate

Materials Research Foundations 191 (2026) 343-363
https://doi.org/10.21741/9781644904213-11

that separates the gas and liquid phases [57]. When employing a diethanolamine solution as the absorbent, they discovered >90% of the CO_2 could be extracted from a mixed amount containing CO_2 (25%) within <10 seconds of gas residence time.

3.2.5 Engineering projects

Several industrial companies in the US and Europe have initiated engineering projects to commercialize gas-liquid microreaction technology, even though most research in this field is still at the bench scale. One such project is being undertaken by UOP, which plans to establish a microreactor facility that will directly combine H_2 and O_2 in a micro-packed bed reactor [58]. This facility is expected to produce approximately 150000 metric tons of hydrogen peroxide annually. A laboratory test using IMM has verified that at 3 MPa pressure and a $H_2 : O_2$ ratios of 1:3, it is possible to obtain 2 grams yield of H_2O_2 for every gm of catalyst/hour. To examine the reaction conditions and kinetics, as well as the capital expenditures, UOP conducted pilot-scale tests at higher pressures. It is expected that the full-scale factory will be built in the coming years.

3.3 Liquid–Gas–Solid Reactions

3.3.1 Hydrogenation

An efficient microstructured reactor capable of generating a high interfacial area between phases remains a key goal in the field. Kobayashi et al. demonstrated a microfluidic system for gas-liquid-solid hydrogenation using a palladium-immobilized microchannel [55]. The catalyst was synthesized in situ by functionalizing the channel surface and introducing a microencapsulated Pd precursor, followed by thermal treatment to induce cross-linking and immobilization. The system enabled continuous hydrogenation of various substrates at room temperature, achieving quantitative yields within minutes. Notably, the space-time yield exceeded conventional batch processes by over 140,000-fold, with negligible palladium leaching. The reactor maintained performance over repeated use, highlighting its potential for broader application in multiphase reactions.

3.3.2 Photocatalytic reactions

L-pipecolinic acid has been synthesized efficiently from L-lysine through photocatalysis using a titanium-modified microchannel chip (TMC) (Figure 1a) [59]. The TMC is cosist of 2 Pyrex glass substrates patterned with branched microchannels, which are coated with titanium dioxide (TiO_2) thin films prepared via a sol-gel process using titanium tetrabutoxide. Upon thermal fusion of the substrates, a uniform TiO_2 film with an anatase crystal structure is formed, as confirmed by scanning electron microscopy (SEM) and X-ray diffraction (XRD) analysis (Figure 1b). The system supports potential-controlled photocatalytic reactions, achieving high conversion rates and allowing continuous product separation without the need for catalyst removal. This enhances selectivity and operational simplicity. However, the relatively low specific surface area of the TiO_2 film limits the overall reaction efficiency. Moreover, while the system demonstrates promise, the broader application of potential-controlled photocatalytic microreactors in organic synthesis remains underdeveloped and warrants further investigation.

Materials Research Foundations 191 (2026) 343-363 https://doi.org/10.21741/9781644904213-11

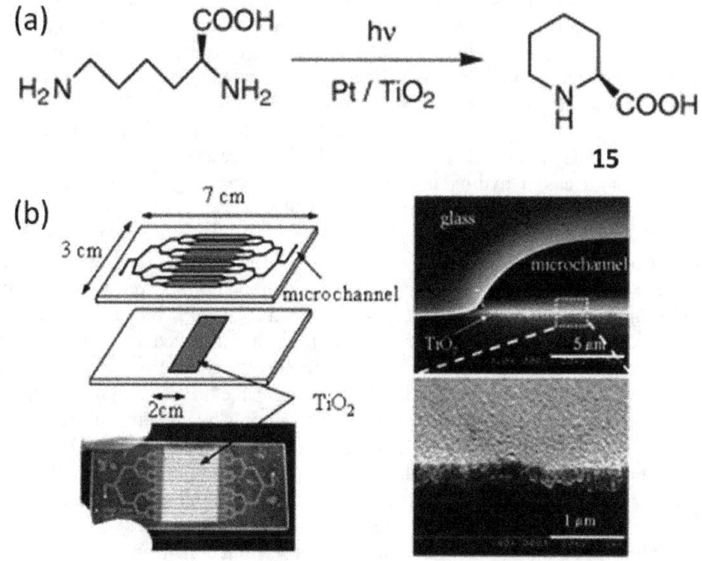

Figure 1. (a) Synthesis of L-pipecolinic acid; (b) TiO_2-modified microchannel chip along with SEM images showing the cross-section of the channel. Reprinted with permission from Ref. [59]. Copyright 2005 Elsevier.

3.4 Solid Substrates and Reagents Bound to Monoliths in Continuous Flow

There are only a limited number of studies that have reported on solution-phase organic synthesis in microreactors utilizing solid-supported catalysts or reagents. This is primarily due to the challenges associated with creating microreactors that can effectively utilize supported reagents.

3.4.1 Solid-supported Reagents

Among the various solid-supported reagents used in chemistry, supported catalysts are particularly advantageous because they allow for the use of excess immobilized catalysts to push reactions to completion. Recent research has shown that solid supports experience less physical damage in flow systems in comparison to bulk reactions, which often require intense stirring. N. Nikbin et al. present an innovative technique for achieving effective and dependable solid-supported synthesis inside a microreactor that operates in a continuous flow [60]. This method merges the supremacy of constant flow synthesis with the microreactor format. They demonstrate the application of electroosmotic flow in a microreactor for Knoevenagel reactions.

3.4.2 Solid-supported Catalysts

The recoverability and reusability of catalysts, along with their ability to be extracted from the reaction stream, make supported catalysis in flow systems very appealing. When comparing homogeneous catalysis to catalyst loading, the latter is generally less significant. Consequently, the overall concentrations of catalysts may be quite elevated, resulting in quicker conversion times. This is one reason why continuous flow reactors have become widely adopted in organic chemistry, especially with the expanded chemical space that has emerged since the widespread use of transition metals. Kirschning and colleagues developed a palladium catalyst supported by a monolith that exhibited strong performance across a variety of cross-coupling reactions [61]. They employed an exchange method to apply the Pd catalyst on the monolithic support, followed by a reduction process. This supported catalyst enabled various reactions, including Heck–Mizoroki reactions, involving transfer hydrogenation reductions of various functional groups, as well as Suzuki–Miyaura and Sonogashira couplings. The Pd(0) functionalized monolith was successfully reapplied for the Heck reaction over several runs. After seven runs, the catalytic activity decreased significantly from 490% to 71%. Notably, the Sonogashira coupling could be completed without the need for a copper co-catalyst, which reduced the likelihood of reactor blockage. Although the lengthy reaction times (ranging from 2.5 to 24 hours) and the need to recycle the reaction solution are minor drawbacks in this continuous flow system, they resulted in very low throughputs. Additionally, Ley and colleagues developed a monolithic cartridge derivatized with palladium (0) to facilitate Heck reactions in ethanol (EtOH) [62]. Under extremely high temperatures, these reactions were completed in approximately twenty-five minutes, yielding substantial amounts of Heck products. The monolithic reactor demonstrated remarkable durability, enabling up to 25 reuses without requiring catalyst revival. The incorporation of porous monolithic materials as supports offers benefits because of the comparatively swift mass transfer that occurs between these materials and the liquid phase. In flow-through processes for ring-closing and ring-opening metathesis polymerization, a monolithic support that is functionalized with an N-heterocyclic carbene/ruthenium metathesis catalyst was utilized.

4. Industrial uses and Perspectives

The transition from laboratory-scale reactions to industrial production traditionally involves scaling up through intermediate pilot plant stages, a process that is often resource-intensive and time-consuming. This conventional scale-up strategy is challenged by changes in the surface-to-volume ratio, which directly impact heat and mass transfer, necessitating re-optimization of reaction conditions at each stage. As a result, the scale-up pathway from bench to production can be economically and logistically demanding. Microreactor technology presents a promising alternative by enabling process development and optimization on a single, small-scale device. Scale-up is then achieved through a "numbering-up" or "scale-out" strategy, wherein multiple microreactors operate in parallel to increase production capacity. This approach maintains consistent reaction conditions across scales, thereby reducing the need for repeated process redesign and improving overall efficiency. Furthermore, the emergence of commercially available parallel reactor systems reflects the growing feasibility of this methodology for industrial applications. In addition to shortening the development timeline, microreactors offer enhanced safety and performance features, including improved thermal control, rapid reaction rates, and the ability to operate safely within hazardous or explosive environments [63].

The adoption of microfluidic and flow-through systems has made it possible to convert a number of intricate and unsafe industrial processes into controlled, scalable, and efficient ones. Microreactors allow for standardized reaction conditions through the accurate control of reaction parameters like temperature, pressure, and mixing rates, which create homogeneous reaction environments that are generally impossible in traditional batch reactors. Their use covers a broad

Materials Research Foundations 191 (2026) 343-363 https://doi.org/10.21741/9781644904213-11

spectrum of industries—pharmaceuticals and specialty chemicals to polymer processing—where safety, reproducibility, and sustainability are key. By combining automation with inline monitoring, these systems reduce human intervention to a minimum, minimize waste, and enable fast optimization of reaction conditions, thus closing the gap between laboratory-scale development and industrial-scale manufacture.

Pharmaceutical Synthesis: The nitration of pyrazole-5-carboxylic acid, a crucial intermediate in the production of Sildenafil, exemplifies microreactors' exceptional ability to handle highly exothermic reactions with technical precision and enhanced safety. By employing a stainless steel CPC CYTOS microreactor, the simultaneous addition of reagents at meticulously controlled temperatures effectively eliminated the significant risk of thermal runaway, a common and dangerous issue in traditional batch nitrations. This precise control over reaction parameters ensures not only safety but also optimized reaction pathways. Industrially, this highly controlled process achieved a stable throughput of 5.5 g/h with a high 73% yield, robustly demonstrating the feasibility and technical advantages of microreactor technology for pharmaceutical-scale operations requiring stringent control and reproducibility [64].

Polymer Production: The methyl methacrylate and butyl acrylate radical polymerizations were successfully and continuously used to produce 4 kg/week of the polymer PMMA in a stainless steel pilot-scale microreactor. The microreactor design inherently achieves automatic highly uniform polymerization conditions within the reaction compartment. Such precision engineering is important to obtaining a highly tight molecular weight distribution control, of great significance to polymer quality as well as performance. This application illustrates microreactors high potential in the polymer field, allowing precise polymer construction with consistently high, predictable quality of the product in the batch, with difficulty to accomplish with less-controlled processes [65].

Fine Chemical Manufacturing: The continuous production of this highly explosive and highly volatile reagent, diazomethane, is one of the most significant breakthroughs in safe, technically reproducible industrial processing. Industry has generated spectacular annual throughput rates of 50–60 tonnes without hazardous accumulation of the highly volatile intermediate through the marriage of in situ synthesis with on-stream reaction in a microflow setup. This new method relies on the precise control of flow with the quick mixing potential of microreactors, eschewing safety issues at high levels of efficiency. This innovation plays a pivotal role in the scalable, safe production of intermediates of crucial antiretroviral drugs such as Nelfinavir Mesylate, as a manifestation of safety as well as chemical conversion in precision [66,67].

Synthesis of Heterocyclic Compound: Microreactor heterocycle-preparation systems, including challenging reactions such as Grignard additions as well as oxidative dehydrogenations, have made possible high-yielding, moisture-sensitive conversions under highly controlled microenvironments. The systems' inherent continuous mode substantially reduces unwanted side reactions and considerably enhances reproducibility. Such technical precision is precisely crucial to the fine chemical industry, wherein high purity as well as precise product specifications are mandatory specifications. The microenvironment-controlled conditions allow precise control over reaction conditions, achieving high-quality product as well as high yields [68,69].

High-throughput Drug Discovery: Microreactor networks that are integrated revolutionized drug discovery with the potential to execute parallel synthesis as well as screening of reactions quickly, considerably lowering the optimization time. Microflow chemistry is highly amenable to

Materials Research Foundations 191 (2026) 343-363 | https://doi.org/10.21741/9781644904213-11

automation as well as to miniaturization, with unrivalled technical accuracy in screening enormous chemical spaces. This enables chemists to experiment quickly as well as efficiently with a variety of reaction conditions with minimum volumes of precious reagents. This very precise as well as automated methodology is redefining drug research as well as development pipelines within the pharmaceutical industries, speeding up lead identification as well as considerably anything else Increase the accuracy as well as the efficiency of studies of structure–activity relationship. Such highly advanced microfluidic systems, frequently with modular chips for molecular analysis, facilitate systematic investigations of reaction parameters as well as quick development of new synthetic pathways to active pharmaceuticals as well as to valuable chemical products [70].

Table 1 highlights the diverse range of industrial reactions and processes that have been successfully translated from batch to microfluidic systems.

Table 1. Key Industrial Applications and Examples

Application Area	Reaction / Process Type	Microfluidic Approach & Device	Industrial Significance / Output	Reference
Pharmaceutical synthesis	Electrophilic nitration of aromatic compounds	Stainless steel CPC CYTOS lab system (70 mL internal volume)	Nitration of pyrazole-5-carboxylic acid (intermediate in Sildenafil synthesis); throughput 5.5 g/h, 73% yield	[64]
Polymer production	Radical polymerization of methyl methacrylate (MMA) and butyl acrylate (BA)	Stainless steel pilot plant microreactor using numbering-up	Produced 4 kg/week of PMMA with controlled molecular weight and dispersity	[65]
Fine chemical manufacturing	Continuous diazomethane generation and methylation reactions	Continuous microreactor with steady-state control	Safe large-scale diazomethane generation (~50–60 tonnes/year) for HIV protease inhibitor synthesis	[66,67]
Heterocyclic compound synthesis	Grignard additions and oxidative dehydrogenations	Metal and glass microflow systems enabling continuous operation	Scalable and safe handling of moisture-sensitive reagents	[68,69]
High-throughput drug discovery	Parallel synthesis and screening	Integrated microreactor networks	Rapid reaction optimization and library generation using small reagent volumes	[70]

Recent industrial disasters and evolving regulatory needs have refined the interest in safer and environmentally friendly industrial procedures among chemical producers. Hazardous industrial accidents brought the importance of enhancing safety procedures into the limelight, and regulatory requirements now demand environmentally friendly procedures. Microfluidic technology presents five key benefits to commercial chemical manufacturing to resolve this. The traits typically improve environmental sustainability, safety, and efficiency.

Improved Process Safety: Microreactors also have low internal volumes and thus reduce the risk of explosion or thermal runaway. The highly hazardous reaction intermediates that are peroxides and diazo compounds are formed and consumed in seconds within the microchannels. An example is the continuous synthesis of diazomethane, where the operator exposure as well as the accident risk is lowered considerably.

Materials Research Foundations 191 (2026) 343-363 https://doi.org/10.21741/9781644904213-11

Efficient Material and Heat Transfer: Microscale channel diameter offers the high surface-to-volume ratio necessary to achieve efficient heat dissipation with a consistent temperature profile. This property avoids the possibility of hot spots as well as increased yield consistency that is preferred in highly exothermic reactions such as oxidations as well as nitrations.

Scale-up Using Numbering-Up: Rather than scale-up of the size of the reactor, with potential to cause mixing and thermal non-uniformities, microreactors can be scaled up by operating some identical units in parallel. The example of the polymerization of the poly(methyl methacrylate) (PMMA) shows that pilot plants can be steadily operated for the long term through numbering-up of microreactor modules.

Reduced Waste and Cost: Continuous-flow systems enable precise stoichiometry and efficient reagent use, thereby reducing solvent consumption and minimizing waste generation. Reactions such as Suzuki–Miyaura couplings benefit from continuous purification and inline monitoring, which eliminate post-synthesis waste streams.

Accelerated Research and Development Cycles: Microreactors enable real-time reaction optimization and kinetic studies at the milligram scale. This capability shortens the interval between laboratory discovery and pilot-scale validation, supporting the current emphasis on seamless laboratory-to-plant transitions in sustainable manufacturing.

Microfluidic devices have transformed industrial organic synthesis by facilitating continuous, safe, and automated chemical processing. These systems enable the transition from laboratory-scale experimentation to industrial-scale manufacturing, thereby advancing sustainable and environmentally responsible chemistry.

Although microfluidic technology offers significant benefits and is increasingly adopted, several obstacles hinder its broad industrial implementation. These challenges include complex equipment setup, limited flow capacity, difficulties with cleaning and dismantling, high pressure drops, and frequent blockages in specific systems. Additionally, further research is needed to address automation, system integration, optimal configuration design, process optimization, and comprehensive cost analysis.

Conclusions

Microreactors, initially developed as analytical tools, have demonstrated exceptional versatility across diverse applications, including the synthesis of nanoparticles and organic compounds. Their potential in chemical and pharmaceutical engineering lies in their ability to exploit the unique properties of micro- and nanostructured materials. Beyond synthesis, microstructured reactors have shown promise as micromixers and microseparators, with tunable channel porosity enhancing mass transfer and species dispersion.

To further improve reactor performance—particularly yield and selectivity—ongoing research is essential for optimizing structural parameters such as channel architecture, connectivity, and material composition. High-performance microreactors fabricated from a wide range of materials, including metals, ceramics, and polymers, have found applications in sectors like automotive and aerospace. Incorporating multifunctional nanomaterials and surfactant-mediated wall coatings can further elevate their efficiency, especially in processes like hydrogenation.

Recent advancements have enabled the direct fabrication of composite-based microreactors, including those made from zeolites and ceramic foams, allowing operations under extreme conditions with enhanced mixing control. Despite technical challenges related to integration, monitoring, and cost, microreaction technology continues to progress toward becoming a sustainable, scalable, and widely adopted solution for modern chemical processing.

References

[1] N. Fay, C. Kouklovsky, A. de la Torre, Natural product synthesis: The endless quest for unreachable perfection, ACS Org. Inorg. Au. 3 (2023) 350-363. https://doi.org/10.1021/acsorginorgau.3c00040

[2] S. Das, B. Choudhury, S. Chauhan, M. M. Balamurali, K. Chanda, An overview of the synthesis of biheterocyclic compounds featuring four-membered rings containing nitrogen, oxygen, or sulfur, Tetrahedron. 181 (2025) 134679. https://doi.org/10.1016/j.tet.2025.134679

[3] D. C. Blakemore, L. Castro, I. Churcher, D. C. Rees, A. W. Thomas, D. M. Wilson, A. Wood, Organic synthesis provides opportunities to transform drug discovery, Nat. Chem. 10 (2018) 383-394. https://doi.org/10.1038/s41557-018-0021-z

[4] A. Dey, A. Maity, T. Singha Mahapatra, E. Suresh, A. K. Mandal, A. Das, A tuneable hierarchical self-assembly of a C3-symmetric triaminoguanidinium-derivative into a rhombic dodecahedral morphology, CrystEngComm. 22 (2020) 5117-5121. https://doi.org/10.1039/D0CE00909A

[5] R. Porta, M. Benaglia, A. Puglisi, Flow chemistry: Recent developments in the synthesis of pharmaceutical products, Org. Process Res. Dev. 20 (2016) 2-25. https://doi.org/10.1021/acs.oprd.5b00325

[6] C. J. Taylor, A. Pomberger, K. C. Felton, R. Grainger, M. Barecka, T. W. Chamberlain, R. A. Bourne, C. N. Johnson, A. A. Lapkin, A brief introduction to chemical reaction optimization, Chem. Rev. 123 (2023) 3089-3126. https://doi.org/10.1021/acs.chemrev.2c00798

[7] B. C. Roy, V. R. Ramlal, D. Basak, S. Basak, S. Roy, S. Ghosh, T. Singha Mahapatra, Light-emitting coordination polymers: stimuli-responsive gels, PMMA-based composite films, and UV-shielding, Inorg. Chem. 63 (2024) 23044-23056. https://doi.org/10.1021/acs.inorgchem.4c04333

[8] V. Hessel, S. Mukherjee, S. Mitra, A. Goswami, N. N. Tran, F. Ferlin, L. Vaccaro, F. M. Galogahi, N. -T. Nguyen, M. Escribà-Gelonch, Sustainability of flow chemistry and microreaction technology, Green Chem. 26 (2024) 9503-9528. https://doi.org/10.1039/D4GC01882F

[9] P. Laurino, A. Odedra, X. Y. Mak, T. Gustafsson, K. Geyer, P. H. Seeberger, Microfluidic Devices for Organic Processes, in: S. V. Luis, E. Garcia-Verdugo (Eds.) Chemical Reactions and Processes under Flow Conditions, RSC, 2009, pp. 118-162. https://doi.org/10.1039/9781847559739-00118

[10] L. Chen, C. Yang, Y. Xiao, X. Yan, L. Hu, M. Eggersdorfer, D. Chen, D. A. Weitz, F. Ye, Millifluidics, microfluidics, and nanofluidics: manipulating fluids at varying length scales, Mater. Today Nano. 16 (2021) 100136. https://doi.org/10.1016/j.mtnano.2021.100136

[11] B. K. Chegeni, Chapter 10 - Microfluidics in organic chemistry, in: M. R. Hamblin, M. Karimi (Eds.) Biomedical Applications of Microfluidic Devices, Academic Press, 2021, pp. 235-255. https://doi.org/10.1016/B978-0-12-818791-3.00011-5

[12] B. C. Roy, S. Ghosh, T. Singha Mahapatra, A. Das, Ultrathin lanthanide-based 2D-coordination nanosheets: A versatile class of 2D materials, Coord. Chem. Rev. 518 (2024) 216058. https://doi.org/10.1016/j.ccr.2024.216058

[13] S. Gimondi, H. Ferreira, R. L. Reis, N. M. Neves, Microfluidic devices: A tool for nanoparticle synthesis and performance evaluation, ACS Nano. 17 (2023) 14205-14228. https://doi.org/10.1021/acsnano.3c01117

[14] A. L. Yarin, Novel nanofluidic and microfluidic devices and their applications, Curr. Opin. Chem. Eng. 29 (2020) 17-25. https://doi.org/10.1016/j.coche.2020.02.004

[15] S. Siavashy, M. Soltani, S. Rahimi, M. Hosseinali, Z. Guilandokht, K. Raahemifar, Recent advancements in microfluidic-based biosensors for detection of genes and proteins: Applications and techniques, Biosens. Bioelectron. X. 19 (2024) 100489. https://doi.org/10.1016/j.biosx.2024.100489

[16] D. Zhang, S. Jia, Z. Jin, J. Bu, G. Guo, J. Deng, X. Wang, Microfluidic technology: A new strategy for controllable synthesis of metal nanomaterials, ChemElectroChem. 12 (2025) e202400666. https://doi.org/10.1002/celc.202400666

[17] A. Dey, V. R. Ramlal, S. S. Sankar, T. S. Mahapatra, E. Suresh, S. Kundu, A. K. Mandal, A. Das, Crystalline free-standing two-dimensional zwitterionic organic nanosheets for efficient conduction of lithium ions, ACS Appl. Mater. Interfaces. 12 (2020) 58122-58131. https://doi.org/10.1021/acsami.0c17683

[18] L. Yang, S. Pijuan-Galito, H. S. Rho, A. S. Vasilevich, A. D. Eren, L. Ge, P. Habibović, M. R. Alexander, J. de Boer, A. Carlier, P. van Rijn, Q. Zhou, High-throughput methods in the discovery and study of biomaterials and materiobiology, Chem. Rev. 121 (2021) 4561-4677. https://doi.org/10.1021/acs.chemrev.0c00752

[19] B. C. Roy, A. Kundu, P. Biswas, S. Roy, T. Singha Mahapatra, Recent advances in stimuli-responsive luminescent supramolecular lanthanide-based metallogels, ChemistrySelect. 9 (2024) e202304755. https://doi.org/10.1002/slct.202304755

[19] D. E. Fitzpatrick, C. Battilocchio, S. V. Ley, Enabling technologies for the future of chemical synthesis, ACS Cent. Sci. 2 (2016) 131-138. https://doi.org/10.1021/acscentsci.6b00015

[20] T. Singha Mahapatra, B. C. Roy, B. Dutta, J. Lengyel, M. Shatruk, D. Ray, Structures and magnetic properties of a trinuclear angular [Ni3] and a heptanuclear wheel-like [Ni7] complexes with a Schiff base ligand, Polyhedron. 249 (2024) 116782. https://doi.org/10.1016/j.poly.2023.116782

[21] C. J. Taylor, A. Pomberger, K. C. Felton, R. Grainger, M. Barecka, T. W. Chamberlain, R. A. Bourne, C. N. Johnson, A. A. Lapkin, A brief introduction to chemical reaction optimization, Chem. Rev. 123 (2023) 3089-3126. https://doi.org/10.1021/acs.chemrev.2c00798

[22] D. S. Mattes, N. Jung, L. K. Weber, S. Bräse, F. Breitling, Miniaturized and automated synthesis of biomolecules-overview and perspectives, Adv. Mater. 31 (2019) 1806656. https://doi.org/10.1002/adma.201806656

[23] T. Singha Mahapatra, A. Roy, S. Chaudhury, S. Dasgupta, S. L. Shrivastava, V. Bertolasi, D. Ray, Trapping of a methanoato bridge in μ-1,1,3,3 mode for [Cu4] aggregate formation: synthesis, steric control on nuclearity, antimicrobial activity, and DNA-interaction properties, Eur. J. Inorg. Chem. 2017 (2017) 769-779. https://doi.org/10.1002/ejic.201601092

[24] T. Singha Mahapatra, M. Gan Choudhuri, S. Ghosh, S. Roy, Advances in functional photoisomerisable low-molecular-weight π-gelators, Chem. Commun., 62 (2026) 8374-8393. https://doi.org/10.1039/D6CC00923A

[25] G. Nunziata, A. Borroni, F. Rossi, Advanced microfluidic strategies for core-shell nanoparticles: the next-generation of polymeric and lipid-based drug nanocarriers, Chem. Eng. J. Adv. 22 (2025) 100759. https://doi.org/10.1016/j.ceja.2025.100759

[26] S. Khizar, N. Zine, A. Errachid, N. Jaffrezic-Renault, A. Elaissari, Microfluidic-based nanoparticle synthesis and their potential applications, Electrophoresis. 43 (2022) 819-838. https://doi.org/10.1002/elps.202100242

[27] Y. Bi, S. Xie, Z. Li, S. Dong, L. Teng, Precise nanoscale fabrication technologies, the "last mile" of medicinal development, Acta Pharm. Sin. B. 15 (2025) 2372-2401. https://doi.org/10.1016/j.apsb.2025.03.040

[28] C. Sun, S. T. Pham, S. L. Boyall, B. Douglas, A. J. Britton, S. Micklethwaite, T. W. Chamberlain, M. O. Besenhard, R. Drummond-Brydson, K. -J. Wu, S. M. Collins, Ultrasound-assisted continuous aqueous synthesis of sulfonate, imidazolate, and carboxylate MOFs with high space time yield, Commun. Chem. 8 (2025) 154. https://doi.org/10.1038/s42004-025-01548-5

[29] C. M. Hussain, G. Hussain, R. Keçili, Smart analytical chemistry: Integrating green, sustainable, white and AI-driven approaches in modern analysis, TrAC - Trends Anal. Chem. 191 (2025) 118295. https://doi.org/10.1016/j.trac.2025.118295

[30] A. Ratanpara, Y. Li, M. Kim, A review of microfluidic approaches for carbon capture and storage research, Lab on a Chip. (2025) https://doi.org/10.1039/D5LC00208G

[31] A. I. Osman, A. Ayati, P. Krivoshapkin, B. Tanhaei, M. Farghali, P.-S. Yap, A. Abdelhaleem, Coordination-driven innovations in low-energy catalytic processes: Advancing sustainability in chemical production, Coord. Chem. Rev. 514 (2024) 215900. https://doi.org/10.1016/j.ccr.2024.215900

[32] T. -Y. Chen, Y. W. Hsiao, M. Baker-Fales, F. Cameli, P. Dimitrakellis, D. G. Vlachos, Microflow chemistry and its electrification for sustainable chemical manufacturing, Chem. Sci. J. 13 (2022) 10644-10685. https://doi.org/10.1039/D2SC01684B

[33] H. Xue, Z. -H. Zhao, M. Yuan, G. Zhang, Microfluidic reactors for paired electrosynthesis: Fundamentals, applications and future prospects, Green Energy & Environment. 10 (2025) 471-499. https://doi.org/10.1016/j.gee.2024.06.006

[34] P. Watts, C. Wiles, Micro reactors: a new tool for the synthetic chemist, Org. Biomol. Chem. 5 (2007) 727-732. https://doi.org/10.1039/b617327f

[35] A. Šalić, A. Tušek, B. Zelić, Application of microreactors in medicine and biomedicine, J. Appl. Biomed. 10 (2012) 137-153. https://doi.org/10.2478/v10136-012-0011-1

[36] A. E. Guber, M. Heckele, D. Herrmann, A. Muslija, V. Saile, L. Eichhorn, T. Gietzelt, W. Hoffmann, P. C. Hauser, J. Tanyanyiwa, A. Gerlach, N. Gottschlich, G. Knebel, Microfluidic lab-on-a-chip systems based on polymers-fabrication and application, Chem. Eng. J. 101 (2004) 447-453. https://doi.org/10.1016/j.cej.2004.01.016

[37] R. Knitter, D. Göhring, P. Risthaus, J. Haußelt, Microfabrication of ceramic microreactors, Microsyst. Technol. 7 (2001) 85-90. https://doi.org/10.1007/s005420100107

[38] K. Geyer, J. D. C. Codée, P. H. Seeberger, Microreactors as tools for synthetic chemists-the chemists' round-bottomed flask of the 21st century?, Chem Eur J. 12 (2006) 8434-8442. https://doi.org/10.1002/chem.200600596

[39] P. Žnidaršič-Plazl, Biotransformations in microflow systems: bridging the gap between academia and industry, J. Flow Chem. 7 (2017) 111-117. https://doi.org/10.1556/1846.2017.00021

[40] B. Ahmed-Omer, J. C. Brandt, T. Wirth, Advanced organic synthesis using microreactor technology, Org. Biomol. Chem. 5 (2007) 733-740. https://doi.org/10.1039/B615072A

[41] P. Watts, C. Wiles, Micro reactors: a new tool for the synthetic chemist, Org. Biomol. Chem. 5 (2007) 727-732. https://doi.org/10.1039/b617327f

[42] N. Hoffmann, Photochemical reactions as key steps in organic synthesis, Chem. Rev. 108 (2008) 1052-1103. https://doi.org/10.1021/cr0680336

[43] E. E. Coyle, M. Oelgemöller, Micro-photochemistry: photochemistry in microstructured reactors. The new photochemistry of the future?, Photochem. Photobiol. Sci. 7 (2008) 1313-1322. https://doi.org/10.1039/b808778d

[44] T. Fukuyama, Y. Hino, N. Katama and I. Ryu, Quick execution of [2+2] type photochemical cycloaddition reaction by continuous flow system using a glass-made microreactor, Chem. Lett. 33 (2004) 1430. https://doi.org/10.1246/cl.2004.1430

[45] B. D. A. Hook, W. Dohle, P. R. Hirst, M. Pickworth, M. B. Berry, K. I. Booker-Milburn, A practical flow reactor for continuous organic photochemistry, J. Org. Chem. 70 (2005) 7558-7564. https://doi.org/10.1021/jo050705p

[46] D. R. J. Acke, C. V. Stevens, A HCN-based reaction under microreactor conditions: industrially feasible and continuous synthesis of 3,4-diamino-1H-isochromen-1-ones, Green Chem. 9 (2007) 386-390. https://doi.org/10.1039/b615186h

[47] M. Movsisyan, M. M. A. Moens, C. V. Stevens, Flow Synthesis of Heterocycles, in: E. F. V. Scriven, C. A. Ramsden (Eds.) Advances in Heterocyclic Chemistry, Academic Press, 2016, pp. 25-55. https://doi.org/10.1016/bs.aihch.2015.10.006

[48] C. Wiles, P. Watts, S. J. Haswell, E. Pombo-Villar, The application of microreactor technology for the synthesis of 1,2-azoles, Org. Process Res. Dev. 8 (2004) 28-32. https://doi.org/10.1021/op034125a

[49] D. M. Ratner, E. R. Murphy, M. Jhunjhunwala, D. A. Snyder, K. F. Jensen, P. H. Seeberger, Microreactor-based reaction optimization in organic chemistry-glycosylation as a challenge, ChemComm. 5 (2005) 578-580. https://doi.org/10.1039/B414503H

[50] D. Grant, R. Dahl, N. D. P. Cosford, Rapid multistep synthesis of 1,2,4-oxadiazoles in a single continuous microreactor sequence, J. Org. Chem. 73 (2008) 7219-7223. https://doi.org/10.1021/jo801152c

[51] A. Odedra, K. Geyer, T. Gustafsson, R. Gilmour, P. H. Seeberger, Safe, facile radical-based reduction and hydrosilylation reactions in a microreactor using tris(trimethylsilyl)silane, ChemComm. 26 (2008) 3025-3027. https://doi.org/10.1039/b803715a

[52] G. Chen, J. Yue, Q. Yuan, Gas-liquid microreaction technology: Recent developments and future challenges, Chin. J. Chem. Eng. 16 (2008) 663-669. https://doi.org/10.1016/S1004-9541(08)60138-X

[53] K. Jähnisch, M. Baerns, V. Hessel, W. Ehrfeld, V. Haverkamp, H. Löwe, C. Wille, A. Guber, Direct fluorination of toluene using elemental fluorine in gas/liquid microreactors, J. Fluor. Chem. 105 (2000) 117-128. https://doi.org/10.1016/S0022-1139(00)00300-6

[54] R. D. Chambers, M. P. Greenhall, J. Hutchinson, Direct fluorination of 1,3-dicarbonyl compounds, Journal of the Chemical Society, ChemComm. 1 (1995) 21-22. https://doi.org/10.1039/c39950000021

[55] J. Kobayashi, Y. Mori, K. Okamoto, R. Akiyama, M. Ueno, T. Kitamori, S. Kobayashi, A microfluidic device for conducting gas-liquid-solid hydrogenation reactions, Science 304 (2004) 1305-1308. https://doi.org/10.1126/science.1096956

[56] T. Inoue, M. A. Schmidt, K. F. Jensen, Microfabricated multiphase reactors for the direct synthesis of hydrogen peroxide from hydrogen and oxygen, Ind. Eng. Chem. Res. 46 (2007) 1153-1160. https://doi.org/10.1021/ie061277w

[57] W. E. TeGrotenhuis, R. J. Cameron, V. V. Viswanathan, R. S. Wegeng, Solvent Extraction and Gas Absorption Using Microchannel Contactors, in: W. Ehrfeld (Ed.) Microreaction Technology: Industrial Prospects, Springer Berlin Heidelberg, Berlin, Heidelberg, 2000, pp. 541-549. https://doi.org/10.1007/978-3-642-59738-1_56

[58] H. Pennemann, V. Hessel, H. Löwe, Chemical microprocess technology-from laboratory-scale to production, Chem. Eng. Sci. 59 (2004) 4789-4794. https://doi.org/10.1016/j.ces.2004.07.049

[59] G. Takei, T. Kitamori, H. -B. Kim, Photocatalytic redox-combined synthesis of l-pipecolinic acid with a titania-modified microchannel chip, Catal. Commun. 6 (2005) 357-360. https://doi.org/10.1016/j.catcom.2005.02.010

[60] N. Nikbin, P. Watts, Solid-supported continuous flow synthesis in microreactors using electroosmotic flow, Org. Process Res. Dev. 8 (2004) 942-944. https://doi.org/10.1021/op049857x

[61] W. Solodenko, H. Wen, S. Leue, F. Stuhlmann, G. Sourkouni-Argirusi, G. Jas, H. Schönfeld, U. Kunz, A. Kirschning, Development of a continuous-flow system for catalysis with Palladium(0) particles, Eur. J. Org. Chem. 2004 (2004) 3601-3610. https://doi.org/10.1002/ejoc.200400194

[62] N. Nikbin, M. Ladlow, S. V. Ley, Continuous flow ligand-free heck reactions using monolithic Pd [0] nanoparticles, Org. Process Res. Dev. 11 (2007) 458-462. https://doi.org/10.1021/op7000436

[63] C. Wille, H. P. Gabski, T. Haller, H. Kim, L. Unverdorben, R. Winter, Synthesis of pigments in a three-stage microreactor pilot plant-an experimental technical report, Chem. Eng. J. 101 (2004) 179-185. https://doi.org/10.1016/j.cej.2003.11.007

[64] G. Panke, T. Schwalbe, W. Stirner, S. Taghavi-Moghadam and G. Wille, A practical approach of continuous processing to high energetic nitration reactions in microreactors, Synthesis. 18 (2003) 2827-2830. https://doi.org/10.1055/s-2003-42491

[65] T. Iwasaki and J. Yoshida, Free radical polymerization in microreactors. significant improvement in molecular weight distribution control, Macromolecules. 38 (2005) 1159-1163. https://doi.org/10.1021/ma048369m

[66] L. D. Proctor and A. J. Warr, Development of a continuous process for the industrial generation of diazomethane, Org. Process Res. Dev. 6 (2002) 884-892. https://doi.org/10.1021/op020049k

[67] M. Struempel, B. Ondruschka, R. Daute and A. Stark, Making diazomethane accessible for R&D and industry: generation and direct conversion in a continuous micro-reactor set-up, Green Chem. 10 (2008) 41-43. https://doi.org/10.1039/B710554A

[68] J. Choe, J. H. Seo, Y. Kwon and K. H. Song, Lithium-halogen exchange reaction using microreaction technology, Chem. Eng. J. 135 (2008) S17-S20. https://doi.org/10.1016/j.cej.2007.07.015

[69] N. Kockmann, M. Gottsponer, B. Zimmermann and D. M. Roberge, Enabling continuous-flow chemistry in microstructured devices for pharmaceutical and fine-chemical production, Chem. Eur. J. 14 (2008) 7470-7477. https://doi.org/10.1002/chem.200800707

[70] H. Wakami and J. Yoshida, Grignard exchange reaction using a microflow system: from bench to pilot plant, Org. Process Res. Dev. 9 (2005) 787-791. https://doi.org/10.1021/op0501500

Chapter 12

Microfluidic Materials for Energy Applications

Guocai Tian[1]*

[1]State Key Laboratory of Complex Non-Ferrous Metal Resource Clean Utilization, Faculty of Metallurgical and Energy Engineering, Kunming University of Science and Technology, Kunming 650093, China

tiangc@kust.edu.cn, tiangc01@gmail.com

Abstract

Microfluidic materials and methods are rapidly expanding and achieving significant breakthroughs in the field of energy due to their precise manipulation of microscale fluids, efficient mass and heat transfer, and integrability. The research progress of microfluidic materials in energy applications, including an introduction to microfluidics and their properties, their applications such as in capacitors, fuel-cells, and other batteries were discussed carefully in this chapter. The current challenges and future development directions of microfluidic materials for energy applications also discussed.

Keywords

Microfluidic, Microfluidic Materials, Energy Storage, Fuel Cells, Capacitors, Sodium-Ion Batteries, Lithium Ion Batteries, Vanadium Battery

Contents

1. Introduction

Energy is the driving force behind all life, and all human activities are closely related to the utilization of energy. The degree of human dependence on energy has also deepened with the development of industrialization. Nowadays, the main energy sources used by humans are still the

Materials Research Foundations 191 (2026) 364-395 https://doi.org/10.21741/9781644904213-12

fuels of fossil such as natural gas, coal, and oil [1,2]. The consumption of fossil fuels is high and their reserves are limited, so there is a critical need for new energy sources or the development of new energy utilization technologies. The International Energy Agency (IEA) proposed in 2021 that vigorously developing renewable energy is an important turning point for achieving net zero emissions by the middle of this century. Scholars predict that green and renewable energy will lead the world's third industrial revolution. At present, new sources of energy mainly include wind energy, biomass energy, solar energy, and the geothermal energy. These new sources of energy are still in the development and research stage and have not been widely applied to people's lives. Developing new energy utilization technologies still face significant challenges [3,4].

The utilization of energy usually refers to the conversion of the energy contained in energy into mechanical and kinetic energy for humans through a series of means. The traditional energy conversion method is generally achieved through heat engines, which usually convert the stored chemical energy in fuel into the internal energy, and then convert the internal energy into mechanical energy or kinetic energy. However, there is energy loss in each process, so the energy conversion efficiency of heat engines is usually only 40% -60%. In addition, during the operation of heat engines, a large amount of CO_2 and nitrogen oxides are generated, which not only causes greenhouse effect but also has a serious impact on the global ecological environment and climate. To achieve green and low-carbon energy transformation, it is necessary to cut down on proportion of the consumption of fossil energy and vigorously develops efficient and clean utilization technologies of energy. The advancement of renewable sources of energy (such as wind, hydro-power, solar, nuclear, biomass, etc.) is an important area of in-depth global research of energy. However, the large-scale utilization of renewable energy still faces some challenges, such as high initial investment costs, intermittent and unstable power supply, high requirements for energy storage technology, and geographical limitations [5,6]. Therefore, the development of renewable energy requires a new generation of storage, transportation, and conversion methods and technologies for energy to fulfill the practical needs of users.

Among numerous clean energy technologies, electrochemical storage method or technology is an important method for energy conversion that reserves electrical energy as chemical energy and directly chemical converts its energy stored into the electrical energy needed, with high-energy conversion efficiency. It has the advantages of environmental friendliness, sustainable operation, and renewability, solving problems such as energy storage, transportation, and intermittent use. At the same time, it is not limited by geographical factors and is currently one of the most commonly used energy utilization technologies. Numerous new electrochemical technologies of energy conversion, such as super-capacitors, batteries (such as lithium-ion batteries (LIBs), metal-ion batteries, redox flow batteries), fuel cells, and so on, have been extensively researched and applied. Since the 21st century, the explosive growth in the consumption of portable and wireless electronic products has accelerated the advancement of new power technologies, with large power and high-energy density being the current focuses in the advancement of new power products. With the emergence of mobile phones, laptops, portable biosensors, and wearable electronic products, the demand for high-performance power sources has put significant pressure on portable power supplies. Miniaturization, light-weighting, and integration are the development trends of portable electronic product power supplies in the future.

The increasing requirement for new energy has greatly stimulated the advancement of storage technologies of energy, such as super-capacitors, sodium ion batteries, LIBs, vanadium redox flow batteries, and the fuel cells. Being one of the most significant branches, the electrode materials can

Materials Research Foundations 191 (2026) 364-395 https://doi.org/10.21741/9781644904213-12

endow these storage devices of energy with expected performance/functional characteristics. Microfluidic technology can provide a continuous platform for precise manipulation of fluid states through integrated micro-reactors with channel sizes ranging from 10 to 1000 μm [6-10]). Thanks to the limited micro-scale effects, the mass/heat transfer coefficients in micro-reactors are multiple orders of magnitude greater than those in traditional high-pressure reactors, enabling enhanced processes of precursor mixing, thermodynamic equilibrium, and nucleation/growth reactions. Microfluidic technology, as an emerging interdisciplinary technology, combines knowledge from multiple fields such as microelectronics, fluid mechanics, and materials science, providing new ideas and methods for energy applications. So far, microfluidic synthesis has become a promising technology for bulk production of high value-added micro/nano scale materials with advanced micro-structure, clear morphology, uniform size distribution, and multifunctional properties, bringing new development opportunities to the energy field. Microfluidic materials and technology have been widely employed in various fields such as physics [11], chemistry [12], and biology [13,14], especially in the nano-materials synthesis [15,16]. It has pioneered a new field of material synthesis.

2. Microfluidics and their properties

The concept of the microfluidic materials and method was first systematically described by Professor Manz in the 1990s [17]. From the perspective of disciplinary terminology, the concept of "microfluidics" can usually be used interchangeably with "microfluidics," "microfluidic chips," and "microfluidic technology". Its core connotation includes three progressive levels: micro or Nano-scale fluid behavior as a physical phenomenon, microfluidic processes as a technical method, and microfluidic chip devices as implementation carriers. This technology essentially utilizes microfabrication techniques to construct micro reaction systems with feature sizes ranging from sub-micron to sub millimeter, enabling precise control, detection, and analysis of fluids.

In micro-scale channels, the dynamic behavior of fluids exhibits significantly different characteristics from macroscopic scales (as shown in Figure 1): 1) viscous force replaces inertial forces as the dominant factor; 2) The sharp increase in a surface volume ratio significantly enhances surface effects, including capillary action, double-layer formation, etc. This scale effect leads to a series of unique physical phenomena, such as the ability of liquids to flow against gravity under capillary action, and the formation of double layers near charged surfaces that can be used for electric propulsion. More importantly, the gradient changes of fluid physical parameters (temperature, pressure, concentration, etc.) within micro-channels are significantly amplified, which can increase the coefficient of the heat transfer by various orders of magnitude and significantly shorten the mass transfer reaction time, bringing revolutionary efficiency improvements to fields such as chemical reactions and biological detection.

Microfluidic technology, as one of the important groundbreaking reaction engineering technologies and methods of the 21st century, is reshaping the traditional paradigm in the chemical industry and analytical testing fields. The revolutionary nature of this technology is mainly reflected in three dimensions [7-10]: firstly, it's astonishing mass and heat transfer efficiency, achieved by reducing the fluid channel size to the micrometer level, resulting in an astonishing specific surface area of 1×10^6 - 5×10^5 m^2/m^3, which is hundreds of times higher than conventional reactors. Secondly, there is excellent safety performance, with stainless steel micro-reactors capable of withstanding extreme operating pressures up to 60MPa. The most distinctive feature is its precise fluid control ability, which solves the interface control problem that is difficult to achieve with traditional stirring through laminar flow operation. These characteristics collectively drive the development of chemical processes towards greater efficiency, safety, and precision.

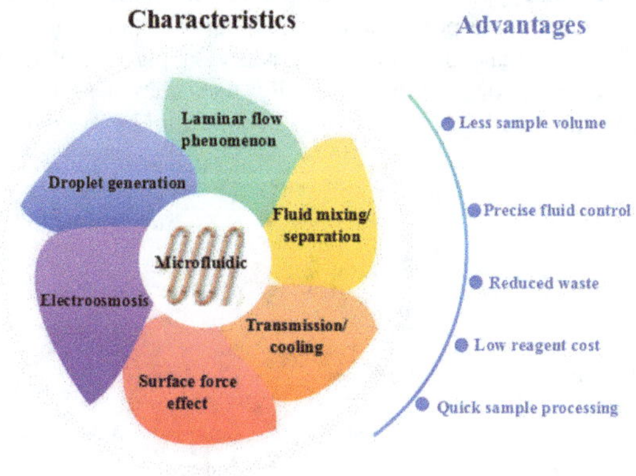

Figure 1 Characteristics and advantages of microfluidics

Microscopic scale fluid manipulation (microfluidics) is showing revolutionary application prospects in many fields such as chemistry and bio-medicine. Microfluidic chip is one of the scientific technologies behaved by the manipulation of fluids in the scale of micrometer space. It can shrink the basic operations of biological, chemical, and other labs onto a tiny chip (just a few cm^2), hence the name "lab on a chip." [18-22]. The advantages it brings are also reflected in: due to the increase in reaction speed, reaction equipment and reaction systems can be greatly reduced, making the safety of the reaction process greatly improved, land investment greatly saved, and material consumption sharply reduced; For toxic reactions in raw materials, processes, or products, a small-scale distributed production method can be adopted to avoid the transportation risks of toxic and harmful raw materials or products; Its processing capability can be improved by prompting the number of functional units without the need to scale up the reaction equipment step by step. It is a typical cutting-edge interdisciplinary field, with its upstream involving at least physics and mechanics, parallel to micro-mechanics and microelectronics, and downstream including biomedical, pharmaceutical, analytical chemistry, synthetic chemistry, and other disciplines [11-27].

The core of microfluidic technology or method is to use microscale channels and structures to manipulate fluids, achieving precise control and manipulation of fluids. Compared to the usage of specific solutions isolated, microfluidic platforms provide a group of fluid operational units designed to combination easier in well-defined manufacturing techniques. It enables for fast, easy, and cost-effective achievement of different applications, achieving specific (bio) chemical reaction processes. With the rapid development of micro-fabrication technology, microfluidic technology has achieved significant breakthroughs. Compared with traditional batch preparation methods, the main advantages of microfluidic synthesis include: (a) Rapid synthesis. Microfluidic technology can significantly shorten reaction time and achieve rapid material synthesis. (b)Precise controlling.

Materials Research Foundations 191 (2026) 364-395 | https://doi.org/10.21741/9781644904213-12

It can precisely control conditions of reaction such as pressure, temperature, and concentration, improving the controllability of the synthesis process. (c)High yield and purity. The fluid mixing and quality/heat transfer efficiency in microfluidic channels are high, which helps to prompt the yield and purity of the product. (d) The structure and morphology control. By designing micro-channel structures and fluid states, the microstructure and morphology of materials can be precisely controlled. (e)Continuous production. Microfluidic technology supports continuous flow synthesis, suitable for large-scale production. (f)Energy saving and efficient. Compared with traditional methods, microfluidic synthesis usually consumes less energy and has higher efficiency. (g)Safety. The small reaction volume in microfluidic systems reduces the risk of dangerous reactions.

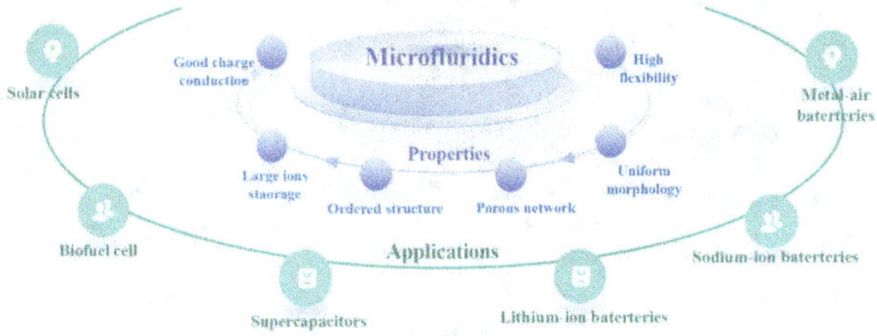

Figure 2 Properties of microfluidics and their applications in the energy sector

Compared with traditional macroscopic fluid manipulation, microfluidic technology or method has the following significant advantages (as shown in Figure 2). (a) Specific surface area. Microfluidic devices typically have the channel sizes in the micrometer range, resulting in a very large specific surface area. This characteristic is particularly important in energy applications as it can significantly improve reaction efficiency and mass transfer rate. For example, in solar cells, by designing microfluidic channels and structures, the absorption area of light will be enhanced, thereby improving the conversion efficiency of solar cells. (b) Low consumption of energy. Since the small size of microfluidic devices, the flow resistance of fluids in them is correspondingly reduced, thereby reducing energy consumption. This advantage is particularly evident in fuel cells. By designing microfluidic channels and structures, the transport resistance of hydrogen and oxygen can be reduced, thereby prompting the efficiency power and density of fuel cells. (c) High integration. Microfluidic devices can achieve the integration of multiple functional units, thereby enhancing the compactness and efficiency of the system. In battery applications, by designing microfluidic channels and structures, rapid electrolyte transfer and efficient charge storage can be achieved, thereby improving the charge discharge rate and cycle life of the battery. (d)Strong controllability. By precisely designing the structure and parameters of microfluidic devices, precise control over fluid flow, mixing, reactions, and other processes can be achieved. This advantage is particularly important in thermal energy management. By designing microfluidic channels and structures, efficient heat exchange between fluids and heat sources can be achieved, thereby improving the heat dissipation effect of heat sinks.

Based on the above advantages, microfluidic technology has been successfully used in multiple disciplines such as chemistry, materials, physics, and biology. Microfluidic technology has shown great potential energy applications for fuel cells, capacitors, and batteries (as shown in Figure 3[23]). Microfluidic method and technology have been applied in electrochemistry of energy and has made progress, which can be used to prepare efficient fuel cells. By designing microfluidic channels and structures, the transport efficiency of hydrogen and oxygen can be improved, thereby enhancing the efficiency and the power density of fuel cells. Microfluidic technology can be used to prepare efficient batteries. By designing microfluidic channels and structures, the transport efficiency of lithium ions can be improved, thereby enhancing the charge discharge rate and cycle life of lithium-ion batteries. Figure 3 shows its application case [23].

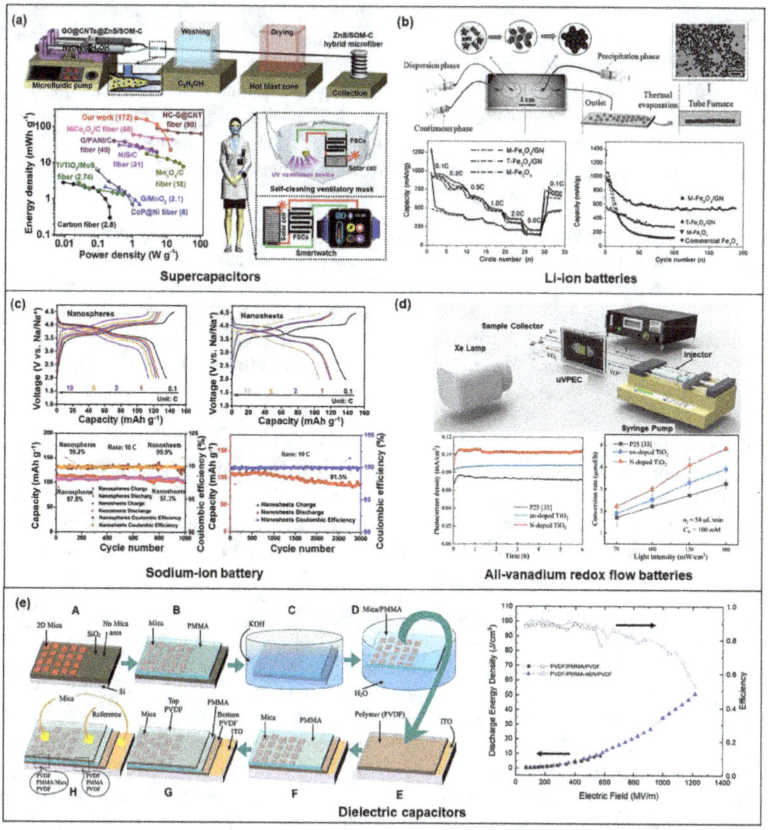

Figure 3 Application of microfluidics in super-capacitors and batteries. Reprinted with permission from [23]. Copyright 2024 American Chemical Society.

 Materials Research Forum LLC
https://doi.org/10.21741/9781644904213-12

3. Applications of microfluidic materials in super-capacitors

Super-capacitors are regarded as highly potential energy storage devices since their excellent power density, fast discharging and charging speed, long-lasting cycling performance, and reliable safety characteristics. The principle of charge storage includes two categories: double-layer capacitors (EDLC) based on electrostatic interactions at the surface/interface between the electrolyte and electrodes. The pseudo-capacitors relying on reversible redox reactions of electroactive materials [24]. Carbon materials with a large conductivity, and a broad specific surface area and a stable electrochemical performance are most commonly applied for electrode materials in EDLC [25]. However, the lower theoretical capacity (25 μm.cm^{-2}) caused by electrostatic interactions keeps a challenging problem need to be solved. Compared to EDLC, the theoretical capacity of pseudo-capacitors has significantly increased, reaching up to a hundred times that of EDLC. However, insufficient conductivity and poor electrochemical stability often lead to significant polarization phenomena and degradation of cycling performance. Therefore, in electrode materials' design, strengthening the interfacial interaction between EDLC and active materials has become a widely studied optimization scheme [26]. Microfluidic technology provides a new possibility (as shown in Figure 4) [27].

Zhang et al. [28] designed a tube in a tube microfluidic chip using continuous phase shear graphene oxide/polyvinylpyrrolidone (GO/PVP) microspheres with dimethyl silicone oil. The prepared graphene microspheres have a good capacitance of 243 F.g^{-1} and mass energy storage stability (with a 97.5% capacitance retention rate after 3000 cycles). Xu et al. [29] prepared all fiber super-capacitors with calcium alginate polyvinyl alcohol (ACa PVA)/rGO CNT structure using a universal spinning and simple method based on microfluidic spinnerets. A 35 mF.cm^{-2} of capacitance for this fiber super-capacitor is founded, and it has good mechanical flexibility and electrochemical stability after 1000 bending cycles. Wu et al. [30] proposed a microfluidic guided strategy for synthesizing a uniform nitrogen-doped fiber with porous graphene. A porous fiber micro SC was constructed using the polyvinyl alcohol/solid-state phosphoric acid (PVA/H_3PO_4), polyvinylidene fluoride co-hexafluoropropene/[Emim]BF_4 (PVDF-HFP/[Emim]BF_4) electrolyte. Based on H_3PO_4/PVA, a super large capacitance (1132 mF.cm^{-2}), long-term bending durability, and large cycling stabilities were achieved. PVDF-HFP/[Emim]BF_4 achieved a 95.7-46.9 μWh.cm^{-2} of energy densities, and power densities of 1.5-15 W.cm^{-2}.

Tong et al. [31] successfully designed the robust and conductive MCNTs/rGO/TPU/PANI composite fibers wrapped in polyaniline on a large scale by combining a simple process of microfluidic with an in-situ aniline polymerization. The specific capacitance of this constructed flexible super-capacitor enhanced sharply from 42.1 mF.cm^{-2} to 155.5 mF.cm^{-2}, where the energy density can attain 3.45 μWh.cm^{-2}. Wu et al. [32] developed a microfluidic guided core-shell polyaniline nanorod array/graphene (PNA/G) composite fiber electrode for high-performance micro-super-capacitors (micro SCs), which exhibits excellent bending stability. This electrode demonstrates a capacitance of 230 mF.cm^{-2}, maintains 86.9% retention capacitance after about 8000 cycles, shows long-term longevity under bending, and achieves a 37.2 μWh·cm^{-2} of energy density. Wu et al. [33] synthesized a black phosphorus-based heterostructure material. Through microfluidic spinning technology, a mixture of black phosphorus and carbon Nano-tubes was built into non-woven fiber fabrics, serving as good-behavior electrodes for super-capacitors. The resulting device exhibits a 96.5 mWh.cm^{-3} of energy density, a 308.7 F·cm^{-3} of volumetric capacitance, good cycle stability, and resistance to deformation. Chen et al. [13] synthesized CD-doped graphene hetero-structures via a microfluidic strategy. The as-prepared CD/graphene

Materials Research Foundations 191 (2026) 364-395 https://doi.org/10.21741/9781644904213-12

hetero-structure features a large specific area of surface (SSA=435 $m^2 \cdot g^{-1}$), excellent electrical conductivity (11000 $S \cdot m^{-1}$), abundant active ion channels, substantial energy storage capacity (607 $mF.cm^{-2}$), and high cycling durability (96% of capacitance retention through a 10,000 cycles). Chen et al. [33] designed a tube in tube micro-reactor and prepared core-shell graphene/polyaniline microfibers, which exhibited high conductivity of 18734 $S.m^{-1}$, significant electrochemical activity, large pseudo-capacitance of 230 $mF.cm^{-2}$, and good cycling durability (with a retention rate of 86.9% for capacitance after about 8000 cycles). Davor Copic et al. [34] constructed mono-disperse CNT micro-spheres with a higher SSA of 351.6 m^2. g^{-1} by continuously shearing emulsified CNT dispersion into spherical droplets within a tubular micro-channel.

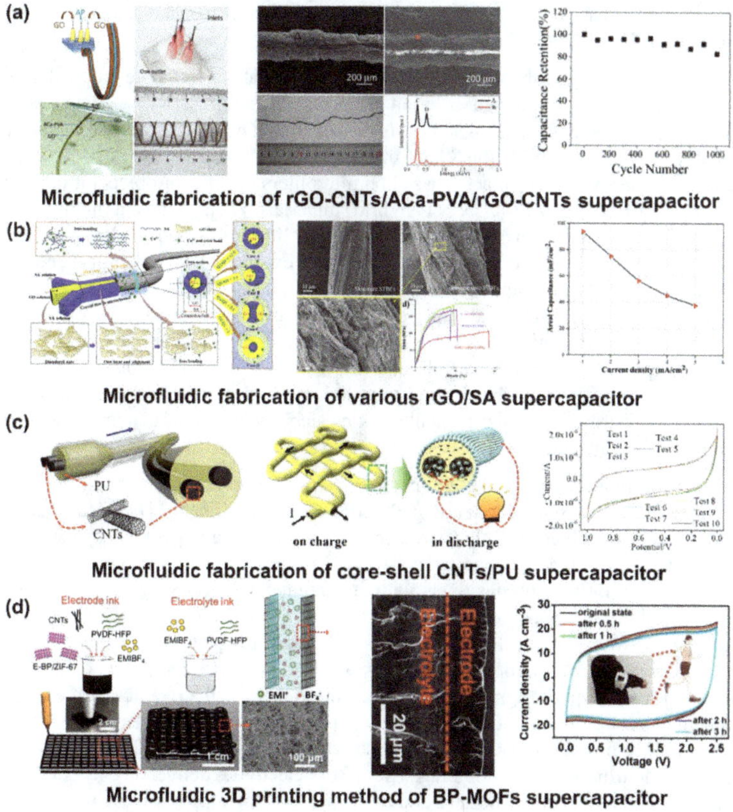

Figure 4 Application of microfluidics in super-capacitors. Reprinted with permission from [27]. Copyright 2022 American Chemical Society.

Chen et al. [35] developed a novel one-dimensional (1D) carbon Nano-material decorated with grape-shaped carbon dots (CDs) via direct pyrolysis of a green starch precursor. Using customized

Materials Research Foundations 191 (2026) 364-395 https://doi.org/10.21741/9781644904213-12

microfluidic spinning technology to fabricate conductive CNB fibers, they constructed a micro-super-capacitor (micro SC) exhibiting a high specific capacitance of 201.4 $mF.cm^{-2}$, a 4.5 $\mu Wh.cm^{-3}$ of energy density, and superior cycling stability. Chen et al. [35] synthesized a hierarchical metal-organic framework/graphene/carbon nanotube hybrid (MOF/G/CNT) via a microliter-droplet-based method. They also produced a microfluidic blow-spinning method to continuously generating nanofiber-based (MOF/G/CNT) flexible super-capacitor electrodes. The resulting super-capacitor exhibits a large volumetric energy density of 147.5 $mWh.cm^{-3}$, a 472 $F.cm^{-3}$ of specific capacitance, and stable deformation-resistant energy supply. Hu et al. [36] constructed rGO/SA fiber super-capacitors with cross-sectional combinations of sandwich, skin-core, and reverse core-shell architectures. This was achieved by regulating sodium alginate electrolyte precursors and graphene oxide electrode within microfluidic micro-channels. The rGO/SA fiber super-capacitor achieves a 93.75 $mF\cdot cm^{-2}$ of capacitance.

Du et al. [37] developed a T-shaped micro-channel to continuously generate droplet based micro-reactors, which contained a mixed solution of black phosphorus (BP) skeleton and ZIF-67 precursor. The prepared ZIF-67/BP hetero-structure exhibited high-energy storage capacity (1347 $F.g^{-1}$) and good cycling durability (an 89.3% of capacitance retention with 12000 cycles). Chen et al. developed a T-shaped micro-reactor based on microliters of liquid droplets for the continuous preparation of 0D/2D/1D ZIF-8/G/CNT hybrid materials [11]. It has a highly conductive framework (236 $S.m^{-1}$), uniform porous channels (SSA=1206 $m^{2}\cdot g^{-1}$), and an ultra-high mechanical flexibility 0D/1D/2D heterojunction structure exhibiting excellent electrochemical energy storage performance (energy storage capacity of 472 $F.cm^{-3}$, with a 94.2% retention rate of capacitance after about1000 cycles). Guo et al. [38] developed a core-shell conductive polymer/sodium alginate (PEDOT/Na Alg) microfiber by in-situ injection of PEDOT core layer into a hollow Na Alg shell through in tube micro-channels and multi flow microfluidic spinning. The prepared PEDOT/Na Alg microfiber exhibited good morphology, high conductivity (3$S.m^{-1}$), adjustable composition/size distribution, and persistent working stability in various bending states (resistance remained almost unchanged after 20 cycles of measurement).

Qiu et al. [39] introduced an innovative magneto-caloric microfluidic method to design graded porous graphene/carbon polyhedral (HG/CP) core-shell microfibers. The resulting capacitors demonstrate a 335.8 $\mu Wh.cm^{-2}$ of ultra-large energy density and a large-area capacitance of 2760 $mF.cm^{-2}$. Meng et al. [40] adopted an innovative microfluidic approach to prepare an anisotropic core-shell as well as ordered porous fibers with the graphene/nickel oxide arrays Nano-materials. These fibers exhibit large specific capacitance (605.9 $mF.cm^{-3}$) and ultra-high energy density of 120.3 $\mu Wh.cm^{-2}$ within microfluidic channels. Through template etching and electroactive coupling, electrically active porous graphene fibers were fabricated. The nickel oxide arrays were coupled in situ onto the porous graphene surface. This yields a 120.3 $\mu Wh\cdot cm^{-3}$ of energy density and excellent electrochemical stability. Wu et al. [41] synthesized a novel two-dimensional boron heterogeneous Nano-sheet (ABCN) and constructed it into highly flexible super-capacitor electrodes via microfluidic electrospinning. The ABCN electrodes deliver an exceptionally 167.05 $mWh.cm^{-3}$ of volumetric energy density and a 534.5 $F.cm^{-3}$ of superior capacitance, due to their large interlayer conductivity, abundant ion channels, and accessible surface area. Guo et al. [42] fabricated core-shell structured CNT/polyurethane (PU) super-capacitors using coaxial microfluidic spinning technology. The devices exhibit areal and volumetric capacitances of 231.89 C/m^2 and 1.34×10^5 C/m^3, respectively. After 50 cycles of testing, the capacitance variation in carbon nanotube microfibers remained below 10 C/m^2, demonstrating excellent durability.

Materials Research Foundations 191 (2026) 364-395 https://doi.org/10.21741/9781644904213-12

Jia et al. [43] fabricated graphene fibers via microfluidic self-assembly, showcasing good flexibility and electrical conductivity. super-capacitors made with these graphene fibers achieve a 3.9 mWh·cm^{-3} of energy density. Wu et al. [44] designed a metal-organic framework/black phosphorus (BP-MOF) hetero-nanostructure with a Co-O-P covalent bond, using a microfluidic droplet strategy involving ultrafast and confined droplet reactions. The hybrid material denotes a large capacitance of 1347 F·g^{-1}. Additionally, a flexible super-capacitor fabricated via 3D printing delivers a 506 F.cm^{-3} of capacitance, an ultra-high volumetric energy density of 109.8 mWh.cm^{-3}, as well as maintains good long-time stability over 12,000 cycles.

Wu et al. [45] employed efficient micro-channel reactions and microfluidic assembly to graft in situ the ZIF-L (Zn) Nano-sheets onto MXene fibers, fabricating ZIF-L(Zn)@MXene fibers. This fiber-based super-capacitors exhibit a large capacitance of 1700 F·cm^{-3}. Again, the ZIF-L(Zn)@$Ti_3C_2T_x$ fibers asymmetric super-capacitors deliver a 19 mWh·cm^{-3} of energy density, cycling stability over a 20,000 cycles, and excellent capacitance of 854 F·cm^{-3}. Wu et al. [26] developed a one-step micro-reactor strategy with in-tube micro-channel in situ phase separation to construct core-shell graphene/multi-metal oxide composites. The material features abundant ion channels (432.6 m^2.g^{-1}), high active mass loading (13.26 wt%), and superior electronic conductivity (3028 S.m^{-1}), showcasing an ultra-high areal specific capacitance of 2959 mF.cm^{-2} and significantly improved cycling durability (a 95.8% retention of capacitance through 20,000 cycles). The vertically arranged graphene/ZIF-67 prepared by Zhu et al. [46] through microfluidic droplet printing exhibits ordered diffusion channels (SSA=903 m^2.g^{-1}), fast electron conduction pathways (200 Ω.cm^{-1}), and excellent redox activity, providing a high energy storage capacity of 1674 F.g^{-1} and good cycling durability (a 86.7% of capacitance retention after 10000 cycles). The vertically arranged $Ti_3C_2T_x$/COF has high SSA (19.3 m^2.g^{-1}), high conductivity (24200 S.m^{-1}), and strong synergistic effects between vertically ordered active channels. It can provide a large energy storage capacity of 787 F.g^{-1}, with a 20000 of capacity. After the second charge discharge measurement, 89% of the initial capacitance was still maintained. Wu et al. [47] constructed a layered graphene/MoS_2 core-shell microfiber based on a tube type micro-channels and microfluidic micro-reaction strategies, achieving a highly conductive, interconnected porous network (364.2 m^2.g^{-1}) and abundant redox active sites. It has a high capacity of energy storage of 1093 mF.cm^{-2} and excellent cycling durability (with a 91% of the capacitance retention rate after the cycles of 8000).

Xu et al. [48] developed a tube-in-tube micro-channel system to controllably fabricate ordered macro-porous/microporous MOF microspheres. These microspheres feature ordered macro-porous channels (the pore size is 154 nm), a large specific surface area of 377.5 m^2.g^{-1}, and ultrafine active sites (the average size is about 9 nm), demonstrating a high-energy storage capacity of 1158 F.g^{-1} and robust cycling durability (82% capacitance retention after 10,000 cycles). Subsequently, a micro-reaction approach was adopted to synthesize ordered macro/microporous carbon/ZnS nanoclusters superstructure materials (SOM-C/ZnS). SOM-C/ZnS fiber super-capacitors assembled via microfluidic spinning technology exhibit both high-energy storage capacity and durable cycling performance (82% of capacitance retention through 10,000 cycles).A commercial-grade energy density of 172 mWh.g^{-1}, good cycling durability (initial capacity of 82% after 10000 cycles), and excellent flexibility. Tawade et al. [49] in-situ grew the PMMA on the nanoparticles $BaTiO_3$ surface through a surface initiated atomic transfer polymerization, forming a core-shell structure $BaTiO_3$@PMMA The polymer dielectric capacitor has a large energy density of capacitance with 5.2 J.cm^{-3}. Singh et al. prepared oriented PVDF/PMMA-2D fillers/PVDF

Materials Research Foundations 191 (2026) 364-395 https://doi.org/10.21741/9781644904213-12

polymers as dielectric materials through microfabrication technology. The polymer dielectric capacitor has a high-energy density of 75 $J.cm^{-3}$ and a 79% of efficiency. Liu et al. [50] synthesized nanoscale Ir/carbon hetero-structures using a continuous flow T-shaped micro-reactor with rapid heat/mass transfer. The structure exhibited ultrafine average size (less than 1 nm), high conductivity (113 $\Omega.cm^{-2}$), and high electrochemical activity, with long-term cycling stability (slight changes in current density after 12 hours of cycling).

4. Application microfluidic materials in fuel cells

As the devices of energy conversion that directly transform the chemical energy from the fuel into electrical energy, the fuel cells have shown as a pivotal form of energy utilization, driven by their environmental benefits, large energy conversion efficiency, and stable operation [51-53]. The core architecture of a fuel cell comprises three essential components: anode, cathode, and electrolyte. This system achieves energy conversion by facilitating oxidation of fuel oxidant reduction at the cathode and at the anode, thereby translating chemical energy into electrical energy. Electrolytes have a dual role in the fuel cells: they act as a physical barrier between fuel and oxidant, as well as a medium for ion transport between anode and cathode, thus constructing a complete electrochemical reaction circuit. With the advancement of technology, fuel cells are developing towards light weighting, integration, portability, and miniaturization to meet the diverse application needs in fields such as aerospace, individual equipment, biochemical testing, and medical diagnosis [54-56]. However, traditional fuel cells still face technological bottlenecks such as complex structures and large volumes, which restrict their miniaturization development.

The key factor currently restricting its commercial application is the high cost of proton exchange membranes, which may account for 40% of the total system cost, and can also cause operational failures such as electrode displacement and membrane dehydration [57]. In recent years, advancements in microfabrication technology have spurred the emergence of microfluidics, leading to the proposal of a novel membraneless microfluidic fuel cell (MFC) [58-61]. Microfluidics encompasses the technology and science of manipulating or processing small-volume fluids within micro-channels (feature sizes: 1–1000 μm). As an interdisciplinary field integrating chemistry, fluid mechanics, engineering physics, and biological engineering, it is distinguished by its miniaturization and integration capabilities. Typical applications of microfluidics include Lab-on-a-Chip systems, biomedical diagnostics, drug discovery, and conversion of energy.

Based on the unique properties of microfluidics, research has found that two or more fluids with similar density and viscosity in micro-channels can form parallel laminar flow [62, 63]. Based on this characteristic, fuel and oxidant flows with similar viscosity and density can form parallel laminar flow in micro-channels, achieving natural separation of oxidant and fuel without the need for ion -exchange membranes. In 2002, the Ferrigno et al. [57] at Harvard University first proposed membrane free microfluidic fuel cells (MMFCs) based on vanadium redox pairs. The basic principle and structure are shown in Figure 5a. Its prominent advantages include: simplified structure and convenient assembly; Avoiding the Ohmic losses, high costs, and lifespan limitations caused by membrane materials; Vanadium can be electrochemically regenerated, eliminating the step of product separation; Strong compatibility with microfluidic systems such as biochemical analysis, drug detection, etc; Miniature design (volume < 1.0 cm^3) brings high specific surface area and achieves excellent power density output. This membraneless design not only solves the

Materials Research Foundations 191 (2026) 364-395 https://doi.org/10.21741/9781644904213-12

technical problems related to membranes, but also meets the high requirements of portable electronic devices for reliable micro power sources, opening up new development directions of fuel cells [10-12]. In recent years, global scholars have conducted increasingly in-depth research on MMFCs. According to Web of Science statistics (as of May 20, 2025), the number of literature using the keyword "Microfluidic fuel cells" continues to grow (as shown in Figure 5 (b)), reflecting the increasing research enthusiasm in this field [63-66].

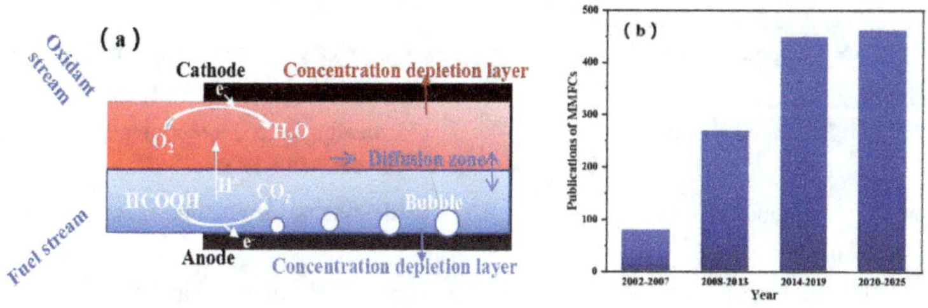

Figure 5 Schematic of the working principle and structure membraneless microfluidic fuel cell (a) and publications of MMFCs versus the year in Web of Science (b).

MMFCs can be systematically classified based on different features [63-72]. From the perspective of micro-channel geometry, it mainly includes I-type, H-type, Y-type, F-type, and T-type channel structures (as shown in Figure 6 (a-e)). According to electrode configuration, it can be divided into longitudinal swept electrodes, permeable electrodes, three-dimensional cylindrical electrodes, and air breathing cathodes (as shown in Figure 6 (f-j)). The behaviors of various types of MMFCs are shown in Table 1. According to the electrode placement position, it can be classified into parallel flow and vertical stratified flow. Based on the characteristics of the flow field, it can divide into four flow patterns: laminar flow and reverse flow. The microfluidic fuel cell system is mainly composed of three key components: in terms of fuel, the main fuels include hydrogen (H_2), ethanol (C_2H_5OH), methanol (CH_3OH), hydrogen peroxide (H_2O_2), formic acid (HCOOH), hydrazine (N_2H_4), sodium borohydride ($NaBH_4$), and vanadium oxide. Liquid fuels such as hydrogen and methanol are highly favored since their large energy density. The main oxidants include air, oxygen, hydrogen peroxide, potassium permanganate, and vanadium oxide. However, there is an inherent defect in the slow reaction kinetics of dissolved oxygen in oxidants, and the behavior needs to be prompted by enhancing the concentration. Electrolytes can enhance ion conductivity and reduce Ohmic losses. Due to the absence of proton exchange membranes, MMFCs can operate in various media environments [67-72,13-14], providing flexibility for electrode kinetics optimization.

Materials Research Foundations 191 (2026) 364-395 https://doi.org/10.21741/9781644904213-12

Table 1. The MMFCs types and their behaviors [60].

Category	MMFCsType	Behavior
Channel Configuration	F-shaped channel	Vertically layered streaming, supports flow over and flow through type electrodes
	H-shaped channel	Side-by-side streaming, supports flow over and flow through type electrodes
	I-shaped channel	No crossover issues, supports flow over and flow through type electrodes
	T-shaped channel	Vertically layered streaming, supports flow over and flow through type electrodes
	Y-shaped channel	Side-by-side streaming, supports flow over type electrodes
	Flow-over	Lower fuel utilization and mass transfer
Electrode Architecture	Air-breathing cathode	Resolve diffusivity and solubility issues of oxygen and enhance mass transfer
	Flow-through electrodes	Improve mass transfer and fuel utilization
	Single layer monolithic design with side-by-side streaming	On-chip integration, complicated fabrication, lower fuel utilization, disapprove stacking
Electrode Placement	Co-laminar flow	Higher mass transer losses
	Multi-layer sandwiched design with vertically layered streaming	Easy fabrication, higher fuel utilization, liquid-gaseous interface amplify mixing region, stackable
Flow configurations	Counter flow	Minimize diffusive mixing region
	Gaseous fuel and oxidant	Flow-over electrodes
	Orthogonal flow	Higher mass transfer losses and complex assembly
	Recirculation	Enhances fuel utilization
	Single Stream	No crossover issues
Fluidic Phase of Fuel and oxidant	Liquid fuel and gaseous oxidant	MMFCs with air-breathing cathode
	Liquid fuel and liquid oxidant	MMFCs with flow-through electrodes

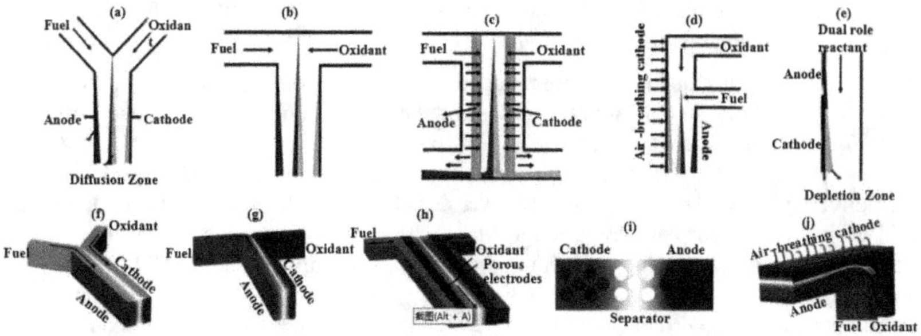

Figure 6 Microchannel geometries of MMFCs

Cohen et al. [67] first implemented an acid-base dual-electrolyte system in H_2/O_2 membraneless microfluidic fuel cells (MMFCs), achieving a 1.4 V of open-circuit voltage and a 0.96 mW·cm^{-2} of power density. Compared with the acidic electrolyte system, the battery's OCV and maximum power increased by 57.3% and 433.3%, respectively. Relative to the alkaline electrolyte system, these values improved by 50.5% and 235.6%, respectively. Jayashree et al. [58] proposed MMFCs featuring porous air-self-breathing cathodes, which consist of a diffusion layer of gas (GDL) and a catalytic layer. By exposing the GDL to air, atmospheric oxygen serves directly as the oxidant. Notably, oxygen's diffusion coefficient in air (2×10^{-5} m^2·s^{-1}) is four orders of magnitude larger than that of dissolved oxygen (2×10^{-9} m^2·s^{-1}), and its concentration in air (10 mM) exceeds dissolved oxygen levels (2–4 mM). This enables enhanced oxygen mass transfer, accelerated oxygen reduction reaction kinetics, and significant performance improvements in fuel cell.

Choban et al. [68] designed a flow-over-electrode cell, using a deposition method to fabricate catalysts: the anode comprised Pt/Ru alloy particles with 2 mg·cm^{-2}, and the cathode was loaded with Pt nanoparticles about 2 mg·cm^{-2}. Using 1 mol/L methanol dissolved in H_2SO_4 as fuel and oxygenated KOH solution as the oxidant, the resulting fuel cell with microfluidic attained a 12 mW·cm^{-2} of power density. Jayashree et al. [69] developed an air-breathing electrode battery using gas diffusion electrodes. With 1 mol/L formic acid dissolved in 0.5 mol/L H_2SO_4 as the anolyte and 0.5 mol/L H_2SO_4 as the catholyte (blank electrolyte), and a 0.3 mL/min rate of flow for each stream, the battery delivered a 26 mW·cm^{-2} of power density. Through electrode and catalyst optimization, along with methanol flow rate and concentration tuning, the density of power was further enhanced to 100 mW·cm^{-2}.

In 2007, Kjeang et al. [70] first developed a fully vanadium MMFC with an electrode of three-dimensional graphite rod array. The device utilized graphite rods with diameter as of 0.5 mm as the cathode and the anode, respectively. Enhanced mass transfer enabled the battery to achieve a 52 mW·cm^{-3} of volumetric power density and a 78% of fuel utilization rate. By fabricating a MFC with multiple graphite rod electrodes—independent arrays for the anode and cathode—it was demonstrated that a 2000 μL/ of minimal flow rate yielded a maximum current of 86 mA, 28 mW of output power, and a 78% of peak rate of fuel utilization. Optimizing the spacing and diameter of graphite rods to further improve fuel utilization efficiency. Additionally, the team applied a flow-over battery configuration to vanadium microfluidic fuel cells, employing high-specific-surface-area porous carbon electrodes. At a 1 mL/min of flow rate, the maximum of power density reached 70 mW·cm^{-2}, though fuel utilization was only 5%. The fuel utilization rose to 55% at 1 μL/min, but power density dropped to 10 mW·cm^{-2}.

Kjeang et al. [71] subsequently proposed a fully vanadium MMFC featuring a flow-through electrode of three-dimensional. Both the anode and the cathode utilized commercial carbon paper as flow-through components, achieving a 131 mW·cm^{-2} of power density and 94% of fuel utilization. When applying flow-through porous electrodes to vanadium MFC at 1 μL/min, the device delivered 0.8 V, a 20 mW·cm^{-2} of power density, and 94% of fuel utilization. Salloum et al. [72] designed a co-current flow-through MFC with porous disc-shaped electrodes. Reducing the fuel flow rate increased fuel utilization from 4% (5 mL/min) to 58% (100 μL/min), while increasing electrolyte concentration, the power density boosted from 1.5 mW·cm^{-2} to 3 mW·cm^{-2}. In a subsequent counter-current flow-through design [73], the battery achieved maximum rates of fuel utilization of 24.9% (50 μL/min) and 8.3% (300 μL/min), with output power increasing but utilization dropping significantly at higher flow rates. Hollinger et al. [74] developed a microfluidic fuel cell incorporating Nano-pore separators to mitigate fuel-electrolyte cross-contamination. This configuration attained a 70 mA·cm^{-2} of power density of and a 661.7 mA·cm^{-2} of peak current density.

Salloum and Posner [75] developed an all vanadium MMFC stack, in which the single cells use vanadium redox pairs as oxidant and fuel, and separate the fuel and oxidant by adding additional supporting electrolyte flow. The fuel cell stack is connected in parallel. Research has found that fuel utilization, the efficiency, and peak power density of fuel stacks are almost twice that of single cells. Moore et al. [76] prepared a plate frame type MMFC with three-dimensional permeable electrodes and constructed successfully an MMFC stack by arranging two single cells in parallel vertically. The test results indicate that the maximum output stacks performance of the fuel cell is twice that of a single cell. Yang et al. [77] invented a unique microfluidic assisted patterning method for battery materials. By transferring the grid pattern from the silicon mold to the PDMS and evaporating from the current collector of gold onto the substrate of PDMS, the electrode material of the battery is filled using microfluidic assisted methods, and then the gold film on the substrate of PDMS is peeled off. The transparency of the battery is 60%, 78%, and 30%, with the energy densities of 10, 5, and 20 Wh^{-1}, respectively.

Zhang et al. [78] employed electrochemical deposition to load a Pd catalyst onto graphite rod surfaces as the battery anode, paired with the air-breathing cathode. The device can achieve a 21.5 mW·cm^{-3} of power density with a 118.3 mA·cm^{-3} of current density and an 87.6% of fuel utilization rate. The three-dimensional graphite rod array electrodes enabled simultaneous high fuel utilization efficiency and output power. Additionally, increasing the number of graphite rod electrode arrays facilitated battery scaling. Hur et al. [79] designed a passive MMFC driven by CO_2 bubbles, utilizing the fuel of formic acid and the air-breathing type cathode. Prior to discharge, the battery fills gradually expanding micro-channels with fuel/electrolyte. As discharge proceeds, CO_2 bubbles generated at the anode undergo generation, growth, coalescence, and are ultimately directed toward the micro-channel outlet. When CO_2 bubbles are expelled through the vent, fresh fuel/electrolyte rapidly flows into the channel, initiating a new reaction and pumping cycle. However, this bubble-driven passive MMFC is constrained by reaction conditions, suitable only for systems generating bubbles. Moreover, bubble growth's periodic nature causes non-constant reactant flow rates, leading to periodic fluctuations in discharge current. Shaegh et al. [80] utilized porous carbon as the anode, allowing fuel to fully permeate the anode and effectively increasing fuel-catalyst contact area. At a 100 μL/min of flow rate, 1 mmol/L formic acid yielded a 26.5

Materials Research Foundations 191 (2026) 364-395 https://doi.org/10.21741/9781644904213-12

mW·cm^{-2} of power density and a 21.4% of fuel utilization rate, compared to 19.4 mW·cm^{-2} and 15.7% for the flat anode, respectively.

Xuan et al. [81] analyzed the limiting factors of MMFCs using air self-breathing cathodes through numerical simulations and studied whether the battery cathode could obtain sufficient oxygen from the air. Research has found that in the vast majority of operating conditions, air breathing cathodes can obtain sufficient oxygen from the air for cathodic half-cell reactions. Zhu et al. [82] proposed a direct formic acid MMFC with a three-dimensional graphite rod array anode.

Esquivel et al. [83] first proposed a paper-based MMFC in 2014. Methanol is used as fuel, and the cathode adopts air self-breathing cathode. Electrolyte and fuel flow through paper-based channels driven by capillary force, achieving passive transportation. When the battery uses different concentrations of electrolyte and fuel, the maximum battery power density can attain 1-5 mW.cm^{-2}. Arun et al. [84] proposed a paper-based MMFC based on pencil lead scratching graphite electrodes. By repeatedly scratching the pencil lead on filter paper, an integrated electrode was prepared, simplifying the battery structure. HCOOH was used as fuel for the anode of the battery, and the cathode is self-breathing air. A 32 mW.cm^{-2} can be obtained for power density. Zhang et al. [85] designed an air-breathing electrode battery utilizing multiple columnar graphite arrays loaded with Pd catalyst particles as anodes to enhance anodic reaction kinetics. Results showed that with 1 mol.L^{-1} formic acid dissolved in 1 mol/L H_2SO_4 as the electrolyte/fuel and a 0.3 mL/min of flow rate, the device achieved a power density of up to 21.5 mW·cm^{-2} with a 118.3 mA·cm^{-2} of current density. López-González et al. [86] proposed a hybrid MFC using fuel of glucose, where laccase and 3-ethylbenzothiazole-6-sulfonic acid (ABTS) and were immobilized on Vulcan XC-72 carbon black as the bio-cathode, and Au/Ag nanoparticles were deposited on XC-72 carbon black as the anode. This configuration yielded a 0.45 mW·cm^{-2} of power density.

In 2015, Wang et al. [87] developed a novel fuel cell system. A methanol solution with high-concentration stored in a fuel pool is treated to generate gaseous methanol, which diffuses across the anode surface and reacts with the catalyst layer. Pt-Ru black and Pt/C served as the anode catalysts and the cathode catalysts, respectively. This cell achieves a 55.4 mW·cm^{-2} of power density and a 9.4% of energy efficiency, which are 1 and 27.5 times larger than those of traditional liquid-phase methanol MFC. Zhang et al. [88] investigated fuel cell performance using potassium hydroxide as an alkaline medium under identical conditions. The device attained a 74.9 mW·cm^{-2} of power density and a 478.8 mA·cm^{-2} of current density, significantly surpassing performance in acidic media. Dector et al. [89] studied the impact of glucose sources on fuel cell performance, using a glucose oxidase as the catalyst of cathode and Pt as the catalyst of the anode. Notably, blood-derived glucose as fuel yielded superior performance, with a 0.52 V of open-circuit potential and a maximum power density of 0.20 mW·cm^{-2}. Renaud et al. [90] constructed microfluidic biofuel cells via X-ray imaging technology. These fuel cells feature parallel micro-channels and metal electrodes, generating electricity from oxygen and glucose solutions under catalysis by specific enzyme catalysts. Vertical and horizontal fluid manipulation endows them with robust power output capabilities. Zhang et al. [91] prepared a microfluidic all-vanadium solar cell that harvests solar energy through photoelectric conversion. The microfluidic platform enables efficient mass and photon transfer, small internal resistance, and uniform light distribution, enhancing light response stability. By tuning micro-channel dimensions, light intensity, and fluid concentration, precise optimization of photocurrent density, proton transfer, and the efficiency of the photo-electrochemical conversion is achieved.

Materials Research Foundations 191 (2026) 364-395 https://doi.org/10.21741/9781644904213-12

Zou et al. [92] developed an H_2/O_2 MMFC with an acid-alkali dual-electrolyte system, employing gas diffusion electrodes (GDEs) at both the anode and cathode. This configuration achieved a 1.76 V of open-circuit voltage and a maximum 145 $mW \cdot cm^{-2}$ of power density, three times the performance of single-electrolyte system batteries. Wang et al. [93] proposed a circular-structured MMFC stacks using pure methanol as fuel, where methanol evaporates into vapor and diffuses to the anode, while the cathode employs air for self-breathing. The stack sequentially connects six single cells in series. Experimental results show the stack's open-circuit voltage ranges from 5 to 5.5 V, with a maximum 108.7 $mW \cdot cm^{-2}$ of power density and a 98.4% of fuel cell efficiency. Wang and Leung [94] designed a six-cell series stack utilizing in situ hydrogen generation from Al-H_2O reactions as fuel, paired with an air-breathing cathode. At room temperature, the stack exhibits a 6 V, of open-circuit voltage, and a 531 mW of peak power, and maximum current of 248 mA. They also developed an aluminum-powered (MFC) six-cell MFC stacks in series, using the electrolyte of kitchen foil aluminum and 2 M sodium hydroxide as fuel. This stack achieves a nearly 6 V of open-circuit voltage at room temperature, a 180.6 $mW \cdot cm^{-2}$ of peak power density, as well as a 1 $Wh \cdot g^{-1}$ of energy density.

Zhang et al. [95] developed a passive MMFC based on the evaporative pump drive. The battery uses polyvinyl chloride (PVC) tubing as the evaporation tube. The ethyl acetate (EA) is used as evaporation solvent. When the evaporating solvent evaporates in the evaporation tube, a negative pressure is formed in the pipeline connected to the battery outlet, and the oxidant and fuel are passively transported under the drive of the negative pressure. The battery uses vanadium redox pairs as oxidant and fuel, and the peak current density of 22 $mA.cm^{-2}$ and a 10.2 $mW.cm^{-2}$ of power density are obtained, respectively. The disadvantages of using an evaporative pump as the driving force are: 1) the volume occupied by the evaporative pump is relatively large (length: 80 cm); 2) The evaporating solvent used is flammable and toxic; 3) The evaporated solvent is difficult to recover and needs to be replenished regularly. Ehteshami et al. [96] proposed a single chamber paper-based H_2O_2 MMFC, in which aluminum and Prussian blue and were applied as catalysts for the anode and the cathode, respectively. H_2O_2 was used as both oxidant and fuel. After testing, the maximum of power density can reach 0.81 $mW.cm^{-2}$ and a 0.61 V of open circuit voltage of the battery is obtained.

Goulet et al. [97] loaded carbon nanotubes onto carbon paper through in-situ flow deposition and applied them to vanadium MMFC. The maximum battery power density is as large as 2010 $mW.cm^{-2}$, which is also the highest performance that MMFCs have been reported to achieve so far. Abrego-Martínez et al. [98] also applied Ag/Pt/CP cathodes with methanol tolerance to MMFC stacks. The fuel cell stacks only needs to be fed with a single stream of anolyte (0.5 M KOH+5 M MeOH) during operation, so the two cells share the electrolyte channel. When two single cells are linked, the voltage of the open circuit (0.89 V) of the stack increases by nearly twice, and the absolute stack power increases by 75%. Lee et al. [99] successfully constructed a novel planar fuel cell by achieving multiple laminar flows within a micro-channel. When the stack of fuel cell is linked in series and parallel, its power and volumetric density increases by 39.3% and 97.5%, respectively, and the fuel utilization efficiency increases by 30.7% and 209.3%, respectively. Domalaon et al. [100] proposed a fabric based MMFC. The battery uses shoelaces made of cotton polyester as the channel, cling film as the separator, folded Kimwipe wiping paper as the absorbent pad, and passive transport of fuel and oxidizer under capillary force. When the battery uses 30% H_2O_2 and 5 M HCOOK as oxidant and fuel, a 4.40 $mW.cm^{-2}$ of power density and maximum current density of 22.83 $mA.cm^{-2}$ can be obtained, respectively.

Materials Research Foundations 191 (2026) 364-395
https://doi.org/10.21741/9781644904213-12

Kim et al. [101] proposed a passive MMFC driven by an osmotic pump, which connects an osmotic pump at the battery outlet to achieve continuous flow of oxidant and fuel in the micro-channel under the action of osmotic pressure drop. Due to the low molecular weight of water, it can easily pass through semi permeable membranes, while high molecular weight polyethylene glycol (PEG) solutions cannot pass through semi permeable membranes, resulting in osmotic pressure drop at the upper and lower ends of the membrane. However, the fuel driven by the osmotic pump has a small flow rate (< 0.5 $mL.h^{-1}$), and only 0.3 $mW.cm^{-2}$ of the maximum battery power density can obtain. In addition, the infiltration pump occupies a large volume in the entire system, which goes against the original intention of miniaturization of battery systems. Yan et al. [102] proposed a paper-based H_2O_2 MMFC cell using a mixed solution of 1.5 M H_2SO_4 and 2 M H_2O_2 and as the sole reaction solution, which serves as both oxidant and fuel. The anode catalysts and cathode catalysts were Prussian blue and silver nanowires, respectively. The maximum battery power density can be obtained as 0.88 $mW.cm^{-2}$.

Wen et al. [103] developed a paper-based FC of microfluidic with a microporous array in an anode channel. By constructing a paper-based microporous array on the anode side, the convection/diffusion transport of the main solution inside the micropores to the anode catalytic layer is enhanced. The results showed a 41.2% improvement in battery performance. Wu et al. [104] proposed a wire based passive microfluidic fuel cell, which uses cotton thread as a channel to attain passive transport of fuel and oxidant under the drive of gravity and capillary force. Using a two-dimensional carbon paper electrode, due to the consumption of electrochemical reactions, reactants will form a boundary layer of concentration on the surface of a planar electrode. The dissolved oxygen was used as an oxidant, which results in limited mass transfer at a cathode. Mu et al. [105] used the lattice Boltzmann method to reveal the characteristics of mass transfer of the anodic coupled electrochemical reactions in cotton thread based MFC. The distributions of concentration and velocity of the fuel in the flow channel were obtained, and the system discussed the effect of fuel inlet flow concentration and rate on the characteristics of mass transfer and anode performance of the battery. It was found that the fuel flow rate inside the cotton thread is low, and its concentration minimizes along the direction of flow under different over-potentials. The greater concentration, the larger average current density and the greater the anode performance.

Materials Research Foundations 191 (2026) 364-395 https://doi.org/10.21741/9781644904213-12

5. Applications microfluidic materials in batteries

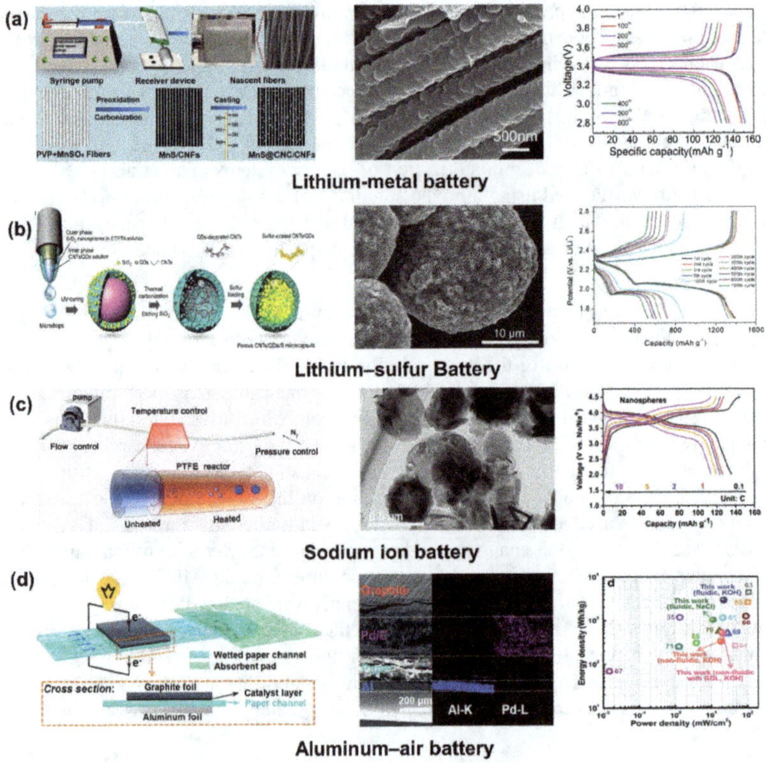

Figure 7 Application of microfluidics in electrochemical energy storage devices. Reprinted with permission from [27]. Copyright 2022 American Chemical Society.

Compared to super-capacitors, the application of microfluidics in electrochemical devices of energy storage has its own unique features, represented by sodium ion batteries (SIBs), lithium-ion batteries(LIBs), and all vanadium redox flow batteries The LIBs have controlled the fields of consumer electronics and electric vehicles since their large energy density (150-250 Wh/kg) and stable cycling performance. SIBs have become a new choice for large-scale storage of energy since their abundant raw materials and low cost (expected to be 30-40% lower than lithium batteries).The all vanadium flow battery exhibits unique advantages in grid-level storage of energy since its ultra-long cycle life (>20000 cycles) and flexible power/capacity design. The performance core of these battery systems lies in the materials' design of the electrode for SIBs or LIBs (See Figure 7 [27]). Their energy storage mechanisms mainly include three forms: 1) reversible insertion/extraction of ions in layered materials, such as the graphite (lithium battery) and hard carbon (sodium battery); 2) Alloying reactions, such as silicon-based materials (the theoretical capacity is up to 4200 mAh/g); 3) Conversion reactions, such as metal oxides. Carbon-based materials have become the mainstream choice due to their excellent conductivity and structural stability, while alloy materials, although having high-capacity characteristics, face volume expansion problems of over 300%. The all vanadium flow battery uses $VO^{2+}/VO_2^{+}$and V^{2+}/V^{3+} dissolved in the electrolyte as the active material, and achieves charge storage through an ion-exchange membrane. The current research focus is on following aspect [106-109]. 1) Developing new electrode materials (such as Prussian blue analogues for sodium electrolysis); 2) Optimize electrolyte system (especially wide temperature range, high concentration electrolyte); 3) Improve interface stability (such as a solid electrolyte interface facial mask design); 4) Reduce system costs (such as optimizing bipolar plates for flow batteries). These innovations will drive the advancement of next-generation storage technologies of energy towards larger energy density, lower cost and longer cycle life.

Lee et al. [110] fabricated continuous rGO/MnO_2 fibers via a one-pot wet method, integrating manganese oxide (MnO_2) nanowires and spun graphene oxide. The new fiber electrode exhibits excellent cycling behavior, large reversible capacity, and robust rate capability, maintaining 560 $mAh{\cdot}g^{-1}$ through 100 cycles. Liu et al. [111] synthesized monodisperse Fe_3O_4/graphene (GN) microspheres using droplet microfluidic technology. These microspheres feature excellent sphericity, uniform composition distribution, high activity utilization, and stable structural integrity, delivering a 950 $mAh{\cdot}g^{-1}$ of discharge capacity—significantly larger than traditional batch reactions (900 $mAh{\cdot}g^{-1}$). Through discharge/charge 200 cycles, they retain a 550 $mAh{\cdot}g^{-1}$ of discharge capacity. Shen et al. [112] constructed a new aluminum-air battery with microfluidic by integrating aluminum ion thermodynamics. Through micro-channel configuration optimization, the battery achieves a 2900 $mWh{\cdot}g^{-1}$ of high energy density and a1.6 V of operating voltage, positioning it as a potential alternative for traditional commercial batteries. Feng et al. [113] developed the N-doped TiO_2 photo anodes for all-vanadium redox flow batteries with microfluidic. The resulting device exhibits a long-term cycling stability and 0.103 $mA{\cdot}cm^{-2}$ of photocurrent density. Liu et al. [114] employed a droplet microfluidic strategy to prepare monodisperse Fe_3O_4/graphene microspheres as lithium-ion battery anodes. The microfluidic-oriented Fe_3O_4/graphene microspheres attain a 550 $mAh{\cdot}g^{-1}$ of discharge capacity, surpassing traditional Fe_3O_4/graphene and pure Fe_3O_4 microspheres. Lin et al. [115] engineered ordered TiO_2 nanotube arrays on titanium foil for use as photo anodes in microfluidic vanadium redox flow batteries, achieving a large photocurrent density of 0.078 $mA{\cdot}cm^{-2}$, rapid photo response, and good operational stability. Zheng et al. [116] utilized a microfluidic approach to continuously synthesize monodisperse vanadium sodium fluorophosphates Nano-sheets and Nano-spheres. Through a polytetrafluoroethylene (PTFE) micro-reactor, they achieved continuous production of $Na_3V_3O_3(PO_4)_2F$ polyanionic compounds. As sodium-ion battery cathodes, the Nano-sheets exhibit stable cycling and a 106 $mAh{\cdot}g^{-1}$ of sodium storage capacity, while the Nano-spheres deliver a large rate performance of 114 $mAh{\cdot}g^{-1}$ at 10 C. Li et al. [117] synthesized self-doped TiO_2 with disordered surface layers and abundant intermediate bands for use as photo anodes in microfluidic vanadium redox flow batteries. The resulting system demonstrates a high vanadium ion conversion rate and a 0.064 $mA{\cdot}cm^{-2}$ of photocurrent density.

Wei et al. [118] fabricated MnS@CNC/carbon fiber fabric via microfluidic spinning technology, serving as an intermediate layer between lithium-ion deposition electrodes and separators in lithium metal batteries. The battery established can achieve a large discharge capacity of 148.2

mAh·g^{-1}. Liu et al. [119] designed microcapsules by embedding alloy-type SnS_2 QD shell layers and conductive CNTs into transformed Cu_8S_5 QD core layers using a droplet micro-reactor. SnS_2@Cu_8S_5 /carbon nanotube microcapsules deliver a stable capacity retention of 386 mAh·g^{-1} through 330 cycles, an ultra-high Columbic efficiency of 99.9%, and excellent rate capability across operating temperatures from −10 to 45 ℃. Liu et al. [120] used microfluidic technology to fabricate S microcapsules/quantum dot (QD)/core-shell CNT, where CNTs provide conductive channels for charge transport and QDs ensure plentiful active sites for adsorption of polysulfide. The lithium-sulfur battery based on S/QD/CNT microcapsules shows a large capacity of 1025 mAh·g^{-1} and a 99.9% of Coulombic efficiency. Lin et al. [121] developed MoS_2-filled micro-capsules as the core for magnesium battery cathodes, which showed a 100 mAh·g^{-1} of capacity passed 100 cycles. The battery kept a large capacity retention at −10 ℃ and 45 ℃, along with stable rate performance at −5 ℃ and 45 ℃.

Liu et al. [122] introduced self-healing microcapsules into vanadium pentoxide-based cathodes for novel self-healing rechargeable magnesium batteries. The microcapsules feature conductive hydrogel and carbon nanosphere cores encapsulated in UV-curable polymer shells. Upon cathode rupture, the microcapsules release their cores to rebuild transport paths of electron, enhancing the electrochemical stable performance. After 500 cycles, the capacity remains 105 mAh·g^{-1} with 100 mA·g^{-1}, and, it stabilizes at 98.3 mAh·g^{-1} through long-term cycles of 1000 with 200 mA·g^{-1}, demonstrating a good temperature tolerance and recoverable rate performance. Do Hyun Quak et al. [123] achieved large-scale production of $NaCrO_2$ nano-particles via a separated-flow micro-reactor, which exhibit high sodium storage capacity and energy density in SIBs. Zhu et al. [124] established a microfluidic technology to encapsulate egg yolk-shell Si@void@SnO_2 nano-sphere-based YSYS micro-capsules. The lithium-ion battery anode using YSYS micro-capsules delivers over 1005 mAh·g^{-1} through 200 cycles and a 99.9% of Coulombic efficiency. It retains 752 mAh·g^{-1} at −10 ℃ passed 900 cycles. It also demonstrates stable rate behavior across varying charge/discharge speeds. Porous S/QDs/CNTs microcapsule composites were developed by encapsulating CNTs/QDs into micro-capsules with SiO_2 nano-sphere sacrificial templates for porous shell formation. As cathodes of lithium-sulfur battery, these micro-capsules maintain a high a 1025 mAh·g^{-1} of capacity though 100 cycles with 0.1C. With a 2.03 mg·cm^{-2} of sulfur loading, the battery exhibits 700 cycles of stable operation and a exceeding 99.9% of Coulombic efficiency, showcasing a good temperature tolerance and performance of recoverable rate .

Microfluidic technology faces multiple challenges in the application of LIBs and SIBs and all vanadium flow batteries: in terms of fluid control, there are problems such as poor stability of electrolyte laminar flow and insufficient accuracy of two-phase flow control. During mass transfer, high viscosity electrolytes have high-flow resistance and significant local concentration polarization phenomena; When integrating the system, there are difficulties in matching microfluidic units with macroscopic batteries, as well as controlling the uniformity of multi-channel flow; In terms of material compatibility, conventional micro-channel materials (such as PDMS) have poor resistance to organic electrolytes and are prone to structural deformation under high pressure environments; In addition, there are technical bottlenecks such as immature micro scale parameter monitoring technology and complex coupling control of multiple physical fields. In the future, research will focus on developing new corrosion-resistant micro-channel materials, designing bio-mimetic micro-structures to enhance mass transfer, and developing intelligent fluid

Materials Research Foundations 191 (2026) 364-395 https://doi.org/10.21741/9781644904213-12

control systems. These breakthroughs will promote the application of microfluidic technology in battery material screening, new structural design, health monitoring, and other fields.

Conclusion and Prospective

Microfluidic technology, characterized by efficient mass/heat transfer, precise fluid manipulation, and low energy consumption, has emerged as a pivotal platform for synthesizing micro/nano-scale electrode materials and tailoring energy storage architectures. Materials prepared at the micro/nano scale exhibit exceptional performance in storage devices of energy such as metal-ion batteries and super-capacitors. This chapter systematically summarizes the advancements in applying this technology to electrochemical energy storage systems, including batteries (LIBs, SIBs, flow batteries, etc.) and capacitors. However, despite significant progress in microfluidic strategies, several challenges persist in the energy storage field:

Development of high-temperature/high-pressure micro-channels. There is a critical need to develop new microfluidic micro-channels capable of withstanding high temperatures and pressures to enhance material adaptability. Further research is required to construct high-conductivity, high electrochemical activity materials via droplet microfluidics for accelerated charge transfer and ion storage.

2) Microfluidic control theory and mechanism exploration. The scarcity of microfluidic control theories for energy storage devices with excellent performance necessitates in-depth studies on fluid dynamics models and reaction mechanisms. At the micro-scale, structural and operational parameters (flow rate, reactant concentration) and catalyst distribution profoundly influences battery performance and material transport. Additionally, the behavior of two-phase flow of gas-liquid in the micro-channels and its impact on battery performance remain poorly understood. Exploring the structure-activity relationships between reaction kinetics and material architecture will be pivotal for advancing microfluidic theory.

3) Micro-channel design for uniform mass/heat transfer. Designing micro-channels to enhance uniform mass and heat transfer in highly conductive super-capacitors is critical for industrial production. Structural optimization can further boost energy density, while micro-channel miniaturization significantly improves mass transfer efficiency. Developing multidimensional integrated reactors represent a key pathway to achieve large-scale material synthesis and industrialization.

4) Scaling for Commercial Energy Storage Devices. In practical applications, constructing storage devices of energy with high-power density and commercial-level energy density is essential for wearable energy supply systems, representing a future development trend. With technological advancements, microfluidic technology is on the verge of becoming the dominant force in the energy storage domain—from material synthesis to industrial production of super-capacitors and batteries— thereby driving progress in wearable electronics.

5) The deep integration with artificial intelligence will promote the development of microfluidic technology towards intelligence and universality. It can extend and combine its practical applications such as biological detection, chemical analysis, pharmaceuticals, separation and purification, enriching interdisciplinary fields. With the advancement of micro-fabrication technology, microfluidic platforms will undoubtedly provide innovative solutions for the large-

scale preparation of functional materials, effectively promoting technological innovation in the energy storage industry.

Acknowledgement

This work is partially supported by Yunnan Fundamental Research Projects (NO. 202501AS070127) and supported by the National Natural Science Foundation of China (No.52364042)

References

[1] J.L. Li, G.H. Li, D.X. Liang, S.L. Ma, Review and prospect of hydrogen production technology from renewable energy under targets of carbon peak and carbon neutrality, Distributed Energy 6 (2021) 1-9.

[2] B. Zhang, X. Sun, Y. Liu, S. Peng, Development trends and strategic countermeasures of China's emerging energy technology industry toward 2035, Strategic Study of CAE (SSCAE) 22 (2020) 38-46. https://doi.org/10.15302/J-SSCAE-2020.02.006

[3] S. Bouckaert, A.F. Pales, C. McGlade, R. Uwe, W. Brent, V. Laszlo, D. Davide, S. Thomas, Net zero by 2050: A roadmap for the global energy sector, 2021. https://trid.trb.org/View/1856381 (accessed 4 June 2025).

[4] L. Li, H. Huang, X.M. Lin, X. Fan, Y. Sun, W. Zhou, T. Wang, S. Bei, K. Zheng, Q. Xu, X. Wang, M. Ni, Enhancing the performance of paper-based microfluidic fuel cell via optimization of material properties and cell structures: A review, Energ. Convers. Manage. 305 (2024) 118255. https://doi.org/10.1016/j.enconman.2024.118255

[5] M. Hightower, S.A. Pierce, The energy challenge, Nature 452 (2008) 285-286. https://doi.org/10.1038/452285a

[6] N. Hickey, P. Fornasiero, M. Graziani, Renewable resources and renewable energy, in: A Global Challenge, CRC Press, 2006, pp. 225-272. https://doi.org/10.1201/9781420020861

[7] G.M. Whitesides, The origins and the future of microfluidics, Nature 442 (2006) 368-373. https://doi.org/10.1038/nature05058

[8] S. Marre, K.F. Jensen, Synthesis of micro and nanostructures in microfluidic systems, Chem. Soc. Rev. 39 (2010) 1183-1202. https://doi.org/10.1039/b821324k

[9] Y. Geng, S.D. Ling, J. Huang, J. Xu, Multiphase microfluidics: fundamentals, fabrication, and functions, Small 16 (2020) 1906357. https://doi.org/10.1002/smll.201906357

[10] J. Peng, S. Ju, Microfluidic solvent extraction technology, Metallurgical Industry Press, Beijing, 2015. In Chinese.

[11] H. Cheng, J. Meng, G. Wu, S. Chen, Hierarchical micro-mesoporous carbon-framework-based hybrid nano-fibres for high density capacitive energy storage, Angew. Chem., Int. Ed. 58 (2019) 17465-17473. https://doi.org/10.1002/anie.201911023

[12]Z. Ding, Y. Yang, Z. Li, Synthesis and super-capacitor performance of histidine-functionalized carbon dots/graphene aerogel, Journal of Inorganic Materials 35 (2020) 1130-1136. https://doi.org/10.15541/jim20190630

[13]Q. Li, H. Cheng, X. Wu, C.-F. Wang, G. Wu, S. Chen, Enriched carbon dots/graphene microfibers towards high-performance micro-super-capacitors, J. Mater. Chem. A 6 (2018) 14112-14119 https://doi.org/10.1039/C8TA02124D

[14] Y. Zhang, X. Cao, G. Zhuang, X. Chen, In situ anchoring CuS nanoparticles on vertical aligned graphene nanosheets supported on carbon cloth for high-performance super-capacitors, J. Electroanal. Chem. 928 (2023) 117033. https://doi.org/10.1016/j.jelechem.2022.117033

[15] X. Zhu, Y. Zhang, Z. Man, W. Lu, W. Chen, J. Xu, N. Bao, W. Chen, G. Wu, Microfluidic-assembled covalent organic frameworks@$Ti_3C_2T_x$ MXene vertical fibers for high-performance electrochemical super-capacitors, Adv. Mater. 35 (2023) 2307186. https://doi.org/10.1002/adma.202307186

[16] F. Su, X. Lv, M. Miao, High-performance two-ply yarn super-capacitors based on carbon nanotube yarns dotted with Co3O4 and NiO Nanoparticles, Small 11 (2015) 854-861. https://doi.org/10.1002/smll.201401862

[17] A. Manz, N. Graber, H.M. Widmer, Miniaturized total chemical analysis systems: a novel concept for chemical sensing, Sens. Actuators B Chem. 1 (1990) 244-248. https://doi.org/10.1016/0925-4005(90)80209-I

[18] J. Novotný, F. Foret, Fluid manipulation on the micro-scale: Basics of fluid behavior in microfluidics, J. Sep. Sci. 40 (2017) 383-394. https://doi.org/10.1002/jssc.201600905

[19] B. Lin, J. Qin, Microfluidic chip laboratory, Science Press, Beijing, 2006. In Chinese.

[20] Z. Li, Fluid flow in microfluidic chips, Science Press, Beijing, 2012. In Chinese.

[21] D. Mark, S. Haeberle, G. Roth, F. Von Stetten, R. Zengerle, Microfluidic lab-on-a-chip platforms: Requirements, characteristics and applications, in: S. Kakaç, B. Kosoy, D. Li, A. Pramuanjaroenkij (Eds.), Microfluidics based microsystems, NATO science for peace and security series a: Chemistry and biology, Springer, Dordrecht, 2010, pp. 281-304. https://doi.org/10.1007/978-90-481-9029-4_17

[22] J.M. Lu, H.F. Wang, J.Z. Pan, Q. Fang, Research progress of microfluidic technique in synthesis of micro/nano materials, Acta Chim. Sin. 79 (2021) 809-819. https://doi.org/10.6023/A21030086

[23] X. Wu, S. Sun, X. Yu, X. Zhu, M. Xu, G. Wu, J. Xu, Review on microfluidic construction of advanced nanomaterials for high-performance energy storage applications, Energy Fuels 36 (2022) 4708-4727. https://doi.org/10.1021/acs.energyfuels.2c00576

[24] B. Richard, S.A. Thomas, M.A. Reddy, M.R Pallavolu, J.Cherusseri, Mini-review on fluid manipulation techniques for the synthesis and energy applications of two-dimensional MXenes: Advances, challenges, and perspectives, Energy Fuels 37 (2023) 6999-7013. https://doi.org/10.1021/acs.energyfuels.3c00589

[25] S. Liu, C. Zhang, Y. Wang, X Wei, Synthesis of energetic materials by microfluidics, Def. Technol. 44, (2025), 306-319. https://doi.org/10.1016/j.dt.2024.06.002

[26] G. Wu, Z. Ma, X. Wu, X. Zhu, Z. Man, W. Lu, J. Xu, Interfacial polymetallic oxides and hierarchical porous core-shell fibers for high energy-density electrochemical super-capacitors, Angew. Chem. Int. Ed. 61 (2022) e202203765. https://doi.org/10.1002/anie.202203765

Materials Research Forum LLC
https://doi.org/10.21741/9781644904213-12

[27]X. Wu, A. Chen, X. Yu, Z. Tian, H. Li, Y. Jiang, J. Xu, Microfluidic synthesis of multifunctional micro-/nanomaterials from process intensification: structural engineering to high electrochemical energy storage, ACS Nano 18 (2024) 20957-20979. https://doi.org/10.1021/acsnano.4c07599

[28]L. Zang, X. Cao, Y. Zhang, L. Sun, C. Qin, C. Wang, Microfluidic generation of graphene beads for super-capacitor electrode materials, J. Mater. Chem. A 3 (2015) 22088-22093. https://doi.org/10.1039/C5TA05095B

[29] T. Xu, X.T. Ding, Y. Liang, Y. Zhao, N. Chen, L.T. Qu, Direct spinning of fiber super-capacitor, Nanoscale 8 (2016) 12113-12117. https://doi.org/10.1039/C6NR03116A

[30] G. Wu, P.F. Tan, X.J. Wu, L. Peng, H.Y. Cheng, C.F. Wang, W. Chen, Z.Y. Yu, S. Chen, High-performance wearable micro-super-capacitors based on microfluidic-directed nitrogen doped graphene fiber electrodes, Adv. Funct. Mater. 27 (2017) 1702493. https://doi.org/10.1002/adfm.201702493

[31] Y.L. Tong, B. Xu, X.F. Du, H.Y. Cheng, C.F. Wang, G. Wu, S. Chen, Microfluidic-spinning-directed conductive fibers toward flexible micro-super-capacitors, Macromol. Mater. Eng. 303 (2018) 1700664. https://doi.org/10.1002/mame.201700664

[32] X.J. Wu, G. Wu, P.F. Tan, H.Y. Cheng, R. Hong, F.X. Wang, S. Chen, Construction of microfluidic-oriented polyaniline nanorod arrays/graphene composite fibers for application in wearable micro-super-capacitors, J. Mater. Chem. A 6 (2018) 8940-8946. https://doi.org/10.1039/C7TA11135E

[33] X.J. Wu, Y.J. Xu, Y. Hu, G. Wu, H.Y. Cheng, Q. Yu, K. Zhang, W. Chen, S. Chen, Microfluidic-spinning construction of black-phosphorus-hybrid micro-fibres for non-woven fabrics toward a high energy density flexible super-capacitor, Nat. Commun. 9 (2018) 4573. https://doi.org/10.1038/s41467-018-06914-7

[34] D. Copic, L. Maggini, M. De Volder, Monodisperse CNT microspheres for high permeability and efficiency flow-through filtration applications, Adv. Mater. 30 (2018) 1706503. https://doi.org/10.1002/adma.201706503

[35] Q.L. Chen, X.J. Wu, H.Y. Cheng, Q. Li, S. Chen, Facile synthesis of carbon nano-branches towards cobalt ion sensing and high-performance micro-super-capacitors, Nanoscale Adv. 1 (2019) 3614-3620. https://doi.org/10.1039/C9NA00181F

[36] X.L. Hu, M.W. Tian, N. Pan, B. Sun, Z.Q. Li, Y.L. Ma, X.S. Zhang, S.F. Zhu, Z.H. Chen, L.J. Qu, Structure-tunable graphene oxide fibers via microfluidic spinning route for multifunctional textiles, Carbon 152 (2019) 106-113. https://doi.org/10.1016/j.carbon.2019.06.010

[37] X.Y. Du, Q. Li, G. Wu, S. Chen, Multifunctional micro/nanoscale fibers based on microfluidic spinning technology, Adv. Mater. 31 (2019) 1903733. https://doi.org/10.1002/adma.201903733

[38] J. Guo, Y. Yu, H. Wang, H. Zhang, X. Zhang, Y. Zhao, Conductive polymer hydrogel microfibers from multi-flow microfluidics, Small 15 (2019) 1805162. https://doi.org/10.1002/smll.201805162

[39] H. Qiu, H.Y. Cheng, J.K. Meng, G. Wu, S. Chen, Magnetothermal microfluidic-assisted hierarchical microfibers for ultrahigh-energy-density super-capacitors, Angew. Chem. Int. Ed. 59 (2020) 7934-7943. https://doi.org/10.1002/anie.202000951

[40] J.K. Meng, G. Wu, X.J. Wu, H.Y. Cheng, Z. Xu, S. Chen, microfluidic-architected nanoarrays/porous core-shell fibers toward robust micro-energy-storage, Adv. Sci. 7 (2020) 1901931. https://doi.org/10.1002/advs.201901931

[41] T.Y. Wu, X.J. Wu, L.H. Li, M.M. Hao, G. Wu, T. Zhang, S. Chen, Anisotropic boron-carbon hetero-nanosheets for ultrahigh energy density super-capacitors, Angew. Chem. Int. Ed. 59 (2020) 23800-23809. https://doi.org/10.1002/anie.202011523

[42] J.H. Guo, Y.R. Yu, L.Y. Sun, Z.H. Zhang, Y.J. Zhao, R.J. Chai, K.Q. Shi, Bio-inspired multicomponent carbon nanotube microfibers from microfluidics for super-capacitor, Chem. Eng. J. 397 (2020) 125517. https://doi.org/10.1016/j.cej.2020.125517

[43] Y. Jia, X. Jiang, A. Ahmed, L. Zhou, Q. Fan, Microfluidic spinning of core-shell α-MnO2@graphene fibers with porous network structure for all-solid-state flexible super-capacitors, J. Electrochem. Soc. 168 (2021) 070514. https://doi.org/10.1149/1945-7111/ac0f85

[44] T.Y. Wu, Z.Y. Ma, Y.Y. He, X.J. Wu, B. Tang, Z.Y. Yu, G. Wu, S. Chen, N.Z. Bao, A covalent black phosphorus/metal-organic framework hetero-nanostructure for high-performance flexible super-capacitors, Angew. Chem. Int. Ed. 60 (2021) 10366-10374. https://doi.org/10.1002/anie.202101648

[45] G. Wu, S.Y. Sun, X.L. Zhu, Z.Y. Ma, Y.M. Zhang, N.Z. Bao, Microfluidic fabrication of hierarchical-ordered ZIF-L(Zn)@$Ti_3C_2T_x$ Core-sheath fibers for high-performance asymmetric super-capacitors, Angew. Chem. Int. Ed. 61 (2022) e202115559. https://doi.org/10.1002/anie.202115559

[46] X. Zhu, Y. Zhang, Z. Man, W. Lu, W. Chen, J. Xu, N. Bao, W. Chen, G. Wu, Microfluidic-assembled covalent organic frameworks@$Ti_3C_2T_x$ MXene vertical fibers for high-performance electrochemical super-capacitors, Adv. Mater. 35 (2023) 2307186. https://doi.org/10.1002/adma.202307186

[47] X. Wu, H. Liu, Y. Geng, X. Liu, G. Wu, J. Xu, Interface-engineered molybdenum disulfide/porous graphene microfiber for high electrochemical energy storage, Energy Storage Mater. 54 (2023) 30-39. https://doi.org/10.1016/j.ensm.2022.10.012

[48] X. Wu, X. Yu, Z. Zhang, H. Liu, S. Ling, X. Liu, C. Lian, J. Xu, Anisotropic zns nanoclusters/ordered macro-microporous carbon superstructure for fibrous super-capacitor toward commercial-level energy density, Adv. Funct. Mater. 33(2023) 2300329. https://doi.org/10.1002/adfm.202300329

[49] B.V. Tawade, M. Singh, I.E. Apata, J. Veerasamy, N. Pradhan, A. Karim, J.F. Douglas, D. Raghavan, Polymer-grafted nanoparticles with variable grafting densities for high energy density polymeric nanocomposite dielectric capacitors, JACS Au (2023) 1365-1375. https://doi.org/10.1021/jacsau.3c00022

[50] H. Liu, X. Wu, Y. Geng, X. Li, J. Xu, Microfluidic-oriented synthesis of enriched iridium nanodots/carbon architecture for robust electrocatalytic nitrogen fixation, Green Energy Environ. (2024) 544-555. https://doi.org/10.1016/j.gee.2022.09.001

[51] Y. Zhou, Fuel Cell, Power Technol. (1996) 161-164.

[52] M. Hou, B. Yi, Progress and perspective of fuel cell technology, J. Electrochem. (2012) 1-13. https://doi.org/10.61558/2993-074X.2872

[53] Y. Zhang, W. Wang, L. Tian, X. Liu, L. Li, Size effect of microfluidics, Devices Technol. (2005) 33-37.

[54] A.E. Khabbazi, A.J. Richards, M. Hoorfar, Numerical study of the effect of the channel and electrode geometry on the performance of microfluidic fuel cells, J. Power Sources (2010) 8141-8151. https://doi.org/10.1016/j.jpowsour.2010.06.094

[55] C. Zhu, T. Fu, X. Gao, W. Liu,Y. Ma, Study on the liquid-liquid two-phase flow patterns in microchannels, Chem. Ind. Eng. Prog. (2011) 65-69. In Chinese.

[56] M. Nie, L. Zhang, Q. Li, B. He, Research status of proton exchange membrane fuel cells, Surf. Technol. (2012) 109-111.In Chinese.

[57] R. Ferrigno, A.D. Stroock, T.D. Clark, M. Mayer, G.M. Whitesides, Membraneless vanadium redox fuel cell using laminar flow, J. Am. Chem. Soc. (2002) 12930-12931. https://doi.org/10.1021/ja020812q

[58] R.S. Jayashree, L. Gancs, E.R. Choban, A. Primak, D. Natarajan, L.J. Markoski, P.J.A. Kenis, Air-breathing laminar flow-based microfluidic fuel cell, J. Am. Chem. Soc. (2005) 16758-16759. https://doi.org/10.1021/ja054599k

[59] S.A.M. Shaegh, N.T. Nguyen, S.H. Chan, A review on membraneless laminar flow-based fuel cells, Int. J. Hydrogen Energy (2011) 5675-5694. https://doi.org/10.1016/j.ijhydene.2011.01.063

[60] M. Tanveer, E.S. Lim, K.Y. Kim, Effects of channel geometry and electrode architecture on reactant transportation in membraneless microfluidic fuel cells: A review, Fuel (2021) 120784. https://doi.org/10.1016/j.fuel.2021.120818

[61] M. Tanveer, K.Y. Kim, Flow configurations of membraneless microfluidic fuel cells: A review, Energies (2021) 3479. https://doi.org/10.3390/en14123381

[62] F.R. Brushett, R.S. Jayashree, W.P. Zhou, P.J.A. Kenis, Investigation of fuel and media flexible laminar flow-based fuel cells, Electrochim. Acta (2009) 7099-7105. https://doi.org/10.1016/j.electacta.2009.07.011

[63] J.C. Shyu, P.Y. Wang, C.L. Lee, C. Chang, T.S. Sheu, C.H. Kuo, K.L. Huang, Z.Y. Yang, Fabrication and test of an air-breathing microfluidic fuel cell, Energies (2015) 2082-2096. https://doi.org/10.3390/en8032082

[64] L. Liu, D. Ye, J. Li, Performance of carbon felt permeable anode air self-breathing microfluidic fuel cell, Sci. Bull. (2017) 3821-3828. In Chinese. https://doi.org/10.1360/N972017-00304

[65] T. Zhang, Transport characteristics and performance enhancement of air-breathing microfluidic fuel cells based on the regulation of anode catalyst distribution, Chongqing University, Chongqing, 2022. In Chinese.

[66] Y. Zhou, Transport and performance enhancement of electrode/electrolyte interface in membrane-less microfluidic fuel cell, Chongqing University, Chongqing, 2022.In Chinese.

[67] J.L. Cohen, D.J. Volpe, D.A. Westly, A. Pechenik, H.D. Abruña, A dual electrolyte H_2/O_2 planar membraneless micro-channel fuel cell system with open circuit potentials in excess of 1.4 V, Langmuir (2005) 3544-3550. https://doi.org/10.1021/la0479307

[68] E.R. Choban, J.S. Spendelow, L. Gancs, A. Wieckowski, P.J.A. Kenis, Membraneless laminar flow-based micro fuel cells operating in alkaline, acidic and acidic/alkaline media, Electrochim. Acta (2005) 5390-5398. https://doi.org/10.1016/j.electacta.2005.03.019

[69] R.S. Jayashree, L. Gancs, E.R. Choban, A. Primak, D. Natarajan, L.J. Markoski, P.J.A. Kenis, Air-breathing laminar flow-based microfluidic fuel cell, J. Am. Chem. Soc. (2005) 16758-16759. https://doi.org/10.1021/ja054599k

[70] E. Kjeang, J. McKechnie, D. Sinton, N. Djilali., Planar and three-dimensional microfluidic fuel cell architectures based on graphite rod electrodes, J. Power Sources (2007) 379-390. https://doi.org/10.1016/j.jpowsour.2007.02.087

[71] E. Kjeang, R. Michel, D.A. Harrington, N. Djilali, A microfluidic fuel cell with flow-through porous electrodes, J. Am. Chem. Soc. (2008) 4000-4006. https://doi.org/10.1021/ja078248c

[72] K.S. Salloum, J.R. Hayes, C.A. Friesen, C.A. Friesen, J.D. Posner, Sequential flow membraneless microfluidic fuel cell with porous electrodes, J. Power Sources (2008) 243-252. https://doi.org/10.1016/j.jpowsour.2007.12.116

[73] K.S. Salloum, J.D. Posner, Counter flow membraneless microfluidic fuel cell, J. Power Sources (2010) 6941-6944. https://doi.org/10.1016/j.jpowsour.2010.03.096

[74] A.S. Hollinger, R.J. Maloney, R.S. Jayashree, D. Natarajan, L.J. Markoski, P.J.A. Kenis, Nanoporous separator and low fuel concentration to minimize crossover in direct methanol laminar flow fuel cells, J. Power Sources (2010) 357923-357928. https://doi.org/10.1016/j.jpowsour.2009.12.063

[75] K.S. Salloum, J.D. Posner, A membraneless microfluidic fuel cell stack, J. Power Sources (2011) 1229-1234. https://doi.org/10.1016/j.jpowsour.2010.08.069

[76] S. Moore, D. Sinton, D. Erickson, A plate-frame flow-through microfluidic fuel cell stack, J. Power Sources (2011) 9481-9487. https://doi.org/10.1016/j.jpowsour.2011.07.024

[77] Y. Yang, S. Jeong, L. Hu, H. Wu, S.W. Lee, Y. Cui, Transparent lithium-ion batteries, Proc. Natl. Acad. Sci. USA (2011) 13013-13018. https://doi.org/10.1073/pnas.1102873108

[78] B. Zhang, D.D. Ye, J. Li, X. Zhu, Q. Liao, Electrodeposition of Pd catalyst layer on graphite rod electrodes for direct formic acid oxidation, J. Power Sources (2012) 277-284. https://doi.org/10.1016/j.jpowsour.2012.04.007

[79] J.I. Hur, D.D. Meng, C.J. Kim, Self-pumping membraneless miniature fuel cell with an air-breathing cathode, J. Microelectromech. Syst. (2012) 476-483. https://doi.org/10.1109/JMEMS.2011.2176920

[80] S.A.M. Shaegh, N.T. Nguyen, S.H. Chan, Air-breathing membraneless laminar flow-based fuel cell with flow-through anode, Int. J. Hydrogen Energy (2012) 3466-3476. https://doi.org/10.1016/j.ijhydene.2011.11.051

[81] J. Xuan, D.Y.C. Leung, H.Z. Wang, M.K.H. Leung, B. Wang, M. Ni, Air-breathing membraneless laminar flow-based fuel cells: Do they breathe enough oxygen, Appl. Energy (2013) 400-407. https://doi.org/10.1016/j.apenergy.2012.10.063

[82] X. Zhu, B. Zhang, D.D. Ye, J. Li, Q. Liao, Air-breathing direct formic acid microfluidic fuel cell with an array of cylinder anodes, J. Power Sources (2014) 346-353. https://doi.org/10.1016/j.jpowsour.2013.08.119

[83] J.P. Esquivel, F.J. Del Campo, J.L.G. de la Fuente, S. Rojas, N. Sabaté, Microfluidic fuel cells on paper: meeting the power needs of next generation lateral flow devices, Energy Environ. Sci. (2014) 1744-1749. https://doi.org/10.1039/C3EE44044C

[84] R.K. Arun, S. Halder, N. Chanda, S Chakraborty, A paper-based self-pumping and self-breathing fuel cell using pencil-stroked graphite electrodes, Lab Chip (2014) 1661-1664. https://doi.org/10.1039/C4LC00029C

[85] X. Zhu, B. Zhang, D.D. Ye, J. Li, Q. Liao, Air-breathing direct formic acid microfluidic fuel cell with an array of cylinder anodes, J. Power Sources (2014) 346-353. https://doi.org/10.1016/j.jpowsour.2013.08.119

[86] B. López-González, A. Dector, F.M. Cuevas-Muñiz, N. Arjona, C. Cruz-Madrid, A. Arana-Cuenca, Hybrid microfluidic fuel cell based on Laccase/C and Au Ag/C electrodes, Biosens. Bioelectron. (2014) 221-226. https://doi.org/10.1016/j.bios.2014.06.054

[87] Y.F. Wang, D.Y.C. Leung, J. Xuan, A vapor feed methanol microfluidic fuel cell with high fuel and energy efficiency, Appl. Energy (2015) 456-465. https://doi.org/10.1016/j.apenergy.2015.03.028

[88] B. Zhang, D.D. Ye, J. Li, X. Zhu, Q. Liao, Air-breathing microfluidic fuel cells with a cylinder anode operating in acidic and alkaline media, Electrochim. Acta (2015) 264-269. https://doi.org/10.1016/j.electacta.2015.03.080

[89] A. Dector, R.A. Escalona, D. Dector, V. Vallejo-Becerra, A.U. Chávez-Ramírez, Perspective use of direct human blood as an energy source in air-breathing hybrid microfluidic fuel cells, J. Power Sources (2015) 70-75. https://doi.org/10.1016/j.jpowsour.2015.04.089

[90] L. Renaud, D. Selloum, S. Tingry, Xurography for 2D and multi-level glucose/O_2 microfluidic biofuel cell, Microfluid. Nanofluid. (2015) 1407-1416. https://doi.org/10.1007/s10404-014-1539-z

[91] X.H. Jiao, R. Chen, X. Zhu, Q. Liao, D.D. Ye, B. Zhang, L. An, H. Feng, W. Zhang, A microfluidic all-vanadium photo-electrochemical cell for solar energy storage, Electrochim. Acta (2017) 842-849. https://doi.org/10.1016/j.electacta.2017.11.134

[92] H.Y. Zou, J. Chen, Y.N. Fang, J. Ding, W. Peng, R. Liu, A dual-electrolyte based air-breathing regenerative microfluidic fuel cell with 1.76 V open-circuit-voltage and 0.74 V water-splitting voltage, Nano Energy (2016) 619-626. https://doi.org/10.1016/j.nanoen.2016.07.036

[93] Y.F. Wang, D.Y.C. Leung, A circular stacking strategy for microfluidic fuel cells with volatile methanol fuel, Appl. Energy (2016) 659-669. https://doi.org/10.1016/j.apenergy.2016.11.019

[94] Y.F. Wang, D.Y.C. Leung, A high-performance aluminum-feed microfluidic fuel cell stack, J. Power Sources (2016) 427-436. https://doi.org/10.1016/j.jpowsour.2016.11.009

[95] Q.Q. Zhang, H.L. Li, X.J. Liu, W. Zhao, C. Zong, H. Gai, A self-driven miniaturized liquid fuel cell, Chem. Commun. (2016) 12068-12071. https://doi.org/10.1039/C6CC06063C

[96] S.M.M. Ehteshami, M. Asadnia, S.N. Tan, SH Chan, Paper-based membraneless hydrogen peroxide fuel cell prepared by micro-fabrication, J. Power Sources (2016) 392-395. https://doi.org/10.1016/j.jpowsour.2015.10.038

[97] M.A. Goulet, O.A. Ibrahim, W.H.J. Kim, E. Kjeang, Maximizing the power density of aqueous electrochemical flow cells with in operando deposition, J. Power Sources (2017) 80-85. https://doi.org/10.1016/j.jpowsour.2016.11.053

[98] J.C. Abrego-Martínez, A. Moreno-Zuria, F.M. Cuevas-Muñiz, L.G. Arriaga, S. Sun, M. Mohamedi, Design, fabrication and performance of a mixed-reactant membraneless micro direct methanol fuel cell stack, J. Power Sources (2017) 10-17. https://doi.org/10.1016/j.jpowsour.2017.10.026

[99] S.H. Lee, Y. Ahn, A laminar flow-based single stack of flow-over planar microfluidic fuel cells, J. Power Sources (2017) 67-73. https://doi.org/10.1016/j.jpowsour.2017.03.102

[100] K. Domalaon, C. Tang, A. Mendez, F. Bernal, K. Purohit, L. Pham, J. Haan, F.A. Gomez, Fabric-based alkaline direct formate microfluidic fuel cells, Electrophoresis (2017) 1224-1231. https://doi.org/10.1002/elps.201600306

[101] S.H. Kim, K. Kim, M. Go, J.Y. Park, Stand-alone external power-free microfluidic fuel cell system harnessing osmotic pump for long-term operation, J. Micromech. Microeng. (12)(2018) 1-10. https://doi.org/10.1088/1361-6439/aae773

[102] X.H. Yan, A. Xu, L. Zeng, P. Gao, T. Zhao, A paper-based microfluidic fuel cell with hydrogen peroxide as fuel and oxidant, Energy Technol. (2018) 140-143 https://doi.org/10.1002/ente.201700470

[103] J. Wen, D. Ye, X. Zhu, Cell performance of passive paper-based microfluidic fuel cells with mini-hole array in the anode flow channel, Chin. Sci. Bull. (2020) 2527-2534. https://doi.org/10.1360/TB-2020-0126

[104] R. Wu, D.D. Ye, R. Chen, B. Zhang, X. Zhu, H. Guo, Z. Liu, A membraneless microfluidic fuel cell with continuous multi-stream flow through cotton threads, Int. J. Energy Res. (2020) 2243-2251. https://doi.org/10.1002/er.5085

[105] A. Mu, D. Ye, R. Chen, X. Zhu, Q. Liao, LB simulation of anode mass transfer characteristics in cotton thread-based microfluidic fuel cell, CIESC J. (2020) 3278-3287.

[106] S.J. Luo, Y.F. Wang, T.C. Kong, W. Pan, X. Zhao, D.Y.C. Leung, Flexible direct formate paper fuel cells with high performance and great durability, J. Power Sources (2021) 1-8. https://doi.org/10.1016/j.jpowsour.2021.229526

[107] M. Kim, Y. Kwon, Y. Ahn, Paper-based mediatorless enzymatic microfluidic biofuel cells, Biosens. Bioelectron. (2021) 1-7. https://doi.org/10.1016/j.bios.2021.113391

[108] Y. Yang, H.R. Zhang, X. Zhu, D. Ye, R. Chen, Q. Liao, A hollow structure for flow and bendable paper-based zinc-air battery, Int. J. Energy Res. (2022) 1-7.

[109] Z. Liu, Mass Transfer Characteristics and Enhanced Power Generation Performance in Passive Microfluidic Fuel Cells Based on Wires, Chongqing University, Chongqing, 2022. In Chinese.

[110] J.G. Lee, Y. Kwon, J.Y. Ju, S. Choi, Y. Kang, W.R. Yu, D.W. Kim, Fiber electrode by one-pot wet-spinning of graphene and manganese oxide nanowires for wearable lithium-ion batteries, J. Appl. Electrochem. (2017) 865-875. https://doi.org/10.1007/s10800-017-1085-y

[111] N. Liu, C. Li, H. Xie, Y. Liu, J. Liu, J. Gao, J. Han, J. Chen, Y. Wei, B. Lin, Z. Liu, Doping graphene into monodispersed Fe_3O_4 microspheres with droplet microfluidics for enhanced electrochemical performance in lithium-ion batteries, Batteries Supercaps (2019) 49-54. https://doi.org/10.1002/batt.201800072

[112] L.L. Shen, G.R. Zhang, M. Biesalski, B.J.M. Etzold, Paper-based microfluidic aluminum-air batteries: Toward next-generation miniaturized power supply, Lab Chip (2019) 3438-3447. https://doi.org/10.1039/C9LC00574A

[113] H. Feng, X. Jiao, R. Chen, X. Zhu, Q. Liao, D. Ye, B. Zhang, W. Zhang, A microfluidic all-vanadium photoelectrochemical cell with the N-doped TiO_2 photoanode for enhancing solar energy storage, J. Power Sources (2019) 162-170. https://doi.org/10.1016/j.jpowsour.2019.02.074

[114] N. Liu, C.L. Li, H. Xie, Y. Liu, J. Liu, J. Gao, J. Han, J. Chen, Y. Wei, B. Lin, Z. Liu, Doping graphene into monodispersed Fe_3O_4 microspheres with droplet microfluidics for enhanced electrochemical performance in lithium-ion batteries, Batteries Supercaps (2019) 49-54. https://doi.org/10.1002/batt.201800072

[115] Y. Lin, H. Feng, R. Chen, B. Zhang, L. An, One-dimensional TiO_2 nanotube array photoanode for a microfluidic all-vanadium photoelectrochemical cell for solar energy storage, Catal. Sci. Technol. (2020) 4352-4361. https://doi.org/10.1039/D0CY00342E

[116] L.M. Zheng, D.T. Zhang, X.Y. Wang, G Guo, Continuous-flow rapid and controllable microfluidic synthesis of sodium vanadium fluorophosphate as a cathode material, Appl. Mater. Today (2021) 101032. https://doi.org/10.1016/j.apmt.2021.101032

[117] J. Li, Y. Lin, R. Chen, X. Zhu, D. Ye, Y. Yang, Y. Yu, D. Wang, Q. Liao, Solar energy storage by a microfluidic all-vanadium photoelectrochemical flow cell with self-doped TiO_2 photoanode, J. Energy Storage (2021) 103228. https://doi.org/10.1016/j.est.2021.103228

[118] L.Y. Wei, N.P. Deng, X.X. Wang, H. Zhao, J. Yan, Q. Yang, W. Kang, B. Cheng, Flexible ordered MnS@CNC/carbon nanofibers membrane based on microfluidic spinning technique as interlayer for stable lithium-metal battery, J. Membr. Sci. (2021) 119615. https://doi.org/10.1016/j.memsci.2021.119615

[119] J. Liu, M. Zhu, K. Mu, T. Han, Z. Pan, Y. Gan, H. Zhang, T. Si, Engineering a novel microcapsule of Cu_8S_5 core and SnS_2 quantum dot/carbon nanotube shell as a Li-ion battery anode, Chem. Commun. (2021) 13397-13400. https://doi.org/10.1039/D1CC05657C

[120] J. Liu, M. Zhu, Z. Shen, T Han, T Si, C Hu, H Zhang, A polysulfides-confined all-in-one porous microcapsule lithium-sulfur battery cathode, Small (2021) 2103051. https://doi.org/10.1002/smll.202103051

[121] X. Lin, J. Liu, H. Zhang, Y. Zhong, M. Zhu, T. Zhou, X. Qiao, H. Zhang, T. Han, J. Li, General liquid-driven coaxial flow focusing preparation of novel microcapsules for rechargeable magnesium batteries, Adv. Sci. (2021) 2002298. https://doi.org/10.1002/advs.202002298

[122] J. Liu, Y. Wang, C. Yang, K. Mu, S. Wang, Z. Pan, M. Zhang, T. Si, A microcapsule-assistant self-healing magnesium battery cathodes, Energy Technol. (2021) 2100393. https://doi.org/10.1002/ente.202100393

[123] D.-H. Quak, M. Sarif, P. Opitz, M. Lange, O. Jegel, D.H. Pham, M. Koziol, L. Prädel, M. Mondeshki, Generalized synthesis of $NaCrO_2$ particles for high-rate sodium ion batteries prepared by microfluidic synthesis in segmented flow, Dalton Trans. (2022) 10466-10474. https://doi.org/10.1039/D1DT04333A

[124] M. Zhu, Preparation of novel capsule energy storage materials based on microfluidic technology and study on their battery performance, Anhui Normal University, Anhui, 2022. In Chinese.

Materials Research Foundations 191 (2026) 396-410
https://doi.org/10.21741/9781644904213-13

Chapter 13

Microfluidic Devices for Automated Remote Analytics

Kannan Deepa[1]*

[1]Department of Chemical Engineering, College of Engineering and Technology, Kattankulathur, Tamil Nadu - 603203, India

kdeepa86@gmail.com

Abstract

Microfluidic devices are finding widespread applications in automated remote analytics due to their rapid and precise handling of small volumes of fluids at microscale dimensions. These compact systems integrate sample preparation, chemical or biological analysis, and data transmission, enabling real-time monitoring in remote, on site, and resource-limited environments. Their advantages include low reagent consumption, fast response times, portability, and compatibility with wireless technologies. Microfluidic devices are used in healthcare diagnostics, environmental monitoring, and food safety sensing, to name a few. Recent advancements in microfabrication, biosensor integration, and Artificial Intelligence-assisted data interpretation have significantly enhanced the performance and reliability of these systems. By minimizing the need for manual intervention and centralized laboratories, microfluidic platforms facilitate decentralized analysis and continuous monitoring, making them a promising solution for various field-deployable analytical tasks.

Keywords

Microfluidics, Microchannels, Lab-on-a-chip, Sensors, Point-of-Care Diagnostics, Environmental Sensors

Contents

Materials Research Foundations 191 (2026) 396-410
https://doi.org/10.21741/9781644904213-13

1. Introduction

Microfluidic devices offer the advantage of requiring only minimal sample volumes, eliminating the need for sophisticated analytical equipment while offering additional advantages such as accuracy, portability, rapid response, and convenient on site analysis. The integration of microchannels, valves, chambers, and sensors capable of analyzing just a few microliters of sample on a single compact platform has led to the term "lab-on-a-chip." Miniaturization of a sophisticated laboratory is achieved in milliscale dimensions reducing the amount of chemical reagents, time, and efforts [1]. Such microfluidic devices have found extensive applications in remote analytics facilitating detection of diseases, biological compounds, and chemical pollutants such as pesticides, dyes, and heavy metals, to name a few [1-4].

The integration of artificial intelligence, especially machine learning, has proven essential for interpreting complex biosensor signals. Although AI often operates as a black box, its critical role in handling the vast data generated by microfluidic systems is well-established and supported by numerous applications [5]. This chapter provides an overview of the developments in automated remote sensing of target analytes using microfluidic platforms. There is a separate section focusing on the realm of Artificial Intelligence which has made its inevitable entry into the field of microfluidics leading to several scientific breakthroughs.

2. Automated Remote Analytics in Microfluidic platforms

Automated remote analytics in microfluidic platforms have revolutionized the field of analytical chemistry and biomedical engineering by merging the precision of microfluidics with the power of automation and connectivity. Microfluidic devices can process and manipulate minute fluid volumes with high accuracy, by revolutionizing in vitro diagnostics, drug development, and environmental monitoring. Integrating automation into these platforms enhances reproducibility and reduces human intervention, thereby minimizing errors and labor-intensive processes. Furthermore, coupling microfluidics with remote analytics allows real-time data acquisition and processing from the user's own locations, opening new avenues for point-of-care diagnostics, personalized medicine, and environmental surveillance. By integrating automation and remote connectivity, microfluidic platforms are expected to become indispensable tools for real-time, accessible, and precise analysis across diverse fields. Fig. 1 illustrates the various strategies used in sensing of analytes using microfluidics.

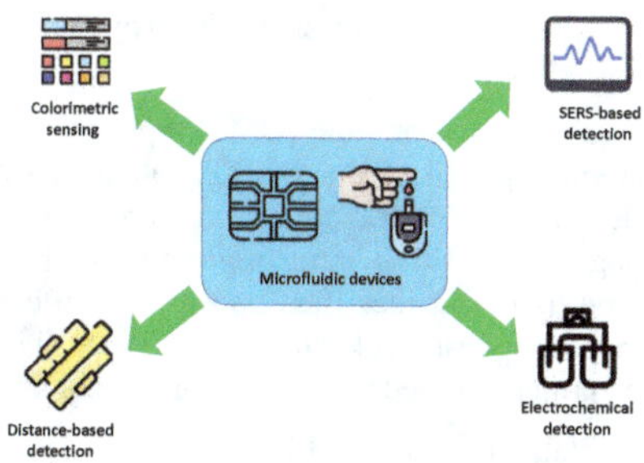

Figure 1. Various Techniques of Sensing in Microfluidic Devices

Materials Research Foundations 191 (2026) 396-410 https://doi.org/10.21741/9781644904213-13

3.1 Healthcare, Diagnostics and Biosensing

In recent years, the FDA has observed a rise in microfluidic medical devices, driven by recent progress in microfluidic technologies and boosted by their use in COVID-19 diagnostic test kits [6]. Microfluidics has revolutionized the field of healthcare by serving as a 'sample-in-answer-out' system that identifies biomarkers of diseases. Extensive and laborious processing of samples is avoided when using microfluidic devices. All the stages in bioanalysis such as sample pretreatment, sample/reagent addition, reaction, detection, and signal analysis are executed on a single microfluidic device [7]. Epidermal electronics and miniaturized wearable diagnostic devices have advanced significantly, enabling continuous real-time biological data transmission to the Internet for clinical analysis. Non-wearable devices also use Bluetooth and WiFi to send data through phones or computers to centralized healthcare systems. However, widespread clinical implementation faces challenges, including the need to integrate detection components with fluid-handling elements at the microscale, where fluid behavior is dominated by viscous forces instead of inertial forces [4].

A fully integrated, one-button microfluidic analyzer has been developed to automate serum separation, reagent mixing, and biochemical analysis from finger-prick blood. It enables simultaneous detection of glucose, cholesterol, and triglycerides within 15 minutes, supporting self-monitoring of chronic diseases at home. The results from volunteer samples were comparable to those obtained using commercial hospital analyzers [8]. A fully automated, label-free platform has been developed for leukocyte activation analysis to monitor immune responses in sepsis, using just 50 μL of blood and delivering results in 25 minutes. It integrates multidimensional double spiral (MDDS) and isodielectric separation (IDS) subplatforms, enabling high-purity leukocyte separation and electrical profiling without human intervention or ex vivo activation. The system effectively distinguishes leukocytes from sepsis patients and healthy individuals, offering potential for real-time bedside immune monitoring in critical care settings [9].

A ferrobotic microfluidic system was developed to automate complex fluidic tasks at small scales, aiming to overcome challenges in diagnostics, drug development, and material synthesis. Inspired by automated guided vehicle systems, the platform uses individually addressable robots actuated by electromagnetic fields to manipulate droplets containing magnetic nanoparticles. These "ferrobots" operate contactlessly and with high speed, repeatability, and long-term stability, supporting diverse fluidic functions like droplet dispensing, merging, and filtering. Demonstrating system-level integration, the ferrobotic platform successfully conducted a fully automated assay to quantify matrix metallopeptidases which are cancer-related biomarkers present in human plasma [10].

A fully automated all-in-one digital microfluidic (DMF) pipeline was developed for processing trace protein samples, integrating steps like reduction, alkylation, digestion, isotopic labeling, and analysis. The system includes thermal control, optimized droplet handling, and direct interfacing with HPLC-MS/MS, enabling high-sensitivity proteomic analysis at the nanogram level. Applied to breast cancer cell lines and tissues, it identified known and differentially expressed proteins linked to metabolism, hormone response, and cancer sub-typing. This integrated systen offers a robust solution for analyzing protein-limited samples for diverse biomedical applications [11].

Centrifugal microfluidics, or lab-on-a-disc, enables the integration of complex assays onto a rotating platform, offering a compact and efficient diagnostic solution. By scaling down reaction volumes to sub-microliter levels, it minimizes reagent use and allows for parallel processing,

Materials Research Foundations 191 (2026) 396-410
https://doi.org/10.21741/9781644904213-13

increasing the speed of the process [12]. This platform allows implementing and standardizing unit operations like fluid movement, mixing, and metering. The centrifugal force enables integrated sample preparation steps such as blood separation and debris sedimentation, and can be combined with capillary forces to automate sequential fluidic actions. By adjusting the rotor speed, various assay requirements can be met, making this technology ideal for in-vitro diagnostics and point-of-care applications [13]. For example, understanding cancer heterogeneity requires isolating all types of circulating tumor cells (CTCs) to accurately detect genetic mutations in metastatic cancers. A fully automated, negative depletion-based continuous centrifugal microfluidic (CCM) system was developed which demonstrated over 90% recovery and 99.9% white blood cell depletion across diverse CTC phenotypes and cancer types, including lung, breast, and bladder. In lung cancer patients, CTC analysis using CCM showed perfect genetic concordance with cell-free DNA and correlated with disease progression and treatment response, highlighting its clinical potential [14]. In another study, a centrifugal microfluidic system was developed to automate and conduct parallel evaluation of bacterial endotoxins using Limulus amebocyte lysate (LAL), the standard reagent for detecting endotoxins in injectable drugs. The platform reduces LAL consumption by over 90 % and enables precise, contamination-free liquid handling with 104 reactions integrated into a single microplate. It demonstrated high accuracy, with a coefficient of variation below 0.75 %, and delivered results consistent with conventional testing methods [15].

Microfluidic devices, which provide efficient platforms for reproducing biologically relevant 3D in vitro models, have been developed. These environments are typically formed using hydrogels and controlled chemical gradients. In addition, validated computational models can simulate such conditions, eliminating the need for physical experiments thereby reducing both time and costs. A web-based application to evaluate chemical gradients within 3D extracellular matrices in microfluidic devices, has been developed based on a pre-existing numerical model. Users can input chip geometry and hydrogel diffusion parameters to simulate chemical distributions, with results delivered in under three minutes. The tool includes four predefined microfluidic geometries and demonstrates how both physical diffusion and device geometry influence gradient formation. It simplifies device design and reduces experimental workload since it is freely available online as well [16].

3.2 Environmental Monitoring

Microfluidic devices are increasingly used nowadays for detecting environmental contaminants due to their portability, affordability, and quick response. They are applied in monitoring air, water, soil, and agricultural conditions, with recent advances focusing on detecting substances like heavy metals, pesticides, nutrients, microorganisms, and polyfluoroalkyl substances. Significant advancements have been made in microfluidic systems utilizing techniques such as absorbance, electrochemical, fluorescence, and chemiluminescence detection [17]. Table 1 gives the various applications of microfluidics in detecting environmental contaminants.

A portable, low-cost, and fully integrated microfluidic system was developed for rapid, point-of-source detection and quantification of nitrite using a disposable PDMS chip and a controlled Griess reaction. The setup includes a 3D-printed syringe pump, LED-photodiode photometric detection, a transimpedance amplifier for signal clarity, and a bluetooth-enabled smartphone app for control and data storage. The system demonstrated good repeatability, with a detection limit of 0.07103 ppm and quantification limit of 0.21524 ppm, rendering real-time nitrite monitoring simple and effective [18]. In another study, a 3D-printed sulfide-selective microfluidic platform (3D-SSMP) was developed using a dual-stage printing process combined with screen printing and long-term Ag_2S deposition. The integrated device demonstrated a strong Nernstian response (−29 ± 1 mV/decade) over a wide concentration range (0.1–1000 mM), with a quick response time of 8 seconds and a low detection limit of 0.03 mM. Validation with real water samples showed results comparable to standard methods, suggesting the potential of this fabrication strategy for creating advanced electrochemical sensors for environmental and biotechnological applications [19].

A microfluidic optical fiber sensor was developed for detecting Hg(II) ions using a selective tripodal chromoionophore which was embedded in a PVC film. The system showed a linear absorbance response at 495 nm for Hg(II) concentrations ranging from 1.0×10^{-6} to 2.5×10^{-4} M, with a detection limit of 0.5 μM. Results from river water samples closely matched those obtained using standard cold vapor atomic absorption spectrometry, demonstrating the accuracy and reliability of the sensor [20]. An autonomous microfluidic analytical system using the vanadomolybdate method was developed for detecting soluble reactive phosphorus in seawater. It integrates a tinted PMMA chip, custom syringe pump, embedded electronics, and onboard calibration, enabling field deployment and comparison with standard lab analyzers. The system achieved a low detection limit of 52 nM, a wide linear range (0.1–60 μM), and accurate performance matching reference methods with no significant differences. Its low reagent and power requirements, combined with high sampling throughput, make it ideal for real-time, high-resolution marine nutrient monitoring [21].

High levels of arsenic in water pose serious health risks globally, necessitating rapid, in-field testing methods for early detection. A modular microfluidic system with an integrated electrochemical sensor—featuring gold nanoparticle-modified electrodes—has been developed for sensitive, on-site detection of As(III) using anodic stripping voltammetry. The system offers automatic calibration and analysis with a detection limit of 0.42 μg/L, showing strong performance in real water samples and meeting WHO safety standards [22].

Etching Inkjet Maskless Lithography (E-IML) is introduced as a low-cost method to fabricate nanoporous gold leaf (NPGL)-based flexible biosensors for detecting organophosphate pesticides like paraoxon. The technique enables precise patterning of NPGL electrodes and Ag/AgCl pseudo-reference electrodes on flexible substrates using inkjet printing. A multi-electrode biosensor developed with E-IML achieved ultra-low detection limits (0.53 pM) and a broad sensing range (1 nM–10 mM), showing over 90% signal recovery in spiked soil samples. This method shows strong potential for use in electrochemical sensing, energy devices, and flexible electronics [23].

Monitoring antibiotic concentrations in drinking water is crucial to prevent the development of resistant bacteria, as even trace amounts can contribute to antibiotic resistance and significantly endanger public health [24]. A passive micromixer was designed and simulated to enhance mixing efficiency for detecting aminoglycosidic antibiotics such as gentamicin in water using a PDMS microfluidic chip. Optimized geometries were fabricated through 3D printing and PDMS casting, and an automated stepper motor-based fluid delivery system was developed for precise reagent control. The integrated platform demonstrated a detection range of 8 to100 μM/mL with a 13 μM/mL limit, showcasing its potential for real-time, automated environmental and biological monitoring [24].

An autonomous in situ microfluidic analyzer was developed for measuring total alkalinity in seawater, a key parameter for understanding the ocean carbon cycle and monitoring carbon dioxide removal via ocean alkalinity enhancement. The lightweight, reagent-efficient device uses closed-cell titrations driven by stepper-motor syringe pumps, enabling precise pH control and measurement across a range of ocean temperatures (5–25 °C). Each titration uses minimal sample and energy, completing in under two minutes. Field trials over 25 days demonstrated high precision (16 μmol kg^{-1}) and good accuracy (−0.17 ± 24 μmol kg^{-1}), supporting long-term, remote deployments for high-resolution ocean monitoring [2].

Table 1. Microfluidic Sensors applied to Environmental Monitoring

Target Compound/ Parameter	Method of Analysis	Limit of Detection (LoD)	Reference
Nitrite	LED-photodiode photometric detection in a disposable PDMS chip	0.07103 ppm	[18]
Sulfide	3D-printed microfluidic platform with integrated potentiometric detection	0.03 mM	[19]
Hg (II) ions	Microfluidic device containing a selective tripodal chromoionophore -PVC film	0.5 μM	[20]
Phosphorus	Tinted poly (methyl methacrylate) (PMMA) microfluidic chip for measuring soluble reactive phosphorus using the vanadomolybdate method	52 nM	[21]
As (III)	Modular microfluidic system with an integrated electrochemical sensor—containing gold nanoparticle-modified electrodes	0.42 μg/L	[1]
Aminoglycosidic antibiotics such as gentamicin	3D-printed PDMS microfluidic chip with automated fluid pumping	13 μM/mL	[24]
Total alkalinity	Closed-cell titrations driven by stepper-motor syringe pumps	Accuracy: −0.17 ± 24 μmol kg^{-1}; Precision: 16 μmol kg^{-1}	[2]

3.3 Food Safety

Due to major food safety issues occurring frequently, it is necessary to develop fast, accurate, and on site sensing techniques to assess the quality of food. The presence of harmful additives, pathogens, pesticides, and microplastics in food has been documented [25, 26]. Extensive monitoring of food products from the point of harvest to processing and distribution demands quick, accurate, safe and on site sensing devices. Microfluidic devices offer all the aforementioned advantages in addition to their compact sizes and robustness. Microfluidic lab-on-a-chip platforms have also offered promising means for the efficient detection of several mycotoxins such as aflatoxins, ochratoxin A, fumonisin B1 and citrinin from fungi like *Aspergillus* spp., *Fusarium* spp., and *Penicillum* spp., entering the food chain in any stage from processing to transportation and storage [27]. Fig. 2 shows the major food safety concerns affecting food quality.

Materials Research Foundations 191 (2026) 396-410
https://doi.org/10.21741/9781644904213-13

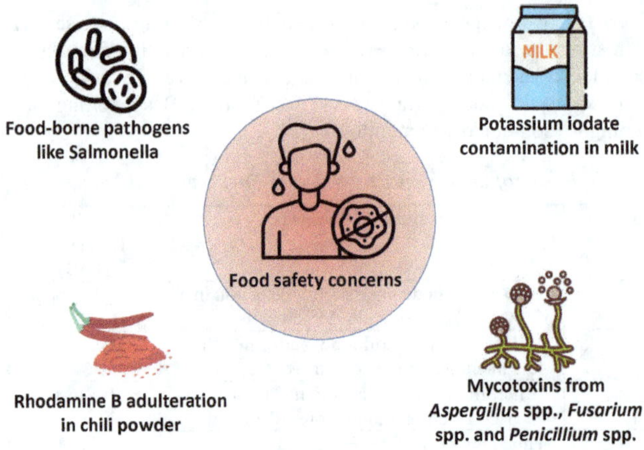

Figure 2. The Major Food Safety issues affecting the Quality of Food

Multiplexed microfluidic technologies enable advanced detection of foodborne pathogens through integrated sample preparation and diverse detection methods. Multiplexed detection within a single microfluidic sensor is achieved through different reactions taking place in different zones of the device. Techniques like cell enrichment, nucleic acid processing, amplification, immunoassays, and spectroscopy were compared with microfluidic detection for practical applications [28]. Among the food borne pathogens, *Salmonella* is responsible for several disease outbreaks since they thrive well in meat and vegetables. Culturing them in labs require a few days while Polymerase Chain Reaction (PCR) or Enzyme-Linked Immunosorbent Assay (ELISA) tests are tedious and expensive [29]. A microfluidic biosensor using gold nanoparticles on thiolated polystyrene microsphere aptamer conjugates and complementary DNA - magnetic nanoparticles were used to detect *Salmonella* in freshly cut vegetables. The biosensor detected *Salmonella* with up to 60 CFU/mL, showing a high degree of accuracy in spiked salad samples [29]. Another sensor detected *Salmonella* using immunomagnetic separation, fluorescence, labeling, and a smartphone for analysis. It could detect Salmonella within the range of 1.4×10^2 to 1.4×10^6 CFU/mL, with a limit of 58 CFU/mL, and the authors suggest that this method could be applied to identify other foodborne pathogens as well [30]. In another study, a renewable 3D-printed rotating manifold was developed that automates reagent delivery, washing, and colorimetric detection steps on disposable paper-based microfluidic devices (μPADs), enabling semi-automated *Salmonella* detection in 65 minutes without pipettes. The system achieved detection limits of 4.4×10^2 CFU/mL in growth media and 6.4×10^2 CFU/mL in milk, offering a sensitive, user-friendly, and field-deployable solution [31].

Campylobacter spp., one of the major foodborne pathogens, are increasingly showing antibiotic resistance, posing serious public health risks. To address this, a polymeric microfluidic device was developed for rapid recognition and antimicrobial susceptibility testing (AST) of *Campylobacter* in contaminated food, using colorimetric detection. The device showed 100% specificity, high sensitivity, and agreement (91–100%) with conventional methods, delivering results within 24

Materials Research Foundations 191 (2026) 396-410
https://doi.org/10.21741/9781644904213-13

hours compared to several days using standard techniques [32]. A dual-rolling circle amplification (RCA) strategy was developed in a microfluidic device to enhance the sensitivity of whole-cell detection of E. coli O157:H7. The system uses capturing RCA (cRCA) to modify microchannel surfaces with poly-aptamers, improving cell capture by 3-fold, and signaling RCA (sRCA) to boost detection signals up to 250-fold. This dual-RCA platform enables highly sensitive detection in food matrices like milk and juice, achieving a detection limit of 80 cells/mL, offering a simple and effective tool for food safety monitoring [33].

Surface-enhanced Raman spectroscopy (SERS) enabled sensitive detection of illegal Rhodamine B in chili powder using paper-based SERS substrates created via self-assembly. Density functional theory helped to assign the characteristic Rhodamine B peaks, with a very low detection limit of 10^{-6} g/g. The method offers reliable quantification (96.4–108.9 % recovery) and is suitable for on-site analysis [34]. Microfluidic paper-based analytical devices (μPADs) with manually opened wax valves enable faster, more sensitive detection by allowing controlled mixing and incubation. The devices achieved linear detection ranges of 0.05 to 0.5 mM for quantifying potassium iodate within 15 min [35]. Table 2 summarizes the several applications of microfluidic sensors applied to food safety monitoring.

Table 2. Microfluidic Sensors for Food Safety Monitoring

Target Compound/ Microorganism	**Food source**	**Method of Analysis**	**Limit of Detection (LoD)**	**Reference**
Salmonella	Freshly cut vegetables	Thiolated polystyrene microspheres - mediated aggregation of gold nanoparticles	60 CFU/mL	[29]
Salmonella typhimurium	Apple juice	Immunomagnetic separation using magnetic bacteria labeled with the fluorescent microspheres	58 CFU/mL	[30]
Salmonella	Milk	Immuno-magnetic separation (IMS)-based immunoassay coupled with chlorophenol red-β-D-galactopyranoside detection on a μPAD	6.4×10^2 CFU mL^{-1}	[31]
Campylobacter spp.	Artificially contaminated poultry meat and milk	Polymeric microfluidic device for AST	Milk: 1×10^2 CFU/ml Meat: 1×10^4 CFU/25 g	[32]
E. coli O157:H7	Milk and Orange juice	Dual-RCA sandwich format for whole cell detections	80 cells/mL	[33]
Rhodamine B	Chili powder	paper-based SERS substrates prepared using liquid–liquid interface mediated self-assembly	10^{-6} g/g of Rhodamine B in chili powder	[34]
Potassium iodate	Salt/milk	μPADs integrating novel wax valves for distance-based detection	0.05 to 0.5 mM	[35]

4. Challenges in Real Time Applications

As with any technology that makes life easier for human beings, microfluidics has its own challenges in implementation, effective operation, and interpretation of data. The performance of a microfluidic device relies on careful execution across all stages such as design, fabrication, sample handling, and signal processing. Fabrication involves creating microchannels and chambers using materials like glass or polymers, while accurate results depend on proper sample collection and analysis using optical, electrochemical, or biological techniques to detect key analytes [36]. Besides this, single-use point-of-care diagnostic devices are valuable for global health, particularly in areas with limited lab infrastructure, but their widespread use raises serious environmental and health concerns. These devices are typically made from fossil-based polymers and often contain toxic chemicals, requiring incineration for safe disposal, which contributes to CO_2 emissions and pollution. To address these issues, especially in low- and middle-income countries, sustainable design strategies and alternative materials must be adopted early in the development of such diagnostics to reduce their ecological footprint [37]. An analysis of more than 40,000 adverse event reports related to microfluidic technologies for the diagnosis of diseases found that the most frequent failure modes involve issues with flow, device operation, and data output [6].

5. Emerging Trends

Automation in sensing systems is essential to minimize human involvement, enhance reproducibility, and enable remote operations. Various systems incorporating automation, such as flow injection, sequential injection, microfluidics, and robotics offer precise sample handling, increased throughput, and improved safety. Computer technologies, including AI and machine learning, play a key role in real-time data processing, prediction, and integration with IoT for effective remote sensing and data management [38]. Microfluidics is widely used in areas like molecular diagnostics and sequencing but widespread usage is minimal due to the expertise required in device design and control. Overcoming these barriers could enable fully automated platforms that make microfluidics accessible to non-experts. By applying machine learning to capture expert intuition and predict system behavior, the integration of AI could greatly broaden the usability and impact of microfluidic technologies [39].

The integration of microfluidics with cyber-physical systems has revolutionized environmental and agricultural monitoring by enabling compact, on-site detection tools. A miniaturized colorimetric device using PDMS microfluidics and optoelectronic components was developed to quantify soil nutrients like nitrogen and phosphorus. Key to this advancement is the incorporation of Internet of Things (IoT) and Bluetooth modules, along with an Arduino microcontroller, allowing real-time data generation, transmission, and storage even in remote locations. An Android app further supports continuous monitoring, highlighting the potential of IoT-enabled microfluidic systems for low-cost, point-of-care environmental sensing [40].

Real-time, noninvasive monitoring of sweating rate and electrolyte loss can support personalized hydration strategies. Traditional sweat analysis methods are impractical for active use, as they require sweat collection and lab-based analysis. A new skin-interfaced wearable microfluidic device, combined with smartphone image processing, enables on-the-go analysis of regional sweating rate and sweat chloride concentration. Tested on 312 athletes, the system showed strong correlations and uses predictive algorithms to estimate whole-body sweat metrics, offering a low-

Materials Research Foundations 191 (2026) 396-410 https://doi.org/10.21741/9781644904213-13

cost, accessible tool for athletes and sports professionals [41]. In another study by the same research group, a wearable microfluidic patch designed for recreational athletes allowed real-time sweat analysis using a smartphone app with image processing algorithms, accommodating various sweat rates and lighting conditions. Field trials with 148 participants showed strong correlation with standard lab methods, validating the system's effectiveness for remote sweat monitoring [42].

A 3D-printed opto-microfluidic analyzer was developed for autonomous, miniaturized analysis with wireless-controlled auto-calibration. It integrates four key functions namely, control and data acquisition, hydrodynamic fluid handling, optical detection, and wireless communication. The experimental fluoride measurements closely matched laboratory-grade results while using 50% less reagent and generating 80% less waste. Thus, the device is portable and cost-efficient, reducing equipment cost by about 75% compared to conventional laboratory systems. The compact design reduces reagent and energy use, enabling real-time, on-site analysis and potential integration into IoT-based wireless sensor networks [43].

Complete blood counts (CBCs) are crucial for diagnosing various health conditions by analyzing red and white blood cells and platelets. Traditional methods depend on expensive and time-consuming hematology analyzers, highlighting the need for faster, automated, and cost-effective alternatives. A novel approach using deep learning algorithms, specifically the YOLOv3 model, enables rapid and accurate detection of blood cells, making it suitable for integration into microfluidic point-of-care devices. The method was effectively applied to blood samples from patients affected by COVID-19, where clotting is a recurring symptom [44].

Microalgae play a vital role in ecosystems and have significant commercial value in areas like biofuels, pharmaceuticals, and food supplements. Hence, it is necessary for proper identification of microalgae to enable environmental monitoring and have a few industrial applications as well. An automated and intelligent microfluidic platform (AIMP) was developed to enable real-time, on-site detection and monitoring of microalgae, addressing limitations of conventional lab-based methods. The system integrates automated control, intelligent data analysis, and user interaction, offering a portable and cost-effective solution. Using a YOLOv5-based detection network, AIMP accurately identified and classified four microalgal species with a mean average precision of 92.8% and demonstrated long-term monitoring of astaxanthin production, supporting ecological studies and commercial exploitation [45].

Another significant breakthrough is the development of triboelectric nanosensors in the field of sensing. These miniature devices harness the triboelectric effect, generating electrical signals from simple contact and motion between materials to detect pressure, vibration, or movement without the need for external power sources. Known for their flexibility, sensitivity, and low cost, they are increasingly being integrated into wearable electronics, environmental monitors, and biomedical devices. Their compatibility with microfluidic platforms and smart textiles has broadened their real-world applications. Triboelectric nanosensors represent a major step toward sustainable and autonomous sensing solutions for next-generation devices [46].

In some cases, microfluidic extraction processes were involved before processing the samples using sophisticated analytical equipment. One such study involved textile fibers which are frequently encountered as trace evidence in forensic investigations, and analyzing their dye content to provide critical molecular-level insights. While conventional techniques like microscopy and spectroscopy are commonly used, a new approach uses an automated microfluidic dye extraction (MFD) system coupled with Q-TOF mass spectrometry for faster, more precise analysis. This system requires only 10 mL of solvent and completes dye extraction and identification in under 12 minutes, even from single fibers as short as 1 mm and as thin as 10 μm. It enables characterization of multiple dyes along with elemental and structural analysis, enhancing the efficiency of forensic fiber examination [47].

Deploying medical diagnostics in developing countries is often hindered by limited infrastructure, healthcare personnel, and transportation systems. A novel electronics-enabled microfluidic paper-based analytical device (EE-μPAD) has been developed to quantitatively measure micronutrient levels in such settings. This compact, credit card-sized system integrates flexible electronics and optoelectronic sensors, offering features like self-calibration, plasma separation, and wireless data transfer to mobile devices for real-time geo-tagged reporting. Clinical tests on 95 blood samples show sensitivity and specificity comparable to ELISA, enabling rapid, decentralized diagnostics and large-scale health data aggregation [48].

To overcome the cost and complexity of integrating microfluidic devices with multiwell plates, a simple open-source robotic system has been developed that delivers liquids sequentially through a single inlet. This system was validated by reliably dispensing 96 dye solutions, capturing neural responses in *C. elegans*, and identifying compounds that modulate neural activity. It also successfully automated a ten-step, 85-minute cell staining protocol, highlighting its adaptability. Overall, the platform offers an efficient and economical solution for automating high-throughput microfluidic experiments [49].

Conclusions and Future Perspectives

The effective operation of a microfluidic device requires attention to every stage, from design and fabrication to sample handling and signal processing. Accurate sample collection and storage are essential to maintain result reliability, with techniques varying based on sample type. The core analytical processes, often optical, electrochemical, or biological assays enable precise detection and quantification of analytes like biomarkers. Future biosensor and Lab/Organ-on-a-Chip designs are expected to support machine learning by enabling setups that extract rich data from biosensing elements. This synergy creates a feedback loop where data generation and intelligent analysis enhance each other. Sensors and microfluidic devices serve as big data sources for machine learning, which can learn to detect specific analytes or flow patterns. In turn, machine learning can control fluidic responses and monitor critical target molecules, especially in lab-on-a-chip systems, closely mimicking in vivo environments.

References

[1] B. Alhalaili, I. N. Popescu, C.O. Rusanescu, R. Vidu, Microfluidic devices and microfluidics-integrated electrochemical and optical (bio)sensors for pollution analysis: A review, Sustain. 14 (2022) 12844. https://doi.org/10.3390/su141912844

[2] C. Sonnichsen, D. Atamanchuk, A. Hendricks, S. Morgan, J. Smith, I. Grundke, E. Luy, V. J. Sieben, An automated microfluidic analyzer for in situ monitoring of total alkalinity, ACS Sensors 8 (2023) 344-352. https://doi.org/10.1021/acssensors.2c02343

[3] S. Apoorva, N. T. Nguyen, K. R. Sreejith, Recent developments and future perspectives of microfluidics and smart technologies in wearable devices, Lab Chip 24 (2024) 1833-1866. https://doi.org/10.1039/D4LC00089G

[4] J. R. Mejía-Salazar, K. R. Cruz, E.M.M. Vásques, O. N. de Oliveira, Microfluidic point-of-care devices: New trends and future prospects for ehealth diagnostics, Sensors 20 (2020) 1951. https://doi.org/10.3390/s20071951

[5] J. Zheng, T. Cole, Y. Zhang, J. Kim, S.Y. Tang, Exploiting machine learning for bestowing intelligence to microfluidics, Biosens. Bioelectron. 194 (2021) 113666. https://doi.org/10.1016/j.bios.2021.113666

[6] R. Natu, L. Herbertson, G. Sena, K. Strachan, S. Guha, A systematic analysis of recent technology trends of microfluidic medical devices in the United States, Micromachines 14 (2023) 1293. https://doi.org/10.3390/mi14071293

[7] P.P. Behera, N. Kumar, M. Kumari, S. Kumar, P. K. Mondal, R. K. Arun, Integrated microfluidic devices for point-of-care detection of bio-analytes and disease, Sensors and Diagnostics 2 (2023) 1437-1459. https://doi.org/10.1039/D3SD00170A

[8] Y. Zhu, X. Meng, Y. Chen, J. Li, H. Shao, Y. Lu, L. Pan, Y. Xu, J. Cheng, Self-served and fully automated biochemical detection of finger-prick blood at home using a portable microfluidic analyzer, Sensors Actuators, B Chem. 303 (2020) 127235. https://doi.org/10.1016/j.snb.2019.127235

[9] H. Jeon, D.H. Lee, B. Jundi, M. Pinilla-Vera, R.M. Baron, B.D. Levy, J. Voldman, J. Han, Fully automated, sample-to-answer leukocyte functional assessment platform for continuous sepsis monitoring via microliters of blood, ACS Sensors 6 (2021) 2747-2756. https://doi.org/10.1021/acssensors.1c00887

[10] W. Yu, H. Lin, Y. Wang, X. He, N. Chen, K. Sun, D. Lo, B. Cheng, C. Yeung, J. Tan, D. Di Carlo, S. Emaminejad, A ferrobotic system for automated microfluidic logistics, Sci. Robot. 5 (2020) 1-12. https://doi.org/10.1126/scirobotics.aba4411

[11] J. Peng, C. Chan, S. Zhang, A.A. Sklavounos, M.E. Olson, E.Y. Scott, Y. Hu, V. Rajesh, B.B. Li, M.D. Chamberlain, S. Zhang, H. Peng, A.R. Wheeler, All-in-one digital microfluidics pipeline for proteomic sample preparation and analysis, Chem. Sci. 14 (2023) 2887-2900. https://doi.org/10.1039/D3SC00560G

[12] W. Al-Faqheri, F. Ibrahim, T.H.G. Thio, M.M. Aeinehvand, H. Arof, M. Madou, Development of novel passive check valves for the microfluidic CD platform, Sensors Actuators, A Phys. 222 (2015) 245-254. https://doi.org/10.1016/j.sna.2014.12.018

[13] S. Smith, D. Mager, A. Perebikovsky, E. Shamloo, D. Kinahan, R. Mishra, S.M. Torres Delgado, H. Kido, S. Saha, J. Ducrée, M. Madou, K. Land, J.G. Korvink, CD-based microfluidics for primary care in extreme point-of-care settings, 7 (2016) 22. https://doi.org/10.3390/mi7020022

[14] H.J. Woo, S.H. Kim, H.J. Kang, S.H. Lee, S.J. Lee, J.M. Kim, O. Gurel, S.Y. Kim, H.R. Roh, J. Lee, Y. Park, H.Y. Shin, Y. Il Shin, S.M. Lee, S.Y. Oh, Y.Z. Kim, J. Il Chae, S. Lee, M.H. Hong, B.C. Cho, E.S. Lee, K. Pantel, H.R. Kim, M.S. Kim, Continuous centrifugal microfluidics (CCM) isolates heterogeneous circulating tumor cells via full automation, Theranostics 12 (2022) 3676-3689. https://doi.org/10.7150/thno.72511

[15] J.N. Klatt, I. Schwarz, T. Hutzenlaub, R. Zengerle, F. Schwemmer, D. Kosse, J. Vincent, M. Scaer, K. Franaszczuk, D. Wadsworth, N. Paust, Miniaturization, parallelization, and

automation of endotoxin detection by centrifugal microfluidics, Anal. Chem. 93 (2021) 8508-8516. https://doi.org/10.1021/acs.analchem.1c01041

[16] M. Cóndor, T. Rüberg, C. Borau, J. Piles, J.M. García-Aznar, A web-based application for automated quantification of chemical gradients induced in microfluidic devices, Comput. Biol. Med. 95 (2018) 118-128. https://doi.org/10.1016/j.compbiomed.2018.02.001

[17] P. Aryal, C. Hefner, B. Martinez, C.S. Henry, Microfluidics in environmental analysis: advancements, challenges, and future prospects for rapid and efficient monitoring, Lab Chip 24 (2024) 1175-1206. https://doi.org/10.1039/D3LC00871A

[18] S. Dudala, S.K. Dubey, S. Goel, Fully integrated, automated, and smartphone enabled point-of-source portable platform with microfluidic device for nitrite detection, IEEE Trans. Biomed. Circuits Syst. 13 (2019) 1518-1524. https://doi.org/10.1109/TBCAS.2019.2939658

[19] R. Pol, F. Céspedes, D. Gabriel, M. Baeza, Fully integrated screen-printed sulfide-selective sensor on a 3D-printed potentiometric microfluidic platform, Sensors Actuators, B Chem. 290 (2019) 364-370. https://doi.org/10.1016/j.snb.2019.03.132

[20] Nuriman, B. Kuswandi, W. Verboom, Optical fiber chemical sensing of Hg(II) ions in aqueous samples using a microfluidic device containing a selective tripodal chromoionophore-PVC film, Sensors Actuators, B Chem. 157 (2011) 438-443. https://doi.org/10.1016/j.snb.2011.04.084

[21] F.E. Legiret, V.J. Sieben, E.M.S. Woodward, S.K. Abi Kaed Bey, M.C. Mowlem, D.P. Connelly, E.P. Achterberg, A high performance microfluidic analyser for phosphate measurements in marine waters using the vanadomolybdate method, Talanta 116 (2013) 382-387. https://doi.org/10.1016/j.talanta.2013.05.004

[22] P. Giménez-Gómez, A. Baldi, C. Ayora, C. Fernández-Sánchez, Automated determination of As(III) in waters with an electrochemical sensor integrated into a modular microfluidic system, ACS Sensors 4 (2019) 3156-3165. https://doi.org/10.1021/acssensors.9b01286

[23] J.A. Hondred, Z.T. Johnson, J.C. Claussen, Nanoporous gold peel-and-stick biosensors created with etching inkjet maskless lithography for electrochemical pesticide monitoring with microfluidics, J. Mater. Chem. C 8 (2020) 11376-11388. https://doi.org/10.1039/D0TC01423K

[24] N. Sethu, P. Premchandra, S.B. Kolhe, M.B. Kulkarni, R. Vyas, An integrated microfluidic device driven by an automated system for precise detection of antibiotics in water, Sensors Actuators A Phys. 388 (2025) 116474. https://doi.org/10.1016/j.sna.2025.116474

[25] H. Gao, C. Yan, W. Wu, J. Li, Application of microfluidic chip technology in food safety sensing, Sensors 20 (2020) 1792. https://doi.org/10.3390/s20061792

[26] G.E. De-La-Torre, Microplastics: an emerging threat to food security and human health, J. Food Sci. Technol. 57 (2020) 1601-1608. https://doi.org/10.1007/s13197-019-04138-1

[27] L. Guo, J. Feng, Z. Fang, J. Xu, X. Lu, Application of microfluidic "lab-on-a-chip" for the detection of mycotoxins in foods, Trends Food Sci. Technol. 46 (2015) 252-263. https://doi.org/10.1016/j.tifs.2015.09.005

[28] X. Han, Y. Liu, J. Yin, M. Yue, Y. Mu, Microfluidic devices for multiplexed detection of foodborne pathogens, Food Res. Int. 143 (2021) 110246. https://doi.org/10.1016/j.foodres.2021.110246

[29] Y. Man, M. Ban, A. Li, X. Jin, Y. Du, L. Pan, A microfluidic colorimetric biosensor for in-field detection of Salmonella in fresh-cut vegetables using thiolated polystyrene microspheres, hose-based microvalve and smartphone imaging APP, Food Chem. 354 (2021) 129578. https://doi.org/10.1016/j.foodchem.2021.129578

[30] S. Wang, L. Zheng, G. Cai, N. Liu, M. Liao, Y. Li, X. Zhang, J. Lin, A microfluidic biosensor for online and sensitive detection of Salmonella typhimurium using fluorescence labeling and smartphone video processing, Biosens. Bioelectron. 140 (2019) 111333. https://doi.org/10.1016/j.bios.2019.111333

[31] C.S. Carrell, R.M. Wydallis, M. Bontha, K.E. Boehle, J.R. Beveridge, B.J. Geiss, C.S. Henry, Rotary manifold for automating a paper-based: Salmonella immunoassay, RSC Adv. 9 (2019) 29078-29086. https://doi.org/10.1039/C9RA07106G

[32] L. Ma, M. Petersen, Identification and antimicrobial susceptibility testing of campylobacter using a microfluidic lab-on-a-chip device, Appl. Environ. Microbiol. 86 (2020) 1-15. https://doi.org/10.1128/AEM.00096-20

[33] Y. Jiang, Z. Qiu, T. Le, S. Zou, X. Cao, Developing a dual-RCA microfluidic platform for sensitive E. coli O157:H7 whole-cell detections, Anal. Chim. Acta 1127 (2020) 79-88. https://doi.org/10.1016/j.aca.2020.06.046

[34] S. Lin, W.L.J. Hasi, X. Lin, S.Q.G.W. Han, X.T. Lou, F. Yang, D.Y. Lin, Z.W. Lu, Rapid and sensitive SERS method for determination of Rhodamine B in chili powder with paper-based substrates, Anal. Methods 7 (2015) 5289-5294. https://doi.org/10.1039/C5AY00028A

[35] C. Chen, L. Zhao, H. Zhang, X. Shen, Y. Zhu, H. Chen, Novel wax valves to improve distance-based analyte detection in paper microfluidics, Anal. Chem. 91 (2019) 5169-5175. https://doi.org/10.1021/acs.analchem.8b05764

[36] G. Antonelli, J. Filippi, M. D'Orazio, G. Curci, P. Casti, A. Mencattini, E. Martinelli, Integrating machine learning and biosensors in microfluidic devices: A review, Biosens. Bioelectron. 263 (2024) 116632. https://doi.org/10.1016/j.bios.2024.116632

[37] A.E. Ongaro, Z. Ndlovu, E. Sollier, C. Otieno, P. Ondoa, A. Street, M. Kersaudy-Kerhoas, Engineering a sustainable future for point-of-care diagnostics and single-use microfluidic devices, Lab Chip 22 (2022) 3122-3137. https://doi.org/10.1039/D2LC00380E

[38] C.M. Raju, D.P. Elpa, P.L. Urban, Automation and computerization of (bio)sensing Systems, ACS Sensors 9 (2024) 1033-1048. https://doi.org/10.1021/acssensors.3c01887

[39] D. McIntyre, A. Lashkaripour, P. Fordyce, D. Densmore, Machine learning for microfluidic design and control, Lab Chip 22 (2022) 2925-2937. https://doi.org/10.1039/D2LC00254J

[40] Pal, S.K. Dubey, S. Goel, IoT enabled microfluidic colorimetric detection platform for continuous monitoring of nitrite and phosphate in soil, Comput. Electron. Agric. 195 (2022) 106856. https://doi.org/10.1016/j.compag.2022.106856

Materials Research Foundations 191 (2026) 396-410 — https://doi.org/10.21741/9781644904213-13

[41] L.B. Baker, J.B. Model, K.A. Barnes, M.L. Anderson, S.P. Lee, K.A. Lee, S.D. Brown, A.J. Reimel, T.J. Roberts, R.P. Nuccio, J.L. Bonsignore, C.T. Ungaro, J.M. Carter, W. Li, M.S. Seib, J.T. Reeder, A.J. Aranyosi, J.A. Rogers, R. Ghaffari, Skin-interfaced microfluidic system with personalized sweating rate and sweat chloride analytics for sports science applications, Sci. Adv. 6 (2020) 17-19. https://doi.org/10.1126/sciadv.abe3929

[42] L.B. Baker, M.S. Seib, K.A. Barnes, S.D. Brown, M.A. King, P.J.D. De Chavez, S. Qu, J. Archer, A.S. Wolfe, J.R. Stofan, J.M. Carter, D.E. Wright, J. Wallace, D.S. Yang, S. Liu, J. Anderson, T. Fort, W. Li, J.A. Wright, S.P. Lee, J.B. Model, J.A. Rogers, A.J. Aranyosi, R. Ghaffari, Skin-interfaced microfluidic system with machine learning-enabled image processing of sweat biomarkers in remote settings, Adv. Mater. Technol. 7 (2022) 2200249. https://doi.org/10.1002/admt.202200249

[43] R.M. Camarillo-Escobedo, J.L. Flores-Nuñez, G. Garcia-Torales, E. Hernandez-Campos, J.M. Camarillo-Escobedo, 3D printed opto-microfluidic autonomous analyzer for photometric applications, Sensors Actuators A Phys. 337 (2022) 113425. https://doi.org/10.1016/j.sna.2022.113425

[44] T. Xia, Y.Q. Fu, N. Jin, P. Chazot, P. Angelov, R. Jiang, AI-enabled microscopic blood analysis for microfluidic COVID-19 hematology, Proc. - 2020 5th Int. Conf. Comput. Intell. Appl. ICCIA 2020 (2020) 98-102. https://doi.org/10.1109/ICCIA49625.2020.00026

[45] J. Zheng, T. Cole, Y. Zhang, N. Bayinqiaoge, D. Yuan, S.Y. Tang, An automated and intelligent microfluidic platform for microalgae detection and monitoring, Lab Chip 24 (2023) 244-253. https://doi.org/10.1039/D3LC00851G

[46] S. Krishnan, A. Giwa, Advances in real-time water quality monitoring using triboelectric nanosensors, J. Mater. Chem. A 13 (2025) 11134-11158. https://doi.org/10.1039/D4TA08871A

[47] N. Sultana, S. Gunning, S.J. Furst, K.P. Garrard, T.A. Dow, N.R. Vinueza, Direct analysis of textile dyes from trace fibers by automated microfluidics extraction system coupled with Q-TOF mass spectrometer for forensic applications, Forensic Sci. Int. 289 (2018) 67-74. https://doi.org/10.1016/j.forsciint.2018.05.020

[48] S. Lee, A.J. Aranyosi, M.D. Wong, J.H. Hong, J. Lowe, C. Chan, D. Garlock, S. Shaw, P.D. Beattie, Z. Kratochvil, N. Kubasti, K. Seagers, R. Ghaffari, C.D. Swanson, Flexible opto-electronics enabled microfluidics systems with cloud connectivity for point-of-care micronutrient analysis, Biosens. Bioelectron. 78 (2016) 290-299. https://doi.org/10.1016/j.bios.2015.11.060

[49] R.C. Lagoy, D.R. Albrecht, Automated fluid delivery from multiwell plates to microfluidic devices for high-throughput experiments and microscopy, Sci. Rep. 8 (2018) 6217. https://doi.org/10.1038/s41598-018-24504-x

Keyword Index

About the Editors

Inamuddin, Ph.D, is an assistant professor at the Department of Applied Chemistry, Zakir Husain College of Engineering and Technology, Faculty of Engineering and Technology, Aligarh Muslim University, Aligarh, India. He has extensive research experience in multidisciplinary fields of analytical chemistry, materials chemistry, electrochemistry, renewable energy, and environmental science. He has worked on different research projects funded by various government agencies and universities and is the recipient of awards, including the Department of Science and Technology, India, Fast-Track Young Scientist Award and Young Researcher of the Year Award 2020 from Aligarh Muslim University. He has published about 216 research articles in various international scientific journals, 18 book chapters, and 230 edited books with multiple well-known publishers. His current research interests include ion exchange materials, a sensor for heavy metal ions, biofuel cells, supercapacitors, and bending actuators.

Dr. Tariq Altalhi is a Professor of Chemistry at Taif University, Saudi Arabia, where he has served as Head of the Chemistry Department and Vice Dean of the College of Science. He earned his Ph.D. from the University of Adelaide, Australia, receiving the Dean's Commendation for Doctoral Thesis Excellence. In 2015, his research was nominated among the top ten entries for the Green Tech Awards, Germany. His research interests include nanomaterial synthesis, membrane technologies, molecular separation, desalination, drug delivery, and biosensing. He published numerous research articles and has co-edited several scientific books and maintains strong collaborations with major industrial partners in Saudi Arabia.

Eng. Abdalla Salah Mustafa is a Chemical Engineer with over nine years of experience in hazardous waste management, industrial operations, environmental compliance, and sustainable engineering practices in Saudi Arabia. He has worked with major industrial organizations, including Petro Rabigh and the Saudi Electricity Company, managing large-scale environmental and waste treatment projects while maintaining strong safety and regulatory compliance standards.

He earned his Bachelor's degree with Honors in Chemical Engineering from the University of Science and Technology in 2013. Currently serving as Projects Manager at SRB Environmental Services, he oversees integrated waste management operations, including waste treatment, recycling, incineration, wastewater systems, and environmentally compliant disposal practices in coordination with Saudi Aramco and Petro Rabigh.

Eng. Abdalla is recognized for his expertise in process optimization, environmental operations, team leadership, and industrial waste management. He has successfully managed projects involving more than 45,000 m^3 of industrial liquid waste and remains committed to sustainable industrial development and environmental protection.

Dr. Luqman has 17+ years of post-PhD multinational industrial and academic (Research and Teaching) experience in the field of Energy, Materials, and Environment. Dr. Luqman is a dedicated and active researcher, academician and administrator. He successfully served as an Editor to many books, published by world renowned publishers. He published numerous high-quality papers, and book chapters. He has attracted a few important research grants from industry and academia. One of his papers secured its place on the Cover Page of the Journal. Another Paper in "Polymer", was ranked the "First" among the "Top 25 Hottest Articles", among "All" related Journals (108 journals) during "April to June" 2008. His research interests include but are not limited to Development of Ionomer/Polyelectrolyte/non-ionic Polymer Nanocomposites/Blends for Smart and Industrial/Engineering Applications.

Dr. Mohammad Luqman earned his PhD degree in the field of Ionomers (Ion-containing Polymers), from Chosun University, South Korea. He served as a post-doctoral fellow at Artificial Muscle Research Center, Konkuk University, South Korea, in the field of Ionic Polymer Metal Composites for the development of Artificial Muscles, Robotic Actuators and Dynamic Sensors. He has served as a Research Engineer at Samsung Cheil Industries. He earned his MS and BS degrees in Industrial Chemistry from Aligarh Muslim University, India. Currently, he is serving as an Assistant Professor of Chemical Engineering in Taibah University, Saudi Arabia. He has held many administrative positions as well including College registrar, Advisor to Academic Advisors, chair and members to various committees at department, college and university levels. He served very actively in the development of Chemical Engineering and Industrial Chemistry programs.

www.ingramcontent.com/pod-product-compliance
Lightning Source LLC
LaVergne TN
LVHW052220200926
848481LV00001B/162

* 9 7 8 1 6 4 4 9 0 4 2 0 6 *